IFIP Advances in Information and Communication Technology 740

IFIP Advances in Information and Communication Technology

The IFIP AICT series publishes state-of-the-art results in the sciences and technologies of information and communication. The scope of the series includes: foundations of computer science; software theory and practice; education; computer applications in technology; communication systems; systems modeling and optimization; information systems; ICT and society; computer systems technology; security and protection in information processing systems; artificial intelligence; and human-computer interaction.

Edited volumes and proceedings of refereed international conferences in computer science and interdisciplinary fields are featured. These results often precede journal publication and represent the most current research.

The principal aim of the IFIP AICT series is to encourage education and the dissemination and exchange of information about all aspects of computing.

More information about this series at https://link.springer.com/bookseries/6102

Pradorn Sureephong · Christophe Danjou ·
Abdelaziz Bouras
Editors

Product Lifecycle Management

Integrating Digital Technologies for Sustainability and Innovation

21st IFIP WG 5.1 International Conference, PLM 2024
Bangkok, Thailand, July 7–10, 2024
Revised Selected Papers, Part I

Springer

Editors
Pradorn Sureephong
Chiang Mai University
Chiang Mai, Thailand

Christophe Danjou
Polytechnique Montréal
Montréal, QC, Canada

Abdelaziz Bouras
Qatar University
Doha, Qatar

ISSN 1868-4238 ISSN 1868-422X (electronic)
IFIP Advances in Information and Communication Technology
ISBN 978-3-031-93318-9 ISBN 978-3-031-93319-6 (eBook)
https://doi.org/10.1007/978-3-031-93319-6

This Springer imprint is published by the registered company Springer Nature Switzerland AG
The registered company address is: Gewerbestrasse 11, 6330 Cham, Switzerland

Preface

Since 2003 the International Conference on Product Lifecycle Management (PLM) has brought together researchers, developers and users of Product Lifecycle Management. It aims to integrate business approaches to the collaborative creation, management and dissemination of product and process data throughout the extended enterprises that create, manufacture and operate engineered products and systems. The conference aims to involve all stakeholders of the wide concept of PLM, hoping to shape the future of this field and advance the science and practice of enterprise development.

The 21th International Conference on Product Lifecycle Management (PLM 2024) was hosted in Bangkok, Thailand, from July 7th to July 10th, 2024.

For its 21st anniversary, PLM 2024 registered 188 participants from 16 countries who had the chance to attend nine Special Sessions (Digital Transformation) plus one industrial day. In addition, PLM 2024 featured Academic Keynotes, Industrial Keynotes and round tables.

Keynote Speakers

- Sudarsan Rachuri, Department of Energy, USA
- Chidchanok Lursinsap, Chulalongkorn University, Thailand
- Ram D. Sriram, National Institute of Standards and Technology, USA
- Tiranee Achalakul, Big Data Institute, Thailand
- Sittapong Settapat, SCB Data X Co., Ltd., Thailand
- Santisook Limpeeticharoenchot, STelligence Co., Ltd., Thailand
- Passakon Prathombutr, Digital Economy Promotion Agency, Thailand

Industry day

- Data Driven Government Thailand Used Cases – Tiranee Achalakul, Big Data Institute, Thailand
- Generative AI And Knowledge Graph For Digital Twin In Industrial 4.0 – Sittapong Settapat, SCB Data X Co., Ltd., Thailand
- Change Management In The Era Of AI And Fintech: Strategies For Success – Santisook Limpeeticharoenchot, STelligence Co., Ltd., Thailand
- Smart City Lifecycle – Passakon Prathombutr, Digital Economy Promotion Agency, Thailand

Industrial Round Table

BIM and Digital Twin

- Dconstruct Robotics Pte., Ltd.
- Microsoft Partner
- Altotech Global Co., Ltd.

Digital Twin for Sustainable Smart City

- Pcs Thailand Co., Ltd.
- Ai And Robotics Ventures Company Limited
- Siam Inno City Co., Ltd

All submitted papers were double-blind peer-reviewed by at least two reviewers. In total, 105 contributions were submitted, of which 64 were accepted. Program chairs Christophe Danjou and Pradorn Sureephong organized an amazing scientific conference with presentations grouped in 20 thematic sections.

We would like to thank everyone who directly or indirectly contributed to making the PLM 2024 conference a success, particularly by participating in rich scientific discussions and meeting all of us in Bangkok, Thailand.

Pradorn Sureephong

Organization

Conference Chairs

Abdelaziz Bouras	Qatar University, Qatar
Kitipong Promwong	Office of National Higher Education Science Research and Innovation Policy Council, Thailand
Kriengkrai Thiennukul	Federation of Thailand Industries, Thailand

Program Committee Chairs

Pradorn Sureephong	Chiang Mai University, Thailand
Christophe Danjou	Polytechnique Montréal, Canada

Local Organizing Committee

Krittika Kantawong	University of Payao, Thailand
Suepphong Chernbumroong	Chiang Mai University, Thailand
Tama Duangnamol	Chiang Mai University, Thailand
Santichai Wicha	Mae Fah Luang University, Thailand
Wanus Srimaharaj	Payap University, Thailand
Yootthapong Tongpaeng	Mae Fah Luang University, Thailand
Paweena Suebsombut	Mae Fah Luang University, Thailand
Shisa Singkhamfu	Chiang Mai University, Thailand

Steering Committee Members

Abdelaziz Bouras	Qatar University, Qatar
Anderson Luis Szejka	Pontifical Catholic University of Paraná, Brazil
Balan Gurumoorthy	Indian Institute of Science, India
Benoit Eynard	University of Technology of Compiègne, France
Christophe Danjou	Polytechnique Montréal, Canada
Clement Fortin	Skolkovo Institute of Science and Technology, Russia
Darli Rodrigues Vieira	Université du Québec à Trois-Rivières, Canada

Debashish Dutta	University of Michigan, USA
Frédéric Noël	Grenoble Institute of Technology, France
José Rios	Polytechnic University of Madrid, Spain
Klaus-Dieter Thoben	University of Bremen, Germany
Néjib Moalla	University of Lyon, France
Osiris Canciglieri Junior	Pontifical Catholic University of Paraná, Brazil
Paolo Chiabert	Polytechnic of Turin, Italy
Ramy Harik	University of South Carolina, USA
Romeo Bandinelli	University of Florence, Italy
Sebti Foufou	University of Sharjah, UAE
Sergio Terzi	Politecnico di Milano, Italy

Doctoral Workshop Committee

Monica Rossi	Politecnico di Milano, Italy
Yacine Ouzrout	University of Lyon, France

Scientific Committee

Abdelaziz Bouras	Qatar University, Qatar
Abdelhak Belhi	Qatar University, Qatar
Aicha Sekhari	University of Lyon, France
Alain Bernard	Central School of Nantes, France
Alexander Smirnov	Russian Academy of Sciences, Russia
Alison McKay	University of Leeds, UK
Améziane Aoussat	Arts et Métiers ParisTech, France
Anderson Luis Szejka	Pontifical Catholic University of Paraná, Brazil
Balan Gurumoorthy	Indian Institute of Sciences Bangalore, India
Bas Koomen	CADMES, The Netherlands
Benoit Eynard	University of Technology of Compiègne, France
Bob Young	Loughborough University, UK
Brendan Sullivan	Polytechnic of Milan, Italy
Chris McMahon	Technical University of Denmark, Denmark
Chen Zheng	Northwestern Polytechnical University, China
Christophe Danjou	Polytechnique Montréal, Canada
Claudio Sassanelli	Politecnico di Bari, Italy
Clement Fortin	Skolkovo Institute of Science and Technology, Russia
Daniel Schmid	Zurich University of Applied Sciences, Switzerland

Contents – Part I

Contents – Part II

Circular Economy Rebound Effect: Theoretical Implications and a Research Agenda

Mariantonietta Ferrante(✉), Micaela Vitti, Francesco Facchini, and Claudio Sassanelli

Department of Mechanics, Mathematics, and Management, Polytechnic University of Bari, Via Orabona, 4 70126 Bari, Italy
m.ferrante6@phd.poliba.it

Abstract. The Circular Economy (CE) is a paradigm aimed at slowing and closing resource loops, and reducing resource flows, to achieve environmental, economic, and social benefits. Notwithstanding its current widespread adoption in manufacturing, the effectiveness of circular manufacturing (CM) strategies could be hampered by the occurrence of the so-called rebound effects (REs). RE is defined as an undesirable increase in consumption of a resource due to improved efficiency. The identification and prevention of REs' occurrence is crucial to ensure the effectiveness of CM strategies. Although CE's RE is gaining momentum in literature, there is still a limited understanding of how to explore this topic and implement related avoidance, mitigation and/or remediation actions. To this concern, this paper aims to unveil and systematize the main research gaps and to open rooms for further studies in this domain. To address this objective, a systematic literature has been conducted, investigating all of the five previously identified dimensions composing the twofold domain CE and REs (i.e., business models, drivers, product lifecycle management, circular manufacturing ecosystem, and socio-economic aspects). The analysis of the selected sources allowed to identify seven gap categories. Finally, from a criticality prioritization of the gap categories, a research agenda has been developed, indicating as priority industrial applications, technologies potential and performance assessment.

Keywords: Circular economy · Rebound effect · Circular manufacturing · Research gaps · Research agenda

1 Introduction

Circular economy (CE) practices have been proposed to overcome relevant and actual ecological challenges (i.e., climate change, pollution, and loss of biodiversity). CE is a model of production and consumption aimed to narrow, slow, and close economic loops [1]. This model turned out to be successful in optimizing the use of resources and minimize waste generation [2], becoming widespread in the manufacturing industry. Data-driven Circular manufacturing (CM) has been recently introduced, leveraging on the role of technologies to trigger the several possible CM strategies [3]. Nevertheless,

Published by Springer Nature Switzerland AG 2025
P. Sureephong et al. (Eds.): PLM 2024, IFIP AICT 740, pp. 1–14, 2025.
https://doi.org/10.1007/978-3-031-93319-6_1

the environmental, economic, and social benefits achievable by implementing circular strategies could be hampered by the occurrence of the so-called Rebound Effect (RE). A RE is defined in the energy-economics sector as an efficiency improvement that reduces prices and increases consumer demand unexpectedly [4]. The occurrence of a RE in CE refers to an efficiency in production that doesn't reduce overall demand, increasing it in the long run in an undesirable way due to lower prices. From an ecological perspective, the undesirable increase in consumption decreases the expected environmental benefits, generating higher negative impacts [5]. Different mechanisms that generate REs have been investigated [6] (e.g., price and imperfect substitution mechanisms). It is crucial to understand how and why REs occur to remediate, mitigate, or even prevent and avoid their effects, thus maximizing the benefits of CE strategies. Despite the growing interest in the REs of CE, indeed, a range of different definitions have been organised and the mechanisms generating RE have been identified [7], but existing knowledge on this topic is still limited and fragmented, neglecting to provide indications on how to systematize the knowledge and provide directions for future research. Thus, this research aims to identify and systematize the main gaps characterizing the twofold domain CE and REs, also proposing a related research agenda to cope with them. To address this objective, a systematic literature has been conducted, investigating all of the five previously identified dimensions composing the twofold domain CE and Res (i.e., business models (BM) (D1), drivers (D2), CM ecosystem (D3), product lifecycle management (PLM) (D4), and socio-economic aspects (D5)). The analysis allowed to identify seven gap categories. Finally, from a criticality prioritization of the gap categories, a research agenda has been developed, indicating as priority industrial applications, technologies potential and performance assessment.

The paper is structured as follows. Section 2 outlines the research methodology adopted to carry out the SLR and to prioritize the main research gaps. In Sect. 3, the results are described, while Sect. 4 discusses them and concludes the paper.

2 Research Methodology

To address the research objective, a five-step SLR [8] has been conducted (Fig. 1).

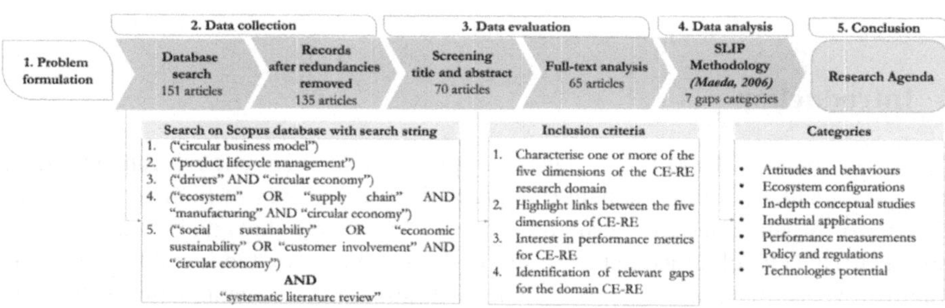

Fig. 1. Research methodology overview [8]

The first one concerned the identification of the problem (i.e., to identify the main gaps to be addressed in the twofold CE-RE domain). The second consisted in the collection of papers, performed through different search strings in the Scopus database (defined combining each of the five CE-RE dimensions previously defined in [9] with the keyword "systematic literature review"). The searches were carried out without constraints in terms of types of papers and time, considering only contributions in English language. After removing 16 redundant contributions, searches led to a set of 135 documents. In the third step, this sample of papers was analysed, selecting the 65 contributions addressing at least one of four inclusion criteria. In detail, papers have been selected if they explain CE-RE dimension's characteristics, if they highlight links among them, if they discuss performance metrics, or if they identify relevant domain research gaps. Finally, taking research gaps declared in selected papers as main driver, the sample was analysed through the Sort, Label, Integrate and Prioritize (SLIP) methodology [10]. Indeed, seven gap categories were identified and prioritized based on the frequency of occurrence of each category within the five dimensions and within the overall sample of selected papers.

3 Results

The 65 papers selected span the years from 2013 to 2023, with an increasing frequency in 2022 and 2023, when the 25% and 23% of the selected papers have been published, respectively. In the selected papers, different industries are involved, mainly textile (44%), chemical (16%), and plastic (16%). After analysing the research gaps highlighted within each of the 65 selected papers, seven key gaps categories (and related number of specific gaps) have been identified: attitudes and behaviours (14); ecosystem configurations (8); in-depth conceptual studies (4); industrial applications (23); performance measurements (14); policy and regulations (4); technologies potential (21). A brief description of each detected gap, classified by dimension and gap category, is provided in Table 1.

Attitudes and behaviours of consumers and supply chain (SC) actors play a central role in generating RE related to CE strategies. In the BM dimension, the interest in consumer actions is emphasized [11], raising in particular the need of focus on the social aspect of sustainability [12]. Also the impact on social wellbeing as well as on the different SC actors involved within the implementation of CE strategies require further investigation [13, 14]. More research is required to investigate the possibility that a consumer group's attitudes contribute to RE [15, 16]. Within drivers dimension, it has been highlighted that the social factors need more consideration for a successful implementation of CE [17, 18], and that, in this context, human behaviour during the adoption of CE practices should be examined [19, 20], also educating consumers to take positive actions [21], such as their behaviour in circular fashion and culture [22]. In the socio-economic aspects dimension, the factors influencing consumers' ecological choices must be examined by assessing their impact [23, 24].

Ecosystem configurations category refer to the interaction dynamics of the actors involved in achieving sustainability. SC configurations, such as the use of reverse logistics (CM ecosystem dimension) [25], need investigation to evaluate the CE effects on the value chain [14, 26]. This category raises the need of fostering information sharing (CM ecosystem dimension) [27] inside eco-friendly industrial systems (drivers dimension) [28] to enact innovative circular strategies (BM dimension) [29]. To achieve the sustainability of the whole ecosystem, an overall approach should be developed, with general approaches of technical quality methods [30, 31].

In-depth conceptual studies category highlights the need of more in-depth theoretical studies. In the PLM dimension, PSSs should be investigated more to manage the lifecycle of a product [32] and to connect them with lean thinking [33]. In the drivers dimension, CE's effects in different geographical areas should be deepened [20]. In addition, further studies should consider the negative impact caused by standard practices [34].

Table 1. Gap classification according to the seven categories identified (BM [D1], Drivers [D2], CM ecosystem [D3], PLM [D4] and Socio-economic aspects [D5])

Ref.	Specific gaps and related gap categories	Dimensions (D)				
		D1	D2	D3	D4	D5
[15]	Attitude-behaviour in a group of people, like customers and governments	X				
[11]	Consumer behaviour research on some CBM	X				
[16]	Social aspects and RE of a circular bioeconomy	X				
[12]	CBM neglects the social dimension of sustainability	X				
[14]	Challenges and consequences of actions among supply chain actors	X				
[13]	Social wellbeing impact of CE	X				
[21]	Educate consumers on the environmental effects of plastic packaging waste		X			
[18]	Insufficient regard for various actors' views and social behaviour		X			
[19]	Restricted examination of macro-level effects on society		X			
[22]	Consumer behaviour in circular fashion and culture and impact		X			
[20]	Human behaviour in adopting CE practices in firms		X			
[17]	Social factors in CE field		X			
[23]	Factors that influence consumer's choice of ecologically friendly packaging					X
[24]	Evaluation study to assess the general social impacts of the CE					X
	Total: Attitudes and behaviours category (14 gaps, 16% of total)	6	6	-	-	2
[26]	Strategic management frameworks and CE effects in value chain categories	X				
[29]	Expand CE and innovation research in other circular fields	X				
[30]	General approach of technical quality methods	X				

(continued)

Table 1. *(continued)*

Ref.	Specific gaps and related gap categories	Dimensions (D)				
[31]	New general categories to achieve sustainable	X				
[14]	Study SC configuration types to meet CE initiatives	X				
[28]	Industrial systems that are eco-friendly and circular		X			
[25]	Framework for circular logistics integration in the lifecycle			X		
[27]	Information sharing between organisations			X		
	Total: Ecosystems configurations category (8 gaps, 9% of total)	5	1	2	-	-
[34]	Standardised practices to consider the negative impact on sustaina- bility		X			
[20]	Understanding of the CE's effects in different geographical areas		X			
[32]	Study on new PSS business model categories				X	
[33]	More studies to recognise the relationship between lean thinking and PSS				X	
	Total: In-depth conceptual analysis category (4 gaps, 4,5% of total)	**-**	**2**	**-**	**2**	**-**
[45]	Examine which BMs are most effective for which sectors.	X				
[38]	More consolidated theory on BM with CE in construction	X				
[48]	Actions to practically promote a BM shift in CE practice	X				
[47]	Benefits and obstacles of CE of WEEE industry	X				
[13]	Need for more practical CE implementation cases in companies	X				
[52]	Specific reverse logistics components in industry		X			
[22]	More practical studies on the application of CE in fashion firms		X			
[46]	Examination of how specific industry sectors adopt CBM		X			
[41]	Sustainability in textile, apparel and fashion is not related to zero waste.		X			
[43]	Focus only on food industry; lack in research on food contact films (made of different materials)		X			
[37]	Lack of empirical evidence		X			
[39]	Achieve net-zero emission: manage better resources in construction projects		X			
[49]	No sector-specific research		X			
[50]	Implement research in other dimensions		X			
[42]	Interactions between packaging business and food industry members			X		
[51]	Further proof from full life cycle assessment studies			X		

(continued)

Table 1. (*continued*)

Ref.	Specific gaps and related gap categories	Dimensions (D)				
[36]	Extend and validate the topic map's use cases, barriers, and drivers				X	
[44]	More results from food industry examination				X	
[32]	Case studies about PSS to meet strategic and operational levels				X	
[35]	Additional use case analysis				X	
[40]	Empirical studies in fashion industry on influence of digital world on PLM				X	
[23]	Further studies on the industry of reusable packaging					X
[24]	More case studies in the built environment, agriculture, and food					X
	Total: Industrial applications category (23 gaps, 26% of total)	5	9	2	5	2
[53]	Metrics for sustainability performance, and social dimensions	X				
[60]	Expected financial performance, evaluation parameters, and design.	X				
[54]	Micro-level for environmental performance; economic and social dimensions not considered	X				
[55]	Implement performance metrics that consider triple bottom line	X				
[13]	Provide CE indicators to evaluate the micro level of analysis	X				
[30]	Develop a quantity method to evaluate circularity	X				
[31]	Generalize results to evaluate sustainable development objectives	X				
[52]	Develop a performance measurement to assess reverse logistic benefits		X			
[57]	Framework must be operationalized with quantitative analysis		X			
[6]	Lack of reliable methods to assess and predict RE's potential magnitude		X			
[20]	Develop performance measures to assess the CE effects in firms		X			
[56]	Integrating with quantitative and qualitative studies			X		
[59]	Additional metrics to track sustainable supply chain activity			X		
[58]	Conduct quantitative analysis along the manufacturing process				X	
	Total: Performance measurements category	**7**	**4**	**2**	**1**	-
[61]	Examine the political barriers	X				
[47]	Implications of policy on CBM adoption		X			
[20]	Examine the policy actions to encourage the adoption of CE	X				

(*continued*)

Table 1. (*continued*)

Ref.	Specific gaps and related gap categories	Dimensions (D)				
[62]	Political initiatives to assist businesses in using green technologies	X				
	Total: Policy and regulations category	**1**	**3**	**-**	-	-
[61]	Investigate the links between CE and digitalization	X				
[64]	Insufficient knowledge of IoT's environmental benefits	X				
[70]	Additional research on reverse logistics and augmented reality	X				
[72]	Digital technology for system simulation/scheduling/control/optiMisation	X				
[57]	Digital technologies to assess the impact of data and to track and manage it		X			
[69]	Digitalization and CE to promote sustainability in the food supply chain		X			
[20]	Analyse how industry 4.0 promotes CE practices		X			
[17]	Investigate the technological effects of applying CE practices.		X			
[27]	Investigate the benefits of CE achievable with information sharing			X		
[67]	Drive for AI adoption; SC and traditional AM restructured			X		
[74]	Reduce the size of waste sorting plants and food waste digesters; enhance cooperative platforms for IS			X		
[59]	Algorithms to turn data into information; digitalization of sustainable SCs			X		
[65]	Investigate the connection between the PLM and IoT platforms				X	
[71]	Framework for the digital transformation and sustainability transition				X	
[73]	Integrated system with multiple product lifecycles for remanufacturing				X	
[66]	Data analytics to integrate conventional design processes for SMEs				X	
[33]	Examine the digital technologies to manage the lean PSS lifetime				X	
[35]	Analysis of big data analytics and the IoT to support the PLM System				X	
[40]	Study about how changes in the developing digital world influence PLM				X	
[63]	More research on industry 4.0 integration with CE and CM	X				
[68]	Technology for quality and safety concerns in sustainable food supply chain	X				
	Total: Technologies potential category	**4**	**4**	**4**	**7**	**2**
	Overall occurrence of gaps in CE-RE dimensions	**28**	**29**	**10**	**15**	**6**

To the category of industrial applications is related to the need of practical cases in firms or of more studies in the same or different industries different from the examined research. In PLM dimension, case studies would be necessary to adopt PSS for operational levels and validate them in strategic approaches [32] and analyse possible barriers and drivers [35, 36]. As for the drivers' dimension, empirical evidences to enable the correct implementation and application of CE in companies are needed [22, 37]. CBMs require practical case studies to investigate their successful and proper implementation [13]. In construction sector, more studies would be needed on how CBMs operate [38] to optimize resource use and achieve net-zero emissions [39]. In fashion and textile industries, empirical studies are needed to investigate how PLM is influenced by digital technologies [40] to minimize waste [41]. The food industry should be concerned about the interrelationships with reusable packaging [23, 42] and the materials from which food contact films are made [43] throughout the entire product lifecycle [44]. Several gaps refer to examining the CBM [45] proposed in other sectors to find the most suitable for the application of this kind of business model [46] and maximize the circular benefits [47]. In some studies, CE practices didn't include action [48] for a specific industrial sector [49], making necessary more research [50] and assessment in other fields [51] such as built environment and agriculture [24]. Finally, some studies should consider specific reverse logistic components in industry [52].

The category of performance measurements generally highlights the lack of a system to evaluate CE and RE. In particular, a lack of metrics for environmental, economic, and social performance is highlighted in many sources [53–55]. Moreover, further studies should develop a quantitative and qualitative method to assess circular strategies and reach sustainable objectives [30-, 31, 56]. Operational CE indicators need to be developed in firms for evaluating manufacturing processes [13, 20, 57, 58]. Other performance metrics should evaluate the sustainability of the value chain, focusing on SC [59] and reverse logistics [52]. Lastly, methods to assess and predict the RE of CE would be required for the proper selection and design of circular strategies [6, 60].

In policy and regulations, gaps focus on political initiatives for the CE. The category consists of political barriers that hamper CE initiatives [61, 62] and policy actions that encourage the adoption of CE strategies (e.g., CBM) [20, 46].

The category of technologies potential relates to the examination of the use of technologies to promote CE. It is confirmed that Industry 4.0 is one of the main enablers for CE. Different gaps underline the relevance of investigating how to integrate industry 4.0 technologies and CE strategies [20, 63]. For example, the Internet of Things (IoT) could lead to environmental benefits [64] and support PLM systems [35, 65]. Data analytics could help design a successful circular process [66], and the adoption of artificial intelligence (AI) and Additive Manufacturing (AM) could modify the traditional SC [67]. The SC needs the support of technologies to manage and monitor the level of sustainability [59, 68], especially in the food industry [69], and increase efficient reverse logistics [70]. More studies should concern the digital transformation to achieve sustainability transition [71] and connecting to CE for sharing information to obtain more benefits [27, 61]. Digital technologies in manufacturing should be a system to manage, schedule, and control [72] the effects of CE practices [17, 57]. In addition, the digital world influences and supports PLM systems [40, 73], and it should be studied how to manage the lean

PSS lifetime [33]. Finally, technologies should be studied for practical application in waste sorting plants and industrial symbiosis (IS) [74].

Finally, gap categories were prioritized, indicating Industrial applications, Technologies potential and Performance measurements as those calling more for further research, followed by Attitudes and behaviours, Ecosystems configurations, In-depth conceptual analysis, and Policy and regulations.

4 Conclusions

The objective of this research was to identify and systematize the main gaps characterizing the twofold domain CE and RE, also proposing a related research agenda to cope with them. According to this purpose, a SLR was conducted, and seven gap categories were detected. These seven gap categories have been prioritized to provide a research agenda related to RE in CE. In particular, industrial applications, technologies, and performance measurements are prior categories to be investigated. Several studies have highlighted the importance of investigating the adoption of CE practices in practical applications with companies (Industrial applications category) to assess the benefits reached and identify possible REs, providing an opportunity to examine the progression of CBM and circular strategies adoption, and assess whether they are able to achieve net-zero waste and emissions or cause undesirable effects. Concerning Technologies potential category, many studies suggest digital technologies for supporting the implementation of CE practices (managing, monitoring, and tracking the impact caused and giving the possibility of preventing RE in the long run). The PLM dimension represents the most suitable to foster technology adoption, managing the stages of the product lifecycle, helping to understand the status of a product at the end of its life and to close the related loops. Relating to performance assessment in CE (Performance measurements category), the development of frameworks, methods, tools, and approaches for qualitatively and quantitatively evaluate the benefits of circular practices and identify the possible occurrence of RE should be an important priority. The BM dimension presents several gaps in this category, driven by the interest of companies adopting CBM to scrutinize the sustainable results achieved. Moreover, concerning Attitudes and behaviors category, there is need to understand consumers' sustainable choices, to detect those leading to REs, and to educate consumers on the effects of their actions.

The research provides several contributions to knowledge and practice. Firstly, the SLR identifies the gaps in each of the five dimensions of the CE-RE research context, clusterizing them into seven categories and prioritizing them to propose a research agenda. Furthermore, this paper wants to raise practitioners' awareness about the main gaps to explore and overcome to support the CE transition. This research offers managers and policymakers decision support for future work, such as industrial applications of circular strategies and policy actions to reach sustainability. However, this research presents some limitations. It focuses only on the research gaps in the five dimensions of the twofold domain CE and REs in manufacturing due to the research string adopted during the SLR. In addition, a limited number of papers were collected for some gap categories and dimensions (i.e., CM ecosystem and socio-economic aspects).

Fundings. Project funded under the National Recovery and Resilience Plan (NRRP), Mission 4 Component 2 Investment 1.3 - Call for tender No. 341 of 15/03/2022 of Italian Ministry of University and Research funded by the European Union – NextGenerationEU. Award Number: PE00000004, Concession Decree No. 1551 of 11/10/2022 adopted by the Italian Ministry of University and Research, CUP D93C22000920001, MICS (Made in Italy - Circular and Sustainable)

References

1. Geissdoerfer, M., Savaget, P., Bocken, N.M.P., Hultink, E.J.: The circular economy – a new sustainability paradigm? J. Clean. Prod. **143**, 757–768 (2017). https://doi.org/10.1016/J.JCLEPRO.2016.12.048
2. Gusmerotti, N.M., Testa, F., Corsini, F., Pretner, G., Iraldo, F.: Drivers and approaches to the circular economy in manufacturing firms. J. Clean. Prod. **230**, 314–327 (2019). https://doi.org/10.1016/J.JCLEPRO.2019.05.044
3. Acerbi, F., Sassanelli, C., Taisch, M.: A conceptual data model promoting data-driven circular manufacturing. Oper. Manag. Res. **15**(3–4), 838–857 (2022). https://doi.org/10.1007/s12063-022-00271-x
4. Wei, T., Liu, Y.: Estimation of global rebound effect caused by energy efficiency improvement. Energy Econ **66**, 27–34 (2017). https://doi.org/10.1016/j.en-eco.2017.05.030
5. Zink, T., Geyer, R.: Circular Economy Rebound. J. Ind. Ecol. **21**(3), 593–602 (2017). https://doi.org/10.1111/JIEC.12545
6. Metic, J., Pigosso, D.C.A.: Research avenues for uncovering the rebound effects of the circular economy: A systematic literature review. J. Clean Prod. **368** (2022). https://doi.org/10.1016/J.JCLEPRO.2022.133133
7. Castro, C.G., Trevisan, A.H., Pigosso, D.C.A., Mascarenhas, J.: The rebound effect of circular economy: Definitions, mechanisms and a research agenda. J. Clean. Prod. **345**, 131136 (2022). https://doi.org/10.1016/J.JCLEPRO.2022.131136
8. Denyer, D., Tranfield, D.: Producing a Systematic Review, pp. 671–689 (2009)
9. Ferrante, M., Vitti, M., Digiesi, S., Sassanelli, C.: Investigating the circular economy's rebound effect in manufacturing: a systematic literature review (2023). Forthcoming
10. Maeda, J.: The laws of simplicity. MIT Press (2006)
11. Gomes, G.M., Moreira, N., Ometto, A.R.: Role of consumer mindsets, behaviour, and influencing factors in circular consumption systems: A systematic review. Sustain Prod Consum **32**, 1–14 (2022). https://doi.org/10.1016/j.spc.2022.04.005
12. Salvador, R., Barros, M.V., da Luz, L.M., Piekarski, C.M., de Francisco, A.C.: Circular business models: Current aspects that influence implementation and unaddressed subjects. Journal of Cleaner Production, vol. 250. Elsevier Ltd. (2020). https://doi.org/10.1016/j.jclepro.2019.11955
13. Merli, R., Preziosi, M., Acampora, A.: How do scholars approach the circular economy? A systematic literature review. Journal of Cleaner Production, vol. 178. Elsevier Ltd., pp. 703–722 (2018). https://doi.org/10.1016/j.jclepro.2017.12.112
14. Bressanelli, G., Perona, M., Saccani, N.: Challenges in supply chain redesign for the Circular Economy: a literature review and a multiple case study. Int. J. Prod. Res. **57**(23), 7395–7422 (2019). https://doi.org/10.1080/00207543.2018.1542176
15. De Bruyne, M.J., Verleye, K.: Realizing the economic and circular potential of sharing business models by engaging consumers. J. Serv. Manag. **34**(3), 493–519 (2023). https://doi.org/10.1108/JOSM-08-2021-0318/FULL/XML

16. Salvador, R., et al.: Key aspects for designing business models for a circular bioeconomy. Journal of Cleaner Production, vol. 278. Elsevier Ltd. (2021). https://doi.org/10.1016/j.jclepro.2020.124341
17. Jia, F., Yin, S., Chen, L., Chen, X.: The circular economy in the textile and apparel industry: a systematic literature review. Journal of Cleaner Production, vol. 259. Elsevier Ltd. (2020). https://doi.org/10.1016/j.jclepro.2020.120728
18. Mies, A., Gold, S.: Mapping the social dimension of the circular economy. Journal of Cleaner Production, vol. 321. Elsevier Ltd. (2021). https://doi.org/10.1016/j.jclepro.2021.128960
19. Acerbi, F., Taisch, M.: A literature review on circular economy adoption in the manufacturing sector. J. Clean. Prod. **273**, 123086 (2020). https://doi.org/10.1016/J.JCLEPRO.2020.123086
20. Silvério, A.C., Ferreira, J., Fernandes, P.O., Dabić, M.: How does circular economy work in industry? Strategies, opportunities, and trends in scholarly literature. Journal of Cleaner Production, vol. 412. Elsevier Ltd. (2023). https://doi.org/10.1016/j.jclepro.2023.137312
21. Fogt Jacobsen, L., Pedersen, S., Thøgersen, J.: Drivers of and barriers to consumers' plastic packaging waste avoidance and recycling – A systematic literature review. Waste Management, vol. 141. Elsevier Ltd., pp. 63–78 (2022). https://doi.org/10.1016/j.wasman.2022.01.021
22. Hugo, A. de A., de Nadae, J., Lima, R. da S.: Can fashion be circular? A literature review on circular economy barriers, drivers, and practices in the fashion industry's productive chain. Sustainability (Switzerland), vol. 13, no. 21. MDPI (2021). https://doi.org/10.3390/su132112246
23. Pålsson, H., Olsson, J.: Current state and research directions for disposable versus reusable packaging: a systematic literature review of comparative studies. Packaging Technology and Science, vol. 36, no. 6. John Wiley and Sons Ltd., pp. 391–409 (2023). https://doi.org/10.1002/pts.2722
24. Valencia, M., Bocken, N., Loaiza, C., De Jaeger, S.: The social contribution of the circular economy. J. Clean Prod., vol. 408 (2023). https://doi.org/10.1016/j.jclepro.2023.137082
25. Ding, L., Wang, T., Chan, P.W.: Forward and reverse logistics for circular economy in construction: a systematic literature review. Journal of Cleaner Production, vol. 388. Elsevier Ltd. (2023). https://doi.org/10.1016/j.jclepro.2023.135981
26. Eisenreich, A., Füller, J., Stuchtey, M., Gimenez-Jimenez, D.: Toward a circular value chain: Impact of the circular economy on a company's value chain processes. Journal of Cleaner Production, vol. 378. Elsevier Ltd. (2022). https://doi.org/10.1016/j.jclepro.2022.134375
27. Jäger-Roschko, M., Petersen, M.: Advancing the circular economy through information sharing: A systematic literature review. Journal of Cleaner Production, vol. 369. Elsevier Ltd. (2022). https://doi.org/10.1016/j.jclepro.2022.133210
28. Despeisse, M., Acerbi, F.: Toward eco-efficient and circular industrial systems: ten years of advances in production management systems and a thematic framework. Prod Manuf Res **10**(1), 354–382 (2022). https://doi.org/10.1080/21693277.2022.2088634
29. Sehnem, S., de Queiroz, A.A.F.S.L., Pereira, S.C.F., dos Santos Correia, G., Kuzma, E.: Circular economy and innovation: A look from the perspective of organizational capabilities. Bus Strategy Environ, vol. 31, no. 1, pp. 236–250 (2022). https://doi.org/10.1002/bse.2884
30. Salvador, R., et al.: How to advance regional circular bioeconomy systems? Identifying barriers, challenges, drivers, and opportunities. Sustainable Production and Consumption, vol. 32. Elsevier B.V., pp. 248–269 (2022). https://doi.org/10.1016/j.spc.2022.04.025
31. Kuzma, E.L., Sehnem, S., Machado, H.P.V., Campos, L.M.D.S.: The new business is circular? Analysis from the perspective of the circular economy and entrepreneurship. Production **31**, 1–17 (2021). https://doi.org/10.1590/0103-6513.20210008
32. Reim, W.: STRATEGY, BUSINESS MODELS, OR TACTICS: WHAT IS PRODUCT SERVICE SYSTEM (PSS) LITERATURE TALKING ABOUT? (2013). https://www.researchgate.net/publication/285985202

33. Sassanelli, C., et al.: Defining lean product service systems features and research trends through a systematic literature review (2019)
34. Blüher, T., Riedelsheimer, T., Gogineni, S., Klemichen, A., Stark, R.: Systematic literature review-Effects of PSS on sustainability based on use case assessments. Sustainability (Switzerland), vol. 12, no. 17. MDPI (2020). https://doi.org/10.3390/su12176989
35. Sakao, T., Nordholm, A.K.: Requirements for a product lifecycle management system using internet of things and big data analytics for Product-as-a-Service. Frontiers in Sustainability, vol. 2. Frontiers Media S.A. (2021). https://doi.org/10.3389/frsus.2021.735550
36. Brandt, N., Ahlemann, F., Rehring, K., Reining, S.: Internet of Things in Product Lifecycle Management-A Review Internet of Things in Product Lifecycle Management-A Review on Value Creation through Product Status Data on Value Creation through Product Status Data (2021). https://aisel.aisnet.org/amcis2021
37. Hina, M., Chauhan, C., Kaur, P., Kraus, S., Dhir, A.: Drivers and barriers of circular economy business models: Where we are now, and where we are heading. J. Clean Prod. **333** (2022). https://doi.org/10.1016/j.jclepro.2021.130049
38. Munaro, M.R., Tavares, S.F., Bragança, L.: Towards circular and more sustainable buildings: A systematic literature review on the circular economy in the built environment. Journal of Cleaner Production, vol. 260. Elsevier Ltd. (2020). https://doi.org/10.1016/j.jclepro.2020.121134
39. Chen, Q., Feng, H., Garcia de Soto, B.: Revamping construction supply chain processes with circular economy strategies: A systematic literature review. Journal of Cleaner Production, vol. 335. Elsevier Ltd. (2022). https://doi.org/10.1016/j.jclepro.2021.130240
40. Conlon, J.: From PLM 1.0 to PLM 2.0: the evolving role of product lifecycle management (PLM) in the textile and apparel industries. Journal of Fashion Marketing and Management, vol. 24, no. 4. Emerald Group Holdings Ltd., pp. 533–553 (2020). https://doi.org/10.1108/JFMM-12-2017-0143
41. Abbate, S., Centobelli, P., Cerchione, R., Nadeem, S.P., Riccio, E.: Sustainability trends and gaps in the textile, apparel and fashion industries. Environment, Development and Sustainability. Springer Science and Business Media B.V. (2023). https://doi.org/10.1007/s10668-022-02887-2
42. Ada, E., Kazancoglu, Y., Gozacan-Chase, N., Altin, O.: Challenges for circular food packaging: Circular resources utilization. Applied Food Research, vol. 3, no. 2 (2023). https://doi.org/10.1016/j.afres.2023.100310
43. Tanveer, U., Ishaq, S., Oqueli, T.: An insight into the application of gradations of circularity in the food packaging industry: a systematic literature review and a multiple case study. Sustainability (Switzerland) **15**(4) (2023). https://doi.org/10.3390/su15043007
44. Corallo, A., Latino, M.E., Menegoli, M., Pontrandolfo, P.: A systematic literature review to explore traceability and lifecycle relationship. International Journal of Production Research, vol. 58, no. 15. Taylor and Francis Ltd., pp. 4789–4807 (2020). https://doi.org/10.1080/00207543.2020.1771455
45. Mhatre, P., Panchal, R., Singh, A., Bibyan, S.: A systematic literature review on the circular economy initiatives in the European Union. Sustainable Production and Consumption, vol. 26. Elsevier B.V., pp. 187–202 (2021). https://doi.org/10.1016/j.spc.2020.09.008
46. Assmann, I.R., Rosati, F., Morioka, S.N.: Determinants of circular business model adoption—a systematic literature review. Bus Strategy Environ (2023). https://doi.org/10.1002/bse.3470
47. Rosa, P., Sassanelli, C., Terzi, S.: Circular Business Models versus circular benefits: An assessment in the waste from Electrical and Electronic Equipments sector. Journal of Cleaner Production, vol. 231. Elsevier Ltd, pp. 940–952 (2019). https://doi.org/10.1016/j.jclepro.2019.05.310

48. Rosa, P., Sassanelli, C., Terzi, S.: Towards Circular Business Models: A systematic literature review on classification frameworks and archetypes. Journal of Cleaner Production, vol. 236. Elsevier Ltd. (2019). https://doi.org/10.1016/j.jclepro.2019.117696
49. Mehmood, A., Ahmed, S., Viza, E., Bogush, A., Ayyub, R.M.: Drivers and barriers towards circular economy in agri-food supply chain: A review. Business Strategy and Development **4**(4), 465–481 (2021). https://doi.org/10.1002/bsd2.171
50. Aloini, D., Dulmin, R., Mininno, V., Stefanini, A., Zerbino, P.: Driving the transition to a circular economic model: a systematic review on drivers and critical success factors in circular economy. Sustainability (Switzerland), vol. 12, no. 24. MDPI, pp. 1–14 (2020). https://doi.org/10.3390/su122410672
51. Amicarelli, V., Bux, C., Spinelli, M.P., Lagioia, G.: Life cycle assessment to tackle the take-make-waste paradigm in the textiles production. Waste Management, vol. 151. Elsevier Ltd. pp. 10–27 (2022). https://doi.org/10.1016/j.wasman.2022.07.032
52. Mallick, P.K., Salling, K.B., Pigosso, D.C.A., McAloone, T.C.: Closing the loop: Establishing reverse logistics for a circular economy, a systematic review. Journal of Environmental Management, vol. 328. Academic Press (2023). https://doi.org/10.1016/j.jenvman.2022.117017
53. Försterling, G., Orth, R., Gellert, B.: Transition to a circular economy in europe through new business models: barriers, drivers, and policy making. Sustainability (Switzerland), vol. 15, no. 10. MDPI (2023). https://doi.org/10.3390/su15108212
54. van Loon, P., Diener, D., Harris, S.: Circular products and business models and environmental impact reductions: Current knowledge and knowledge gaps. J. Clean Prod. **288** (2021). https://doi.org/10.1016/j.jclepro.2020.125627
55. Vegter, D., van Hillegersberg, J., Olthaar, M.: Supply chains in circular business models: processes and performance objectives. Resour Conserv Recycl. **162** (2020). https://doi.org/10.1016/j.resconrec.2020.105046
56. Marić, I., Opazo-Basáez, M., Vlačić, B., Dabić, M.: Innovation management of three-dimensional printing (3DP) technology: Disclosing insights from existing literature and determining future research streams. Technol. Forecast Soc. Change **193** (2023). https://doi.org/10.1016/j.techfore.2023.122605
57. Acerbi, F., Sassanelli, C., Terzi, S., Taisch, M.: A systematic literature review on data and information required for circular manufacturing strategies adoption. Sustainability **13**(4), 2047 (2021). https://doi.org/10.3390/SU13042047
58. Karkasinas, A., Rentizelas, A.: Verification and validation: a novel product lifecycle management-based approach (2019)
59. Liu, I., Song, W., Liu, Y.: Leveraging digital capabilities toward a circular economy: Reinforcing sustainable supply chain management with Industry 4.0 technologies. Comput. Ind. Eng. **178**(2023). https://doi.org/10.1016/j.cie.2023.109113
60. Kanzari, A., Rasmussen, J., Nehler, H., Ingelsson, F.: How financial performance is addressed in light of the transition to circular business models - A systematic literature review. Journal of Cleaner Production, vol. 376. Elsevier Ltd. (2022). https://doi.org/10.1016/j.jclepro.2022.134134
61. Chauhan, A., Parida, V., Dhir, A.: Linking circular economy and digitalisation technologies: A systematic literature review of past achievements and future promises. Technol. Forecast Soc. Change **177** (2022). https://doi.org/10.1016/j.techfore.2022.121508
62. Silvestri, A., Silvestri, L., Forcina, A., Di Bona, G., Falcone, D.: Green chemistry contribution towards more equitable global sustainability and greater circular economy: A systematic literature review. J. Clean Prod. **294** (2021). https://doi.org/10.1016/j.jclepro.2021.126137
63. Agrawal, R., Wankhede, V.A., Kumar, A., Upadhyay, A., Garza-Reyes, J.A.: Nexus of circular economy and sustainable business performance in the era of digitalization. Int. J. Product. Perform. Manage. **71**(3), 748–774 (2022). https://doi.org/10.1108/IJPPM-12-2020-0676

64. Ding, S., Tukker, A., Ward, H.: Opportunities and risks of internet of things (IoT) technologies for circular business models: a literature review. Journal of Environmental Management, vol. 336. Academic Press (2023). https://doi.org/10.1016/j.jenvman.2023.117662
65. Barrios, P., Danjou, C., Eynard, B.: Literature review and methodological framework for integration of IoT and PLM in manufacturing industry. Comput. Ind. **140** (2022). https://doi.org/10.1016/j.compind.2022.103688
66. Meyer, I., Wiederkehr, I., Koldewey, C., Dumitrescu, R.: Understanding usage datadriven product planning: A systematic literature review. In: Proceedings of the Design Society, pp. 3289–3298. Cambridge University Press (2021). https://doi.org/10.1017/pds.2021.590
67. Sood, S.K., Rawat, K.S., Sharma, G.: Role of enabling technologies in soft tissue engineering: a systematic literature review. IEEE Eng. Manage. Rev. **50**(4), 155–169 (2022). https://doi.org/10.1109/EMR.2022.3195923
68. Kumar, A., Mangla, S.K., Kumar, P.: An integrated literature review on sustainable food supply chains: Exploring research themes and future directions. Science of the Total Environment **821** (2022). https://doi.org/10.1016/j.scitotenv.2022.153411
69. Adams, A., Donovan, J., Topple, C.: Achieving sustainability in food manufacturing operations and their supply chains: Key insights from a systematic literature review. Sustainable Production and Consumption, vol. 28. Elsevier B.V., pp. 1491–1499 (2021). https://doi.org/10.1016/j.spc.2021.08.019
70. Abideen, A.Z., Pyeman, J., Sundram, V.P.K., Tseng, M.L., Sorooshian, S.: Leveraging capabilities of technology into a circular supply chain to build circular business models: a state-of-the-art systematic review. Sustainability (Switzerland) vol. 13, no. 16. MDPI AG (2021). https://doi.org/10.3390/su13168997
71. De Oliveira, I.M., Andreatta, L.G., Stjepandić, J., Junior, O.C.: Product lifecycle management and sustainable development in the context of industry 4.0: A systematic literature review. In: Advances in Transdisciplinary Engineering, pp. 213–222. IOS Press BV (2021). https://doi.org/10.3233/ATDE210100
72. Cioffi, R., Travaglioni, M., Piscitelli, G., Petrillo, A., Parmentola, A.: Smart manufacturing systems and applied industrial technologies for a sustainable industry: A systematic literature review. Applied Sciences (Switzerland), vol. 10, no. 8. MDPI AG (2020). https://doi.org/10.3390/APP10082897
73. Li, K., Liu, J., Fu, H., Zhao, N.G.: An integrated system with multiple product lifecycles for remanufacturing (IS-MPLR): new opportunities and challenges. Int. J. Comput. Integr. Manuf. **34**(1), 20–40 (2021). https://doi.org/10.1080/0951192X.2020.1858498
74. Kerdlap, P., Low, J.S.C., Ramakrishna, S.: Zero waste manufacturing: A framework and review of technology, research, and implementation barriers for enabling a circular economy transition in Singapore. Resour Conserv Recycl. **151** (2019). https://doi.org/10.1016/j.resconrec.2019.104438

Multi-aspect Ontologies as Support for Quantified Products Lifecycle Management

Kurt Sandkuhl[1,3](✉), Nikolay Shilov[2], Alexander Smirnov[2], and Ulf Seigerroth[3]

[1] University of Rostock, Albert-Einstein Street 22, 18059 Rostock, Germany
kurt.sandkuhl@uni-rostock.de

[2] SPC RAS, 39, 14th Line, St. Petersburg 199178, Russia
{nick,smir}@iias.spb.su

[3] School of Engineering, Jönköping University, 55111 Jönköping, Sweden
ulf.seigerroth@ju.se

Abstract. Quantified products (QP) collect data about themselves as a basis for data-driven services. They consist of fleets of physical products, IT components built into the products, IT services and business services. The product lifecycle management of QP has to integrate the individual lifecycles for the different components and faces the challenge of a large variability of possible customer solutions. Multi-aspect ontologies can serve as the fundament for an approach to capture interface specifications and formal dependencies of the different QP components and variants. Motivated by an industrial case study, the paper investigates the use of multi-aspect ontologies as support for QP engineering. The main contributions of this work are (1) an approach to integration of multi-aspect ontologies into QP engineering, (2) two industrial case studies showing the actual multi-aspect ontology, and (3) experiences from the case studies.

Keywords: Multi-aspect ontology · quantified product · digital transformation

1 Introduction

In many industrial areas, the digitalization of products resulted in new business models [1] and digital ecosystems [2] that do not only enhance products with accompanying services but also exploit the data as such [3]. Examples are smart connected products [4], Internet-of-Things applications [5], new kinds of product/service systems [6], or quantified products [7]. Quantified products (QP) are products collecting data about themselves that can be measured for the purpose of data analysis, data sharing and data-driven services (cf. Section 2.1).

QP consist of physical products (as single device or fleet of devices), IT components built into the product or implementing IT services, and business services using physical products or IT components. As all constituents of QP require systematic development and maintenance processes, the lifecycle management of QP consists of several integrated lifecycles for the different types of components (see Section 2.1) and different

Published by Springer Nature Switzerland AG 2025
P. Sureephong et al. (Eds.): PLM 2024, IFIP AICT 740, pp. 15–29, 2025.
https://doi.org/10.1007/978-3-031-93319-6_2

engineering disciplines involved in the lifecycles (e.g., mechanical, electrical, software and service engineering).

One of the frequent challenges observed in QP engineering is the large variability of possible customer solutions [7]. This variability usually is a consequence of the fact that physical products, IT services and business services all exist in many different variants. Thus, their combination in QP leads to an even higher number of variants. In this context, the engineering process and QP lifecycle would benefit from an approach to capture interface specifications and formal dependencies of the different components while at the same time maintaining the established structures that the different disciplines have. Our conjecture is that a combination of feature modelling and multi-aspect ontologies can serve this purpose. Feature models can be used to specify what characteristics the QP is supposed to offer to the customers and multiaspect ontologies to formalize the relation between these features and the components available for their implementation. Thus, this paper investigates the use of multi-aspect ontologies as support for QP engineering.

Motivated by an industrial case study, we present our approach and develop a multi-aspect ontology for verification purposes. This is followed by a discussion how to integrate the multi-aspect ontologies into QP engineering. The main contributions of this work are (1) an approach how to integrate multi-aspect ontologies into QP engineering, (2) two industrial case studies showing the actual multi-aspect ontology, and (3) experiences from the case studies.

The remainder of this paper is structured as follows: Section 2 provides the background for our work on quantified products, multi-aspect ontologies and previous results. Section 3 briefly discussed the research method applied. Section 4 introduces the industrial case studies and the multi-aspect ontology. Section 5 tackles the question how to integration the ontology into the QP engineering lifecycle. Section 6 summarizes the work and gives an outlook on future research.

2 Theoretical Background

The theoretical background is briefly summarized in this section and originates from multi-aspect ontologies (section 2.2) and quantified products (section 2.1).

2.1 Quantified Products

"Turning data into value" is the motivation for several research streams that evolved in the past decade and form the context for research on quantified products (QP). Smart connected products (SCP) is a category of IoT application and also forms a business model category that is characterized by IT-based and data-driven services [4]. SCP are equipped with sensors, e.g., for tracking geographic positions or energy consumption, for capturing usage profiles or other information relevant to operation and maintenance of the product.

In individual self-management the term of Quantified Self [8] was coined for people collecting physical and biometric data to "quantify" their life. The Quantified Vehicle adapts this idea for the automotive industry as a digital ecosystem establishing data-driven services [9] and generating business opportunities. Quantified vehicles also collect additional data, which could be of interest for third parties.

The service sectors related to SCP and Quantified Vehicles "collect data on everything" [10] to achieve automation and increased efficiency. For example, product manufacturers and component suppliers want to understand how their components are used in the field [11]. In case of vehicles, insurance industry is interested to learn how the insured persons drive [12] and traffic planners how vehicle use the road infrastructure [13].

When defining the term quantified product, we separate between product data produced and captured in traditional product lifecycle management (PLM) and data collected by product instances, i.e., manufactured individual copies of a product: "A Quantified Product is a product whose instances collect data about themselves that can be measured, or, by design, leave traces of data, including operational, physical, behavioural, or environmental information for the purpose of data analysis and (optional) data sharing and services." [14]

Business models for quantified products often build upon (a) collecting data from single instances and the entire fleet of instances of a product while operating in the field, (b) applying this data for fleet and device management, including monitoring, real-time control and maintenance, and (c) aggregating and analysing data for defined target groups and offering them on marketplaces.

Compared to traditional PLM, the PLM for QP has to coordinate and integrate different kinds of products and their lifecycles:

- the actual physical device or product,
- services built upon data from connected physical product instances, and services using data from a fleet of products, i.e., all instances operating in the field,
- services using the fleet data as basis for data-driven ecosystem services, often in combination with other services.

Implementation of QP in enterprises is accompanied by substantial changes in the companies' PLM, as the required coordination of these different lifecycles requires both organizational, structural, and technical solutions.

2.2 Feature Modelling

A feature is a "distinctive and user-visible aspect, quality, or characteristic of a software system or systems" [15]. The purpose of a feature model is to capture, structure and visualize the commonality and variability of a domain or a set of products. Commonalities are the properties of products shared among all the products in a set, placing the products in the same category or family. Variability is the elements of the products that differentiate and show the configuration options, choices and variation points that are possible between variants of the product, aimed to satisfy customer needs and requirements. The variability and commonality are modelled as features and organized into a hierarchy of features and sub-features, sometimes called feature trees, in the feature model. The hierarchy and other properties of the feature model are visualized in a feature diagram. Feature diagrams express the relation between features with the relation types mandatory, optional, alternative, required and mutually exclusive.

Different methodical approaches in the field and the exact syntax of feature diagrams were analyzed and compared in [16], and an overview of methods for feature model development is provided in [17].

2.3 Multi-aspect Ontologies

The concept of multi-aspect ontology is aimed to describe complex domains integrating several aspects (points of view) to the same object or problem [18–20]. The most common problem for such cases is the complexity of the problem domain, certain level of independence between the aspects, and potential usage of different formalisms for description of the aspects. As a result, application of a single ontology usually is not an efficient solution due to its maintenance complexity and the necessity of transferring updates of the aspects into it.

There exist several approaches to integrate different ontologies, for example, multilingual [21], granular [22, 23], temporal [24], and layered ontologies [25]. The carried out analysis [18] showed that the hierarchical approach is the most suitable for the considered problem since it enables preserving the original ontologies or knowledge models unchanged setting links between them.

Following the hierarchical approach multi-aspect ontology has three interconnected levels:

- Local level represents the original ontologies in their original formalisms.
- Aspect level represents fragments of the original ontologies that interact with other aspects.
- Global level represents ontology elements, which are used in two or more aspects.

The aspect and global levels are described using a common formalism. To select the formalism, several ones have been investigated including Distributed Description Logic (DDL) [26], Context OWL [27], Multi-Viewpoint OWL (MVP-OWL) [28], epsilon-relations [29], Package-based Description Logic (P-DL) [30]. It was concluded that the OWL-compatible MVP-OWL was the most suitable for the problem at hand. However, it considered less connected viewpoints to a problem domain unlike aspects of the same object or problem. Thus, the MVP-OWL was modified resulting in an original OWL-compatible formalism for multi-aspect ontologies [18].

3 Research Approach

The results presented in this paper are part of research work that investigates how the lifecycle of quantified products can be supported with technological and methodical means. The overall research approach is based on the paradigm of design science research (DSR) [31]. DSR contributes to the theory and practice of solving real-world problems by offering a systematic way to generate prescriptive knowledge about the design of artifacts, such as methods, technologies, or models. Within the different

phases of our DSR project, the content of this paper has its focus on the design and initial evaluation of the envisioned artefact, a methodical and technological support for integrating the lifecycles of physical product, data-driven services and ecosystems services.

This paper builds on previous research on QP lifecycle management that also investigated common features of QP [7]. However, the feature models developed in this context (see Section 4) only visualized the building blocks for implementing features but did not specify the dependencies between building blocks and constraints across features. This resulted in the following research question for the work presented in this paper: In the context of quantified product engineering, how can multi-aspect ontologies support coordinated development of product features and building blocks?

The research approach used is a combination of descriptive case study and conceptual modelling. The reason for making case studies part of the empirical basis for our work is the lack of scientific work on quantified products. In previous work [14], we conducted a systematic literature analysis (SLA) with the result that hardly any work is published in the field. The SLA investigated publications in Scopus, IEEE Xplore, AISeL and ACM DL. An update of the SLA when preparing this paper confirmed this finding. Thus, we need to explore the nature and phenomenon of QP in real-world environments designing and developing QP, which is possible in case studies.

Based on the research question, we identified industrial case studies suitable to show the challenges related to developing product features and building blocks of QP. The analysis of the case studies and the use of the feature models resulting from the case studies form the basis for investigating the use of multi-aspect ontologies in QP development. The actual development of a multi-aspect ontology essentially is conceptual modelling (see Section 5). According to [32] the case studies in this paper are exploratory, as they are used to explore engineering process of QP in industrial practices and how to integrate multi-aspect ontologies.

4 Industrial Case Studies

This section summarizes the design and content of two case studies on QP engineering performed in the manufacturing industry. The purpose of the case studies was twofold: on the one hand side we intended to study if there are commonalities in the QP engineering processes when it comes to defining building blocks for QP implementation. We use the term "building block" in this section as superordinate concept for components, modules and product parts to reflect the practice in the case study companies. On the other hand, we analyzed the artefacts (e.g., models, interface definitions or specification documents) used to specify dependencies between features and building blocks, between features, and between building blocks.

4.1 Case Study Design

To investigate QP engineering process with respect to practices and artefacts for feature and building block specification, we had the possibility to study two developers of quantified products from the manufacturing sector: the first company (case 1) is a manufacturer of power garden products and the second one (case 2) a manufacturer of pumps. Within both enterprises, we had the possibility to collect information in several ways:

- One researcher worked during several months as part-time member of the development team for power garden products at the enterprise. The researcher maintained a work diary and collected information about the work processes, technologies used and practices by taking notes.
- Another researcher conducted interviews with different members of the development team for pumps during a period of two weeks on the premises of the enterprise.
- Both enterprises offered access to the documentation accompanying the development process (including specifications and models).

We defined two propositions P1 and P2. P1: The cases show commonalities in the QP engineering processes when it comes to defining building blocks for QP. P2: Artefacts documenting features and building block specifications can be observed in the case studies. To set clear boundaries for the cases we defined that only data collected during the work in the development team in the organization units responsible for power garden product (case 1) or during the interviews in the department responsible for pumping system design (case 2) shall be taken into account. Not subject of the case studies were other organization units of the enterprises, product and services provided to other market segments, the clients of the enterprise and sub-suppliers used for QP engineering.

4.2 Summary of Case Study Data

Case 1 is related to a research project conducted with a producer of garden and outdoor products for both, the consumer and business market (robotic and conventional lawn mowers, various sizes and types of chainsaws, garden tractors, watering systems, trimmers, and other devices). The business line "outdoor products" was the main partner in this research project that focused on the shift from products without connectivity to the Internet to smart and networked ones (QPs). The core objective of the business line was to develop an ecosystem of products and services that extends the conventional (physical) product lines and offers a common application architecture and service infrastructure. The ecosystem is meant to include different profit centers of the business line with their services and products, as well as product and services from external companies. The products in most cases have built-in electronic components implementing networking and communication capabilities where the product components connect to a common services architecture and infrastructure. The embedded systems and infrastructure services primarily are controlling the mechatronic components in the product, but also used for collecting data when the product is in operation. The data includes, e.g., status of mechatronic components, performance data and service information, or information from the product's operational environment.

The QP in focus of our investigation is the robotic lawn mower (RLM). RLMs are connected via a wireless connection to a base station installed at the point of operation. The base station in turn is connected to the Internet and the service infrastructure of the case company for providing software updates and services to the RLM. When installed at end-consumers, the RLM also provides operations data to the base station. When installed at business customers, like, e.g., garden service providers, RLMs commonly are operated as a fleet of robotic mowers. The case company provides fleet management services and the required data collection through a cloud-based platform. **Case 2** concerns a medium-sized manufacturer of pumping technologies and products for use in industrial environments (e.g., factory sewage pumps) and in the public infrastructures (e.g., pumps for flood protection). The main business originates from sales of actual pumps, facilities with pumps and integrated control systems, and their installation and maintenance. The company decided to expand their market offer to services based on IoT technologies in combination with the established product portfolio. The development of a "black-box datalink" facilitated the integration of sensors

into pumps and provision of information from the device to the back-office.

In combination with a new pump-controller, a "smart connected pump" was developed that can be monitored and controlled from remote. The black-box datalink provides information about, e.g., energy consumption, alerts about malfunctions, or the volume of fluid pumped. This allows for the new business model of "pumping as a service" where the company is paid for pumped cubic meters or hours of pumping. This also facilitates the offer of fleet management services, as many municipalities or industrial customers operating a fleet of pumps require an integrative view on the status of devices in their fleet.

In both cases, the features and building blocks of the quantified products were captured in a feature model. Excerpts from the feature models are presented in Figures 1 and 2. Each model shows on its first level the features offered to the customers. The second level shows the building blocks required for implementing features and the third level implementation options for the building blocks, which basically constitute variants of the building blocks. The feature models show dependencies between features and building blocks, but only if a building block is optional (o), mandatory (m) or alternative to another building block (a). Dependencies between the building blocks and technical information about the kind of dependency are not part of the model, as a feature model is not designed to include technical specifications. However, the knowledge about dependencies is important for the engineering process, in particular for the building blocks originating from different engineering disciplines as such dependencies are only captured in Excel (see Section 4.3) but in no formal model.

4.3 Case Study Analysis

The analysis of the case study data showed that both propositions can be confirmed.

P1 (commonalities when defining building blocks for QP): In the development of both QP, several disciplines cooperated, like mechanical engineers for the physical components (e.g., chassis, cutting unit, clutch, or wheels of the RLM; casing, impeller, pipes of pumps), electrical engineers for embedded systems and electrical parts (e.g., motor, control unit, battery, or sensors of the RLM; black-box datalink of the pumps;), software engineers for integration with back-office software and cloud, and business engineers for ecosystems service design and business process integration. In both cases, the building blocks of the QP were in a first step defined by the individual disciplines separately. This eased the reuse of already existing building blocks from previous product versions. Building blocks integrating components from different disciplines were mostly a composition of basic building blocks, not a re-design. In case 2 we observed that new QP-focused versions of the building blocks were created to avoid the danger of side-effects in case the original versions were modified in the product lifecycle for other products. Only in case 1, new building blocks were created that integrate building blocks from different disciplines. An example is the "anti-theft component" that integrates a sensor required for geo-location with an IT service for geo-fencing and a business service notifying the user in case of suspicious activity.

P2 (Artefacts documenting features and building block specifications): In both cases, we found a variety of artefacts for specifying and documenting building blocks, and we found that the different engineering disciplines maintained their traditional toolsets. For on-board electronic components, circuit board design software and specification tools were used producing native CAD or ODB++ formats. For the IT parts, we found UML modelling tools and software development environments for module and interface specifications. In mechanical engineering, we found several CAD tools and 3D CAD file formats. The business services in both cases were visualized as PowerPoint presentations that showed the costumer journey (case 1) and analysis of required IT and functional support.

However, the only "integrative" artefact that captured the dependencies between building blocks of different disciplines in both cases was an Excel file. In one case this file contained different sheets with separate matrixes of features and building blocks indicating dependencies by matrix entries or using identifiers for cross referencing. In the other case, the file used the rationale of quality function deployment (QFD) to trace from customer demands to product functions in several steps to building blocks and their costs, again using separate sheets of the file. In both cases, technical dependencies between building blocks or features were incomplete and difficult to reconstruct.

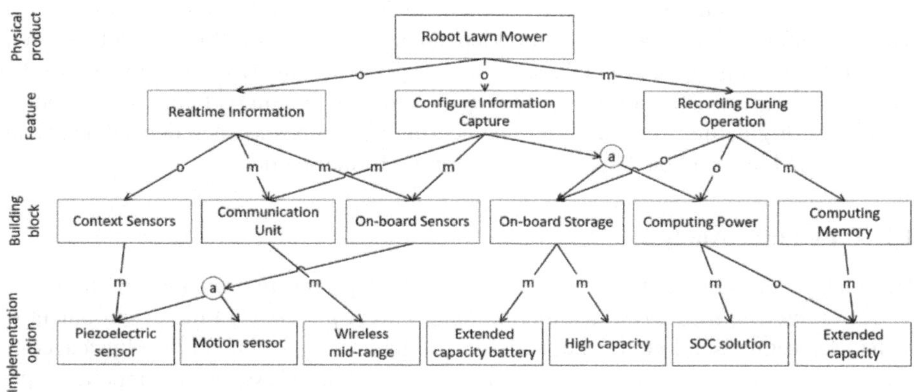

Fig. 1. Feature model excerpt for robot lawn mower product.

5 Multi-aspect Ontology for Quantified Product Design

The built multi-aspect ontology has two aspects: (1) feature aspect describing features and building blocks that are common for the considered case studies, and (2) implementation option aspect that is case study-specific. This division makes it possible to design

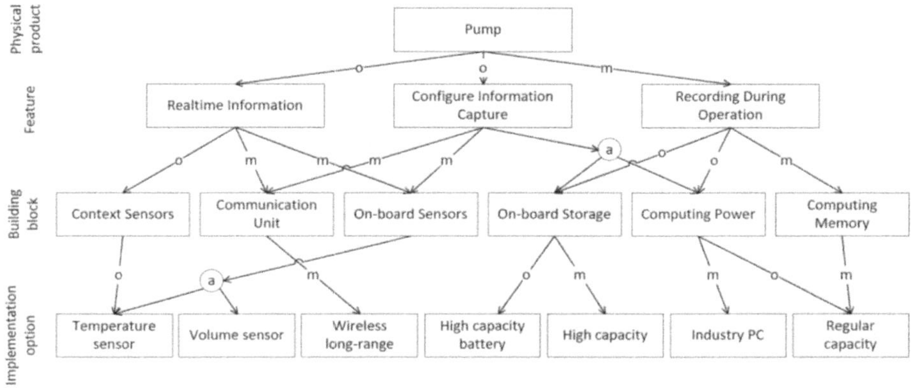

Fig. 2. Feature model excerpt for pump product.

the aspects separately and "plug in" the implementation option aspect for a considered case study without affecting the rest of the ontology. Below, the classes of different levels are described.

- Global level:
  - `Thing` (the root class of the ontology);
  - `Physical_product`;
  - `Feature`;
  - `Implementation_option`;
  - `Building_block` and its subclasses `Context_sensors`, `Communication_unit`, `On-board_sensors`, `On-board_storage`, `Computing_power`, `Computing_memory`, `Configure_information_capture_bb`.

The classes `On-board_storage` and `Computing_power` are disjoint subclasses of the `Configure_information_capture_bb` class.

- Aspect level for the feature aspect (subclasses of the global class `Feature`):
  - `Realtime_information`;
  - `Configure_information_capture`;
  - `Recording_during_operation`.

- Aspect level for the implementation option aspect of the Robot Lawn Mower case study (subclasses of the global class `Implementation_option`):
  - `Piezoelectric_sensor`;
  - `Motion_sensor`;
  - `Wireless_mid_range`;
  - `Extended_capacity_battery`;

- High_capacity;
- SOC_solution;
- Extended_capacity.

- Aspect level for the implementation option aspect of the Pump case study (subclasses of the global class Implementation_option):

  - Temperature_sensor;
  - Volume_sensor;
  - Wireless_long_range;
  - High_capacity_battery;
  - High_capacity;
  - Industry_PC;
  - Regular_capacity.

In order to describe the feature model relationships, the following rules have been implemented within the ontology at aspect levels:

- Aspect level rules for the feature aspect:

  - Realtime_information(?x) - > Communication_unit(?x)

(communication unit is mandatory if real-time information is chosen);

- Realtime_information(?x) - > On-board_sensors(?x) (onboard sensors are mandatory if real-time information is chosen);
- Configure_information_capture(?x) - > Communication_unit(?x) (communication unit is mandatory if configure information capture is chosen);
- Configure_information_capture(?x) - > Onboard_sensors(?x) (on-board sensors are mandatory if configure information capture is chosen);
- Recording_during_operation(?x) - > Computing_memory(?x) (computing memory is mandatory if recording during operation is chosen);
- Configure_information_capture(?x), (not (Computing_power))(?x), (not (On-board_storage))(?x)

->Configure_information_capture_bb(?x) (the auxiliary class Configure_information_capture_bb is mandatory if configure information capture is chosen)

The auxiliary class Configure_information_capture_bb serves as a flag that one of its subclasses has to be chosen, however, both of them cannot be chosen simultaneously since they are disjoint.

The same approach is used for describing the relationships between building blocks and implementation options for both case studies.

The built ontology provides the following functionality. When an instance of the Physical_product class is created and classified via required features, it is automatically classified to the mandatory building blocks and implementation options, except the "alternative" relationship, which is (as it was mentioned above) indicated with the auxiliary class, what means that one of its subclasses has to be chosen. Fig. 3 depicts the Protégé ontology editor window fragment illustrating this functionality. The manually defined classifications are in bold with white background and automatically derived classifications are highlighted with shaded background (see the dashed rectangle in the lower part of the figure).

Fig. 3. A fragment of the Protégé ontology editor window illustrating the functionality of the developed multi-aspect ontology.

6 Integration of Multi-aspect Ontology in QP Lifecycle

Engineering processes of products and services involving different disciplines and module types often consist of specific processes for the individual engineering disciplines and an overarching process managing the integration of discipline-specific tasks. For the engineering of QP, the integration of three lifecycles has been observed [14]: the lifecycle of the physical product, of the data-driven services and the ecosystem consisting of the business services. The three lifecycles are managed by a common QP management process. In both cases analyzed in Section 4, the models and information developed in these lifecycles are available in a joint file system accessible to all engineers involved.

Figure 4 shows the three lifecycle models and our approach for their integration. On the level of the (overarching) QP lifecycle management, the joint product and service development process for all engineering disciplines involved has been added. We propose to use the multi-aspect ontology as an additional artefact in the QP models file storage or repository to interlinks the specific building blocks for physical product, IT-services and services. This is visualized by showing the multi-aspect ontology in the quantified product models layer in grey color.

In addition to the integration into lifecycle processes an integration into engineering tool environment is important. Figure 5 shows an excerpt of the software architecture for this tool support with focus of the multi-aspect ontology integration. The multi-aspect ontology integrates various QP models and their aspects and provides for certain level of automation with the help of OWL reasoning software and neuro-symbolic machine learning models integrating neural networks and ontological rules [33]. This automation is in turn used by the engineering and other PLM support tools.

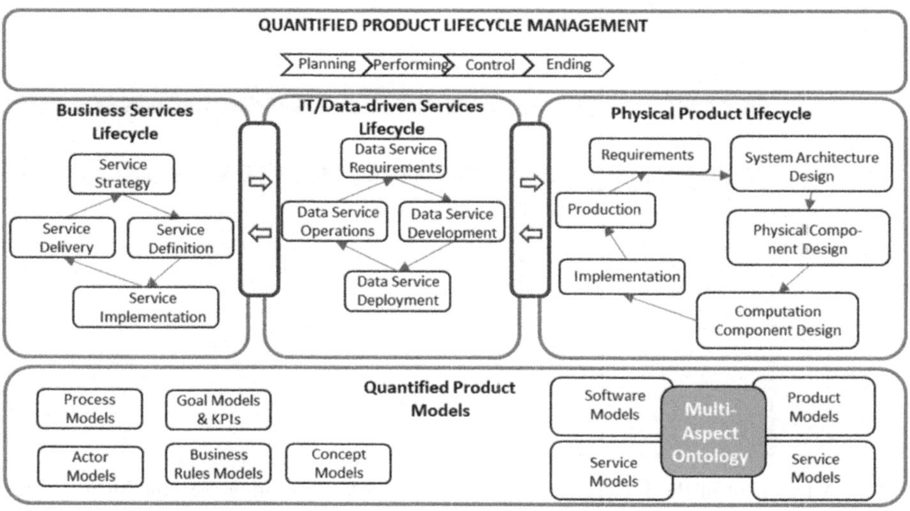

Fig. 4. QP Lifecycle Processes

The use cases described in Section 4 indicated that tool support applicable by all engineering disciplines would ease the management of dependencies between buildings blocks and to features. From a functional perspective, tool support would have to cover

- the actual development of the QP with models for different disciplines including the definition of building blocks and interfaces,
- the storage or linking of all documents related to feature models and building blocks (e.g., interface specification, functional description, implementation documentation), and

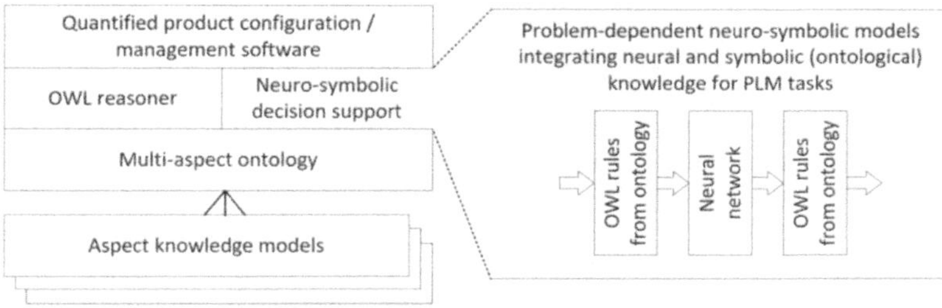

Fig. 5. Software architecture based on the multi-aspect ontology.

- retrieval/search possibilities in building block and interface specifications to support reuse and new product designs.

7 Summary and Future Work

Using two industrial cases as motivation and research context, our work proposed to use feature models and multi-aspect ontologies as means to support integration of the different lifecycles present in QP engineering. More concretely, the different ontologies underlying the multi-aspect ontology are used to capture implementation options for building blocks and the relation of features and building blocks in a more formal way than possible in a feature model. The multi-aspect ontology adds the formal integration of these underlying ontologies. By allowing for separate foci on features and implementation options in ontologies, the established structures of the different engineering disciplines can be maintained, while at the same time the multi-aspect ontology captures the formal dependencies. The multi-aspect ontology also forms an "integrative" artefact between the existing models used for specifying requirements or defining implementation in the different engineering disciplines.

Future work will have to include conceptual and practical work. The conceptual work should focus on integration of more ontologies in the multi-aspect ontology, for example for supporting modularization in the different disciplines or for capturing customer value in relation to features. Furthermore, the linking and integration to other model types used in the engineering process should be defined in a more formal way. The practical work will consist of applying the multi-aspect ontology approach in real-world cases to study acceptance of the engineers and effects on the overall QP engineering process. We expect to gain experience from such case that support improving and extending the approach.

Acknowledgment. The paper is due to the research funded by the Russian Science Foundation (project 22-11-00214).

References

1. Schallmo, D., Williams, C.A., Boardman, L.: Digital transformation of business models--best practice, enablers, and roadmap. Int. J. Innov. Manag. **21**, 1740014 (2017)

2. Gordijn, J., Wieringa, R.: The business model of digital ecosystems: why and how you should do it. In: Enterprise Engineering Working Conference, pp. 3–16 (2022)
3. Hansen, E.B., Bøgh, S.: Artificial intelligence and internet of things in small and medium-sized enterprises: a survey. J. Manuf. Syst. **58**, 362–372 (2021)
4. Porter, M.E., Heppelmann, J.E.: How smart, connected products are transforming competition. Harv. Bus. Rev. **92**, 64–88 (2014)
5. Asghari, P., Rahmani, A.M., Javadi, H.H.S.: Internet of things applications: a systematic review. Comput. Networks. **148**, 241–261 (2019)
6. Yang, M., Evans, S.: Product-service system business model archetypes and sustainability. J. Clean. Prod. **220**, 1156–1166 (2019)
7. Sandkuhl, K.: Quantified products: case studies, features and their design implications. Informatica **34**, 825–845 (2023)
8. Swan, M.: The quantified self: Fundamental disruption in big data science and biological discovery. Big data. **1**, 85–99 (2013)
9. Kaiser, C.: Quantified vehicles: data, services, ecosystems (2021)
10. Mayer-Schönberger, V., Cukier, K.: Big data: A revolution that will transform how we live, work, and think. Houghton Mifflin Harcourt (2013)
11. Farahani, P., Meier, C., Wilke, J.: Digital supply chain management agenda for the automotive supplier industry. Shap. Digit. Enterp. Trends use cases Digit. Innov. Transform. 157–172 (2017)
12. Tselentis, D.I., Yannis, G., Vlahogianni, E.I.: Innovative insurance schemes: pay as/how you drive. Transp. Res. Procedia. **14**, 362–371 (2016)
13. Kong, X., Song, X., Xia, F., Guo, H., Wang, J., Tolba, A.: LoTAD: Longterm traffic anomaly detection based on crowdsourced bus trajectory data. World Wide Web. **21**, 825–847 (2018)
14. Sandkuhl, K., Shilov, N., Seigerroth, U., Smirnov, A.: Towards the quantified product-product lifecycle support by multi-aspect ontologies. IFAC-PapersOnLine. **55**, 187–192 (2022). https://doi.org/10.1016/j.ifacol.2022.04.191
15. Kang, K.C., Kim, S., Lee, J., Kim, K., Shin, E., Huh, M.: FORM: A feature-; oriented reuse method with domain-; specific reference architectures. Ann. Softw. Eng. **5**, 143–168 (1998)
16. Thörn, C., Sandkuhl, K.: Feature modeling: managing variability in complex systems. Complex Syst. Knowledge-based Environ. Theory, Model. Appl. 129–162 (2009)
17. Li, L., et al.: A survey of feature modeling methods: Historical evolution and new development. Robot. Comput. Integr. Manuf. **61**, 101851 (2020)
18. Smirnov, A., Levashova, T., Ponomarev, A., Shilov, N.: Methodology for multi-aspect ontology development: ontology for decision support based on human-machine collective intelligence. IEEE Access. **9**, 135167–135185 (2021). https://doi.org/10.1109/ACCESS.2021.3116870
19. Sandkuhl, K., Shilov, N., Smirnov, A.: Facilitating digital transformation: success factors and multi-aspect ontologies. Int. J. Integr. Supply Manag. **13**, 376 (2020). https://doi.org/10.1504/IJISM.2020.110739
20. Smirnov, A., Shilov, N.: Multi-aspect Ontology for Semantic Interoperability in PLM: Analysis of Possible Notations. In: IFIP International Conference on Product Lifecycle Management. PLM 2018: Product Lifecycle Management to Support Industry 4.0, pp. 314–323. Springer (2018). https://doi.org/10.1007/978-3-030-01614-2_29
21. Espinoza, M., Gómez-Pérez, A., Montiel-Ponsoda, E.: Multilingual and localization support for ontologies. In: Aroyo, L., Traverso, P., Ciravegna, F., Cimiano, P., Hyvönen, E., Mizoguchi, R., Oren, E., Sabou, M., Simperl, E. (eds.) The Semantic Web: Research and Applications. ESWC 2009, pp. 821–825. Springer, Berlin, Heidelberg (2009). https://doi.org/10.1007/978-3-642-02121-3_63
22. Priya, M., Aswani Kumar, C.: An approach to merge domain ontologies using granular computing. Granul. Comput. **6**, 69–94 (2021). https://doi.org/10.1007/s41066-019-00193-3

23. Calegari, S., Ciucci, D.: Granular computing applied to ontologies. Int. J. Approx. Reason. **51**, 391–409 (2010). https://doi.org/10.1016/j.ijar.2009.11.006
24. Song, S., Zhang, X., Qin, G.: Multi-domain ontology mapping based on semantics. Cluster Comput. **20**, 3379–3391 (2017). https://doi.org/10.1007/s10586-017-1087-x
25. Cruz, I.F., Xiao, H.: Ontology driven data integration in heterogeneous networks. In: Complex Systems in Knowledge-based Environments: Theory, Models and Applications. Studies in Computational Intelligence, pp. 75–98. Springer Berlin Heidelberg, Berlin, Heidelberg (2009). https://doi.org/10.1007/978-3-540-88075-2_4
26. Borgida, A., Serafini, L.: Distributed description logics: assimilating information from peer sources. J. Data Semant. Lect. Notes Comput. Sci. **2800**, 153–184 (2003). https://doi.org/10.1007/978-3-540-39733-5_7
27. Bouquet, P., Giunchiglia, F., van Harmelen, F., Serafini, L., Stuckenschmidt, H.: C-OWL: contextualizing ontologies. Lect. Notes Comput. Sci. **2870**, 164–179 (2003). https://doi.org/10.1007/978-3-540-39718-2_11
28. Hemam, M., Boufaïda, Z.: MVP-OWL: a multi-viewpoints ontology language for the Semantic Web. Int. J. Reason. Intell. Syst. **3**, 147 (2011). https://doi.org/10.1504/IJRIS.2011.043539
29. Cuenca Grau, B., Parsia, B., Sirin, E.: Combining OWL ontologies using Epsilon-Connections. J. Web Semant. **4**, 40–59 (2006). https://doi.org/10.1016/j.websem.2005.09.010
30. Bao, J., Slutzki, G., Honavar, V.: A semantic importing approach to knowledge reuse from multiple ontologies. In: 2nd AAAI Conference on Artificial Intelligence, pp. 1304–1309. AAAI Press (2007)
31. Hevner, A., Chatterjee, S.: Design research in information systems: theory and practice. Springer Science & Business Media (2010)
32. Yin, R.K.: Case study research and applications. Sage Thousand Oaks, CA (2018)
33. Smirnov, A., Shilov, N., Ponomarev, A.: Facilitating enterprise model classification via embedding symbolic knowledge into neural network models. Commun. Comput. Inf. Sci. **1875**, 269–279 (2023). https://doi.org/10.1007/978-3-031-39059-3_18

A Digital Twin Implementation Framework for Collaborative Robots

Giulia Bruno(✉), Khurshid Aliev, Ahmed Awouda, and Paolo Chiabert

Politecnico di Torino, Corso Duca Degli Abruzzi 24, 10129 Torino, Italy
{giulia.bruno,khurshid.aliev,ahmed.awouda,
paolo.chiabert}@polito.it

Abstract. In the era of Industry 4.0/5.0, collaborative robots (cobots) play a pivotal role in transforming manufacturing processes by enabling human-robot collaboration and adaptability. However, realizing the full potential of cobots necessitates advanced digital technologies. This paper introduces a digital twin framework specifically tailored for collaborative robots within the context of Industry 4.0 and 5.0. The framework leverages digital twin technology to create a virtual representation of physical cobots, enabling real-time monitoring and optimization of their performance. Emphasizing data flow, communication protocols, and visualization tools, the framework facilitates seamless integration of cobots into smart manufacturing environments. The digital twin application is implemented in the FlexSim software and it has been exploited to perform dynamic task allocation among UR3e cobots. Through the use case, the paper demonstrates the effectiveness and versatility of the proposed framework, offering a roadmap for harnessing the potential of digital twin for cobots management.

Keywords: Digital Twins · Collaborative robots · Cobots · Industry 4.0 · Industry 5.0

1 Introduction

The use of collaborative robots in the manufacturing industry has revolutionized the way tasks are performed, offering increased productivity, flexibility, and safety. Collaborative robots, also known as cobots, have the ability to work alongside human workers in a shared workspace, allowing for physical interaction and collaboration. The convergence of collaborative robots and digital twins represents a transformative paradigm in the manufacturing landscape. As collaborative robots redefine traditional workflows by seamlessly integrating with human operators, digital twins emerge as instrumental tools in harnessing the full potential of these collaborative systems.

Digital twins are virtual representations of physical systems, created through the use of advanced technologies such as sensors, data analytics, and simulation models [1]. Digital twins allow manufacturers to simulate and optimize the performance in a virtual environment before they are implemented in the real world.

Published by Springer Nature Switzerland AG 2025
P. Sureephong et al. (Eds.): PLM 2024, IFIP AICT 740, pp. 30–42, 2025.
https://doi.org/10.1007/978-3-031-93319-6_3

By bridging the physical and virtual realms, digital twins facilitate iterative refinement and optimization of collaborative robot designs, ensuring enhanced performance, adaptability, and safety in real-world deployment scenarios. This symbiotic relationship between collaborative robots and their digital twins underscores a paradigm shift towards agile, data-driven manufacturing processes, where innovation and efficiency converge to drive sustainable competitive advantage in an increasingly dynamic marketplace.

By utilizing digital twins of collaborative robots, manufacturers can design, test, and refine the functionality and capabilities of the robots without the need for physical prototypes [2]. This not only saves time and resources but also ensures that the robots meet the necessary safety requirements for human collaboration. Furthermore, digital twins enable manufacturers to remotely monitor and control the performance of collaborative robots in real-time. This level of visibility and control enhances operational efficiency and enables timely adjustments to be made, improving overall productivity.

The article is structured as follows: Sect. 2 presents the state of the and explores the current literature regarding digital twins and collaborative robot processes. Section 3 defines the case study where the digital twin framework has been applied. Section 4 describes the digital twin design and implementation by describing the framework and the virtual entities involved. Section 5 highlights the main results and findings from implementation of the framework in the case study. Finally, Sect. 6 draws conclusions and provides recommendations for future development.

2 State of the Art

In the pursuit of comprehensively exploring the integration of digital twins and cobots, a systematic approach was undertaken to survey the existing literature. Initially, a targeted keyword search was conducted across prominent academic databases to retrieve pertinent scholarly works pertaining to digital twins and collaborative robotics. This initial endeavor yielded a corpus of 113 documents. Subsequently, abstract-level scrutiny was performed to filter out documents lacking substantive implementations of digital twins, particularly those that merely discussed digital models or shadows without emphasizing real-time synchronization and bidirectional communication – essential attributes of authentic digital twins. Through this filtration process, 58 documents meeting the specified criteria were identified for further examination.

Following the initial screening, a refined analysis was conducted to discern between documents centered on process-oriented digital twins and those predominantly focused on cobot-specific digital twins. Given the primary thematic emphasis on process digital twins within the scope of this work, papers primarily addressing cobot digital twins were systematically excluded. Consequently, a curated subset of 22 scholarly works remained, each deemed pertinent to the core objectives of the literature review. These articles consisted of 12 conference proceedings and 10 journal articles.

The first work in this sector was conducted in 2018 by Malik and Bilberg [3]. They presented a digital twin framework to support the design and control of human-machine cooperation. The framework enables the creation of a simple animation which is extended to an event-based simulation for performance analysis on a system level. A digital twin was developed for an assembly workstation that highlighted the frameworks'.

ability to visualize changes in production behavior without risk or injury to the humans involved in the process.

Several works since then have been conducted to develop frameworks for digital twins of cobots processes. The focus of some of these frameworks has been on improved process monitoring and simulation of production processes, such examples include collaborative painting [4], food packaging [5], and analysis of collaborative assembly processes [6–13]. Moreover, the work conducted by Ganesh et al. [14] focuses on a specific subclass of cobots which are AGVs and developed a digital twin for a fleet of AGVs that focuses on providing modularity and concurrent processing capability to the fleet.

Other works were focused on the human aspect of human robot collaboration by applying digital twin for safety assurance and ergonomics conformity of collaborative processes by using the digital twin as a tool for safety analysis and control [15–17]. Furthermore, the work conducted by Vasquez-Hurtado et al. [18] used digital twin to formulate a mixed reality environment where students and operators can train on working with collaborative robots away from the risks of the real physical environment. Moreover, applications on trajectory and movement path planning of cobots in order to minimize human risks have also been found in literature [19–21]. Other authors conducted a study focused on human factor modeling in robotic digital twins [22]. Their work described the human condition and its impact on technological process KPIs with the objective of improving the quality and degree of integration of human factors into robotic process digital twins.

Furthermore, two works were found that deal with digital twins for task allocation within production processes: the first deals with assembly line worker assignment and balancing problems for a collaborative assembly system, with the aim of optimizing the work allocation between the worker and the robot [23], while the second provides an application of dynamic task allocation in the case of failure within the production process [24].

Finally, two review works were also found. One review highlights the challenges of developing digital twins of human robot collaborations. Its findings show that modeling human operators, computational intensity, resource efficiency, safety concerns, and integration with Industry 4.0 technologies are the key challenges for developing digital twins and are considered areas for further research [25]. The other review provides an overview of prospective applications of collaborative robots and digital twin technologies among others within the industry 5.0 paradigm [26]. The authors concluded that digital twins have the potential to identify production errors, decrease downtime, reduce maintenance costs, and improve the efficiency of the manufacturing process. This leads to more resilient and sustainable production processes and products.

The existing research on digital twins of collaborative robot processes has made significant progress in areas such as integration and control [7], dynamic modeling of robot joints [27], and human-robot interaction [28]. However, there are still some key research gaps that need to be addressed. Firstly, there is a need for further development of digital twin models that can accurately simulate the complex dynamics of collaborative robot systems, including the interactions between human and robot [3]. Secondly, the current focus on kinematic-level digital twins needs to be expanded to include the.

integrated system dynamics of collaborative robot processes [27]. Lastly, there is a need for more research on the practical implementation and real-time monitoring of digital twins in collaborative robot processes, particularly in the context of small and medium-sized enterprises (SMEs) [27].

This work addresses the last two gaps by providing a real-world implementation of a digital twin that includes an error handling mechanism that is not limited to the collaborative robot but rather extended to the whole collaborative process. While many theoretical frameworks exist for collaborative digital twins, a practical case study can provide a clearer idea of the tangible benefits obtained in a material handling process. Moreover, while existing research focuses on performance monitoring this work explores how digital twins can actively identify and address process errors in a collaborative setting. Finally moving beyond static task assignment, this article investigates how digital twins can dynamically allocate tasks based on real-time data and error detection, contributing to a more adaptive and optimized workflow.

3 Cobot Use Case

This section describes the cobot physical devices and the scenarios exploited in the use case.

3.1 Cobots in the Mind4Lab Facility

The physical setup includes a robotic workstation with two UR3e models and one UR3 CB series cobot connected on a shared network. The robots, composed mostly of plastic and aluminium with some steel elements, weigh around 11 kg and have a payload capacity of 3 kg. The robot arm is primarily composed of tubes and joints, facilitating precise tool positioning.

Each robot has its own control unit, connected via TCP/IP, with ports for digital/analog I/O to interface with the arms, the teach pendant, and other peripherals. The cobots are equipped with OnRobot RG6 flexible grippers for the UR3e models and an RG2 gripper for the UR3 CB-series model, allowing them to execute Pick and Place tasks.

The robotic task involves programming a robot to repeatedly perform a cycle of picking up a workpiece from one queue, moving it to another, and repeating the process when there is a workpiece. Figure 1 depicts the cobots utilized in the experiments, located at the Mind4Lab facility in the Department of Management and Production Engineering at Politecnico di Torino.

Fig. 1. Cobots in the Mind4Lab facility: two UR3e and a UR3 CB series with grippers.

3.2 Cobots Use Case Scenarios

The experimental case study entails the coordinated operation of two collaborative robotic agents tasked with the retrieval of items from a designated picking location and their subsequent transfer to a designated release point. The nominal time required to complete this operation stands at 25 s.

In adherence to a static scheduling paradigm, each robotic agent is initially tasked with handling seven items concurrently, as illustrated in Fig. 2.

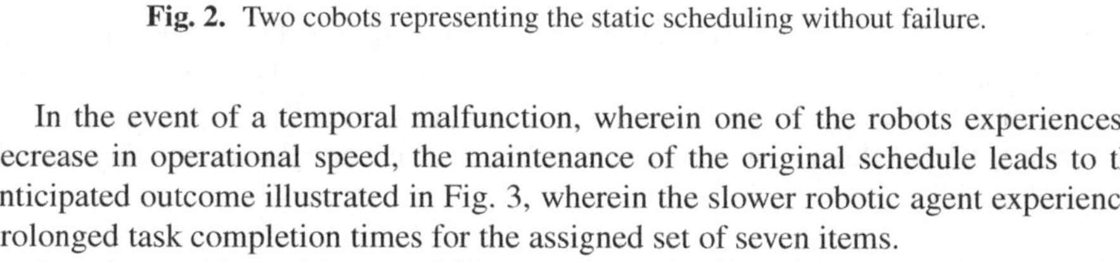

Fig. 2. Two cobots representing the static scheduling without failure.

In the event of a temporal malfunction, wherein one of the robots experiences a decrease in operational speed, the maintenance of the original schedule leads to the anticipated outcome illustrated in Fig. 3, wherein the slower robotic agent experiences prolonged task completion times for the assigned set of seven items.

In order to address this potential issue, a dynamic allocation strategy integrated into the Digital Twin framework autonomously identifies instances of temporal malfunction, prompting a redistribution of tasks to the more expedient robotic agent. This corrective measure is enacted either through the incremental relocation of individual items or the transfer of an optimally calculated batch of items. To facilitate this task reassignment process, a supplementary third robotic agent is introduced, boasting an average task execution time of 28 s.

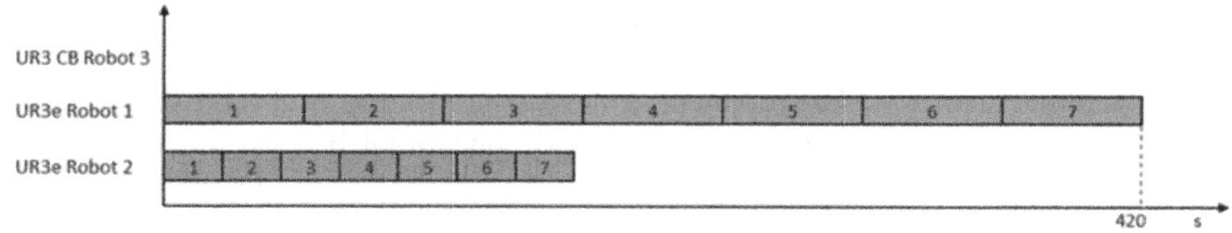

Fig. 3. Two cobots representing the static scheduling with time failure.

Dynamic scheduling with temporal malfunction, wherein the Digital Twin system promptly identifies the operational slowdown of the second robotic agent and automatically undertakes corrective action by reallocating two tasks (individually) to the first robotic agent, as depicted in Fig. 4.

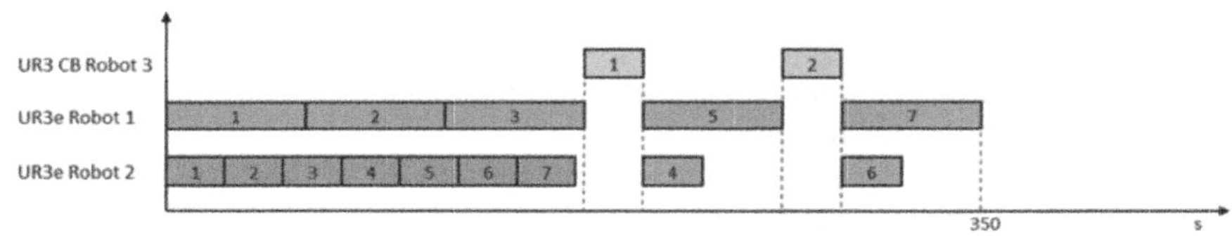

Fig. 4. Three cobots representing dynamic scheduling with temporal malfunction, entailing the relocation of one item at a time.

Dynamic scheduling with temporal malfunction, wherein the Digital Twin system promptly detects the operational slowdown of the second robotic agent and autonomously initiates corrective measures by reallocating three tasks to the first robotic agent, as illustrated in Fig. 5.

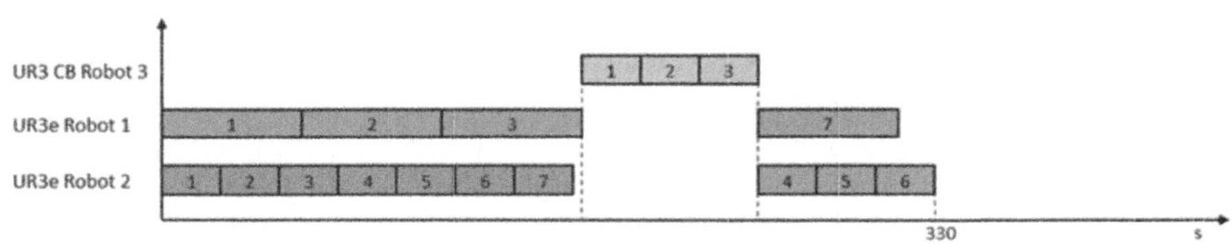

Fig. 5. Three cobots representing dynamic scheduling with temporal malfunction, entailing the relocation of a batch of three units.

4 Digital Twin Design and Implementation

This section describes the developed digital twin. Firstly, the physical entities involved in the process are defined. Then, a brief introduction to the framework is provided with the accompanying domain model of the case study. Furthermore, the virtual model and its environment are presented. Finally, the control logic between the digital and the physical counterparts is described.

4.1 Framework Overview for Cobot Digital Twin

The exponential growth of interconnected devices under the umbrella of the "Internet of Things" (IoT) necessitates a well-defined conceptual framework to bridge the gap between real-world entities and their digital representations. In this context, we use the IoT-DT framework developed by Awouda et al. [29] that is built upon the IoT Architectural Reference Model (IoT-ARM) [30].

The starting point is establishing a comprehensive domain model that facilitates systematic comprehension and development of complex IoT systems. This model transcends specific technologies and use cases, serving as a lingua franca for diverse stakeholders – engineers, researchers, and domain experts alike – striving to navigate the intricate landscape of the IoT and digital twins. By codifying key concepts like Physical and Virtual Entities, Resources, Devices, and Services, along with their interrelationships, the domain model provides a foundation for designing interoperable, adaptable, and scalable IoT architectures. These key elements are defined as follows:

- Physical Entity: Represents a real-world object equipped with sensors and actuators (e.g., a smart thermostat, a connected car).
- Resource: Captures data or functionality exposed by a Physical Entity (e.g., temperature reading, engine control).
- Device: Acts as the interface between the physical and digital realms, Implements or hosts one or more Resources (e.g., a microcontroller, a gateway).
- Virtual Entity: Digital representation of a Physical Entity within the IoT system (i.e. the digital twin).
- Service: Provides functionalities based on accessed Resources (e.g., data processing, analytics, monitoring).
- User: Person or system interacting with the IoT system.

4.2 Cobot Process Domain Model

Since the focus of the case study is the creation of the digital twin of the pick and place process itself and not its individual components, we consider the focus of the model is on the physical entity representing the process as shown in Fig. 6. Furthermore, since the process is made up of multiple components, the corresponding physical entity also consists of these components, namely the two robots, individual products processed, an output queue for finished products and the robot responsible for the re-allocation task. As shown in the figure, each robot has a device attached to it that represents the robot controller which is responsible for monitoring and controlling the robot.

These devices host several resources that provide specific functionalities to the robots, the main resources of relevance to this work are the sensor drivers used for data acquisition, the elements used for invoking actuation (i.e. moving the robots). Another resource is the control logic used to manage the picking and placing of items in the case of robots 1 and 2 and in the case of robot 3, the logic used for task allocation. These resources are exposed by specific services that allow access to the resources.

The digital twin (designated as virtual entity "process") represents the process physical entity in the digital realm. This virtual entity can access specific services to interact with its physical counterpart, in this case the VE accesses the dashboard of the digital

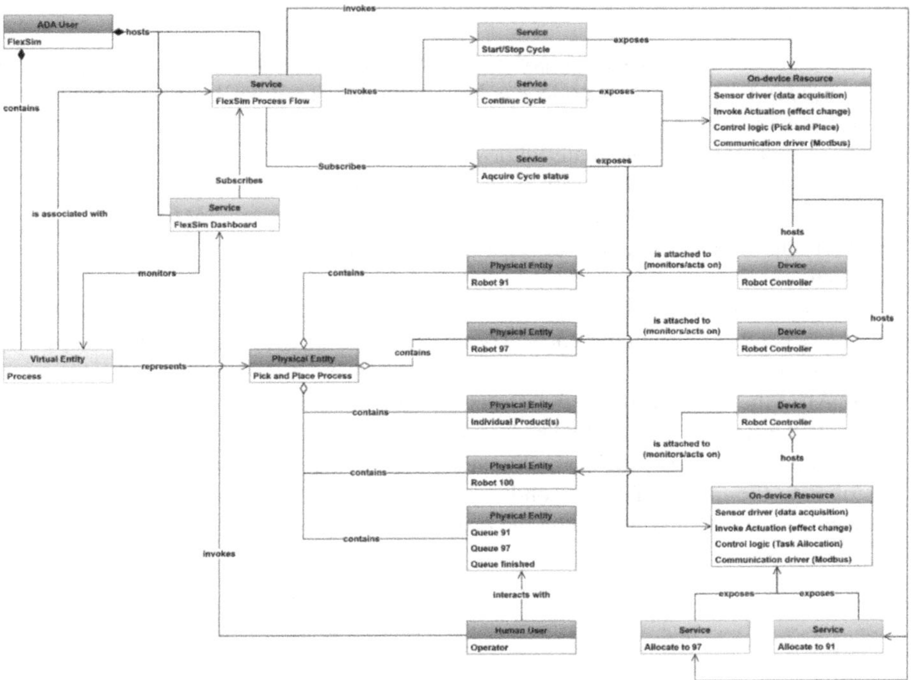

Fig. 6. Process domain model.

model (FlexSim dashboard) to monitor and receive data about the process. Furthermore, it is also associated with the process flow of the digital model (FlexSim process flow), which is modelled as a service that manages the digital process and provides the dynamic task allocation. The dashboard subscribes to the process flow in order to receive data about the process. These two services along with the VE are hosted in the digital environment (FlexSim) that is modelled as a user. It is noteworthy to mention that this type of user is an active digital artifact (a digital object or software that exhibits dynamic behaviors or possesses the ability to react to its environment and user input and has the ability to access services) that is different from the human user.

The process flow in turn subscribes and invokes other lower-level services to access the resources hosted on the devices. It subscribes to the service acquire cycle status to retrieve the current status of the robot's cycle. Moreover, the process flow invokes the two services start/stop cycle and continue cycle in the robots 1 and 2 and it also invokes the services allocate to 2 and allocate to 1. The latter two are used by the dynamic allocation to move an item from the queue of robot 1 to the queue of robot 2 and vice versa depending on the needed re-allocation.

Last but not least is the human user that interacts with the process through the queues of robot 1 and 2, the user also interacts with the finished goods queue at the end of the.

process. This user also utilizes the dashboard to receive insights and information about the ongoing process.

4.3 Virtual Model

The virtual model of the process was developed in FlexSim, which is a versatile platform widely used in the realm of discrete event simulations [31]. It has been deemed suitable for this implementation due to its 3D modeling environment that allows the recreation of the robots, workstations, and material flow, fostering a comprehensive understanding of the process dynamics. The FlexSim environment consists of 3 main parts; the 3D model for building and visualizing the system under study, the process flow for managing and implementing model logic, and the dashboard for extracting and visualizing data from the model.

The virtual model is shown in Fig. 7. It consists of two source elements used to generate a batch of ten items each. These items are then passed to their respective queues that represent a storage area where the items await to be processed by the two robots. After the robots process the items, they are passed along to the final queue that represents the storage area for finished goods. Moreover, the third robot placed between the two queues performs the task allocation by moving items from the queue of the first robot to the queue of the second robot and vice-versa depending on the specific reallocation needed and dictated by the process flow logic.

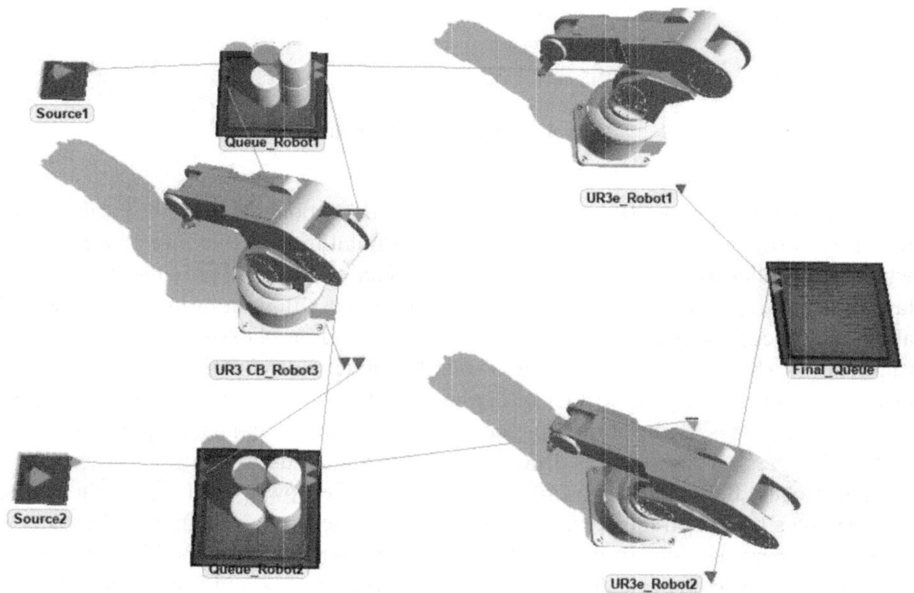

Fig. 7. Virtual model developed in FlexSim.

4.4 Digital Twin Process Flow

Open-source industrial communication protocols like Modbus TCP simplify digital twin development, enhancing interoperability and enabling real-time data exchange between virtual and physical systems. In the proposed DT, Modbus TCP facilitates communication between FlexSim (client) and three robots (servers), allowing seamless data transactions [31].

The information flow between the virtual models and robots is illustrated in Fig. 8 through an activity diagram. The system is initiated by the user, who activates both the virtual and physical models and initializes the item count for each pair of robots. In our scenario, the item count is initialized using variables K1 and K2, both set to 7. As the robots commence their tasks, they adhere to a predetermined scheduling. These robots are programmed to repeat the process upon receiving signals from the virtual models and transmit signals back upon task completion using Modbus registers. Virtual models incorporate logic governing the number of items, time, and speed for each iteration of the process. During each iteration, physical robots send and receive signals to and from virtual models to represent and monitor the process. Consequently, the virtual models synchronize their process time with that of the physical machines. If one physical robot operates slower, the virtual model signals a faster third robot to perform the task, ensuring timely task completion within the system.

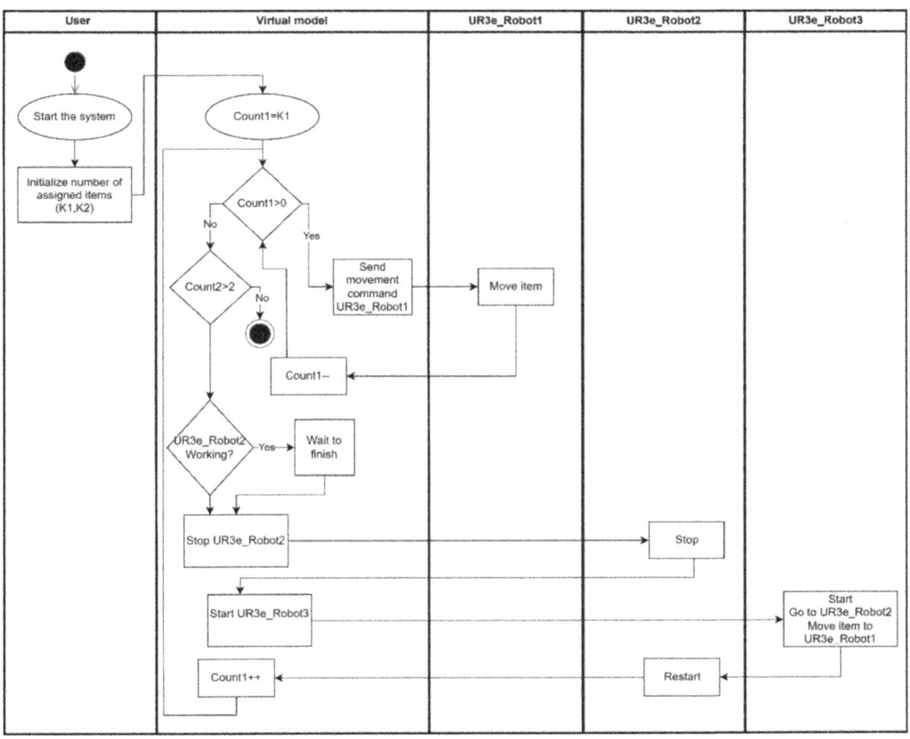

Fig. 8. Activity diagram of the Digital Twin.

5 Results

The findings of this investigation underscore the profound implications of system failures and the efficacy of DT implementation in enhancing system performance optimization. Under ideal conditions, where failures are absent, the system attains notable efficiency metrics: a throughput of 288 items/h, a cycle time of 12.5 s, and full utilization (100%). Conversely, when confronted with failures in the absence of DT intervention, there is a substantial degradation in performance, evidenced by a reduced throughput of 120 items/h, an increased cycle time of 30 s, and a utilization decline to 70.8%.

The introduction of DT-enabled intervention represents a significant improvement in response to failures. Through proactive measures such as item relocation, the system achieves a throughput of 144 items/h, a cycle time of 25 s, and an increase in utilization to 89.3%. Furthermore, the implementation of batch relocation further enhances system response, resulting in the highest performance observed. With a throughput of 152.7 items/h, a cycle time of 23.6 s, and utilization peaking at 97%, the system demonstrates an impressive capacity to adapt and optimize performance. These findings underscore the value of DT-enabled intervention in mitigating failures and maximizing operational efficiency, as depicted in Table 1.

Table 1. Results of indicators in the different scenarios.

Indicator	First scenario (no failure)	Second scenario (failure without DT)	Third scenario (failure with DT, item relocation)	Fourth scenario (failure with DT, batch relocation)
Throughput (TH)	288 items/h	120 items/h	144 items/h	152.7 items/h
Cycle time (CT)	12.5 s	30 s	25 s	23.6 s
Utilization (U)	100%	70.8%	89.3%	97%

6 Conclusions

The paper proposed an automatic data flow DT framework alongside the efficient communication protocols, and intuitive visualization tools. By emphasizing seamless integration of cobots into smart manufacturing environments, the framework empowers manufacturers to achieve agility, flexibility, and efficiency. The use case of the DT application developed in FlexSim showcases dynamic task allocation between UR3 cobots, illustrating the effectiveness and versatility of the proposed framework. The scenario involved three cobots, where the DT detected slow movement in one robot and reallocated the task to a faster robot. This reallocation resulted in greater efficiency compared to other scenarios.

Looking ahead, future research could extend this framework by incorporating various types of resources, such as mobile cobots and/or robotized CNC machines, to broaden manufacturing scenarios. Enhancing its capabilities to manage more complex tasks and environments, exploring innovative communication protocols and visualization techniques, and integrating machine learning algorithms for predictive maintenance and performance optimization are promising avenues for further exploration.

References

1. Lu, Y., Liu, C., Kevin, I., Wang, K., Huang, H., Xu, X.: Digital Twin-driven smart manufacturing: Connotation, reference model, applications and research issues. Robotics and computer-integrated manufacturing **61**, 101837 (2020)
2. Barenji, A.V., Liu, X., Guo, H., Li, Z.: A digital twin-driven approach towards smart manufacturing: reduced energy consumption for a robotic cellular. Int. J. Comput. Integr. Manuf. **34**, 1–16 (2020)
3. Malik, A.A., Bilberg, A.: Digital twins of human robot collaboration in a production setting. In: Sormaz, D., Chen, F.F., Suer, G. (eds.) Procedia Manuf, pp. 278–285. Elsevier B.V. (2018)
4. Chancharoen, R., et al.: Digital Twin for a Collaborative Painting Robot. Sensors **23** (2023). https://doi.org/10.3390/s23010017
5. Lu, X., et al.: A generic and modularized Digital twin enabled human-robot collaboration. In: Proc. - IEEE Int. Conf. e-Bus. Eng., ICEBE, .pp 66–73. Institute of Electrical and Electronics Engineers Inc. (2022)
6. Joseph, A.J., Kruger, K., Basson, A.H.: An Aggregated Digital Twin Solution for Human-Robot Collaboration in Industry 4.0 Environments. Springer Science and Business Media Deutschland GmbH (2021)
7. Diachenko, D., et al.: industrial collaborative robot digital twin integration and control using robot operating system. J. Mach. Eng. **22**, 57–67 (2022). https://doi.org/10.36897/jme/148110
8. Pérez, L., Rodríguez-Jiménez, S., Rodríguez, N., Usamentiaga, R., García, D.F.: Digital twin and virtual reality based methodology for multi-robot manufacturing cell commissioning. Appl. Sci. 10 (2020). https://doi.org/10.3390/app10103633
9. Htet Oo, K., Koomsap, P.: Development of Collaborative Multi-Robotics Assembly System with Digital Twin Approach. In: Cooper, A., Koomsap, P., Stjepandic, J. (eds.) Adv, pp. 492–501. IOS Press BV, Transdiscipl. Eng (2023)
10. Lorenzo, R., Negri, E., Macchi, M.: Local Digital Twin-based control of a cobot-as-sisted assembly cell based on Dispatching Rules. In: IFAC-PapersOnLine, pp. 372–377. Elsevier B.V. (2022)
11. Jonek, M., Niermann, D., Petzoldt, C., Manns, M., Freitag, M.: Automated analysis of assembly processes in human-robot collaboration: research approaches and challenges. In: Mpofu, K., Sacks, N., Damm, O. (eds.) Procedia CIRP, pp. 1203–1208. Elsevier B.V. (2023)
12. Duan, J., Gong, X., Zhang, Q., Qin, J.: A digital twin–driven monitoring framework for dual-robot collaborative manipulation. Int. J. Adv. Manuf. Technol. **125**, 4579–4599 (2023). https://doi.org/10.1007/s00170-023-11064-2
13. Mathiesen, S., et al.: Applying Robotics Centered Digital Twins in a Smart Factory for Facilitating Integration and Improved Process Monitoring. Springer Science and Business Media Deutschland GmbH (2022)
14. Ganesh, M., Ahamed Rizvi, M., Anbu, A.: Digital Twin framework for material handling and logistics in Manufacturing: Part 1. In: Thampi, S.M., Trajkovic, L., Gelenbe, E., Atiquzzaman, M., Berretti, S., Balas, V.E. (eds.) Proc. Int. Conf. Connect. Syst. Intell., CSI. Institute of Electrical and Electronics Engineers Inc. (2022)
15. Weistroffer, V., Keith, F., Bisiaux, A., Andriot, C., Lasnier, A.: Using Physics-Based Digital Twins and Extended Reality for the Safety and Ergonomics Evaluation of Cobotic Workstations. Front Virtual Real 3 (2022). https://doi.org/10.3389/frvir.2022.781830

16. Stefanakos, I., Calinescu, R., Douthwaite, J., Aitken, J., Law, J.: Safety Controller Synthesis for a Mobile Manufacturing Cobot. Springer Science and Business Media Deutschland GmbH (2022)
17. Douthwaite, J.A., et al.: A Modular Digital Twinning Framework for Safety Assurance of Collaborative Robotics. Front Robot AI 8 (2021). https://doi.org/10.3389/frobt.2021.758099
18. Vázquez-Hurtado, C., Altamirano-Avila, E., Roman-Flores, A., Vargas-Martinez, A.: Towards a mixed virtual reality environment implementation to enable industrial robot programming competencies within a cyber-physical factory. In: IEEE Global Eng. Edu. Conf., EDUCON. IEEE Computer Society (2023)
19. Zanchettin, A.M., Messeri, C., Cristantielli, D., Rocco, P.: Trajectory optimisation in collaborative robotics based on simulations and genetic algorithms. Int. J. Intel. Rob. Appl. **6**, 707–723 (2022). https://doi.org/10.1007/s41315-022-00240-4
20. Lee, H.: Physics-based cooperative robotic digital twin framework for contactless delivery motion planning. Int. J. Adv. Manuf. Technol. **128**, 1255–1270 (2023). https://doi.org/10.1007/s00170-023-11956-3
21. Sun, X., Zhang, R., Liu, S., Lv, Q., Bao, J., Li, J.: A digital twin-driven human–robot collaborative assembly-commissioning method for complex products. Int. J. Adv. Manuf. Technol. **118**, 3389–3402 (2022). https://doi.org/10.1007/s00170-021-08211-y
22. Gorkavyy, M.A., Tyurina, Y.A., Ivanov, Y.S., Grabar, D.M.: Human factor modelling in the collaborative robotic process control system. In: Proc. - Int. Conf. Ind. Eng., Appl. Manuf., ICIEAM, pp. 954–960. Institute of Electrical and Electronics Engineers Inc. (2023)
23. Pabolu, V.K.R., Shrivastava, D., Kulkarni, M.S.: A digital-twin based worker's work allocation framework for a collaborative assembly system. In: Bernard, A., Dolgui, A., Benderbal, H.H., Ivanov, D., Lemoine, D., Sgarbossa, F. (eds.) IFAC-PapersOnLine, pp. 1887–1892. Elsevier B.V. (2022)
24. Bruno, G., Aliev, K.: Digital Twin Application for Dynamic Task Allocation. Springer Science and Business Media Deutschland GmbH (2023)
25. Ramasubramanian, A.K., Mathew, R., Kelly, M., Hargaden, V., Papakostas, N.: Digital Twin for Human-Robot Collaboration in Manufacturing: Review and Outlook. Appl. Sci. 12 (2022). https://doi.org/10.3390/app12104811
26. Raffik, R., Vaishali, V., Balavedhaa, S., Jyothi, L.N., Sathya, R.R.: Industry 5.0: Enhancing Human-Robot Collaboration through Collaborative Robots - A Review. In: Int. Conf. Adv. Electr., Electron., Commun., Comput. Autom., ICAECA. Institute of Electrical and Electronics Engineers Inc. (2023)
27. Yang, X., et al.: Dynamic Modeling and Digital Twin of a Harmonic Drive Based Collaborative Robot Joint. In: Proc IEEE Int Conf Rob Autom, pp. 4862–4868. Institute of Electrical and Electronics Engineers Inc. (2022)
28. Malik, A.A., Brem, A.: Digital twins for collaborative robots: A case study in human-robot interaction. Rob. Comput. Integr. Manuf. 68 (2021). https://doi.org/10.1016/j.rcim.2020.102092
29. Awouda, A., Traini, E., Bruno, G., Chiabert, P.: IoT-Based Framework for Digital Twins in the Industry 5.0 Era. Sensors **24**, 594 (2024). https://doi.org/10.3390/s24020594
30. Bassi, A., et al.: Enabling Things to Talk. Springer, Berlin Heidelberg, Berlin, Heidelberg (2013)
31. Beaverstock, M.: Applied simulation: modeling and analysis using Flexsim, 2nd ed. FlexSim Software Products, Orem, UT (2011)

A Procedure Model for Systematic Application of Generative Engineering and Design Software Tools

Timo Köring(✉), Detlef Gerhard, and Matthias Neges

Institute for Product and Service Engineering, Digital Engineering Chair, Ruhr University Bochum, Universitaetsstr. 150, 44801 Bochum, Germany
{timo.koering,detlef.gerhard,matthias.neges}@rub.de

Abstract. The latest developments in AI and advanced algorithms for Generative Engineering and Design have led to wide landscape of different approaches and new software tools. These tools differ in their systematics and areas of applications as well as opportunities for establishing new processes and methodologies in engineering design. The consideration and integration of driving technologies for applications such as topology optimization, design space exploration and design automation are incorporated in different ways. Therefore, a clear structure and procedural model is necessary to provide a methodological guideline with respect to potentials and limitations for engineering tasks. Building on previous work and based on an analysis of existing software tools, the consideration of use case requirements, general principles, and technology trends, this paper extends an approach to structure and compare different tools especially for (mechanical) engineering applications. For each stage inside the design process, specific user needs, features and software characteristics are described. The focus is on how and which interactions between users and software are necessary and which functionalities and tasks the software must provide. The analysis is not limited to existing approaches. Future developments streams in the field are considered. The findings are validated using two commercial software applications.

Keywords: Software-based Generative Design · Generative Engineering · procedural model · Computational Design synthesis

1 Introduction

The impact of Generative Engineering and Design (GE&D) software through the integration of artificial intelligence (AI) is equated with the shift from hand drawing to CAD software in the product development process [1]. GE&D software extends or changes existing CAD software by integrating new technologies such as (generative) AI, design automation and optimization. The importance of integrating software applications into creative processes and the opportunities of AI in engineering software in different phases (concept generation, concept evaluation and decision making) is emphasized in the literature to the effect that AI implementation can be extended to a "copilot" that assist

Published by Springer Nature Switzerland AG 2025
P. Sureephong et al. (Eds.): PLM 2024, IFIP AICT 740, pp. 43–57, 2025.
https://doi.org/10.1007/978-3-031-93319-6_4

users and actively supports them in engineering tasks [2]. This not only revolutionizes the product development process, but also poses new challenges for design engineers. There is an increasing number of different GE&D software tools and a growing market while the technology is only in the innovation trigger phase of the Gardner Hype Cycle [3, 4]. In addition to the systematically correct use of the software, the selection from a large number of existing software options for own application fields and requirements is a major challenge for users. This challenge reinforces the need for a comprehensive overview, procedure model, and differentiation system for GE&D tools.

While many scientific papers consider the differences of specific generative design algorithms and models or case studies within specific software tools [5–7], this paper focuses on different characteristics in commercial software tools. It is dedicated to developing an overview, methodological guideline and procedure model that provides a systematic and practical structure for design engineers, product developers, and researchers. This reduces the gap between research and industrial application in the field of generative product design [7]. Three major aspects are covered in this paper: a) how and which interactions between users and software are necessary, b) which functionalities and tasks can be provided by different software tools c) what technology, theoretical principle or approach is applied as foundation of the software tool? The goal of the procedure model is not only to organize the variety of available tools. It is designed to clearly differentiate their functionalities and application areas. This is built upon a comprehensive literature review, and the analysis of existing software products, the consideration of user needs and technology trends such as AI or cloud computing.

2 State of the Art and Related Work

The term Generative Design (GD) is widely used in the literature, mainly due to the large number of different disciplines, but is still not uniformly defined. Generative Engineering (GE) is used less frequently, especially in commercial applications. Computational Design Synthesis (CDS) is a term widely used in academia and research papers. In the context of this paper, GE&D is understood as the umbrella term for the various streams such as generative design, design automation, design space exploration, in order to differentiate in particular from the typical notion of GD in connection with topology optimization and design for additive manufacturing (AM) and to include the academia term CDS. While Singh and Gu [5] see the main task of GD in the architectural context in process automation to support designers and explore larger design spaces for greater efficiency, the focus in (mechanical) engineering is on performance and increasing manufacturability [8, 9]. GD is an approach that automatically generates and evaluates designs using different advanced algorithms. The user's task shifts from explicitly generating shapes to defining constraints such as material, forces, and manufacturing processes. Using algorithms and constraints, a variety of different designs are generated based on the specified boundary conditions. This allows GD to be applied in the earlier stages of the engineering design process, i.e., ideation [10]. The often bioinspired lightweight structure of GD goes hand in hand with the potentials of AM but is not limited to serve as a specialized engineering design tool for this manufacturing technology [11]. In contrast to topology optimization, GD allows to incorporate adjacent downstream application areas such as CAE and manufacturing process evaluation in order to enable a holistic view and the exploration of new solution spaces [12]. In this way, vast solution spaces can be systematically explored and solutions created that were previously unattainable [13]. This principle is also described as Design Space Exploration (DSE). DSE is the systematic exploration and analysis of different design alternatives within a defined design space [14]. The design space contains all the char-acteristics and parameter variations that can be taken into account when developing a product [15]. The aim is to find innovative and optimized solutions. This approach uses algorithmic and optimization-based techniques to generate and evaluate a variety of design possibilities.

The basic principle of the shift from explicit generation to the definition of boundary conditions with an automatic computational design synthesis methods was generalized by Cagan et al. [16] in a flowchart (s. Figure 1).

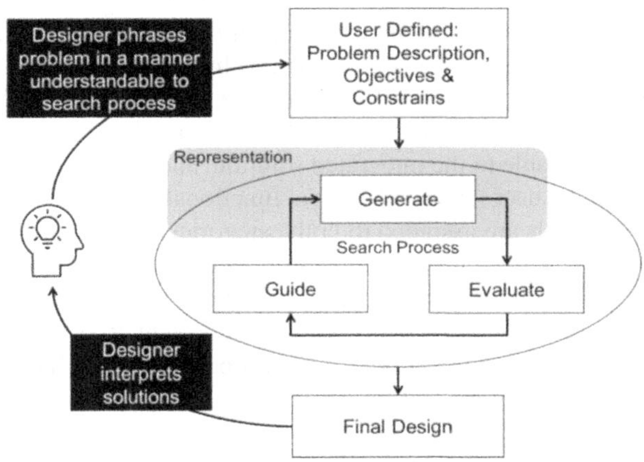

Fig. 1. The generic flowchart for the synthesis of open-ended engineering design problems [16]

The flowchart depicts the interaction between design engineers and the CDS software tool as well as the functioning of the system itself. The process starts with the initial setup of a problem by defining constraints, establishing objective functions and design objectives for the search process. This is accomplished in a way that is comprehensible for the so-called search process, i.e., the process of finding a solution for a given problem. By this, the design engineer formulates the representation to encapsulate the constraints or attributes of the design space. Using this representation, candidate solutions are generated in the "Generate" task. Subsequently, each generated candidate undergoes evaluation ("Evaluate") to assess its alignment with the objectives and constraints of the design problem. Guided by the calculated objectives for the candidates, a guidance strategy ("Guide") is implemented to direct the search process towards identifying improved solutions in subsequent iterations. The finally generated result returned to the design engineer is interpretated for evaluation purposes with respect to requirements, constraints, boundary conditions etc. The approach focuses on the domain of engineering problems with FEM simulations as a practical part for evaluation.

Various state of the art approaches to classify GE&D software identified in the literature are presented below. Mountstephens and Teo [7] conducted an extensive study of various GD system tools to identify significant approaches and trends. This study examined the general aim of the tools, the design phase addressed, the number of solutions, the type/algorithm of generation, the performance (user feedback, objective criteria), and the interactivity (automatic, interactive) of the tools. It was found that many tools (15 out of 37) were generated on a parametric CAD basis, which suggests a transformation. Furthermore, 26 of the 37 systems generated designs automatically, so that the design engineer can focus more on evaluation than on direct and explicit shape generation. Pollák et al. [17] showed the current status of available GD software solutions and carried out a comparison of three tools for practical application. In addition to advantages and disadvantages, yes/no comparison criteria were integrated for the evaluation. Junk and Burkart [18] differentiate between integrated and standalone systems in their specific comparison of GD systems for additive manufacturing. Pilagatti et al. [11] also distinguished between the two and also introduced the possibility of the "fluid study" field as application to differentiate between them. In their comparison, Gerhard et al. [19] differentiated between the general approach (for the application of AI) and GD in software tools. They made a subdivision into "functional CAD programming environment", which is comparable to the integrated solution, and "assisted CAD modeling" in relation to the graphical user interface. Typical functionalities and characteristics based on existing software tools are assigned to both categories. Vlah et al. [12] highlight the possibilities of Autodesk Fusion in a comparative study between topology optimization and generative design. Furthermore, it became clear that the integration of manufacturing processes and the consideration of costs are an important aspect for the development of these tools. Biedermann and Meboldt [20] proposed a CDS framework for additive manufactured multi-flow nozzles under the use of an objectoriented programming environment. They focused on the necessary user inputs for each activity.

Buonamici et al. [21] conducted a case study in Autodesk Generative Design. The workflow within the software for structural applications was examined in more detail and a comparison between GD and topology optimization was made. In addition to the process sequence, they particularly highlight significant features of the software. The focus here was also on the comparison options for decision-making. Furthermore, they concentrated on the manufacturability of the created shapes. They showed that approaches are available for some use cases, while the integration of dynamic and thermal use cases is seen as a first step for improvement. Cloud computing is seen as a driver for many results. Further relevant work and aspects are explained in the following chapters as corresponding evidence for the aspects.

3 Procedural Model for GE&D Software

The generalized flowchart of computational design synthesis methods according to Cagan et al. [16] serves as the basis for the development of a procedural model for today's GE&D software applications. Figure 2 depicts this generic procedural model. It identifies the necessary steps inside the process, the overall structure and highlights the connection to other internal enterprise applications and data sources. The procedural model is structured as follows: As initial work step, before the design engineer starts using a GE&D software, a transfer of the problem in a manner understandable to the problem-solving process inside the software is necessary. This is the definition of the problem space. Afterwards, different application areas of the problem and the software functionalities need to be compared as an initial filter to consider the appropriate use of the software. In the first step of the interaction between design engineer and software, the before defined phrases of the problem need to be converted and inserted into the software as a problem specification. These problem specification and boundary condi-tions are forwarded to the software internal problem-solving process. This mainly iter-ative process of "Generate", "Evaluate" and "Parameter optimization" leads to the so-lution space. The generated results are evaluated regarding the objectives from the boundary conditions and subsequently relevant parameter are optimized. The process is iteratively repeated until the target functions are fulfilled. Afterwards, from the given problem space, a solution space with valid solutions is generated and returned. The design engineer interprets this solution space and, depending on the results, the problem specification is adjusted, or the design engineer carries out further post-processing steps before it leaves the process as a final design. The procedural model, particularly the automated and algorithm driven part is usually supported in certain phases by various integrated data sources of different enterprise applications. This data can originate from PLM, MES, ERP, CAx, Office and other applications. The transfer of historical and live data, design alternatives and industrial know-how are key aspects of this.

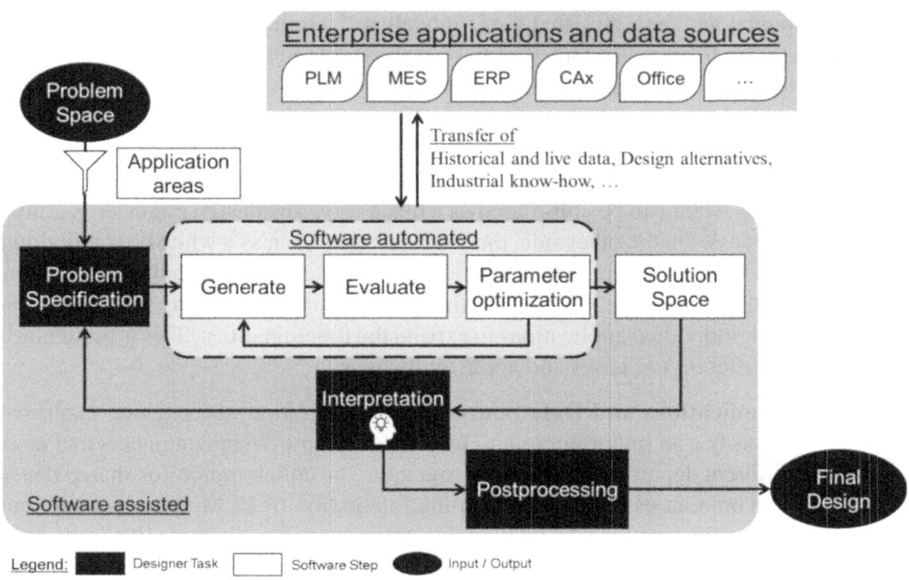

Fig. 2. Generative Engineering and Design procedural model (based on [16])

Based and extended on the previous research, this chapter presents various criteria and characteristics for each process step grouped into the categories "user tasks and inputs", "software functionalities/tasks", and "technology, theoretical principles or approaches". The analysis of existing GE&D software tools (e.g. Synera, Autodesk Fusion,...) determines their functionalities and areas of application. This identifies similarities, differences and specific features that can serve as a basis for the subdivision in the categorizations. The integration of the user perspective is essential in order to enable a practice-relevant procedure model. The consideration of related technology trends, such as AI, machine learning or cloud computing, helps to define and integrate forward-looking characteristics.

The analysis and procedure model of the software is divided into four process steps: a) Software interface to design engineers and enterprise applications b) the interaction between design engineers and software, c) software internal processes and d) the interaction of design engineers and software afterwards.

3.1 Software Interface to Design Engineers and Enterprise Applications

The state of the art and related work have already shown initial approaches to classification and structuring the process inside software tools. As a first step, the software user interface and the integration into existing enterprise applications and data sources are critical aspects, especially regarding further aspects and possibilities.

Software User Interface. The user interface defines the interaction between the design engineer and the software assisted functions as well as the possibles for the software automated part. The basic subdivision into "Functional CAD programming environment", similar to standalone, and "Assisted CAD modelling", similar to integrated CAD, as a superordinate subdivision is adopted. It has a critical role in the implementation of further characteristics and integration into industrial structures. Beyond that, it is directly correlated to the variety of design automation options. On the one hand, the process can be specified by the software manufacturer and cannot be influenced by the users, which primarily occurs with CAD integrated systems. Apart from the input and output parameters, the system can be considered as a black box. The design engineer is guided through the process. On the other side, the process can be seen as a white box. Individual steps can be customized up to a system which is completely flexible and can, for example, have a modular structure [20]. Customizing refers to the possibility of adapting the tools process for individual applications to expand the functionalities. This is particularly important for different use cases and application areas.

Enterprise Applications and Data Sources. The integration into existing enterprise software systems has an important role to path the way how design engineers and engineers from different departments can work together. The collaboration for shared development has become an essential part with the integration of PLM systems and cloud computing in recent years. On second thoughts, this plays a decisive role in how existing systems can be connected and the connection to enterprise databases, information and know how can be realized. For example, GD is understood as process automation of the design creation steps to overcome bottlenecks like a manual export and import for FEM analysis. Besides that, especially for the evaluation tasks the consideration of company knowledge, information and circumstances are mandatory.

In practical applications, the collaboration can take place through manual file exchange, file-based, which limits collaboration. The integration can be implemented through common access to a standardized data format, database integration or middleware, workflow automation or APIs or collaborative (cloud) platforms. Pilagatti et al. [11] stated that some software developers are already working in a cloud-based collaborative environment for product lifecycle management (PLM) optimization inside GE&D software. In a comprehensive cloud-based development environment with simultaneous working, design engineers can work together. In addition to simultaneous working, the environment can enable real-time chats, comments and updates. However, this also leads to advantages on the software side when considering in-house expertise within the algorithms to enable a learning capability based on company specific con-texts and experiences, which will be explained in more detail below.

3.2 Interaction Between Design Engineers and Software

The design engineer's interaction with the software has two main distinguishing criteria: On the one hand, the user input/task and, on the other hand, the software functionality or task. After the design engineer has transferred the problem in a manner understandable to problem-solving process, an analysis of the application areas is necessary for the selection of the software, before transferring the problem specification inside the software.

Application Areas. The design engineer's problem space for the GE&D software can cover many different areas like design space exploration, design automation, modelling, or manufacturing. These areas can range from individual parts to assemblies. Problems such as parameter optimization, geometry optimization, part/variant diversity, construction space, lightweight construction, flow, thermal or functional problems can be addressed. In modern engineering, other domains play a role in addition to optimal design through multidisciplinary analysis. Therefore, the task of the software is to provide multidisciplinary support as well as a multidomain, e.g. through the integration of manufacturing aspects, economic or sustainability aspects.

In practice, software vendors typically applied first functions for structural (static) problems in the sense of the widespread definition of GD for individual parts. An extension to assemblies and dynamic problems is possible. Pilagatti et al. [11] already presented the integration of fluid studies as the first implemented extension, while Buonamici et al. [21] see the integration of dynamic and thermal use cases as the first step. Some tools already included application areas for fluid or thermodynamic problems. A multidisciplinary application is possible, to combined different application areas.

Problem Specification. After the design engineer has mapped the problem to the problem space and the application is within the scope of the software, the user-defined problem must be transferred to the software. The representation of the actual conditions of a system is important in order to create realistic and precise designs. In principle, a distinction is made between boundary conditions (functional and non-functional) and constrains, objectives, and design variables for the system.

The boundary conditions, constrains and design variables for the system can include geometries, spaces, loads, supports, load steps, materials as well as manufacturing processes. The objectives depend on the respective application and can be in the mechanical area e.g. mass reduction, stress reduction or frequency. The functionality and task for the software is to provide user interfaces and/or import functions to define the problem. These inputs are afterwards interpreted and structured in a practical way. At a further level the software provided features to integrate and interpret guidelines for the specific application field.

In practical applications, the large number of different input parameters is compromised to a set of variables. Individual design variables such as dimensions, materials as well as restrictive specifications such as manufacturing processes can be entered. This has a direct influence on the solution space and restricts it accordingly. Usually, the design engineer must specify certain geometries, the so-called non-design spaces, which can no longer be changed. Furthermore, the construction space in which geometries can be created can be specified. The tools support a specific selection of boundary conditions and objectives that can be assigned with values. The focus is on the form, geometric representation, i.e. functional sketches or similar are not addressed.

A decisive role of the problem description are supported manufactured processes / constrains for the development of products that are not only optimally designed but can also be manufactured efficiently and cost-effectively. Vlah et al. [12] emphasize that the integration of manufacturing processes is an important aspect for the development of the tools. Although Briard et al. [22] emphasize the high potential of combining GD and AM, especially through bio-inspired shapes, Wang et al. [9] criticize that commercial software vendors ignore the main manufacturing constraints of AM processes in the generation of solution alternatives. In general, when considering manufacturing processes, a distinction must be made as to whether general parameters can be specified, e.g. the building space of an additive manufacturing machine, or specific machines can be selected based on the companies' own.

In practical applications the integration of manufacturing processes can be broken down into the following aspects: no manufacturing restrictions are taken into account (unresisted). The design process focuses mainly on aesthetic or functional aspects without regard to manufacturability. There is also no integration of manufacturing data or guidelines. Another approach is to consider additive manufacturing processes. This can cover the entire range of additive manufacturing, including different 3D printing technologies, materials and applications. Moreover, traditional 2D manufacturing processes and 3D manufacturing processes are enabled. This will be dedicated to conventional flat manufacturing processes, such as cutting and punching, as well as more advanced three-dimensional manufacturing processes, including CNC machining, precision casting, and other methods that have specific requirements for three-dimensional shaping and material processing. As an advanced approach, the combination of different manufacturing processes is made possible, whereby additive manufacturing, for example, can serve as the basis for further processes.

3.3 Software Automated

The software automated process consists of generate, evaluate and parameter optimization to set up the solution space. This systematic exploration and analysis of different design alternatives within a defined design space is a basic principle of GE, which has also been identified as the primary function of generative product design systems in Mountstephens and Teo [7]. Since no user inputs are required in that steps, the two main distinguishing criteria in this section are: the software functionality or task and the technology, theoretical principle or approach to support this.

Generate. In the first step, the task of the software is to generate one or rather a multitude of different solutions based on the user's specified boundary conditions. One goal is to find ideas and solutions that support human creativity, but also go beyond it to explore alternative solutions. To success this, the software provides different tools and workflows to generate. The most known tool for the design generation is topology optimization. Besides this, classic approaches like search tress or agent based design [16] as well as newer approaches like field driven design, optimization, generic or evolutionary algorithms and AI can be used [23]. In the iterative process, the previous results serve as a guide and are further improved by the various tools. In this field, the knowledge about previous companies' solutions, information and data can be critical. Existing solutions including simulation or live data can be incorporated through machine learning to autonomously controlled generation and optimization of solutions. With respect to the different options, the number of solutions can vary between one solution and multiple solution.

In practical, the number of solutions is primally dominated by the number of design variables from the problem description. In the basic version of GE&D software, this means that the design engineer manually selects one or more corresponding parameters for each variant like materials or manufacturing processes. The solutions are then generated by selecting individual design parameters and the solution space is explored, whereby the determination is very selective. In the second stage, a large number of combinations of the existing design parameter intervals are selected using classic design of experiments methods in order to enable greater coverage of the design space. As an advanced approach, the integration of evolutionary algorithms and generative AI can be used to explore larger spaces and generate independent results.

Evaluate. The analysis and evaluation of the results represents a critical step in the process chain. In general, a distinction must be made between subjective and objective criteria [7]. In addition to aspects that are easy to determine (objective criteria), such as volume, weight or mechanical strength, aspects such as ergonomics, aesthetics, manufacturability, and industrial know-how can also be important (subjective criteria) [21]. Moreover, from the boundary conditions of the user, single and multiple objectives are possible. While in single-objective optimization the solutions can be created and evaluated according to a specific criterion, GE&D results, which are primarily based on multi-objective optimizations, are more difficult [9]. Additionally, the supported kind of criteria can be different from various domain as in the boundary conditions, implemented like costs, manufacturing, and sustainability. The software needs to incorporate all these kinds of criteria and support the internal evaluation as well as the design engineer in their process later. In this area, also the integration inside the companies' knowledge is critical to provide a finer evaluation.

In order to implement these conditions within the software, classic mathematical functions and simulations must be integrated, as well as methods for evaluating single and multi-objective optimizations. The use of machine learning algorithms can provide support in order to make specific evaluations based on historical data within the company [24]. As the internal evaluation has no direct reference to the user, the practical implementation within software tools is postponed for evaluation and interpretation by the design engineer.

Parameter Optimization. In parameter optimization, the software has the task to efficiently compare and evaluate, taking into account the resulting effects of parameter changes, based on the previous evaluation. The results are used to further refine the solution space and generate more precise solutions. In this step, the software focuses on identifying the most promising design options and further optimizing this design parameter.

In addition to optimizing algorithms, this may involve considering trade-offs studies and trade-offs to balance between different performance characteristics and requirements. The application of machine learning techniques can help to identify patterns or correlations in the data to find the most suitable candidate solutions.

3.4 The Interaction Between Design Engineers and Software Afterwards

Interpretation. The different evaluation methods and software functionalities from the evaluation inside the software are also necessary to support the design engineer for the interpretation. The design engineers have the tasks to evaluate between single and multiple criteria inside a pareto front and need to be supported in different domains. The software therefore needs to provide the specific values and represent them in an appropriate way.

In practical, Vlah et al. [12] showed that only simple results such as mass, tension and a visual comparison were possible in the specific comparison, while the consideration of costs is an important aspect for the development of the tools. In general, a simple analysis of (objective) key figures such as strength or weight is possible, which can be determined using mathematical equations or a simulation. In this case, there is no influence from user-specific criteria or specifications and there is no evaluation in the sense of trade-off studies. With extended functions for evaluation, the integration of advanced simulation techniques for more detailed results, whereby several key figures and their interactions (in the sense of trade-off studies) can be taken into account, e.g. through visual forms of presentation is included. Some tools offer evaluation customization through the integration of user-defined criteria and specifications for the evaluation, e.g. cost calculations. The use of machine learning for evaluation based on historical data is possible. In addition to pure design parameters and historical data, other aspects such as machine or material availability or extensive cost calculations and forecasts are also integrated [25]. In this case, AI can be seen as an assistant, especially to support subjective aspects.

Postprocessing. After interpreting the solutions and selecting one or more, the solution must be post-processed. In principle, the process can be automated, manual by the design engineer or a combination of both. Important tasks here are the post-processing of individual geometric components, preparation for production and the overall check with regard to the boundary conditions, which may not have been evaluated beforehand. The software can provide support here through CAD integration for direct modeling and, for example, integration of features or further simulation. Export options to other software systems are also a central task. The software can also help to define the parameters for subsequent downstream processes such as manufacturing processes. This is another area where the integration of machine learning, particularly on the basis of internal company knowledge, is advantageous in order to both define the parameter and evaluate live feedback for the effects of modeling changes without further simulation.

In practical applications, both manual and automatic post-processing options are available. Connections to CAD applications and export options are also included. The integration of AI-supported evaluations of changes has not yet been implemented.

4 Application on Commercial Software Tools

Using the tools Autodesk Fusion and Synera, the presented procedure model will be applied in the sense of an evaluation. In addition to integration, the understanding of the respective vendors with regard to GE&D is also considered.

Autodesk Fusion is a cloud-based GD platform that integrates CAD, CAM and FEM in addition to GD to enable a seamless transfer. For Autodesk, GD requires a close link between the human operator, the underlying technology trends (AI, cloud computing) and the process and manufacturing domain. GD is an approach that "simultaneously generates multiple solutions based on real-world design goals, product performance requirements, and manufacturing constraints" [26].

For Synera, Generative Engineering is the opportunity to overcome the limits of the incremental, traditional product development process through a high focus on design automation and a low-barrier combination of different engineering tools. The aim is to use automatic, customizing algorithms to create shapes and geometry based on boundary conditions in a comprehensive platform [19, 27].

Software Interface to Design Engineers and Enterprise Applications. As a representative of CAD-integrated GE&D software, Autodesk Fusion is specially designed for a guided-through process and only gives the user options for variation within the individual input parameters, but not in the process and connection. The integration within the wider enterprise applications and data sources is therefore limited. Collaborative work is implemented directly through the cloud environment.

For Synera on the other side, the implementation as a low-code platform particularly enhanced the high degree of customizing to create individual, modular composition of algorithms. The resulting free configuration options led to a variety of implementations. A focus is also on the connection to existing enterprise application to include as many tools and information as possible. But the collaboration is primarily file-based, as the so-called workflows are stored on a file basis.

Interaction Between Design Engineers and Software. At present, Autodesk Fusion is limited to structural and fluid application areas. A multidomain is primarily supported in the evaluation phase. The boundary conditions must be entered within the guided process within the software, with a particular focus on the implementation of geometry, materials, manufacturing processes (unrestricted additive manufacturing as well as traditional 2D and 3D manufacturing processes with general parameters) and objectives for the problem specification.

Synera supports both multidisciplinary and multidomain application areas and boundary conditions. The boundary conditions can be incorporated both within the software and by importing them from various sources (problem specification).

Software Automated. In Autodesk Fusion, the automated software processes for optimization are predefined and are not adjustable for the design engineer as a black box. The resulting solution space in structural application results from the combination of materials and manufacturing processes. Further variation of individual parameters is not possible. The evaluation is primarily based on objective criteria through automatically generated (mechanical) simulations of the generated variants and initial cost forecasts for additive manufacturing are provided.

In Synera, the automated software process, which can be freely designed, are primarily support topology optimization approaches for generation. However, it is also possible to integrate machine learning approaches as well as self-written or available optimization algorithms. The integration of manufacturing constrains depends on the optimizer used (e.g. Altair OptiStruct) and its capabilities. In addition to the evaluation of objective criteria using mathematical formulas and simulation, it is also possible to integrate machine learning based on historical data. Through the implementation of a design of experiment with an individual selection of parameters the solution space can be widened up.

The Interaction Between Design Engineers and Software Afterwards. The design engineer is supported in the interpretation by numerous virtual comparison forms, whereby trade-off studies as well as initial AI proposals are integrated inside Autodesk Fusion. Direct implementation in a CAD environment enables manual post-processing without any problems.

On the other side, in Synera the design engineer can be supported by a sensitivity analysis and visual representation. For the postprocessing are automated functions inside the software as well as the connection to other software tools incorporated.

5 Summary and Outlook

The increasing importance and development of Generative AI in the engineering domain leads to an increasing number of software tools that integrate these capabilities into their systems and makes the differentiation and selection of tools challenging. This paper presented a methodology guided procedure model for the differentiation and selection of Generative Engineering & Design software based on an in-depth literature review, the investigation of existing software tools, the consideration of user needs and the underlying technology trends. The procedure model was descripted regarding engineering tasks and inputs, necessary functions of the software and the technology, theoretical principle or approach to enable that. For each main process step, specific characteristics are described in the corresponding areas. A special focus was set to relation to enterprise applications and data sources and the possibilities to integrate industrial knowledge. The procedure model was applied to two commercial software tools.

In the current implementation, the software tools focus on geometric solutions as kind of representation. This often requires non-design space (functional geometries), which means that the solution space is already limited. There is little or no support for the development of active and functional principles or systematic creation without geometries. The synthesis of working principles into realizable geometries is also not supported and forms potential for further integration into software tools. Similarly, subsequent downstream processes are not supported. Manufacturing or design information in the form of tolerances, fits or parameters is also not taken into account. Another aspect is the integration of sustainability aspects, also in relation to a circular economy. This involves not only options for assessing the designs created as part of the evaluation, but also options for creating the design like following rules for the 10R strategy.

In a further step, the evaluation of other commercially available software tools can be carried out. The system can be further broken down and detailed for the individual phases of product development (concept design, detailed design,...). This enables a targeted analysis and evaluation of the available systems for the phase in product development. Furthermore, the procedure model system can be supplemented by the other criteria. In summary, the procedure model presented here offers a first comprehensive option for organizing the available software tools for Generative Engineering and Design for practical work.

Disclosure of Interests. The authors have no competing interests to declare that are relevant to the content of this article.

References

1. Harvard Business Review: The next wave of intelligent design automation. https://hbr.org/resources/pdfs/comm/autodesk/The.Next.Wave.of.Intelligent.Design.Automation.pdf. Accessed 22 Feb 2023
2. Zhang, G., Raina, A., Cagan, J., McComb, C.: A cautionary tale about the impact of AI on human design teams. Des. Stud. (2021). https://doi.org/10.1016/j.destud.2021.100990
3. Generative Design Market Size & Share Analysis - Growth Trends & Forecasts (2024 - 2029). https://www.mordorintelligence.com/industry-reports/generative-design-market. Accessed 21 Feb 2024
4. Gartner. What's New in the 2022 Gartner Hype Cycle for Emerging Technologies. (2022). https://www.gartner.com/en/articles/what-s-new-in-the-2022-gart-ner-hype-cycle-for-emerging-technologies. Accessed 21 Feb 2024
5. Singh, V., Gu, N.: Towards an integrated generative design framework. Design Studies. **33**, 185–207 (2012). https://doi.org/10.1016/j.destud.2011.06.001
6. Jaisawal, R., Agrawal, V.: Generative Design Method (GDM) – A State of Art. IOP Conf. Ser.: Mater. Sci. Eng. (2021). https://doi.org/10.1088/1757-899X/1104/1/012036
7. Mountstephens, J., Teo, J.: Progress and challenges in generative product design: a review of systems. Computers. **9**, 80 (2020). https://doi.org/10.3390/computers9040080
8. Gu, N., Singh, V., Merrick, K.: A framework to integrate generative design techniques for enhancing design automation. In: Dave, B., Li, A.I., Gu, N., Park, H.-J. (eds.) CAADRIA 2010: New Frontiers; 4/7/2010 - 4/10/2010, pp. 127–136. CAADRIA, Hong Kong (2010). https://doi.org/10.52842/conf.caadria.2010.127

9. Wang, Z., Zhang, Y., Bernard, A.: A constructive solid geometry-based generative design method for additive manufacturing. Addit. Manuf. (2021). https://doi.org/10.1016/j.addma.2021.101952
10. Saadi, J., Yang, M.: Observations on the implications of generative design Tools on design process and designer behaviour. Proc. Des. Soc. **3**, 2805–2814 (2023). https://doi.org/10.1017/pds.2023.281
11. Pilagatti, A.N., Atzeni, E., Salmi, A.: Exploiting the generative design potential to select the best conceptual design of an aerospace component to be produced by additive manufacturing. Int. J. Adv. Manuf. Technol. **126**, 5597–5612 (2023). https://doi.org/10.1007/s00170-023-11259-7
12. Vlah, D., Žavbi, R., Vukašinović, N.: Evaluation of topology optimization and generative design tools as support for conceptual design. Proc. Des. Soc.: Des. Conf. **1**, 451–60 (2020). https://doi.org/10.1017/dsd.2020.165
13. Shea, K., Aish, R., Gourtovaia, M.: Towards integrated performance-driven generative design tools. Autom. Constr. **14**, 253–264 (2005). https://doi.org/10.1016/j.autcon.2004.07.002
14. Kang, E., Jackson, E., Schulte, W.: An approach for effective design space exploration. In: Calinescu, R.C., Jackson, E., (eds.) Foundations of computer software: Modeling, development, and verification of adaptive systems; 16th Monterey Workshop 2010, Redmond, WA, USA, March 31-April 2010; revised selected papers, pp. 33–54. Springer, Berlin, Heidelberg (2011). https://doi.org/10.1007/978-3-642-21292-5_3
15. Li, H., Lachmayer, R.: Automated exploration of design solution space applying the generative design approach. Proc. Int. Conf. Eng. Des. **1**, 1085–1094 (2019). https://doi.org/10.1017/dsi.2019.114
16. Cagan, J., Campbell, M.I., Finger, S., Tomiyama, T.: A framework for computational design synthesis: model and applications. J. Comput. Inf. Sci. Eng. **5**, 171–181 (2005). https://doi.org/10.1115/1.2013289
17. Pollák, M., Kočiško, M., Dobránsky, J.: Analysis of software solutions for creating models by a generative design approach. IOP Conf. Ser.: Mater. Sci. Eng. (2021). https://doi.org/10.1088/1757-899X/1199/1/012098
18. Junk, S., Burkart, L.: Comparison of CAD systems for generative design for use with additive manufacturing. Procedia CIRP. **100**, 577–582 (2021). https://doi.org/10.1016/j.procir.2021.05.126
19. Gerhard, D., Köring, T., Neges, M.: Generative engineering and design – a comparison of different approaches to utilize artificial intelligence in CAD software tools. In: Noël, F., Nyffenegger, F., Rivest, L., Bouras, A. (eds.) IFIP International Conference on Product Lifecycle Management, 1st ed., pp. 206–215 Springer Nature Switzerland; Imprint Springer, Cham (2023). https://doi.org/10.1007/978-3-031-25182-5_21
20. Biedermann, M., Meboldt, M.: Computational design synthesis of additive manufactured multi-flow nozzles. Addit. Manuf. **35**, 101231 (2020). https://doi.org/10.1016/j.addma.2020.101231
21. Buonamici, F., Carfagni, M., Furferi, R., Volpe, Y., Governi, L.: "Generative Design": An Explorative Study. In: Computer-Aided Design & Applications, pp. 144–155 (2020). https://doi.org/10.14733/cadaps.2021.144-155
22. Briard, T., Segonds, F., Zamariola, N.: G-DfAM: a methodological proposal of generative design for additive manufacturing in the automotive industry. Int. J. Interact. Des. Manuf. **14**, 875–886 (2020). https://doi.org/10.1007/s12008-020-00669-6
23. Jang, S., Yoo, S., Kang, N.: Generative design by reinforcement learning: enhancing the diversity of topology optimization designs. Comput. Aided Des. **146**, 103225 (2022). https://doi.org/10.1016/j.cad.2022.103225
24. Shin, D., et al.: How to Trade off Aesthetics and Performance in Generative Design. Advances in Structural Engineering and Mechanics (ASEM21) (2021)

25. Moussavi, S.M., Svatoš-Ražnjević, H., Körner, A., Tahouni, Y., Menges, A., Knippers, J.: Design based on availability: Generative design and robotic fabrication workflow for non-standardized sheet metal with variable properties. Int. J. Space Struct. **37**, 119–134 (2022). https://doi.org/10.1177/09560599221081104
26. Autodesk: Autodesk - Generative Design for Manufacturing: How to accelerate your Product Development Process. https://d1.awsstatic.com/partner-network/partner_marketing_web_team/Manufacturing-partner-solutions_Autodesk_brochure.pdf. Accessed 28 Mar 2022
27. Synera | Process Automation for Engineers (2024). https://www.synera.io/. Accessed 21 Feb 2024

Exploring Model-Based Systems Architecting Digital Threads in Virtual Reality

Sergio Camilo Medina Galvis, Romain Pinquié(✉), and Baptiste Paterne

University Grenoble Alpes, CNRS, Grenoble INP, G-SCOP, 38000 Grenoble, France
romain.pinquie@grenoble-inp.fr

Abstract. Adopting a Model-Based Systems Engineering (MBSE) approach generates a seamless flow of linked data that connects all aspects of a product life cycle: digital threads. The consistency of the linked set of digital artefacts must be actively managed over the system life cycle, from design to manufacturing to maintenance and beyond. Immersive analytics has proved helpful in managing large datasets. Still, no research has focused on using advanced interactive visualisation to make use of and manage digital threads. This paper presents GraphXplore, a virtual world that enables systems engineers to explore MBSE digital threads by interacting with a rich, immersive, layered 3D graph visual metaphor where nodes and edges represent MBSE entities (e.g., needs, functions, architecture elements, etc.) and interdependencies, respectively. The MBSE digital threads of a helicopter landing gear system serve as a case study to illustrate and evaluate GraphXplore. Future work will concentrate on the continuous improvement of GraphXplore and an empirical comparison with competitive digital thread solutions.

Keywords: Digital Engineering · Model-Based Systems Engineering · Virtual Reality · Digital Threads · Immersive Analytics

1 Introduction

1.1 Context

Companies strive for Digital Engineering (DE) to become the *de facto* standard strategy for engineering systems. DE is an enterprise approach incorporating data-driven infrastructure, methods and tools, innovation, and workforce transformation centred on an authoritative source of truth for engineering data [18]. It not only aims to integrate all design data related to a system across the entire life cycle but also to integrate design data across the entire system life cycle and "*bridge the business processes across all functional disciplines*" [20]. It describes a holistic approach to designing complex systems using digital models instead of documents and integrating data across models. Thus, the crux of DE is computer-readable models that represent all aspects of the system and support all the activities for its design, development, manufacture, and operation throughout its life cycle. Model-Based Systems Engineering (MBSE), the formalised application

Published by Springer Nature Switzerland AG 2025
P. Sureephong et al. (Eds.): PLM 2024, IFIP AICT 740, pp. 58–72, 2025.
https://doi.org/10.1007/978-3-031-93319-6_5

of modelling to the systems engineering approach, is a subset of DE. Although the proposal presented in this paper can apply to MBSE, the scope is limited to Model-Based Systems Architecting (MBSA). MBSA is often used in place of MBSE, but as Systems Architecting is the front end of Systems Engineering [24], MBSA is the front end of MBSE.

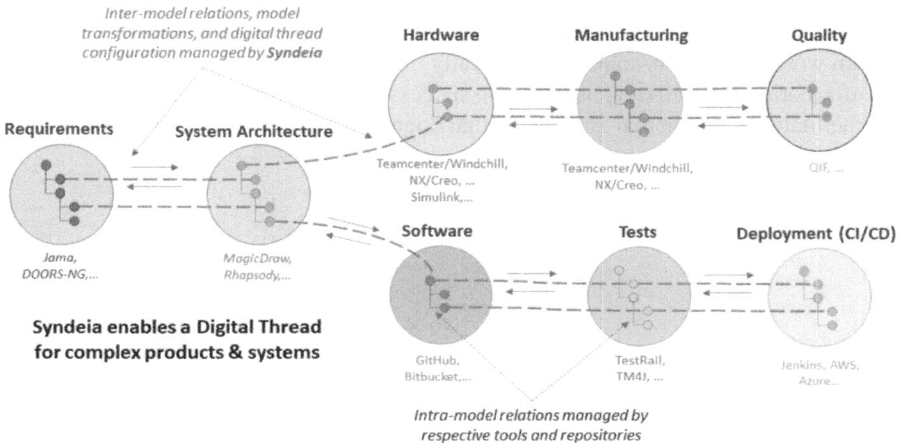

Fig. 1. Illustration of the digital thread concept by Syndeia[1]

A central pillar of DE is the concept of digital thread (Fig. 1), which is a linked set of digital artefacts of a product's lifecycle, from design to manufacturing to maintenance and beyond, providing a seamless flow of data that connects all aspects of the life cycle and whose consistency is actively managed over the system life cycle. Digital threads aim to deploy an integration framework to link heterogeneous information systems and facilitate a continuous, synchronised data flow across the various domains of a system life cycle in dynamic ways without requiring one-to-one data mapping [8]. Digital threads will increase interdepartmental collaboration, streamline product development to reduce production time, and ensure regulatory compliance. A crucial activity of DE, therefore MBSE and MBSA, is developing, using, and maintaining digital threads.

1.2 Practical Problem and Research Gap

Implementing fine-grained end-to-end digital threads remains a technical and organisational challenge for software publishers and professionals. Digital threads of complex model-based engineered systems are, in a way, Big Data. Visual immersive or not-data analytics help to analyse large and complex datasets [4]. Thus, interactive visualisation could help develop, use, and maintain digital threads [25]. However, Figs. 2 and 4 show that DE, MBSE and MBSA software offer 2D planar graph and desktop-based visualisation capabilities with very few interactions like filtering and expandable nodes that do not enable designers to explore a large number of data at once, especially on small, standard screens. A conventional office desk setup usually includes one or two desktop screens, limiting the ability to get a bird's eye holistic view and navigate in a large set of large linked digital artefacts. Consequently, practitioners have no alternative but to navigate multiple models to search for relevant data dispersed among different windows and integrate it into a coherent mental representation. Generating and refining hypotheses about the system-of-interest is such a setting becomes extremely difficult [13]. Understanding the system as a whole is also impossible for complex systems because the number of models exceeds a system architect's span of control. Indeed, the limited capacity of human working memory makes it practically impossible for a system architect to retain all information from all models to reason successfully about the system. MBSE tools do not facilitate the perceptual and conceptual cognitive reasoning processes.

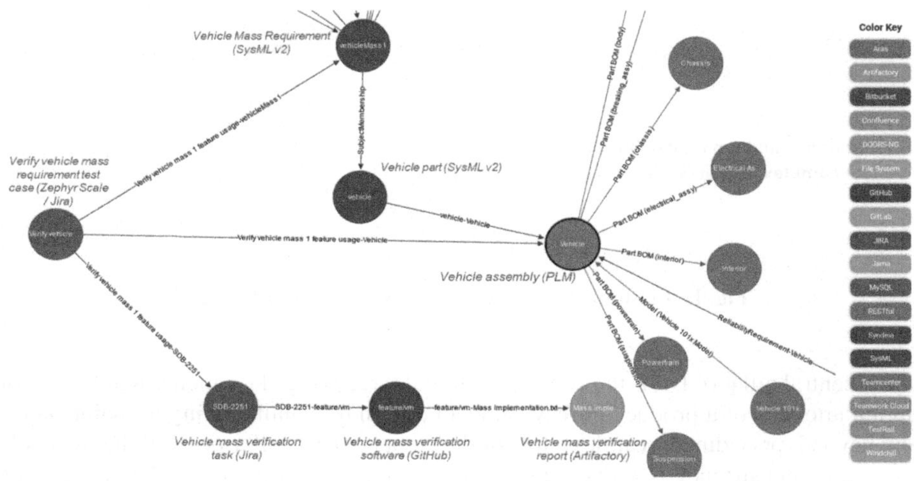

Fig. 2. Digital thread exploration in the Syndeia[2] digital engineering software

This work is problem-based research that seeks to find a solution to the need to explore digital threads in Model-Based Systems Architecting (MBSA). The unlimited and highly interactive virtual 3D space and the sense of 3D depth offered by virtual reality motivated the research question: **What is the potential of virtual reality for exploring digital threads for Model-Based Systems Architecting?** Although the flexibility and genericity of GraphXplore means it can cover any digital threads for digital or systems engineering activities and data, the scope is limited to systems architecting.

1.3 Claim

GraphXplore, a virtual reality-based interactive layered 3D graph, will help system engineers explore and maintain the coherence of MBSA digital threads.

1.4 Outline

Section 2 will provide background knowledge and review related work on digital threads, emphasising digital thread visualisation capabilities and studies exploring the potential of virtual reality for model-based systems architecting. Section 3 presents the immersive virtual environment GraphXplore applied to a helicopter's landing gear system. Section 4 evaluates the proposed solution to the digital threads exploration research problem before discussing the results and suggesting future work in Sect. 5.

2 Related Work

2.1 Digital Thread

A review of digital thread definitions (Table 1) from scientific papers, corporate documentation, and software technologies shows that the concept is seen through two different lenses.

Table 1. Definitions of digital thread.

Source	Definition
IBM[a]	A digital representation of a product's lifecycle, from design to manufacturing to maintenance and beyond, providing a seamless flow of data that connects all aspects of the lifecycle
AIAA[b]	A linked set of digital artifacts whose consistency is actively managed over the lifecycle of a product, process, or system
Syndeia[c]	A graph whose nodes are elements in various enterprise repositories, tools, and version control systems and whose edges are intra-model relationships within each agency and inter-model relationships created and managed by Syndeia between the nodes
DAU[d]	An extensible and configurable analytical framework that seamlessly expedites the controlled interplay of technical data, software, information, and knowledge in the digital engineering ecosystem, based on the established requirements, architectures, formats, and rules for building digital models
Eurostep[e]	A coherent set of integrated product information enriched by data from diverse domains and disciplines. This integrated approach spans the entire value network, from suppliers and partners to customers, across the product's lifecycle. A resilient digital thread is one that is designed to manage evolving links and relationships over time, enabling agile product development and continuous feedback loops

[a] https://www.ibm.com/blog/digital-thread-vs-digital-twin/
[b] https://www.aiaa.org/resources/digital-thread-white-paper
[c] https://intercax.com/products/syndeia/
[d] https://www.dau.edu/glossary/digital-thread
[e] https://www.eurostep.com/what-is-a-digital-thread/

Similarly to PLM, digital threads are seen as a strategic product-centric, lifecycle-oriented and information-driven business approach e.g., Digital thread-enabled Department of Defense acquisition enterprises [5] or an inter-organisational enterprise-level software application (supporting a business strategy) [23].

- Digital thread as a business approach: "*The aim was to lay the foundation for a digital thread enabled acquisitions ecosystem in which the processes (and potentially even organizational structures) are updated to best take advantage of the new digital engineering capabilities*" [5].
- Digital thread as a software application "*A Digital Thread graph is the dig ital blueprint of a system that evolves through its lifecycle. It knows the versions of connected models in various configuration management systems, e.g., Product Lifecycle Management, Application Lifecycle Management, and Requirements Management. Syndeia provides visualization and query services to analyze the digital thread and trace intra and inter-model relationships within and across the boundaries of repositories/tools. For example, start with a SysML block and explore intra and inter-model relationships to requirements in Jama, then relationships to mechanical parts in PLM and CAD systems or issues in JIRA*[1]".

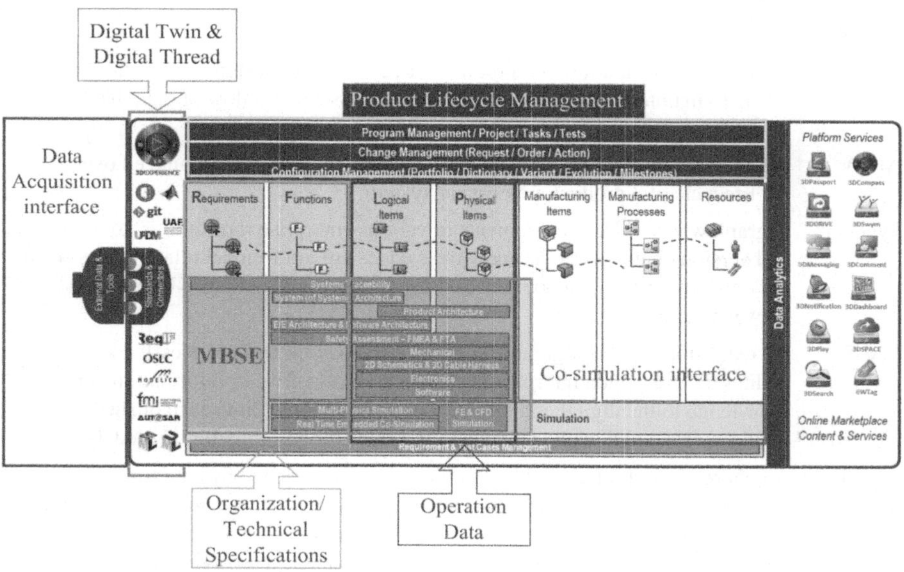

Fig. 3. Digital threads in the 3D Experience PLM software

As Digital thread is a hot topic, numerous software publishers in DE[2] (Fig. 1), PLM (Fig. 3) and MBSE[3] tend to argue that they implement the concept. PLM software

[1] https://intercax.com/products/syndeia/.
[2] https://intercax.com/2020/08/21/syndeia-and-building-the-digital-thread/.
[3] https://fr.mathworks.com/help/slcheck/ug/digital-thread.html.

usually manages data at coarse files or metadata levels rather than fine-grained model elements. However, the recent integration of MBSE capabilities in PLM software offers the opportunity to create and trace atomic data such as requirements, functions, logical units, or parts, provided that they were authored with modelling software that belongs to the same vendor or a partner.

Similarly, some MBSE environments integrated or not in a PLM software propose traceability maps (Fig. 4) similar to digital threads. However, these traceability maps are limited to the model elements represented by architecture diagrams. They do not capture linked data throughout a system's life cycle. Although means for implementing end-to-end digital threads such as the early work on graph-based digital blueprint [1–3] at the origin of the Syndeia software or Open Services for Lifecycle Collaboration (OSLC), ISO 10303–243:2021 [10] Modelling and Simulation information in a collaborative Systems Engineering Context (MoSSEC) and Long Term Archiving and Retrieval[4] (LOTAR) interoperability standards are available or under construction, existing proofs of concept are limited to the management of 3D simulation models [9], data-driven design [26, 27], quality assessment of architecture models [22], or smart manufacturing [8]. As stated [29] "*The challenge ahead is to move beyond the concepts and engineer the framework, processes, and tools to realize the ideal*" and the current implementation level of digital thread is proof of concept [19].

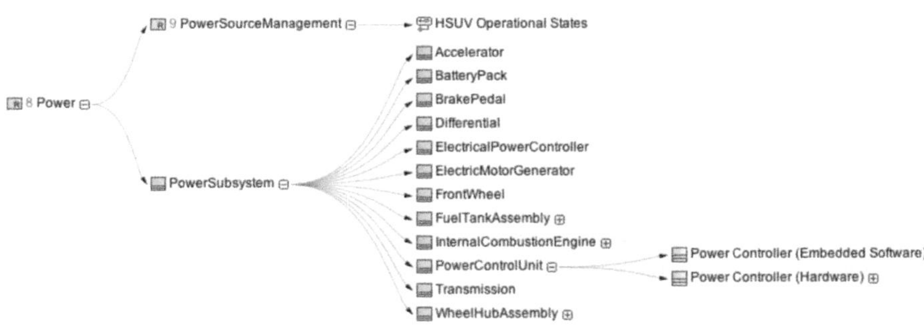

Fig. 4. Digital threads based on relation map in Catia Magic MBSE software

2.2 Digital Thread Visualisation

Digital threads, seen as big data, could be exploited using AI or visual analytics, whether immersive or not. A key technology for implementing digital threads is graph-oriented NoSQL databases such as Neo4j [1, 3, 9, 19, 22]. Neo4j also serves as a means to visualise the graph database (Fig. 5), but it is very limited in terms of graph drawing.

Various research studies [7, 12, 14, 15, 28], mainly in biology and medicine, developed interactive 2D and 3D graph drawing libraries that could be leveraged to explore linked data such as digital threads.

[4] https://lotar-international.org/.

3 Digital Threads Immersive Exploration: GraphXplore

This section presents the solution GraphXplore, an immersive virtual environment for visually exploring MBSA digital threads.

3.1 Tasks Analysis

The main user of GraphXplore is a system engineer who has, among other things, to maintain consistency and ensure traceability during system development. To realise and maintain traceability means to define and keep up to date with all sorts of systems engineering elements (e.g., life cycle concepts, missions, operational activities, functions, business requirements, stakeholder needs, stakeholder requirements, system requirements, architecture elements, interface definitions, analysis results, verification methods, etc.) and their interdependencies.

There are two main types of traceability: horizontal and vertical [16]. In a systems engineering approach, horizontal traceability aims to maintain the coherence of data at a given systemic level (e.g., the digital thread that likes needs, functions, requirements, and architecture elements of a sub-system), whereas vertical traceability aims to maintain the coherence between system and systemelement levels (e.g., System-level requirements, subsystem-level requirements, and component-level requirements). Specific views can combine horizontal and vertical traceability to answer particular questions.

Vertical and horizontal traceability is necessary for a "What-If" impact analysis. System impact analysis is a systems engineering framework and best practice for systemically and systematically identifying, analysing, estimating, and implementing the impact of a significant technical change that ripples across an existing, complex system. Impact analysis enables a system architect to understand the implications of the change and the elements of the system affected, accurately estimate the work required for the change, and make informed decisions that maximise business value.

Exploring digital threads means querying one or multiple databases, which requires learning the query languages. Systems engineers are not software engineers. They do not have to learn query languages. Thus, experts develop custom graphical user interfaces that enable MBSE practitioners to answer their questions without knowing the query languages to hide the complexity of declarative query languages. However, it isn't easy to anticipate all the questions that MBSE practitioners may want to answer by interacting with the database. A more flexible solution is to facilitate visually exploring the linked data.

3.2 Meta-model

As a case study, the meta-model in Fig. 5, which also corresponds to the Neo4j database schema, defines the elements and relationships that digital threads help to maintain. The MBSA of the system of interest, a helicopter landing gear, relies on the definition of life situations, each including one or multiple use cases. Each use case involves at least one actor and a sequence of actions, which.

are transformed into stakeholders needs. Each stakeholder need is mapped to one or multiple landing gear system functions performed by system elements represented by geometrical shapes. Note that the data model is an illustrative example that could be more detailed with requirements, interfaces, etc.

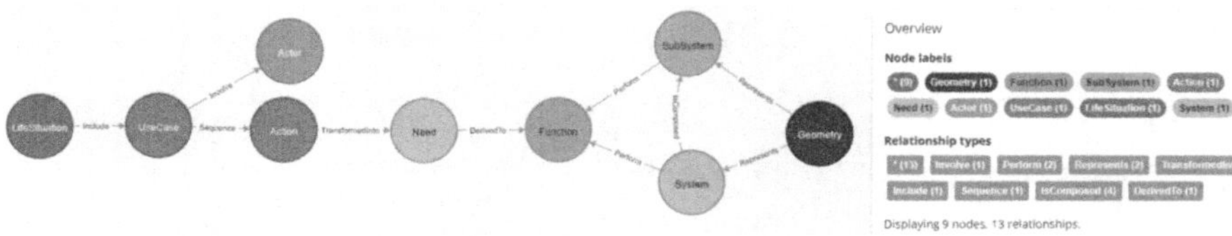

Fig. 5. Neo4j database schema defining the nodes and relationships of the digital thread

The illustrative case study aims to explore digital threads for a helicopter's landing gear system. The database gathers 184 objects and 230 relationships with 6 life situations, 17 use cases, 4 actors, 33 actions, 38 needs, 66 functions, 15 systems, 7 sub-systems, and 5 geometries. It also includes 18 include, 21 involve, 57 sequence, 41 transformedInto, 15 derivedTo, 64 perform, 9 isComposedOf, and 5 represent relationships.

3.3 Visual Metaphors

Systems engineering being a transdisciplinary and integrative approach to enable the successful realisation, use, and retirement of engineered systems, each company develops its own MBSA framework or meta-model tailored to the internal practices (e.g., processes, glossaries, products, etc.). For instance, for some companies, "stakeholder expectations" are communicated in terms of stakeholder needs and requirements; ISO/IEC/IEEE 29148 [11] defines "stakeholder requirements" in terms of "stakeholder needs". For others, stakeholder needs and expectations are communicated in a mission, business goals, use case, user story, user journey, operational scenario, storytelling, functional analysis, or an operational concept. In some domains, "user needs and requirements" and "customer needs and requirements" are specifically addressed instead of, or in addition to, the more generic "stakeholder requirements" designation. The diversity of terms and concepts requires the development of flexible visual metaphors.

The main visual metaphor of a layered 3D graph implemented in GraphXplore was preferred for analysing and managing change propagation [21] without advanced interactive visualisation capabilities. A 3D graph layered or not is a flexible metaphor that enables the visual encoding of the meta-model of any company's MBSA framework by defining new nodes or edges and mapping them to MBSA concepts. In GraphXplore, the user can select a layout algorithm (parallel, zigzag, star-like, cube, circular), which he can apply to a selected set of layers individually (Fig. 6). It also supports four main clustering algorithms (Louvain, Walktrap, Fast-Greedy, and Label propagation) offered by the *igraph* library [6] in combination with various global and local layouts for a set of selected layers and channels to run them separately i.e., per layer or in combination i.e., across layers (Fig. 7).

A systems engineer can interact with nodes and relationships to show attributes, filter by type, explore neighbours and travel paths, etc. Figure 8 illustrates a filtered

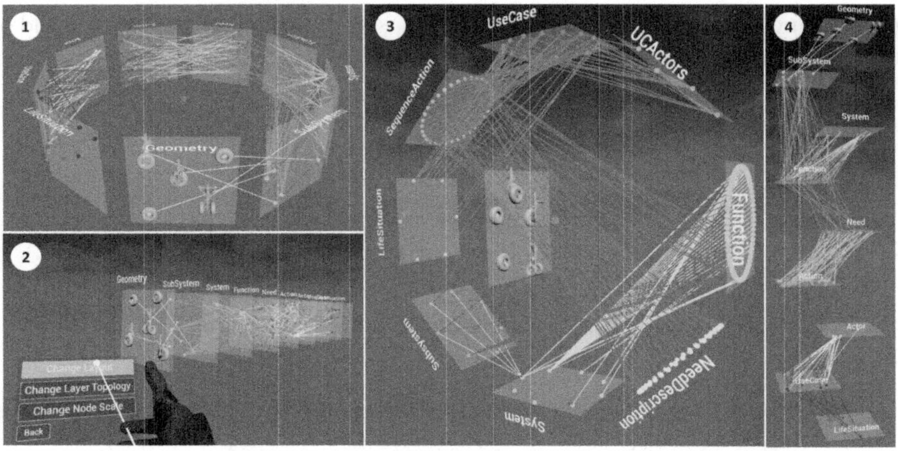

Fig. 6. 3D graph-based immersive visualisation of MBSA digital threads

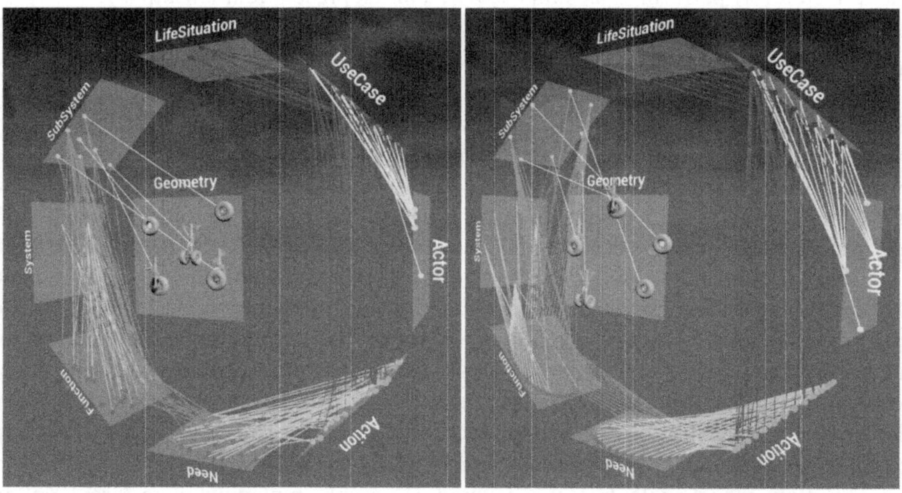

Fig. 7. Circular graph layout algorithm with two different clustering algorithms

view of a subgraph that shows the functions and sub-systems connected to the selected system. The exploration process of linked data can start from any perspective (e.g., life situation, use case, functions, systems, geometries, etc.). So far, GraphXplore does not offer any visual exploration scenario related to the tasks a systems engineer would perform. Consequently, he must have a clear objective to avoid being overwhelmed by information.

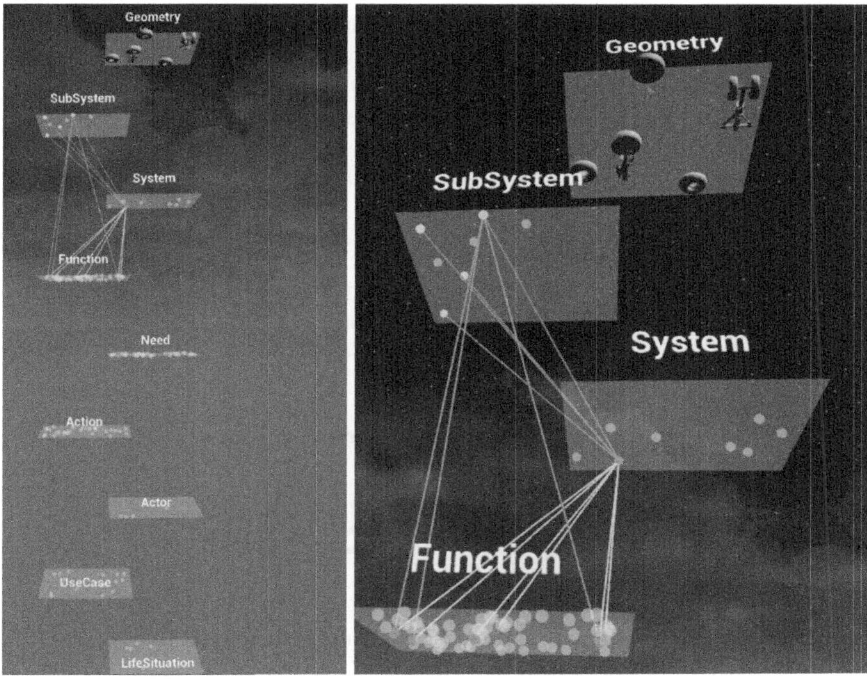

Fig. 8. Filtered view showing elements affected by the modification of a system

3.4 Software Architecture and Implementation

Figure 9 illustrates the software architecture and its functionalities described in this paper. GraphXplore's core functionalities are developed using Unreal Engine (UE), a 3D computer graphics game engine. UE provides an interface between the virtual environment and the hardware for stereoscopic rendering and interaction.

The graph-oriented NoSQL Neo4j database stores the MBSA digital threads. Neo4j is a powerful and reliable database technology that stores and accesses linked data. It also implements various algorithms for graph analytics and provides access to multiple APIs. GraphXplore sends Cypher queries to the Neo4j database and receives results via HTTP requests.

GraphXplore incorporates Arena3Dweb [14], a web-based application that provides a 3D visualization of multilayered graphs built using R and JavaScript. Arena is an intermediary, facilitating access to network algorithms implemented in the *igraph* R library [6]. A custom RESTful API exposes Arena's core functionalities via HTTP requests for communication between UE and Arena.

Although this paper does not elaborate on it, GraphXplore integrates a Generative AI using OpenAI API to translate natural language questions into formal Cypher queries in Neo4j. Being able to interact with GraphXplore effortlessly through natural language voice command recognition is of interest for two main reasons: 1) systems engineers are not software engineers trained to be proficient in the Cypher Neo4j query language, and

2) it is impossible to anticipate the definition and maintenance of all Cypher queries a system engineer can make so as to develop a custom GUI that calls pre-defined queries. Moreover, in some cases, one may explore the 3D graph by asking natural language questions before navigating through the graphical representation.

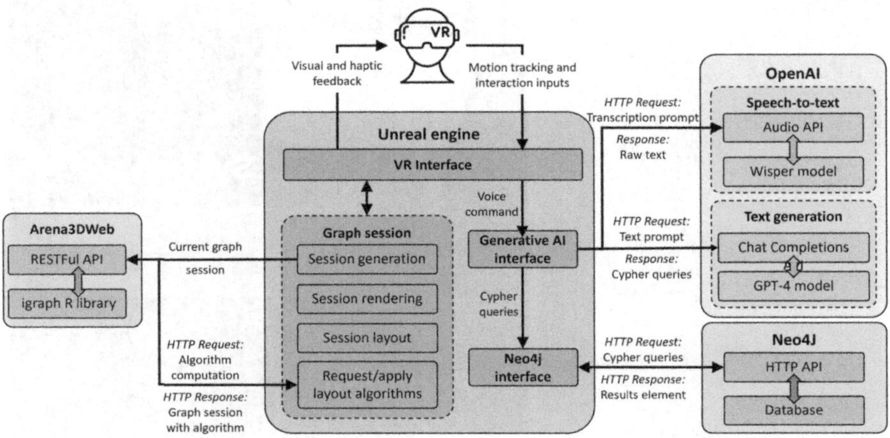

Fig. 9. Software architecture of the immersive virtual environment GraphXplore

When receiving a voice command, UE builds a request containing the audio file and a prompt with a dictionary to ensure correct spellings. This request is dispatched to the Speech-to-text OpenAI Audio API via an HTTP request. As a response, based on the state-of-the-art Whisper model, the API sends the raw text of the user's voice command back to UE. To execute the voice command inside Neo4j, UE generates a new request sent to the OpenAI Chat Completions API. This request incorporates the previous response and a custom prompt to provide a role, domain knowledge, and inputs regarding the definition and restrictions imposed by the meta-model. Upon dispatching this request, the GPT-4 model translates the natural language statement of the original voice command into Cypher queries. After receiving the GPT-4 API's response, UE transmits it to Neo4j via an HTTP request. The result is then deserialised and processed within UE to be executed in the running Neo4j graph session. Generative AI capabilities embedded in GraphXplore are out of the scope of this paper, so details such as few-shot learning, RAG, and […] prompt engineering are not discussed.

4 Qualitative Evaluation

This paper does not propose a quantitative empirical evaluation of GraphXplore's usability and utility. Nor does it offer any fair comparison with the few existing digital thread visualisation solutions, such as the SysML relation map diagram in Catia Magic. The first validation step focused on feasibility and scalability, for which evidence was provided in the real-world landing gear case study. This section will adopt the Physics of Notation [17] as a theoretical framework to provide design justifications and recommend improvements that have to be implemented before designing and conducting controlled laboratory experiments and formative user evaluations based on testable research questions and hypotheses and consider further human-computer interaction criteria such as task correctness, effectiveness, usefulness, usability, user satisfaction and cognitive load.

[Semiotic clarity] A graphical representation is semiotically clear (no synographs and homographs) if there is a 1:1 correspondence between semantic constructs and graphical symbols. When designing a new visual notation, one must maximise expressiveness (by eliminating symbol deficit), precision (by eliminating symbol overload), and parsimony (by eliminating symbol redundancy and excess). There is a natural 1:1 mapping in GraphXplore as MBSA elements correspond to nodes and MBSA relationships correspond to edges. The next version should consider semiotic clarity at a lower level by encoding each type of MBSA element (e.g., requirement and use case) and relationships (e.g., derivedTo and represent) with a different symbol.

[Perceptual discriminability and Visual expressiveness and Graphic economy] Perceptual discriminability is the ease and accuracy with which graphical symbols can be differentiated from each other. While perceptual discriminability measures pairwise visual variation between symbols, visual expressiveness measures visual variation across the entire visual vocabulary. GraphXplore is perceptually direct as it uses shapes and colours that suggest the meaning of their referent concepts. It facilitates the differentiation of graphical symbols as each node type and relationship has its own colour. A limit of two visual variables is voluntary so that the representation is cognitively manageable. Colour was preferred as it is a cognitively effective visual variable. Differences in colour are detected three times faster than shape and more easily remembered. However, colours are sensitive to variations in visual perception and should be used for redundant coding. Thus, each node type and relationship should also be encoded with a different shape. In a future version, GraphXplore will use a greater variety of symbolic shapes. Each sphere object of a node type will have its own texture with signs (e.g., F for function, UC for use case, R for requirement). Redundant coding that suggests using multiple visual variables (e.g., shapes and colours) to distinguish between them will increase the visual distance between symbols, reducing errors and decreasing noise. An additional feature would be to exploit family resemblances among shapes (i.e. system requirement and sub-system requirement or system and sub-system) in the future GraphXplore version.

[Semantic transparency] A visual notation is semantically transparent if the meaning of a symbol can, to some extent, be inferred from its appearance. The graph metaphor of GraphXplore naturally provides some cues to symbols' meaning as form (spherical nodes and line edges) implies content (entities and relationships), but it is semantically conventional as the mapping between symbols' appearance and meaning is purely arbitrary.

[Cognitive integration Complexity management] A visual notation shall include visual cues that facilitate the cognitive integration process. The cognitive integration process involves a "*perceptual integration process characterized by a purposeful search behaviour of transitioning between diagrams and a conceptual integration process characterized by a hypothesis generation and refinement behaviour*" [13]. GraphXplore significantly improves the perceptual and conceptual integration processes since navigating and transitioning between diagrams is unnecessary. All relevant visual items, usually dispersed across various diagrams of different perspectives, are naturally linked in the 3D graph summary (long shot) representation, which provides a view of the system as a whole. To reduce complexity management, GraphXplore represents each concept on a separate layer so as not to overload the human mind.

5 Discussion and Future Work

This paper presents GraphXplore, an immersive virtual environment for exploring and maintaining digital threads in model-based systems architecture. In the current version, the main visual metaphor implemented in GraphXplore is a layered 3D graph. GraphXplore does not intend to replace existing means to support vertical and horizontal traceability or conduct an impact analysis but provides a complementary solution that reduces the cognitive integration process. Indeed, the information usually dispersed across various models of different perspectives is gathered in a layered 3D graph, which, once displayed in an immersive virtual world, provides a holistic view of the digital threads. As a case study, the paper illustrates how GraphXplore enables a system engineer to explore the digital threads of a helicopter landing gear system in virtual reality.

Future work will concentrate on the empirical evaluation of GraphXplore using formative user studies with systems architects and summative comparative evaluation through controlled laboratory experiments. New visual metaphors will be designed and compared with the layered 3D graph. New metaphors will also represent new concepts and support tasks such as vertical traceability, where each layer will correspond to a system level rather than a type of model element. Finally, a generative conversational AI will be fine-tuned with systems architecting data to enable a system architect to interact with GraphXplore effortlessly through natural language voice command recognition.

Aknowledgements. This work was supported by the ArchiTOOL project (grant ANR-20-CE10-0010), MIMESIS joint research lab project (grant ANR-23-LCV1-0002) in partnership with SkyReal, and Equipex+ Continuum (grant ANR-21-ESRE-0030) of the French National Research Agency.

References

1. Bajaj, M., et al.: Graph-based digital blueprint for model based engineering of complex systems **27**(1), 151–169. https://doi.org/10.1002/j.2334-5837.2017.00351.x
2. Bajaj, M., Cole, B., Zwemer, D.: Architecture to geometry integrating system models with mechanical design. In: AIAA SPACE 2016, pp. 1–19. American Institute of Aeronautics and Astronautics. https://doi.org/10.2514/6.2016-5470, bluehttps://arc.aiaa.org/doi/https://doi.org/10.2514/6.2016-5470
3. Bajaj, M., Hedberg, T.: System lifecycle handler spinning a digital thread for manufacturing **28**(1), 1636–1650. https://doi.org/10.1002/j.2334-5837.2018.00573.x
4. Cavallo, M., Dholakia, M., Havlena, M., Ocheltree, K., Podlaseck, M.: Dataspace: A reconfigurable hybrid reality environment for collaborative information analysis, pp. 145–153. https://doi.org/10.1109/VR.2019.8797733, ISBN: 9781728113777

5. Collopy, P., Paredis, C., Nadolski, M., McDermott, T.: Enterprise system-of-systems model for digital-thread enabled acquisition
6. Csárdi, G., Tamas, N.: The igraph software package for complex network research **Complex Systems**, 1695, bluehttps://igraph.org
7. Feyer, S.P., et al.: 2d, 2.5d, or 3d? an exploratory study on multilayer network visualisations in virtual reality, blue http://arxiv.org/abs/2307.10674
8. Hedberg, T.D., Bajaj, M., Camelio, J.A.: Using graphs to link data across the product lifecycle for enabling smart manufacturing digital threads **20**(1). https://doi.org/10.1115/1.4044921
9. Hoppen, M., Rossmann, J., Hiester, S.: Managing 3d simulation models with the graph database neo4j pp. 78–83, blue https://publications.rwth-aachen.de/record/660085, ISBN: 9781612084862
10. International Organization for Standardization: ISO 10303-243:2021 industrial automation systems and integration product data representation and exchange part 243: Application protocol: For modelling and simulation information in a collaborative systems engineering context (MoSSEC)
11. International Organization for Standardization: ISO/IEC/IEEE 29148 systems and software engineering. life cycle processes. requirements engineering
12. Karatzas, E., Baltoumas, F.A., Panayiotou, N.A., Schneider, R., Pavlopoulos, G.A.: Arena3dweb: interactive 3d visualization of multilayered networks **49**, W36– W45. https://doi.org/10.1093/nar/gkab278, blue https://academic.oup.com/nar/article/49/W1/W36/6246395
13. Kim, J., Hahn, J., Hahn, H.: How do we understand a system with (so) many diagrams? cognitive integration processes in diagrammatic reasoning **11**(3), 284–303. https://doi.org/10.1287/isre.11.3.284.12206, publisher: INFORMS Inst. for Operations Res. and the Management Sciences
14. Kokoli, M., et al.: Arena3dweb: interactive 3d visualization of multilayered networks supporting multiple directional information channels, clustering analysis and application integration **5**(2), lqad053. https://doi.org/10.1093/nargab/lqad053, blue https://academic.oup.com/nargab/ article/doi/https://doi.org/10.1093/nargab/lqad053/7185857
15. Koutrouli, M., Karatzas, E., Papanikolopoulou, K., Pavlopoulos, G.A.: NORMA: The network makeup artist — a web tool for network annotation visualization **20**(3), 578–586. https://doi.org/10.1016/j.gpb.2021.02.005, blue https://academic.oup.com/gpb/article/20/3/578/7198173
16. Königs, S.F., Beier, G., Figge, A., Stark, R.: Traceability in systems engineering – review of industrial practices, state-of-the-art technologies and new research solutions **26**(4), 924–940. https://doi.org/10.1016/j.aei.2012.08.002, blue https://linkinghub.elsevier.com/retrieve/pii/S1474034612000766
17. Moody, D.: The physics of notations: Toward a scientific basis for constructing visual notations in software engineering **35**(6), 756–779. https://doi.org/10.1109/TSE.2009.67
18. Office of the Deputy Assistant Secretary of Defense for Systems Engineering: Digital engineering strategy, blue https://ac.cto.mil/wp-content/uploads/2019/06/USA001603-18-DSD.pdf
19. Oroz, J., et al.: Implementing the digital thread - a proof-of-concept. In: AIAA SCITECH 2023 Forum. American Institute of Aeronautics and Astronautics. https://doi.org/10.2514/6.2023-1405, blue https://arc.aiaa.org/doi/10.2514/6.2023-1405
20. Papke, B.L., Wang, G., Kratzke, R., Schreiber, C.: Implementing MBSE – an enterprise approach to an enterprise problem **30**(1), 1550–1567. https://doi.org/10.1002/j.2334-5837.2020.00803.x, blue https://incose.onlinelibrary.wiley.com https://doi.org/10.1002/j.2334-5837.2020.00803.x

21. Pasqual, M.C., De Weck, O.L.: Multilayer network model for analysis and management of change propagation **23**(4), 305–328. https://doi.org/10.1007/s00163-011-0125-6, blue http://link.springer.com
22. Petnga, L.: Graph-based assessment and analysis of system architecture models **29**(1), 922–936. https://doi.org/10.1002/j.2334-5837.2019.00644.x
23. Pinquié, R., Rivest, L., Segonds, F., Véron, P.: An illustrated glossary of ambiguous PLM terms used in discrete manufacturing **8**(2). https://doi.org/10.1504/IJPLM.2015.070580
24. Rechtin, E.: Systems architecting: creating and building complex systems. Prentice Hall
25. Romero, V., Pinquié, R., Noël, F.: An immersive virtual environment for reviewing model-centric designs **1**, 447–456. https://doi.org/10.1017/pds.2021.45, blue https://www.cambridge.org/core/product/identifier/S2732527X21000456/type/journalarticle
26. Singh, V., Willcox, K.E.: Engineering design with digital thread **56**(11), 4515–4528. https://doi.org/10.2514/1.J057255
27. Singh, V., Willcox, K.E.: Decision-making under uncertainty for a digital thread-enabled design process **143**(9), 091707. https://doi.org/10.1115/1.4050108, blue https://asmedigitalcollection.asme.org/mechanicaldesign/article/143/9/091707/1098099/Decision-Making-Under-Uncertainty-for-a-Digital
28. Zhou, G., Pang, Z., Lu, Y., Ewald, J., Xia, J.: OmicsNet 2.0: a web-based platform for multi-omics integration and network visual analytics **50**, W527–W533. https://doi.org/10.1093/nar/gkac376, blue https://academic.oup.com/nar/article/50/W1/W527/6593602
29. Zweber, J.V., Kolonay, R.M., Kobryn, P., Tuegel, E.J.: Digital thread and twin for systems engineering: Requirements to design. In: 55th AIAA Aerospace Sciences Meeting. American Institute of Aeronautics and Astronautics. https://doi.org/10.2514/6.2017-0875, blue https://arc.aiaa.org/doi/10.2514/6.2017-0875

Development and Validation of a Maturity Model for Digital Servitization of Manufacturing Companies

Elena Beducci(✉), Anna de Carolis, and Marco Taisch

Department of Management, Economics and Industrial Engineering, Politecnico di Milano, via Lambruschini 4/B, 20156 Milan, Italy
{elena.beducci,anna.decarolis,marco.taisch}@polimi.it

Abstract. Servitization is considered a path to achieve sustainable competitive advantage. The trend of servitization is particularly relevant for manufacturers, as many companies are moving from selling products to selling product service systems (PSS). Manufacturers are increasingly leveraging digital technologies to support their integrated offerings, a trend called "digital servitization". More in general, to achieve servitization companies are required to undergo great changes. Maturity Models are crucial tools to guide companies in pursuing the transition. Literature proposes many maturity models related to servitization, only few of them are designed to address digital servitization. Moreover, these mostly focus on digital characteristics of the integrated offerings or on the companies' digital architecture, not considering a more encompassing perspective of the companies. This work aims at developing and implementing a descriptive maturity model, which focuses on digital servitization practices, for manufacturing companies. The maturity model is developed following the design science research methodology and has been applied in four manufacturing companies operating in the B2B context. The model appears to be able to effectively assess the maturity of manufacturing companies considering their current digital servitization practices' level. This work contributes to the existing literature on digital servitization, by proposing a dedicated maturity model and identifying opportunities for improvements to be addressed by future researches.

Keywords: Servitization · Product Service System · Digital Servitization · Maturity Model · Assessment

1 Introduction

According to [1], a product service system (PSS) *"is a marketable set of products and services capable of jointly fulfilling a user's needs"*. Servitization is considered a path to lock out competitors, lock in customers, increase the level of differentiation and achieve economic and environmental advantages [2]. The trend of servitization is particularly relevant in the context of manufacturing, as many companies are moving

Published by Springer Nature Switzerland AG 2025
P. Sureephong et al. (Eds.): PLM 2024, IFIP AICT 740, pp. 73–85, 2025.
https://doi.org/10.1007/978-3-031-93319-6_6

from selling products to selling PSS [3]. In particular, manufacturers are increasingly applying digital technologies to enable their integrated offerings [4], this trend is called digital servitization [5].

More in general, the transition towards servitization requires a strong commitment and fundamental changes which are highly dependent from the internal and external context in which the companies operate [6].

These changes encompass all stakeholders covering elements of strategies, organization, management, culture and more [7].

The journey towards servitization, particularly considering digitalization aspects and sustainability objectives, requires a good understanding of the current maturity levels and of the future desired state [8].

Maturity Models (MMs) are crucial to guide companies in pursuing this transition. In fact, MMs are management tools that can be used by companies in the identification, prioritization and development of relevant capabilities [9].

Multiple MMs have been proposed by the literature (e.g. [10]) and a number of them also focus on digitalization. The existing MMs related to digital servitization mostly focus on the digital characteristics of the integrated offerings or on the companies' digital architecture (e.g. [11]). According to the authors, the MM should approach digital servitization with a more holistic perspective, taking into consideration not only the technological elements of PSS but also the company's overall status.

This paper aims at partially addressing the call of [5], which ask for the development of descriptive models, such as assessment, to guide companies in the transformation towards digital servitization.

The objective of the paper is to present a maturity model that specifically address manufacturing companies operating in the B2B context interested in pursuing a digital servitization transformation.

The remainder of the paper is structured as follows. Section 2 presents the theoretical background related to MMs for servitization. Section 3 shows the methodology implemented to develop and evaluate the proposed MM. Section 4 describes the model. Section 5 elucidated the application of the model in four industrial companies. Section 6 concludes the contribution, highlighting limitations and future research opportunities.

2 Theoretical Background

MMs were initially developed in the context of the software industry to increase software process capabilities related [12]. In this context, MMs are defined as *"frameworks representing a path of improvements recommended for software organization that want to increase their software process capabilities"* [13].

MMs can help companies in the identification of strengths and weaknesses to be addressed and in the selection and design of the activities required to improve their current maturity level [13].

MMs are classified according to three distinct objectives [14–16]:

- *Descriptive Maturity Models*: these models are used as a diagnostic tool, and they aim at assessing and describing the as-is status. They assess current capabilities of the entity investigated with respect to given criteria.
- *Prescriptive Maturity Models:* these models aim at identifying the organization desired maturity level and propose approaches to achieve it, they indicate how to achieve maturity improvements to increase business value.
- *Comparative Maturity Models:* these models aim at making benchmarking and comparison between different organizations, both with internal and external boundaries. They can compare similar practices across entities to benchmark them within different industries.

MMs are now being designed and implemented in very different fields [17], such as servitization and digitalization, and receive interest both in academy and industry.

Currently, literature related to MMs for servitization mostly addresses specific aspects of PSS such as service development [18], eco-design [19, 20], digital technologies [21–23] and business model canvas [10].

Few maturity models analyze the PSS maturity across more encompassing dimensions (such as [7, 17]).

The MMs available in literature adopt different approaches to the assessment of the overall maturity level of the companies and select different category of analysis and maturity dimensions to be assessed. Between the most common assessed characteristics, it is possible to identify:

- the relevance of customers (addressed by MMs such as [10, 12, 24]) both in relation to activities such as co-creation and in relation to the understanding of their needs and barriers to the interaction;
- the presence of processes dedicated to servitization, such as the design and deployment of integrated offerings (e.g. [12, 25, 26])
- the relevance of a dedicated strategy to servitization (e.g. [18, 25, 26])
- the organizational changes required to approach effectively the servitization [10, 18, 25])

While elements such as dedicated tools, monitoring strategies, implemented technologies (when not strictly related to servitization) are less commonly considered. This work aims at overcoming this limitation: the presented model considers and assess all the relevant dimension of analysis identified in the literature.

3 Methodology

Following the example of [20] and [19], the authors based the research approach on the Design Science Research Methodology [27] (see Fig. 1). The methodology was applied to develop a MM addressed to manufacturers aiming at increasing their digital servitization level.

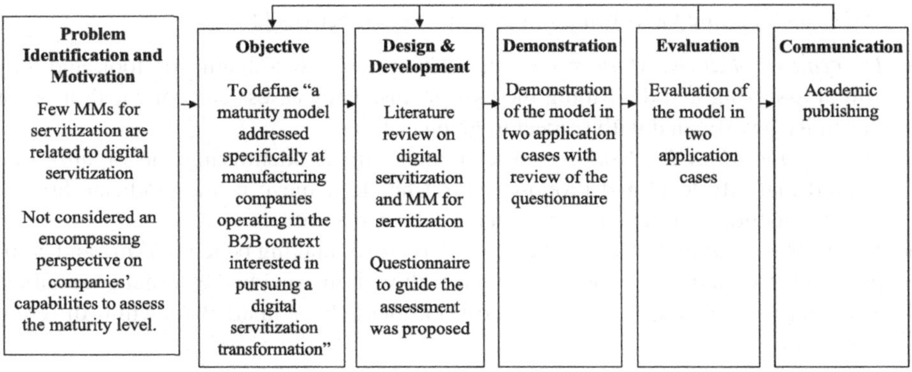

Fig. 1. Implemented Methodology

In the first stage**, problem identification and motivation**, has been identified by reviewing the existing literature related to servitization. As stated in the previous chapters, literature proposes many MMs for servitization and only a few are related to digital servitization. Most of digital servitization MM do not consider an encompassing perspective on companies' capabilities to assess the maturity level.

The **objective**, defined previously, is to define a *"maturity model addressed specifically at manufacturing companies operating in the B2B context interested in pursuing a digital servitization transformation"*.

The **design and development** phases were based on the review of the existing maturity models related to servitization (such as [10]) and digitalization (such as [16]), and by reviewing literature on the transition towards digital servitization. The intended audience of the MM is composed by manufacturing companies aiming at assessing their maturity level in relation to digital servitization. The model is to be applied by third parties, such as researchers. The respondents are those in the companies that manage the development of services and of the servitization strategies. Then the structure of the model was designed: five relevant categories of analysis were identified during the analysis of the literature, while four maturity dimensions and related maturity levels were defined. A questionnaire to guide the assessment of the maturity levels has been developed.

Once the questionnaire standing behind the model was developed, it needed to be tested. In particular, its completeness and the comprehension of its questions were evaluated. Four manufacturing companies have been involved in the **demonstration and evaluation phases**. In the demonstration phase, two companies received the questionnaire in advance and were asked to provide feedback on it. Then the MM was applied. In the evaluation phase, the MM was applied in two other companies. By utilizing the model in four different contexts, it was possible to demonstrate the applicability of the MM, but also to evaluate its adequacy and efficacy, identifying gaps and required modifications. The results will be presented and discussed in Chapter 5.

The **communication** of the model is being currently conducted with this paper.

It is relevant to notice that this paper reports the result of the first evaluation stage, the authors intend to continue the research and adapt of the presented model.

4 Model Description

In this section, the proposed MM has been explained. The model is composed by 5 categories of analysis (see Fig. 2), that represent the main elements required to support the servitization transition, which are investigated across 4 maturity dimensions. This approach has been implemented by [10], that investigate the components of the Business model canvas through maturity dimensions. For each maturity dimension, 5 maturity levels have been developed to assess the companies. The content of the 3 elements will be explained in this chapter. The categories of analysis, maturity dimensions and maturity levels are based on existing maturity models and on the relevant capabilities and elements for servitization identified in the literature.

As anticipated in the introduction, the model does not assess the maturity of the integrated offering, intended only as the union of product and related services sold by the company, but it investigate the maturity at the organization level.

The model is considered a Descriptive Maturity Model, and it aims at describing the companies as-is status.

The first category of analysis is the **Strategic Relevance**. This category aims at analyzing the role of services and of the servitization path in the overall company's strategy. It investigates the presence of a formal strategy to approach servitization and its relevance for the company. It also assesses the changes to organization and processes that have been implemented to adapt to servitization, such as reskilling of the workforce. It investigates the relevance of servitization under the economic perspective. It assesses the presence of processes put in place to define cost and revenues related to services and the activities to monitor and budget costs and revenues.

The second category is related to **Digitalization** of core products and how digital technologies are applied to deploy services. The model is designed to address specifically manufacturing companies, it is expected that the assessed companies will move from offering centered on product to integrated offering. The value of digital technologies as an enabler of servitization is well recognized in the literature [28, 29], also in relation to the sustainability of the offering [30, 31]. Following the consideration of [11], the maturity model does not investigate the hardware or software capabilities, but the roles, capabilities and activities that the digital technologies enable.

The third category is related to **Data**. It investigates the maturity of the company in the collection and analysis of data for servitization. It assesses the processes, structures, and tools available for the collection and analysis of data. This category also investigates the visibility and the exploitation of customer data to deploy services, to improve research and development activities, to define upgrades and improvements of the integrated offering.

The fourth category is **Service Infrastructure.** The service infrastructure category investigates the design and deployment of services included in the integrated offering. This dimension analyses not only Product-Oriented offerings, but also Results and Use

Oriented offerings, according to [32]. This maturity dimension focuses on two main elements of the service infrastructure: the design and executions of PSS.

The first element, related to design, investigate the activities and methods to intercept customers' needs and market's opportunities as well as the related activities of design and development of the PSS. The second element is related to the executions of services: the standardized activities put in place to successfully deliver the PSS, the network created, the adopted tools and methodologies, and the channels exploited by the companies.

The fifth category is related to **Customers**. This category is divided in two main topics of interest: the knowledge of customers and relationship with customers. The first topic investigates the level of knowledge that the company has on its customers. It considers the knowledge related to the customers' processes, the company's ability to cluster them and to monitor the degree of satisfaction in relation to the integrated offering. The second topic is related to how the company keeps in contact with the customers for the deployment and offering of services and to obtain feedback.

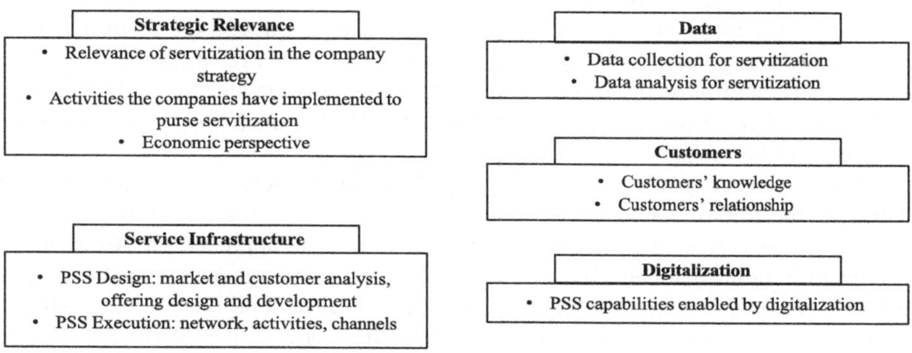

Fig. 2. Categories of analysis

Each category is investigated and assessed through open questions focusing on all the relevant elements. The authors decided to select a questionnaire based on open questions to increase the flexibility of the model. Nonetheless, the questions are structured such as it could be possible to create normative answers for a self-assessment questionnaire. The questions can be descriptive, to have a clearer understanding on the context in which the company operate, or they assess the maturity levels of specific elements. The questions are related to four maturity dimensions that allows to identify the maturity level of the investigated category. The four maturity dimensions are: Processes, Organization, Tools, and Monitoring. Each maturity dimension is assessed through explicit Maturity Scales, which follow a Likert scale 1–5. The selected maturity dimensions and relative levels are adapted from other relevant MM: [10, 16, 18].

The maturity dimension of **processes** assesses how the activities are carried out and the robustness of the processes put in place to support the different categories (see Table 1).

Table 1. Process Maturity Levels (ML)

ML	Process Maturity Levels
1	There are no dedicated processes
2	The process exists but is not formalized and shared
3	The process exists but is only partially formalized and shared
4	The process exists and is formalized and shared
5	The process exists, it is formalized and shared and it is continuously improved and adapted

The **organization** maturity dimension assesses the presence of dedicated roles, areas, and functions with the required skills to carry on specific activities (see Table 2).

Table 2. Organization Maturity Level (ML)

ML	Organization Maturity Levels
1	There is no function dedicated to the activity
2	Ad hoc working groups exist have been created for the function
3	There is a working group responsible, but not fully dedicated to the function
4	There is a working group responsible and fully dedicated to the function
5	There is a working group responsible and fully dedicated to the function and its improvement

The **monitoring** maturity dimension assesses the presence and application of dedicated KPIs (see Table 3).

Table 3. Monitoring Maturity Levels (ML)

ML	Monitoring Maturity Levels
1	There is no KPI and no process to collect data
2	Few data are collected through unstructured processes
3	Data are collected, but used only partially
4	KPIs are elaborated and fully used
5	KPIs are elaborated and fully used for improvement activities

The **tools** maturity dimension is related to artifacts, model, methodologies, or dedicated tools to implement and support activities (see Table 4).

Table 4. Tools Maturity Levels (ML)

ML	Tools Maturity Levels
1	There is no tool used for the activity
2	Non-dedicated/not fully adapted tools are used sporadically for the activity
3	Non-dedicated/not fully adapted tools are used for the activity
4	Dedicated or fully adapted tools are used for the activity
5	Dedicated or fully adapted tools are used for the activity and there is active research for their improvement

To achieve the final maturity level of the company, the average of each maturity level is computed. To increase the readability of the results the maturity level of each category is computed.

5 Model Application and Results

The model was applied through four interviews conducted with the manufacturing companies, both those involved in the demonstration and the evaluation phase.

The involved companies are large enterprises with production plants in Northern Italy. They operate in B2B context in the machinery field. Three companies are machinery producers, while one is an original equipment manufacturer.

The model has been applied through four distinct interviews, spanning from two hours to four hours, following semi-structured interviews. Each companies selected the personnel to be interviewed, the selection criteria were the strict involvement in the servitization process of the companies. The interviews took place in a time frame of 6 months, each interview lasted from 2 to 4 hours. Two interviews took place in presence while the remaining two online.

According to the model's results, the companies have very different maturity levels in relation to servitization (see Fig. 3 and Table 5).

Company A appears to have a homogenous maturity level across all dimensions, around 2. The company has recently started to purse a servitization path and it is introducing in their physical products the digital components to enable intermediate and advanced services. The company, moreover, operates in a field that is currently undergoing a transition towards servitization and the company's competitors have similar or inferior service offerings. The low maturity level was to be expected, and the relative peak in the digitalization field is coherent with the strategy implemented so far.

Company B is highly mature in 3 categories, with maturity levels above 4. This result is coherent with the situation observed: the company created a dedicated operative function, they employ very different digital technologies to offer services and heavily use the collected data. The company adopted and adapted tools from other functions to support the processes related to servitization, such as tool for the analysis of clients. The relevance of servitization is not shared between all levels and functions.

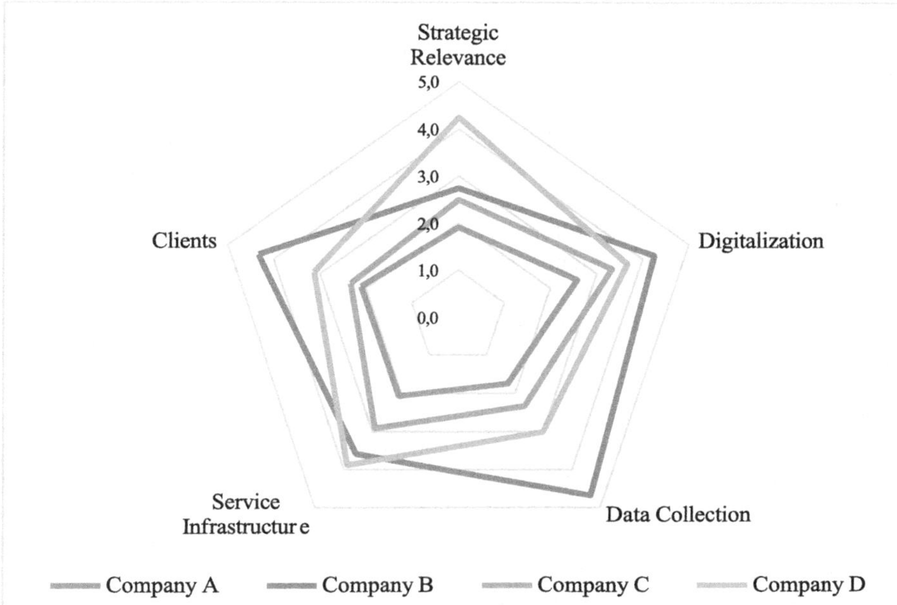

Fig. 3. Maturity Levels of the four companies

Company C is undergoing a servitization process and it is involved in multiple projects addressing the topic. The process is currently focusing on digitalization and how it can enable and support the delivery of integrated offerings.

Company D has started its servitization process since more than 5 years ago, and it appears to be highly mature. The company, in his traditional offers, provide software solutions as well as physical products, thus facilitating the transition towards a servitized approach. Servitization is a strategic priority for the company. It has dedicated processes and methodologies for the development of integrated offerings, often customized for the customers. Nonetheless, it is possible to observe a low attention to the analysis of customer needs and satisfaction.

Table 5. Maturity Levels of the four companies

Dimension	Company A	Company B	Company C	Company D
Strategic Relevance	1,9	2,8	2,5	4,3
Digitalization	2,6	4,3	3,3	3,7
Data Collection	1,8	4,7	2,3	3,0
Service Infrastructure	2,1	3,6	2,9	3,9
Clients	2,1	4,3	2,3	3,1

The model appears to be able to effectively represent the status quo of the companies.

The four companies can be positioned at different stages in their transition towards servitization and the model is able to assess and describe the different conditions of the companies.

Through the questions and all the elements assessed it was possible to map the most relevant characteristics of manufacturing companies embracing digital servitization. During the application of the model, all five maturity levels have been identified in the companies (all five levels of maturities were assessed), representing the ability of the model to map the real different degree of maturity of companies.

The developed questionnaire seems to be able to capture most relevant elements to successfully describe the maturity of the companies. The model is intended to be applied through semi-structured interviews to gain context as well as the company's characteristics.

As stated in Chapter 4, the model is built on both category of interest and maturity dimension. While the categories of interest appear to be well described by the existing model, not every maturity dimension is assessed in each category (ex. The monitoring dimension is not assessed in the Client category). To improve the readability of the assessment results, the questions and investigated elements should be reconfigured to allow the analysis of each category along the four maturity dimensions.

An additional point of interest is the role of digital technologies and digitalization. Currently, the maturity model focus on the role of technologies in relation to servitization. During the evaluation, it was possible to observe that the model could be improved by assessing, not only the digital technologies applied in the contest of servitization, but also the overall level of digital maturity of the organization. Thus, category **Digitalization** should take into consideration all digital technologies available in the companies to identify possible synergies not yet explored that could leveraged to increase the maturity level for servitization.

During the interviews, the respondents often addressed certain elements multiple times in the conversation while answering distinct question.

In order to facilitate the application of the model, and to avoid repetitions during the interviews, the analyzed categories should be rearranged, or the questionnaire should be revised by incorporating questions.

The model is intended to be applied by third party stakeholders, such as researchers, and not as an internal self-assessment tool. Nonetheless, the model could be adapted as a self-assessment MM, by modifying the existing questionnaire. The open questions can be translated to closed answer questions, with normative answers supported by examples to improve the comprehension. This strategy could also increase awareness of industrial practitioners related to the capabilities that impact the degree of success of servitization strategies.

It was possible to observe that the semi-structured interviews allow, not only to cover all elements to be evaluated, but also to acquire the relevant information to identify projects and activities to increase the maturity levels. Nonetheless, this activity is outside the scope of the paper and should be object of further researches.

6 Conclusions

In this paper, the authors set out to design and evaluate a new maturity model focused on digital servitization addressed to manufacturing companies operating in the B2B context.

This paper answers multiple calls asking for tools aimed at supporting industrial practitioners in the development of successful servitization strategies. While multiple Maturity Models addressed to servitization already exist in the literature, there is no consensus on the maturity dimensions that should be addressed. At the same time, most maturity models related to digital servitization focus specifically on the role of technologies.

This paper proposes a new maturity model, build on existing literature, to facilitate manufacturers transition towards servitization.

The main limitation of the study is related to the development of the model. The model has been developed through an analysis of existing maturity models and on the analysis of literature related servitization. While the industrial perspective has been taken into consideration while developing the artifacts, it lacks a robust intervention from industrial practitioners. Future works should focus on increasing the robustness of the tool by validating and corroborating the academic perspective through interviews with relevant industrial practitioners.

A second limitation of the study relies on the validation of the model: the four companies that have been involved are Large Enterprises. Small and Medium Enterprises should be considered in future studies and future validation rounds to demonstrate and evaluate the effectiveness of the tool also in reality with less resources, and generally less awareness.

A third, and last limitation, is that this model does not take into consideration the sustainability dimension. While PSS are recognized as a viable path to achieve sustainability [33, 34], the implementation of servitization does not automatically translate to improved environmental impact [35]. Future studies should focus on the assessment of servitization while taking into consideration the readiness of implementation of PSS with the objective to improve the environmental performances of the company.

Acknowledgments. This study was carried out within the MICS (Made in Italy – Circular and Sustainable) Extended Partnership and received funding from the European Union NextGenerationEU (PIANO NAZIONALE DI RIPRESA E RESILIENZA (PNRR) – MISSIONE 4 COMPONENTE 2, INVESTIMENTO 1.3 – D.D. 1551.11-10-2022, PE00000004). This manuscript reflects only the authors' views and opinions, neither the European Union nor the European Commission can be considered responsible for them.

The authors have no competing interests to declare that are relevant to the content of this article.

References

1. Goedkoop, M.: Product Service systems, Ecological and Economic Basics (1999)
2. Neely, A.: Exploring the financial consequences of the servitization of manufacturing. Oper. Manag. Res. **1**(2), 103–118 (2008). https://doi.org/10.1007/s12063-009-0015-5

3. Schroeder, A., Baines, T., Sakao, T.: Increasing value capture by enhancing manufacturer commitment—managing the servitization process. IEEE Eng. Manag. Rev. **50**(3), 162–170 (2022). https://doi.org/10.1109/EMR.2022.3197075
4. Raddats, A., Naik, P., Ziaee Bigdeli, A.: Creating value in servitization through digital service innovations. Ind. Mark. Manag. **104**, 1–13 (2022). https://doi.org/10.1016/j.indmarman.2022.04.002
5. Paschou, T., Rapaccini, M., Adrodegari, F., Saccani, N.: Digital servitization in manufacturing: A systematic literature review and research agenda. Ind. Mark. Manag. **89**, 278–292 (2020). https://doi.org/10.1016/j.indmarman.2020.02.012
6. Schnürmacher, C., Hayka, H., Stark, R.: Providing product-service-systems - the long way from a product OEM towards an original solution provider (OSP). Procedia CIRP **30**, 233–238 (2015). https://doi.org/10.1016/j.procir.2015.02.124
7. Martinez, V., Neely, A., Velu, C., Leinster-Evans, S., Bisessar, D.: Exploring the Journey to Services. International (2017)
8. González Chávez, C.A., Despeisse, M., Johansson, B., Romero, D., Stahre, J.: Sustainability-as-a-Service: Requirements Based on Lessons Learned from Empirical Studies. In: Alfnes, E., Romsdal, A., Strandhagen, J.O., von Cieminski, G., Romero, D. (eds.) Advances in Production Management Systems. Production Management Systems for Responsible Manufacturing, Service, and Logistics Futures. In: IFIP Advances in Information and Communication Technology. Cham: Springer Nature Switzerland, pp. 181–196 (2023). https://doi.org/10.1007/978-3-031-43666-6_13
9. Häckel, B., Huber, R., Stahl, B., Stöter, M.: Becoming a product-service system provider – a maturity model for manufacturers. presented at the Lecture Notes in Information Systems and Organisation, pp. 169–184 (2021). https://doi.org/10.1007/978-3-030-86790-4_13
10. Adrodegari, F., Saccani, N.: A maturity model for the servitization of product-centric companies. J. Manuf. Technol. Manag. **31**(4), 775–797 (2020). https://doi.org/10.1108/JMTM-07-2019-0255
11. Heinz, D., et al.: A maturity model for smart product-service systems." presented at the Procedia CIRP, pp. 113–118 (2022). https://doi.org/10.1016/j.procir.2022.04.019
12. Kimita, K., McAloone, T.C., Ogata, K., Pigosso, D.C.A.: Servitization maturity model: developing distinctive capabilities for successful servitization in manufacturing companies. J. Manuf. Technol. Manag. **33**(9), 61–87 (2022). https://doi.org/10.1108/JMTM-07-2021-0248
13. Paulk, M., Curtis, B., Chrissis, M., Weber, C.: Capability Maturity Model for Software, Version 1.1 (1993)
14. de Bruin, T., Freeze, R., Kulkarni, U., Rosemann, M.: Understanding the Main Phases of Developing a Maturity Assessment Model. Australas. Conf. Inf. Syst. (2005)
15. Poeppelbuss, J., Roeglinger, M.: What makes a useful maturity model? A framework of general design principles for maturity models and its demonstration in business process management. presented at the 19th European Conference on Information Systems, ECIS 2011 (2011)
16. De Carolis, A.A., Macchi, M., Negri, E., Terzi, S.: A Maturity Model for Assessing the Digital Readiness of Manufacturing Companies, pp. 13–20 (2017). https://doi.org/10.1007/978-3-319-66923-6_2
17. Röglinger, M., Pöppelbuß, J., Becker, J.: Maturity models in business process management. Bus. Process. Manag. J. **18**(2), 328–346 (2012). https://doi.org/10.1108/14637151211225225
18. Rapaccini, M., Saccani, N., Pezzotta, G., Burger, T., Ganz, W.: Service development in product-service systems: A maturity model. Serv. Ind. J. **33**(3–4), 300–319 (2013). https://doi.org/10.1080/02642069.2013.747513
19. Exner, K., Zimpfer, R., Stark, R.: Maturity Model and Action Recommendation: A PSS Capability Self-Assessment Tool for Companies." presented at the Procedia CIRP, pp. 175–180 (2017). https://doi.org/10.1016/j.procir.2017.03.050

20. Exner, K., Balder, J., Stark, R.: A PSS maturity self-assessment tool. presented at the Procedia CIRP, pp. 86–90 (2018). https://doi.org/10.1016/j.procir.2018.04.013
21. Paschou, T., Rapaccini, M., Peters, C., Adrodegari, F., Saccani, N.: Developing a Maturity Model for Digital Servitization in Manufacturing Firms. Lect. Notes Multidiscip. Ind. Eng. **Part F201**, 413–425 (2020). https://doi.org/10.1007/978-3-030-43616-2_44
22. Neff, A.A., Hamel, F., Herz, T.P., Uebernickel, F., Brenner, W.: Fostering efficiency in information systems support for product-service systems in the manufacturing industry. presented at the 19th Americas Conference on Information Systems, AMCIS 2013 - Hyper-connected World: Anything, Anywhere, Anytime, pp. 1715–1724 (2013)
23. Neff, A.A., Hamel, F., Herz, T.Ph., Uebernickel, F., Brenner, W.: Developing a maturity model for product-service systems in manufacturing enterprises." presented at the Proceedings - Pacific Asia Conference on Information Systems, PACIS 2013 (2013)
24. Pigosso, D.C.A., McAloone, T.C.: Supporting the Development of Environmentally Sustainable PSS by Means of the Ecodesign Maturity Model. Procedia CIRP **30**, 173–178 (2015). https://doi.org/10.1016/j.procir.2015.02.091
25. Alvarez, R.L.P., Martins, M.R., Da Silva, M.T.: Applying the maturity model concept to the servitization process of consumer durables companies in Brazil. J. Manuf. Technol. Manag. **26**(8), 1086–1106 (2015). https://doi.org/10.1108/JMTM-04-2014-0050
26. Schroderus, J., Lasrado, L.A., Menon, K., Kärkkäinen, H.: Towards a Pay-Per-X Maturity Model for Equipment Manufacturing Companies. Procedia Comput. Sci. **196**, 226–234 (2022). https://doi.org/10.1016/j.procs.2021.12.009
27. Blessing, L.T.M., Chakrabarti, A.: DRM, a Design Research Methodology. Springer, London (2009). https://doi.org/10.1007/978-1-84882-587-1
28. Alcayaga, A., Wiener, M., Hansen, E.G.: Towards a framework of smart-circular systems: An integrative literature review. J. Clean. Prod. **221**, 622–634 (2019). https://doi.org/10.1016/j.jclepro.2019.02.085
29. Ren, S., Zhang, Y., Liu, Y., Sakao, T., Huisingh, D., Almeida, C.M.V.B.: A comprehensive review of big data analytics throughout product lifecycle to support sustainable smart manufacturing: A framework, challenges and future research directions. J. Clean. Prod. **210**, 1343–1365 (2019). https://doi.org/10.1016/j.jclepro.2018.11.025
30. Bressanelli, G., Adrodegari, F., Perona, M., Saccani, N.: Exploring How Usage-Focused Business Models Enable Circular Economy through Digital Technologies. Sustainability **10**(3), Art. no. 3 (2018). https://doi.org/10.3390/su10030639
31. Neligan, A., Baumgartner, R.J., Geissdoerfer, M., Schöggl, J.-P.: Circular disruption: Digitalisation as a driver of circular economy business models. Bus. Strategy Environ. **32**(3), 1175–1188 (2023). https://doi.org/10.1002/bse.3100
32. Tukker, A.: Eight types of product–service system: eight ways to sustainability? Experiences from SusProNet. Bus. Strategy Environ. **13**(4), 246–260 (2004). https://doi.org/10.1002/bse.414
33. Annarelli, A., Battistella, C., Nonino, F.: Product service system: A conceptual framework from a systematic review. J. Clean. Prod. **139**, 1011–1032 (2016). https://doi.org/10.1016/j.jclepro.2016.08.061
34. Matschewsky, J.: Unintended Circularity?—Assessing a Product-Service System for its Potential Contribution to a Circular Economy. Sustainability **11**(10), Art. no. 10 (2019). https://doi.org/10.3390/su11102725
35. Tukker, A.: Product services for a resource-efficient and circular economy – a review. J. Clean. Prod. **97**, 76–91 (2015). https://doi.org/10.1016/j.jclepro.2013.11.049

Mapping the Utilisation of I4.0 and I5.0 Technologies Along the Product Life Cycle to Enable Circular Economy Endeavours

Marco Dautaj[1,2](✉), Monica Rossi[1], and Sergio Terzi[1]

[1] Politecnico di Milano, via Lambruschini 4b, 20156 Milan, Italy
{marco.dautaj,monica.rossi,sergio.terzi}@polimi.it
[2] University School for Advanced Studies IUSS Pavia, Palazzo del Broletto, Piazza Vittoria 15, 27100 Pavia, Italy

Abstract. With the pressing goals of sustainability and continuous technological advancement, enterprises need to change their strategy regarding their production processes. Furthermore, even the industrial fabric is influenced by this technological disruption, and new paradigms have been introduced. Amongst these new concepts, it is important to point out the new paradigm of Industry 5.0, introduced in 2018, which defines the working environment as characterized by a high integration between humans and technologies. Another important concept to consider is the ideation of the Circular Economy model, introduced by the Ellen MacArthur Foundation, which is a system where materials and products are maintained in a closed-loop process through the 10R strategies. This study aims to understand how companies are exploiting technologies towards the achievement of circular economy strategies. In order to retrieve this information, the authors conducted a literature review, considering only application cases and case studies. For this study, Scopus has been used as a search engine to collect all the useful articles. The authors collected all the relevant data regarding the region of the application, the technologies employed, which circular economy strategies have been used for the specific production process, which part of the product life cycle firms are frequently focused on and what companies need to ameliorate to improve their processes. A synthesis of the findings will be shown and some suggestions to foster and promote circular economy strategies within the production environment will be provided.

Keywords: Industry 5.0 · Circular Economy · Skills

1 Introduction

There are many attempts to reach Industry 5.0, many of which leverage the emerging challenge of introducing technologies (industrial, digital, collaborative, etc.) that have emerged with the paradigms of Industry 4.0 [1] and Industry 5.0 [2]. In fact, Industry 5.0, besides embracing resilience and promoting a human-centred approach, highlights

Published by Springer Nature Switzerland AG 2025
P. Sureephong et al. (Eds.): PLM 2024, IFIP AICT 740, pp. 86–99, 2025.
https://doi.org/10.1007/978-3-031-93319-6_7

sustainability with a particular focus on the circular economy. It is essential to understand, on one hand, the essential role of technologies in facilitating the implementations of the circular economy and, on the other hand, to understand the optimal timing for their adoption to maximize desired outcomes. This implies not only an understanding of the technologies themselves but also a thorough assessment of market dynamics, government regulations, and consumer preferences that influence the adoption and effectiveness of these technologies.

Furthermore, it is important to recognize that implementing the circular economy often requires systemic as well as technological transformation, involving changes in business models, product design practices, and resource management. In this context, the involvement of key stakeholders (e.g., entrepreneurs, and academics) becomes crucial in developing effective strategies and policies that foster widespread adoption of enabling technologies and promote a smooth transition towards a circular economy. This requires close collaboration and open dialogue amongst all stakeholders to overcome challenges and maximize the opportunities offered by emerging technologies for a more sustainable and equitable future.

Moreover, there are numerous technologies, but it is not clear how and when (at which stage of the lifecycle) their use is conducive to achieving the sustainability results that Industry 5.0 aims for. Many researchers have addressed this issue conceptually; however, the study aims to map the technologies through applied practical cases. After a brief theoretical background on I5.0, technologies, and circular economy, the paper introduces the objectives and the methodology, designed through a literature review, and afterwards, the data will be collected and shown, following a results section.

2 Theoretical Background

In the following paragraph, they will be introduced the most important concepts that characterize recent studies in the field of industry, in particular Industry 5.0, Circular Economy and Technologies. This new approach represents a significant shift towards a more human and interactive vision of industrial operations, highlighting the synergy between technology and the human being. Through an overview of key concepts, the authors will outline the conceptual foundations that drive the evolution of modern industry.

2.1 Industry 5.0

Industry 5.0 [2, 3] is a concept that refers to the subsequent evolution of industry after Industry 4.0 (see Fig. 1). While Industry 4.0 focuses on the integration of digital and innovative technologies into production, Industry 5.0 pushes further towards a more human and sustainable approach. In Industry 5.0, greater emphasis is placed on the interaction between man and machine, trying to combine the efficiency and precision of machines with the creative and decision-making abilities of human beings. This human-centric approach translates into greater customization of products and services, as well as greater attention to environmental and social sustainability.

Industry 5.0 represents a shift in the industry that emphasizes the importance of humans. It involves the integration of technologies with human operations to improve

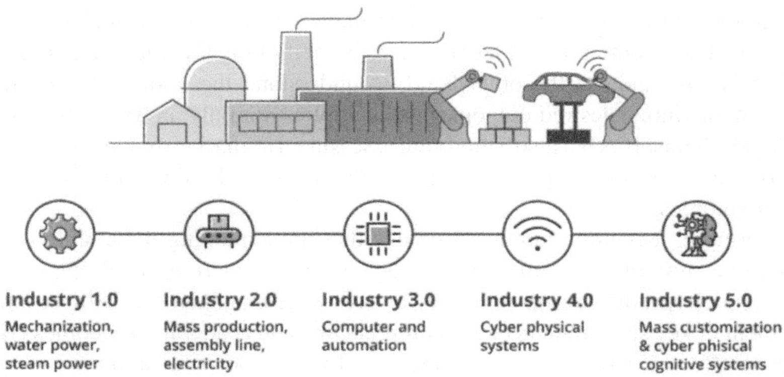

Fig. 1. Industrial Revolutions [4]

human capabilities. The paramount aspect of Industry 5.0 is the balance between machines and humans, allocating mundane tasks to the machines and more creative to the human operators [5].

2.2 Technologies

Technology remains a transformative pressure inside the contemporary technological panorama, reshaping the way companies arrange, manufacture and supply services and products. The wave of new technologies is characterized by high degrees of automation, and more importance is given also to human creativity. At the forefront of this technological development, there is Artificial Intelligence (AI) [6], a multifaceted device that empowers human people with remarkable insight and decision-making strength. Using machine-gaining knowledge of algorithms and cognitive computing, AI will increase performance and accuracy throughout engineering operations. At the same time, the Internet of Things (IoT) is creating and spreading a new generation of connectivity inside the industrial environment. Strong networks of sensors, devices, and related gadgets that facilitate actual-time tracking, predictive renovation, and adaptive response techniques to optimize resource consumption and commercial enterprise performance.

Furthermore, additive manufacturing, typically referred to as 3D printing, has emerged as a disruptive force in modern production. By permitting circuit-based and scalable manufacturing, 3D printing fastens the prototype cycle and enables the advent of custom products to meet individual desires. These are not the only technologies which companies are exploiting, in the following sections the authors will highlight which are the most used technologies by companies operating mostly in the manufacturing sector. In this dynamic environment, cutting-edge technologies are paving the way for a future where innovation and sustainability are meeting human creativity in order to create a human-centric environment.

2.3 Circular Economy

The circular economy (CE) (see Fig. 2) embodies a sustainable system where materials are continually reused, recycled and regenerated. Through 10R [7] practices, products and materials remain in circulation, contributing to resource conservation and waste reduction. By decoupling economic growth from the consumption of finite resources, the circular economy tackles pressing global issues such as climate change, biodiversity loss, waste, and pollution, offering a comprehensive approach to sustainable development [8]. The CE is a global economic model that aims to minimize the consumption of finite resources by focusing on the intelligent design of materials, products, and systems. It seeks to replace the traditional linear economy model, which is characterized by a linear approach 'take, make, and dispose', with a more sustainable, closed-loop model.

The relevance of the CE has been amplified worldwide in recent years, and it has replaced old industrial practices with closed-loop patterns that focus on balancing economic, environmental, and societal impacts [9]. Geissdoerfer et al. [10] defined circular economy as "*a regenerative system in which resource input and waste, emission, and energy leakage are minimized by slowing, closing, and narrowing material and energy loops. This can be achieved through long-lasting design, maintenance, repair, reuse, remanufacturing, refurbishing, and recycling. Second, we define sustainability as the balanced integration of economic performance, social inclusiveness, and environmental resilience, to the benefit of current and future generations.*"

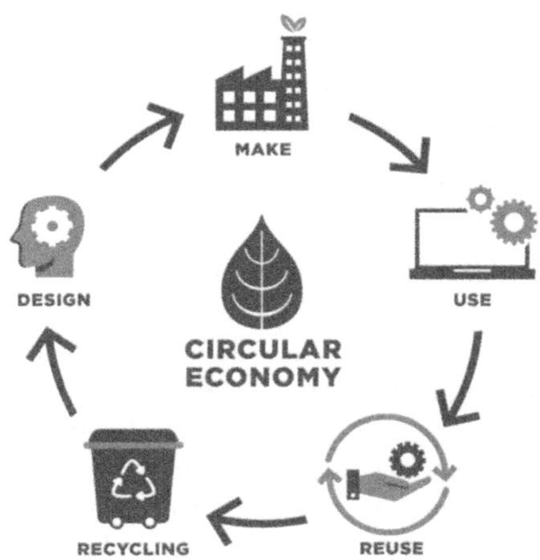

Fig. 2. Circular Economy [11]

3 Methodology

A literature review approach has been used. The literature review is a method to review important literature in a specific field with a meticulous approach.

The main steps of the review are going to be described. Firstly, the research question is the following: "What are the technologies and in what phases of the product life cycle they have been used?". In order to answer the previous question, the authors tailored sub-questions as follows: "What are the technologies frequently exploited by companies?", "On what phase of the product life cycle companies are focusing on?", "What are the circular economy strategies companies are applying?". Answering these questions, the authors have been able to list all the technologies that have been exploited and the specific circular economy strategies employed by enterprises.

In order to obtain the articles and expand the literature, the research was conducted using Scopus as database. The keywords researched were (TITLE-ABS-KEY ("Industry 4.0" OR "I4.0" OR "Industry 5.0" OR "I5.0") AND TITLE-ABS-KEY ("Circular Economy" OR "CE") AND TITLE-ABS-KEY ("Practical case*" OR "Case Stud*" OR "Interview*" OR "questionnaire*")).

The Literature review process is shown in Fig. 3.

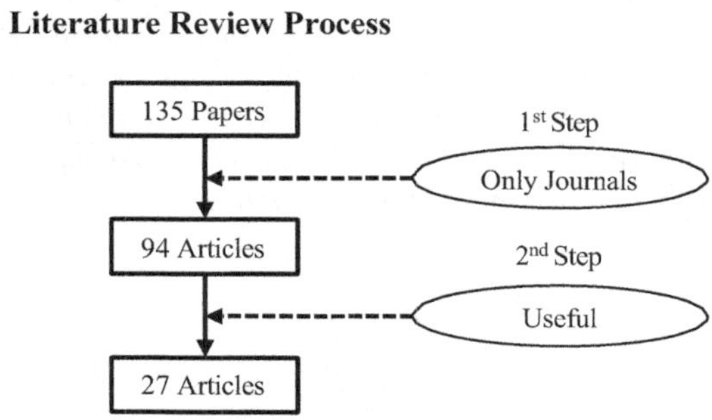

Fig. 3. Literature Review Process

The total amount of articles found using these keywords was 135. A significant percentage of the articles found were "Conference papers" since the concepts of Circular Economy and Industry 5.0 are very recent. The authors tried to consider more journal articles since they pass through a rigorous review analysis before being published in a journal and for this reason, are more reliable. The total number of articles, after considering only articles, was 94. Amongst these 94 articles, only 27 were considered useful for this study because, even though there was a reference to Circular Economy, Industry 5.0, and Industry 4.0, the information based on Case Studies and Practical Cases was not adding value to the analysis, since the authors aimed to operationalize all the data gathered. The publication time interval considered is 2018-2024, and in Fig. 4 the number of publications based on the year is shown.

3.1 Analysis

The analysis conducted is poised to understand the main concepts regarding "Circular Economy", "Industry 5.0" and "Industry 4.0" through the collection of practical cases described in each article. The authors extrapolated the relevant information needed from

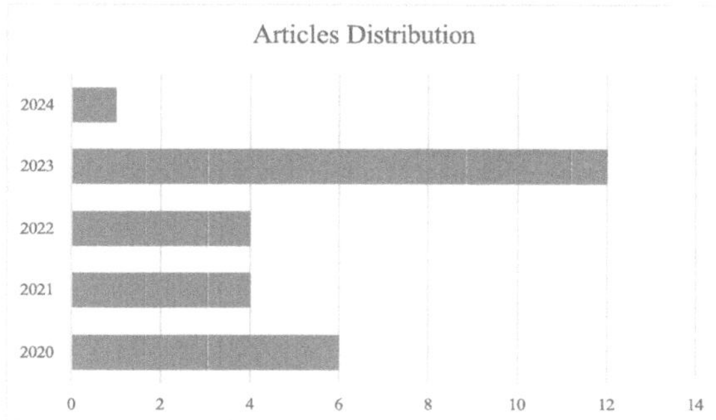

Fig. 4. Articles Distribution

all the articles considered. This was done in order to compare the main concepts of all the articles and consequently to find common ground.

The authors collected all the information in several main categories in order to organize and find commonalities amongst them. Here the main analyzed categories are shown. Firstly, it was extrapolated the information related to the region of application. Consequently, it was analyzed the theory applied in the articles (e.g., resource-based theory, dynamic capabilities theory). Furthermore, information related to the sectors/products, life cycle phases and which 10R has been applied in each case, has been collected. For the sake of knowledge, the authors tried to consider all the circular economy strategies cited in each paper, even though there are some cases where there is not an explicit link between the application of the strategy and the process/product. Lastly, the authors gathered useful information regarding the technologies applied as well as what companies need to overcome criticalities.

4 Results

In this section, all the relevant information gathered throughout the literature review will be highlighted, mostly employing tables. This section is characterized by 6 subsections where the data mentioned in Sect. 3.1 will be shown.

4.1 Region

The information related to the region was not easily findable in all the articles. From the data gathered by the authors (see Fig. 5), it can be stated that India [12–17] is by far the country where the use of circular economy strategies has been exploited in order to comply with the new environmental regulations and to be more sustainable.

4.2 Theory

The authors collected information related to the theory used in the articles. In Fig. 6 the theories collected have been counted. As it is evident from the figure, the most frequent

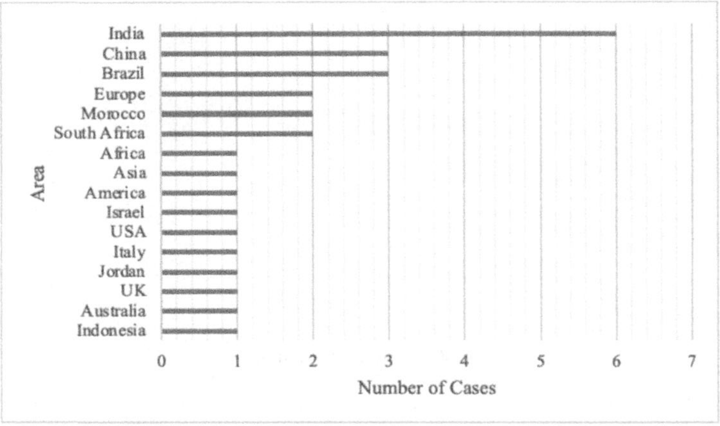

Fig. 5. Application Areas

theories the authors have gathered are Resource-based Theory (RBT) [12, 15, 16, 18–20], which was introduced by Edith Penrose [21, 22] because she "*recognized that internal managerial resources are both drivers and limits to the expansion any one firm can undertake*", and the other frequent theory is the Dynamic Capabilities View [23–29], introduced by Teece et al. [30] as "*the firm's ability to integrate, build, and reconfigure internal and external competences to address rapidly changing environments*".

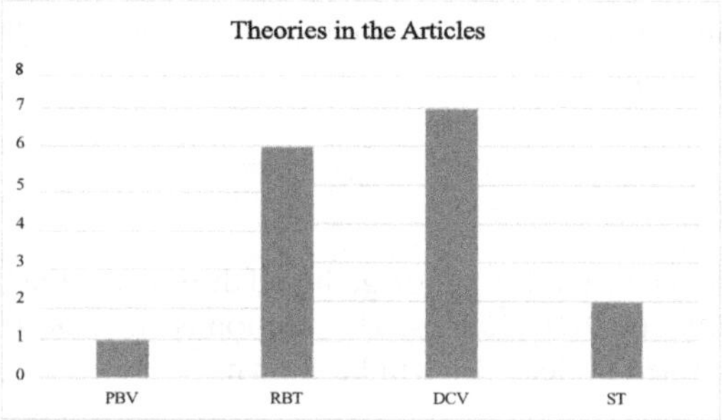

Fig. 6. Theories

5 Product Life Cycle Phases

This section presents the product life cycle phases in which companies are frequently focusing on the product life cycle is characterised by three main phases: Beginning of life (BOL), Middle of life (MOL), and End of life (EOL). The authors have been able to identify the specific processes within each phase.

For the BOL phase [12, 13, 18–20, 23–26, 28, 30–37], four main processes have been identified: "Design and Development", "Material Selection", "Quality Assurance", and "Standardisation".

For the MOL phase [12, 13, 18, 23, 25, 26, 28, 31, 33, 34, 36], five processes have been identified: "Production and Operations", "Supply chain management", "Maintenance", "Sharing Economy", and "Resilience Practices".

For the EOL phase [12, 19, 20, 23–26, 28, 30–38], five processes have been identified "Reverse Logistics", "Regulatory Compliance", "Sustainable Transportation", "Sustainable Practices", and "Customisation"

5.1 Circular Economy Strategies

This section presents the circular economy strategies employed by companies. The circular economy strategies are as follows: Refuse, Rethink, Reduce, Reuse, Repair, Refurbish, Remanufacture, Repurpose, Recycle, Recovering [23].

Almost all the articles focus the attention on Recycling, Reusing and Reducing. The other CE strategies exploited more frequently are Recovering and Remanufacturing. For the other strategies, there is a paucity of practical cases, indeed more examples should be addressed towards Refurbishing, Repurposing, Repairing, Rethinking and Refusing strategies.

5.2 Technologies

A list of the technologies frequently exploited linked with the CE strategies is shown in Table 1.

As shown in Table 1, the most frequent technologies applied and exploited by firms are the Internet of Things (IoT), Artificial Intelligence (AI), Big Data (BD), Blockchain and Robotics. More examples should be focused on the application of those technologies less frequently employed.

5.3 What Companies Need

From the analysis the authors carried out, it is evident to some extent that companies need to improve, ameliorate and integrate some characteristics which hinder the exploitation of circular economy strategies.

Table 2 shows what the authors gathered throughout the analysis of the articles; in particular, it will show useful information extrapolated from the practical cases.

What companies should focus on is to improve skill competencies within their working environment. In doing so, workers and employees know how to interact with a specific technology and, in combination with the knowledge of circular strategies, apply it efficiently to the production context. Furthermore, companies should improve their collaboration with suppliers in order to improve processes. Being more collaborative means creating an optimized supply chain where "supply and demand" are perfectly balanced, with a positive effect on waste reduction.

Table 1. Technologies Exploited by Firms

Technologies	Articles
Artificial Intelligence (AI)	[13, 18–20, 23–25, 28, 31–35, 38–40]
Internet of Things (IoT)	[13, 14, 16–20, 23–28, 30–34, 39]
Big Data Analytics (BDA)	[13, 17, 18, 23–26, 28, 31, 33, 34, 38–40]
Cloud Computing	[12, 16, 25, 30, 38]
RFID	[12, 30]
Cyber Physical System (CPS)	[12, 38]
Additive Manufacturing (AM)	[12, 19, 20, 33, 34, 40]
Simulation	[12, 39, 40]
Blockchain	[17, 19, 24, 25, 28, 32, 33, 35, 38]
Machine Learning (ML)	[24]
Robotics	[14, 20, 26, 31, 35, 37, 40]
5G Technologies	[35]
Building Interface Modeling (BIM)	[30]
Reverse Vending Machine	[14]
Augmented Reality (AR)	[32, 34]
Virtual Reality (VR)	[32, 34]
Digital Twin (DT)	[19, 20, 33]
Material Passport	[33]
Sensors	[39]
Nanoengineering of Materials	[40]

Table 2. What it is needed by Firms

Need	Articles
Skill competencies	[14, 23, 33, 35, 38]
Collaboration with suppliers	[13, 14, 19, 24, 25, 28, 30, 31, 33, 35, 40]
Traceability	[24]
Regulatory compliance	[14, 25, 33, 38, 40]
Community Engagement	[14, 31, 35, 40]
Institutional pressures	[20]

6 Discussion

The industry and the fabric of society have always been able to remodel themselves according to the fast development of cutting-edge technology.

This study is poised to understand the relationship amongst three principals thematic: Circular Economy (CE), which has gained relevance worldwide in recent years, and it has replaced old industrial practices with closed-loop patterns that focus on balancing economic, environmental, and societal impacts [9]. Geissdoerfer et al. [10] defined circular economy as "*a regenerative system in which resource input and waste, emission, and energy leakage are minimized by slowing, closing, and narrowing material and energy loops. This can be achieved through long-lasting design, maintenance, repair, reuse, remanufacturing, refurbishing, and recycling. Second, we define sustainability as the balanced integration of economic performance, social inclusiveness, and environmental resilience, to the benefit of current and future generations*". Technology, which remains a transformative pressure inside the contemporary technological panorama, reshaping the way companies arrange, manufacture and supply services and products. The wave of new technologies is characterized by high degrees of automation, and more importance is given also to human creativity. Industry 5.0 (I5.0), a greater emphasis is placed on the interaction between man and machine, trying to combine the efficiency and precision of machines with the creative and decision-making abilities of human beings. This human-centric approach translates into greater customization of products and services, as well as greater attention to environmental and social sustainability [2].

From the practical cases, it is evident that companies are focusing more and more on sustainable practices. Thanks to the high technological development, such as IoT, Artificial intelligence (AI), and Big Data Analytics (BDA), companies are improving and increasing the efficiency of their processes. This helps to adopt circular economy strategies during almost all the phases of the product life cycle. An important aspect that should not be neglected is the importance of the human impact throughout the production processes. Indeed, in some articles, it is stressed that companies need more competencies and a skilled workforce.

In general, the authors can state that, through the analysis of the practical cases, firms are aware of the production impact on sustainability, and they are trying to be proactive towards the reduction of waste generation, the creation of pollution and all those activities that can have negative repercussion on the environment.

6.1 Future Research

The connection amongst Circular Economy, Technology, and Industry 5.0/Industry 4.0 is not so easily detectable, especially when coping with practical and real applications.

It is fundamental to have more practical cases in the literature since the number of cases considered is not enough to standardize the approaches companies are applying within their production context.

Nowadays, due to continuous technological development and the adoption of circular economy strategies, companies need to rely more often on a skilled workforce. In order to effectively exploit cutting-edge technologies and implement circular economy strategies, companies must invest in a workforce fully equipped with the requisite skills and expertise. The workforce should possess a deep understanding of technologies so being able to interact with them and use them as enablers for the application of circular economy strategies. Companies should know their workforce before the working environment, and one of the ways to do it is to implement study programs in collaboration with universities [41].

Lastly, enterprises should collaborate and work closely with suppliers to create a closed-loop system where resources are continuously circulating, and waste is reduced.

Acknowledgements. This study was carried out within the MICS (Made in Italy – Circular and Sustainable) Extended Partnership and received funding from the European Union NextGeneration EU (PIANO NAZIONALE DI RIPRESA E RESILIENZA (PNRR) –MISSIONE 4 COMPONENTE 2, INVESTIMENTO 1.3 – D.D. 1551.11–10-2022, PE00000004). This manuscript reflects only the authors' views and opinions, neither the European Union nor the European Commission can be considered responsible for them.

This article has received funding from EIT RawMaterials under grant agreement ID 10044 (correspondent to the project shortly entitled "SMILE", "SmartManufacturing Innovation, Learning-Labs, and Entrepreneurship").

This paper and related research have been conducted during and with the support of the Italian inter-university PhD course in sustainable development and climate change (link: www.phd-sdc.it).

This work was produced while attending the PhD programme in PhD in Sustainable Development And Climate Change at the University School for Advanced Studies IUSS Pavia, Cycle XXXVIII, with the support of a scholarship financed by the Ministerial Decree no. 351 of 9th April 2022, based on the NRRP - funded by the European Union - NextGenerationEU - Mis- sion 4 "Education and Research", Component 1 "Enhancement of the offer of educational ser- vices: from nurseries to universities" - Investment 4.1 "Extension of the number of research doctorates and innovative doctorates for public administration and cultural heritage".

References

1. Lasi, H., Fettke, P., Kmper, H.G., Feld, T., Hoffmann, M.: Industry 40. Bus. Inform. Syst. Eng. **6**, 239–242 (2014). https://doi.org/10.1007/s12599-014-0334-4
2. Dautaj, M., Rossi, M.: Towards a new society: solving the dilemma between society 5.0 and industry 5.0. In: Junior, O.C., Noël, F., Rivest, L., Bouras, A. (eds.) Product Lifecycle Management. Green and Blue Technologies to Support Smart and Sustainable Organizations: 18th IFIP WG 5.1 International Conference, PLM 2021, Curitiba, Brazil, July 11–14, 2021, Revised Selected Papers, Part I, pp. 523–536. Springer International Publishing, Cham (2022). https://doi.org/10.1007/978-3-030-94335-6_37
3. Nahavandi, S.: Industry 5.0—a human-centric solution. Sustainability **11**(16), 4371 (2019). https://doi.org/10.3390/su11164371
4. Ekinpublishinggroup: Ready For Industry 5.0 ? https://furniturkonline.com/2018/08/06/ready-for-industry-5-0/
5. Leng, J., et al.: Industry 5.0: prospect and retrospect. J. Manuf. Syst. **65**, 279–295 (2022). https://doi.org/10.1016/j.jmsy.2022.09.017
6. Romero, D., Stahre, J.: Towards the resilient operator 5.0: the future of work in smart resilient manufacturing systems. Procedia CIRP **104**, 1089–1094 (2021). https://doi.org/10.1016/j.procir.2021.11.183
7. Reike, D., Vermeulen, W.J.V., Witjes, S.: The circular economy: New or Refurbished as CE 3.0? — Exploring Controversies in the Conceptualization of the Circular Economy through a Focus on History and Resource Value Retention Options. Resour. Conserv. Recycl. **135**, 246–264 (2018). https://doi.org/10.1016/j.resconrec.2017.08.027

8. What is a circular economy? | Ellen MacArthur Foundation. https://www.ellenmacarthurfoundation.org/topics/circular-economy-introduction/overview. Last accessed 5 Mar 2024
9. Rosa, P., Sassanelli, C., Urbinati, A., Chiaroni, D., Terzi, S.: Assessing relations between Circular Economy and Industry 4.0: a systematic literature review. Int. J. Product. Res/ **58**(6), 1662–1687 (2020). https://doi.org/10.1080/00207543.2019.1680896
10. Geissdoerfer, M., Savaget, P., Bocken, N.M.P., Hultink, E.J.: The circular economy – a new sustainability paradigm? J. Clean. Prod. **143**, 757–768 (2017). https://doi.org/10.1016/j.jclepro.2016.12.048
11. Circular Economy | Exenia: https://www.exenia.eu/en/sustainability/circular-economy/. Last accessed 5 Mar 2024
12. Dev, N.K., Shankar, R., Qaiser, F.H.: Industry 4.0 and circular economy: Operational excellence for sustainable reverse supply chain performance. Resour. Conserv/ Recycl. **153**, 104583 (2020). https://doi.org/10.1016/j.resconrec.2019.104583
13. Jayakumar, J., Kandasamy, J., Vimal, K.E.K., Hasibuan, S.: Modelling of sharing networks in the circuar economy. J. Model. Manage. **15**, 407–440 (2020). https://doi.org/10.1108/JM2-05-2019-0101/FULL/PDF
14. Kanojia, A., Visvanathan, C.: Assessment of urban solid waste management systems for Industry 4.0 technology interventions and the circular economy. Waste Manage. Res.: J. Sustain. Circular Econ. **39**(11), 1414–1426 (2021). https://doi.org/10.1177/0734242X21992424
15. Al-Swidi, A.K., Hair, J.F., Al-Hakimi, M.A.: Sustainable development-oriented regulatory and competitive pressures to shift toward a circular economy: the role of environmental orientation and Industry 4.0 technologies. Bus. Strat. Env. **32**(7), 4782–4797 (2023). https://doi.org/10.1002/bse.3393
16. Despoudi, S., Sivarajah, U., Spanaki, K., Charles, V., Durai, V.K.: Industry 4.0 and circular economy for emerging markets: evidence from small and medium-sized enterprises (SMEs) in the Indian food sector. Ann. Oper. Res. (2023). https://doi.org/10.1007/s10479-023-05404-4
17. Nayal, K., Raut, R.D., Mangla, S.K., Kumar, M., Tuček, D., Gavurova, B.: Achieving market performance via industry 4.0 enabled dynamic marketing capability, sustainable human resource management, and circular product desig. Ind. Market. Manage. **115**, 86–98 (2023). https://doi.org/10.1016/j.indmarman.2023.09.010
18. Chiappetta Jabbour, C.J., et al.: First-mover firms in the transition towards the sharing economy in metallic natural resource-intensive industries: Implications for the circular economy and emerging industry 4.0 technologies. Resourc. Policy **66**, 101596 (2020). https://doi.org/10.1016/j.resourpol.2020.101596
19. Nobre, G.C., Tavares, E.: The role of Industry 4.0 technologies in the transition to a circular economy: a practice perspective. Sustain. Sci. Pract. Policy (2023). https://doi.org/10.1080/15487733.2023.2289260
20. Ali, K., Johl, S.K.: Driving forces for industry 4.0 readiness, sustainable manufacturing practices and circular economy capabilities: does firm size matter? J. Manuf. Technol. Manage. **34**, 838–871 (2023). https://doi.org/10.1108/JMTM-07-2022-0254/FULL/PDF
21. Kor, Y.Y., Mahoney, J.T.: Edith Penrose's (1959) Contributions to the Resource-based View of Strategic Management. J. Manage. Stud. **41**, 183–191 (2004). https://doi.org/10.1111/J.1467-6486.2004.00427.X
22. Zhao, S., Morgan, S.L.: Business to Government Networks in Resource Acquisition. In: Business Networks in East Asian Capitalisms, pp. 69–91. Elsevier (2017). https://doi.org/10.1016/B978-0-08-100639-9.00005-0
23. Bag, S., Gupta, S., Kumar, S.: Industry 4.0 adoption and 10R advance manufacturing capabilities for sustainable development. Int. J. Product. Econ. **231**, 107844 (2021). https://doi.org/10.1016/j.ijpe.2020.107844

24. Bag, S., Wood, L.C., Mangla, S.K., Luthra, S.: Procurement 4.0 and its implications on business process performance in a circular economy. Resour. Conserv. Recycl. **152**, 104502 (2020). https://doi.org/10.1016/j.resconrec.2019.104502
25. Chari, A., et al.: Dynamic capabilities for circular manufacturing supply chains—Exploring the role of Industry 4.0 and resilience. Bus. Strat. Env. **31**(5), 2500–2517 (2022). https://doi.org/10.1002/bse.3040
26. Neri, A., Negri, M., Cagno, E., Kumar, V., Garza-Reyes, J.A.: What digital-enabled dynamic capabilities support the circular economy? A multiple case study approach. Bus. Strat. Environ. **32**, 5083–5101 (2023). https://doi.org/10.1002/BSE.3409
27. AL-Khatib, A.W.: The impact of dynamic capabilities on circular economy: the mediating effect of the industrial Internet of things. J. Manuf. Technol. Manage. **34**, 873–895 (2023). https://doi.org/10.1108/JMTM-01-2023-0003/FULL/PDF
28. Schöggl, J.P., Stumpf, L., Baumgartner, R.J.: The role of interorganizational collaboration and digital technologies in the implementation of circular economy practices—Empirical evidence from manufacturing firms. Bus Strategy Environ. (2023). https://doi.org/10.1002/BSE.3593
29. Dynamic Capabilities — David J. Teece: https://www.davidjteece.com/dynamic-capabilities. Last accessed 6 Mar 2024
30. Xing, K., Kim, K.P., Ness, D.: Cloud-BIM enabled cyber-physical data and service platforms for building component reuse. Sustainability **12**(24), 10329 (2020). https://doi.org/10.3390/su122410329
31. Subramoniam, R., Sundin, E., Subramoniam, S., Huisingh, D.: Riding the digital product life cycle waves towards a circular economy. Sustainability **13**(16), 8960 (2021). https://doi.org/10.3390/su13168960
32. Akram, S.V., Malik, P.K., Singh, R., Gehlot, A., Juyal, A., Ghafoor, K.Z., Shrestha, S.: Implementation of digitalized technologies for fashion industry 4.0: opportunities and challenges. Sci. Programm. **2022**, 1–17 (2022). https://doi.org/10.1155/2022/7523246
33. Talla, A., McIlwaine, S.: Industry 4.0 and the circular economy: using design-stage digital technology to reduce construction waste. Smart Sustain. Built Env. **13**, 179–198 (2024). https://doi.org/10.1108/SASBE-03-2022-0050/FULL/PDF
34. Uriarte-Gallastegi, N., Landeta-Manzano, B., Arana-Landín, G., Laskurain-Iturbe, I.: How do technologies based on cyber–physical systems affect the environmental performance of products? A comparative study of manufacturers' and customers' perspectives. Sustainability **14**(20), 13437 (2022). https://doi.org/10.3390/su142013437
35. Kurniawan, T.A., Meidiana, C., Dzarfan Othman, M.H., Goh, H.H., Chew, K.W.: Strengthening waste recycling industry in Malang (Indonesia): lessons from waste management in the era of Industry 4.0. J. Clean. Product. **382**, 135296 (2023). https://doi.org/10.1016/j.jclepro.2022.135296
36. Satyro, W.C., et al.: Implementing industry 4.0 through cleaner production and social stakeholders: holistic and sustainable model. Sustainability **13**(22), 12479 (2021). https://doi.org/10.3390/su132212479
37. Sehnem, S., Lara, A.C., Benetti, K., Schneider, K., Marcon, M.L., da Silva, T.H.H.: Improving startups through excellence initiatives: addressing circular economy and innovation. Environ. Dev. Sustain. (2023). https://doi.org/10.1007/S10668-023-03247-4/TABLES/4
38. Yu, Z., Khan, S.A.R., Umar, M.: Circular economy practices and industry 4.0 technologies: a strategic move of automobile industry. Bus. Strat. Env. **31**(3), 796–809 (2022). https://doi.org/10.1002/bse.2918
39. Skalli, D., Charkaoui, A., Cherrafi, A., Shokri, A., Garza-Reyes, J.A., Antony, J.: Analysis of factors influencing Circular-Lean-Six Sigma 4.0 implementation considering sustainability implications: an exploratory study. Int. J. Product. Res. **62**(11), 3890–3917 (2023). https://doi.org/10.1080/00207543.2023.2251159

40. Zonnenshain, A., Fortuna, G., Adres, E., Kenett, R.S.: Regional development in the era of industry 4.0. Dyn. Relat. Manage J. **9**, 19–36 (2020). https://doi.org/10.17708/DRMJ.2020.V09N02A02
41. Dautaj, M., Chiriacò, F., Terzi, S., Pero, M., Abdelkafi, N., Callupe, M.: When the learning lab embraces digitalisation: the development of a digital learning lab for the SMILE project. In: Terzi, S., Madani, K., Gusikhin, O., Panetto, H. (eds.) Innovative Intelligent Industrial Production and Logistics: 4th International Conference, IN4PL 2023, Rome, Italy, 15–17 Nov 2023, Proceedings, pp. 270–284. Springer Nature Switzerland, Cham (2023). https://doi.org/10.1007/978-3-031-49339-3_17

Enhancing Data Integrity with CSDT: A Solution to Predict Data Disturbance in IoT Systems

Meriem Smati[1,2(✉)], Jannik Laval[1], Vincent Cheutet[1], and Christophe Danjou[2]

[1] INSA Lyon, Université Lumière Lyon 2, Université Claude Bernard Lyon 1, Université Jean Monnet Saint-Etienne, DISP UR4570, 69621 Villeurbanne, France
{meriem.smati,vincent.cheutet}@insa-lyon.fr,
jannik.laval@univ-lyon2.fr

[2] Laboratoire Poly-Industrie 4.0, Département de Mathématiques Et Génie Industriel, Polytechnique Montréal, Montréal, QC, Canada
christophe.danjou@polymtl.ca

Abstract. With the rapid growth in popularity of Digital Twins (DTs), there is a pressing need to ensure data integrity within Internet of Things (IoT) systems. The complexity of potential anomalies in Internet of Things (IoT) data requires mechanisms that can evaluate these behaviors and understands them in advance without disrupting live systems. DTs offer a solution by replicating physical systems in a virtual environment, allowing new functionalities like monitoring, optimization, and prediction. To improve these functionalities, cognitive capabilities are encompassed, leading to the emergence of Cognitive Digital Twin (CDT). The findings of this research underscore the transformative impact of CDTs in strengthening IoT system security, realized by harnessing the capabilities of cognitive technologies and the additional layer of data generation that introduce the new framework "Cognitive Super Digital Twin" (CSDT).

Keywords: Digital Twins · Internet of Things · Model-Driven Engineering · Machine Learning · Data Integrity

1 Introduction

Amid the ongoing evolution and widespread adoption of IoT devices in today's data-driven world, a key challenge is ensuring uninterrupted functionality in the face of potential disruptions like cyber-attacks, power outages, physical damage, and data anomalies. Many traditional approaches exist to mitigate such risks, for example incorporating redundancy and failover mechanisms into the design of infrastructure and industrial systems [10]. Redundancy entails integrating backup devices, networks, or data centers to enable system continuity in case of component failure. However, implementing this methodology would require a significant investment of both finances and resources.

Published by Springer Nature Switzerland AG 2025
P. Sureephong et al. (Eds.): PLM 2024, IFIP AICT 740, pp. 100–113, 2025.
https://doi.org/10.1007/978-3-031-93319-6_8

Nevertheless, this paper focuses on addressing data anomalies, also referred to as outliers, aberrations or irregularities, which pose a significant hurdle in the current data-centric era. These anomalies represent deviations from expected behavior within interconnected IoT networks. Detecting and understanding these anomalies are vital for ensuring system reliability and integrity across various sectors like manufacturing, healthcare, agriculture, and smart cities. But predicting problems in advance is even better.

Rather than relying solely on conventional methods to tackle anomalies i.e. redundancy, a more cost-effective digital solution emphasizing data integrity preservation is proposed. In this context, DTs technology emerges as a promising strategy, particularly the concept of CDTs [2]. A CDT refers to a virtual representation of a physical or digital entity that possesses the ability to perceive, analyze, learn, and make decisions in a manner similar to human cognition[1] [18]. It leverages Artificial Intelligence (AI) more particularly Machine Learning (ML) to replicate the cognitive processes and behaviors of its real-world counterpart.

Therefore the research objective of this paper is the following:

How can DT be used as a tool for detecting and preventing data disturbance in an IoT system?

To reach this objective, the paper is organized as follows. After reviewing a general introduction in Sect. 1, Sect. 2 covers the reviewed literature on IoT and Data anomalies as well as the use of DT as a prevention tool. The realised framework and its architecture is given in Sect. 3 and its application on a use case is in Sect. 4. And to conclude this article, Sect. 5 allows to see the solved problems as well as the limits and the future works.

2 Related Works

2.1 IoT Data Anomalies

Anomalies in IoT can be caused by multifaceted factors. These factors include not only unintentional occurrences or technical malfunctions, but also deliberate and malicious intrusions orchestrated by external actors who aim to exploit vulnerabilities and disrupt the smooth operation of these IoT systems.

Mishra et al. [11] discuss the detection of data anomalies in pervasive IoT systems, emphasizing the importance of validating input data to prevent severe consequences. The main focus was on two types of anomalies: data bias anomalies, where sensor values are forced to extreme values, and sensor cut-off anomalies, where sensors stop sending data. These anomalies can result from accidents, hardware or software failures, or cyberattacks. The study proposes using statistical techniques, specifically control charts, for anomaly detection in IoT systems. It highlights the challenges in differentiating the heterogeneity of edge networks normal from abnormal behavior in IoT. Another challenge is the importance of standardized datasets for evaluating IoT systems and a huge amount of data, especially abnormal ones, to do the training of the detection models.

[1] https://fuseforward.com/cognitive-digital-twins/.

Fahim et al. [4] presented a literature review and recognized that anomalies in data streams are crucial for optimizing system performance. Challenges include defining normal/abnormal boundaries and scarcity of abnormal observations for model training. Steps for anomaly detection include understanding data nature and structure, identifying anomaly types, and considering availability of training data. Statistical methods, ML techniques, and Deep Learning (DL) models are employed for anomaly detection, with each having its advantages and limitations. Research gaps include the need for easier access to real system data and the development of methods to handle complex data.

Yasaei et al. [17] propose an adaptive context-aware anomaly detection method for IoT systems using sensors and a fog computing platform. The method includes a sensor association algorithm, a predictor model with Long Short-Term Memory (LSTM) neural network and Gaussian estimator, and a consensus algorithm to identify anomalies and their source. The paper addresses challenges such as multi-variant time-series data, resource-constrained sensors, and dynamic characteristics of IoT systems. The contributions include a context-aware sensor association algorithm, a consensus-based strategy for unsupervised anomaly detection, and an adaptive data-driven model.

Riahi et al. [13] consists of a simulation platform for an IoT system that takes into consideration only sensors due to their communication with the real-world and them being a target for many hacks and disturbances. The faulty IoT Network generates many sensor and data disturbance such data alteration and data deletion.

Various facets of anomaly detection and prediction have been addressed and contributions concerning different attacks have been discussed. However, the subject of data anomaly detection poses several significant challenges. The first and foremost challenge involves the complexity of distinguishing between data that is considered normal in one scenario but abnormal in another, as highlighted in Mishra et al. [11]. An even more daunting challenge lies in the prediction and prevention of these outliers. Creating genuine anomalous behaviors is a formidable task due to the resource-intensive nature of the process or, in some cases, due to inherent system limitations, as discussed in Fahim et al. [4]. In Subsect. 2.2, we explore potential solutions, including the utilization of DTs to replicate physical assets into virtual ones.

2.2 The Use of DT and Data Anomalies

DTs have several definitions depending of the domain of application. This concept has been first defined by Michael Grieves in his Product Lifecycle Management (PLM) course in 2002 as a virtual representation of a physical asset that serves as the effective counterpart to the real-world object [6]. And it consists of the physical object, the digital representation of the object and the communication channel between the physical and virtual representations [16]. In the field of IoT, it can be defined as virtual representation of system that is created using data from connected IoT sensors. It allows and facilitate the monitoring of devices, assets, or processes from anywhere[2].

DTs have two types [9]:

[2] https://www.networkworld.com/article/3280225/what-is-digital-twin-technology-and-why-it-matters.html, last access in 21/11/2023.

- *Dynamic DT*: it is a type of DT that is fed by live data flows from the physical asset, it needs real-time updates and impact the Physical Twin (PT) at the same moment.
- *Static DT*: it gives periodic updates to the PT.

Concerning the level of magnification the DT replicates, there are four types[3]:

- *Component Twin*: it is the creation of a model or a twin to each individual small asset.
- *Asset Twin*: it represents when two or more components work together.
- *System Twin/Unit Twin*: this is considered as the next level of magnification, enabling to see how different assets come together to form an entire functioning system.
- *Process Twin*: It is the realization of an entire business process or customer journey and it is considered to be the macro level of magnification.

Regarding the CDT, it is an evolution of DT that integrates advanced learning and self-discovery capabilities using AI functions[4]. They are capable of achieving some elements of cognition, notably memory (encoding and retrieval), and can learn and evolve, integrating different sources of information for a specific purpose [3]. They are distinguishable by six key functionalities[5]:

- Perception: it involves the creation of useful and meaningful representations of data related to the PT and its surroundings for further processing.
- Attention: it is the act of selectively focusing on a task or a specific sensory information.
- Memory: it encompasses three types of memories. The working memory used for active use, the episodic memory for recalling episodes from the PT's life and semantic memory for retaining knowledge about the environment and its interactions with the PT.
- Reasoning: drawing conclusions that align with a starting point or memories.
- Problem-solving: it is the systematic process of identifying a solution for a given problem.
- Learning: a transformative process of converting the experience of the PT into reusable knowledge for new experiences like foreseeing the future and act accordingly.

Below are few works that have utilized DTs for prediction and fault detection:

- Saraeian et al. [14] provide a detailed discussion a fault tolerance approach for predicting and detecting faults in Cyber-Physical Production Systems (CPPS) using a digital twin-based architecture. The proposed approach utilizes an extraction module to efficiently extract fault signatures and a prediction module to generate appropriate responses based on the extracted faults. The results show that the approach can predict reliable fault signatures and maintain CPPS at optimal levels. The proposed component aims to reduce complexity and improve fault detection and maintenance in CPPS. Yet this contribution has limitations since it is highly dependent on historical data and on the complexity of the production system and the accuracy of DTs.

---

[3] https://www.ibm.com/topics/what-is-A-digital-twin, last access in 21/11/2023.
[4] https://fuseforward.com/cognitive-digital-twins/, last access in 21/11/2023.
[5] https://fuseforward.com/cognitive-digital-twins/.

- Flammini [5] discusses the concept of DT as a promising paradigm for modeling and predicting the performance of Cyber-Physical Systems (CPS). It focuses on data-driven evaluation and prediction of dependability attributes, as well as the challenges of addressing resilience in CPS. The contribution consists of a conceptual framework for hosting DT run-time models based on autonomic systems, and discusses the hierarchical architecture and tradeoffs involved in implementing DT at different levels. However, this framework has several limitations due to CPS's sizes, heterogeneity, distribution, and criticality.
- Savaglio et al. [15] introduce the concept of Opportunistic Digital Twin (ODT) which combines AI, Edge Intelligence (EI) and DT technology to enhance applications. A use case of traffic prediction has been presented to demonstrate the improved performance and reduced network resources consumption where an LSTM Recurrent Neural Network (RNN) model trained on traffic intensity, air pollution, and atmospheric data has been used.
- Haidar et al. [8] conduct a review on the DT technology for fault detection in buildings. The study emphasizes that prediction plays a crucial role in Digital Twin technology. AI, such as ML and data analytics, is used to forecast the future behavior and health condition of the physical asset. Predictions are based on measured input and output values, as well as initial circumstances. The study acknowledges the ongoing debate about the maturity level required for accurate predictions and warnings about the asset's performance. It also mentions the importance of big data prediction strategies, which often involve ML or data mining techniques.

As observed in various literature sources and the reviewed studies, they collectively tackle the critical challenge of detecting anomalies in IoT systems either by using statistical methods, DTs or CDTs. Emphasis is placed on validating input data to avert severe consequences which leads to a noteworthy challenges. The DT relies fully and is intricately linked to the data provided by its counterpart, the PT. And like any IoT system, the normal collected data instances are more than the collected abnormal ones resulting an imbalanced dataset, so when aiming to train the prediction model of the CDT, a substantial waiting period is often necessary to achieve accurate results. To circumvent this challenge, a new approach needs to be established to address the data scarcity issue, thus enhancing the efficiency and effectiveness of the prediction model training process.

3 CSDT Proposal

IoT systems rely on a multitude of sensors and interconnected devices, have proven to be invaluable in enhancing efficiency and convenience. However, the seamless operation of IoT systems is not without its challenges. In the pursuit of ensuring the reliable and secure operation of IoT systems, the novel concept of CSDT is created.

3.1 Problematic

Before delving deeper in the CSDT concept, it is imperative to address the problems and issues that led to this contribution and it encompasses the following:

- Data imbalance, since IoT systems generally generate normal data, but anomalies are rare in comparison. This data imbalance poses a significant challenge when it comes to training models to detect abnormal behavior effectively,
- Behavioral Variability concerning what constitutes "normal" or "abnormal" behavior in IoT systems is not always straightforward. Behavior may vary with factors such as season, location, or context, making it difficult to establish clear boundaries,
- Traditional DT reliance on PT data, so data anomaly has to occur first to be predicted the next time after using it for training the model.

3.2 Solution Architecture

As it can be seen in Fig. 1, the chosen definition of a DT is a set of layers illustrated in pink.

The first layer is the **database** which plays a crucial role in storing and managing the vast amount of data required to create and maintain DTs. DTs can have multiple sources of information, such as information from processing the measured data and information collected previously from various data sources along the life cycle of a product [1]. In this context our database is considered to be constituted of two repositories, the "dataset acquisition" repository that collect dynamic time-series data and the "Vault" repository that preprocess the collected data in the first repository and make the archive of the system. The layer that is positioned above of it uses the database layer to **simulate** and replicate exactly what the PT does [12]. The **visualisation** layer in the DT is an important aspect since it allows users to see a virtual and real-time representation of systems or components [19]. And depending on all these layers, the final layer of the DT is the **decision** layer, helping the human being to make the decision. But to make predictions and make the decision-making more automated, a **cognitive** layer is added, presented in orange, making the DT a CDT. Shifting focus on the contribution at hand which answers the problems mentioned in the Subsect. 3.1: the **data generation** layer colored in Blue. This stratum makes a CDT a CSDT.

Since IoT systems are susceptible to a range of vulnerabilities stemming from both software and hardware failures, which can lead to a loss of control. Any disruption in the operation of an IoT system can, at the very least, cause inconvenience and, at worst, pose a life-threatening risk [11].

Finding unusual data points to train models in an IoT system can be quite challenging. In practical applications, instances of abnormal behavior, such as a sudden and significant increase in temperature in a temperature monitoring system, are quite infrequent when contrasted with the frequency of normal behaviors like maintaining a stable temperature within the desired range [4].

Hence, the justification for the generation module is the lack of abnormal data compared to normal ones. This layer fabricates normal and abnormal data, both of those generations are essential to have a balanced global dataset. Therefore, distinguishing those two types of behavior is already a challenge [11]. For instance, a temperature of 10 °C is considered normal during winter, but the same temperature in summer might be regarded as a deviation. This is discussed further more in Sect. 4. But as mentioned in [4], to commence with, we need to know the nature of collected data streams in our IoT system (binary, continuous, discrete) along with the relationship structure (time-series

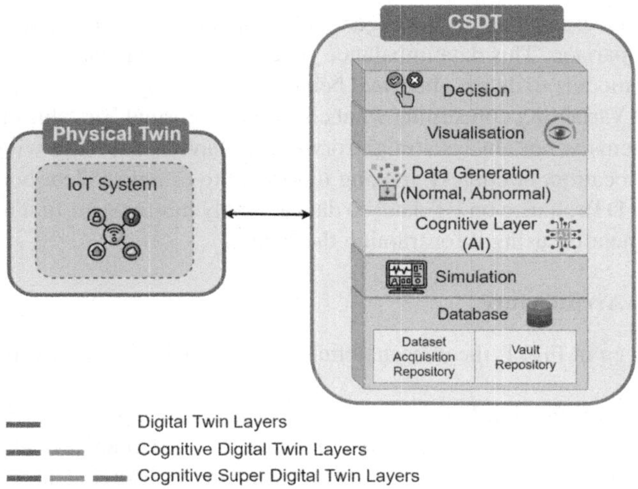

Fig. 1. CSDT's Generation Module

data, spatial data, graph data). Second, the type of anomaly needs to be identified (point anomaly which is an observation point in the data stream that is significantly distant from the rest of the data, contextual anomaly, an observation point normal in one scenario yet abnormal in another and collective anomaly that is when a sequence of observations are analyzed together to know the normal behavior of a data stream. And any deviation is considered an anomaly). And finally know the availability of training data to allow our model to learn.

On the macroscopic view, Fig. 2 represents the versatile architecture required to build a CSDT. As any DT, the architecture must be constituted of the three essential parts: a Physical asset, a communication medium and the digital asset.

In this paper, the considered PT is an IoT system. To facilitate the bidirectional communication between the two assets, an intermediary communication medium plays a pivotal role in establishing a seamless and real-time tunnel between the previously mentioned components. This medium ensures the secure transportation and storage of all interactions, thereby rendering the communication process transparent and immediate.

Moving on to CSDT part, the data generation is represented in the "*Database*" layer. After the data generation, an operation of data acquisition will start, since the origin of the data processed by the CSDT can come from different sources (the generated data, the data collected from the IoT system, etc.) so it serves to collect the data and process it to combine it and make it one global source of data or dataset. The over element of the CSDT is the simulator [12], which is the core of the DT which is a faithful replication of the PT's functionality.

3.3 Class Diagram

The viewed versatile architecture has also been represented in the form of a generic class diagram in Fig. 3 that has been inspired from [13] but the emphasis is on how the

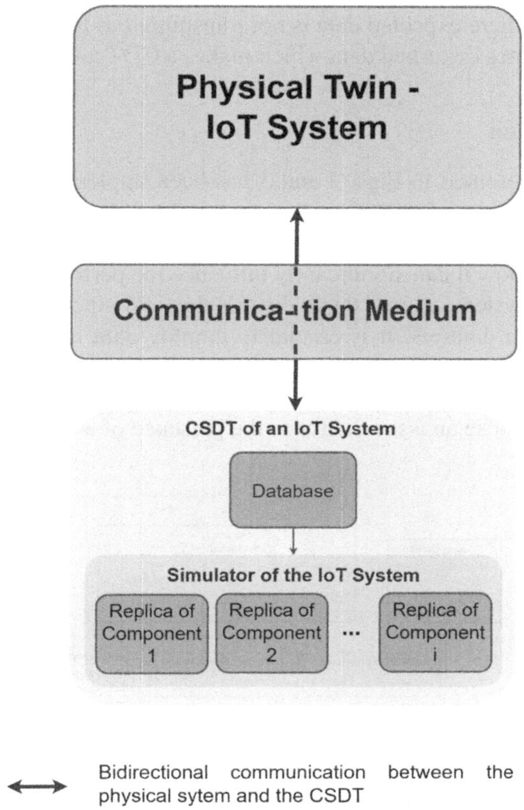

Fig. 2. The Versatile CSDT Architecture

disturbances are made and how the DT is stressed and react to these anomalies. And is described below:

The "Component" class is at the head of the class diagram as it serves as a foundational element, with all classes ultimately deriving from it, ranging from complex to simple. This class gives rise to two specific inheriting classes: "PhysicalTwin" and "DigitalTwin". The "PhysicalTwin" class exhibits a self-contained nature and is composed of one system. It is further equipped with a predictive model trained on normal data.

In contrast, the "DigitalTwin" class is composed of the "DatasetAcquisition" and "Simulation" classes. The "Simulation" class, as the core component of the DT, closely replicates the functionalities of the PT. Consequently, it features a system and has a model trained on normal and perturbed data, eventually attaining the status of a "golden standard model" once sufficient data has been generated and the model has been fully trained. At this juncture, it supersedes the model in the PT. This model makes the DT a CDT. Notably, the inclusion of a "Sensor" class is required since it plays a major role in the data capture process within the "Simulation" class. Various forms of data disturbances are injected into the "Sensor" class, including Data alteration, data deletion, and temporary sensor stoppages, where expected data is not transmitted as anticipated. These injected disturbances generates disturbed data which makes a CDT a CSDT.

3.4 Implementation

The architecture presented in Figs. 2 and 3 has been implemented on a case study to test the ability and limits of CSDT, so it is essential to address the potential impact of disturbances as well that can be in the form of unexpected events such as irregularities in the data and see how it can significantly influence the performance of the system.

To fortify the system against these disturbance and since anomaly detection and prediction works on datasets, it is crucial to employ data mining techniques. It can consider multiple instances at a time, such as Classification-based anomaly detection techniques, which can be applied whenever sufficient amount of labelled data is available [5]. This would not pose an issue thanks to the presence of a data generation module.

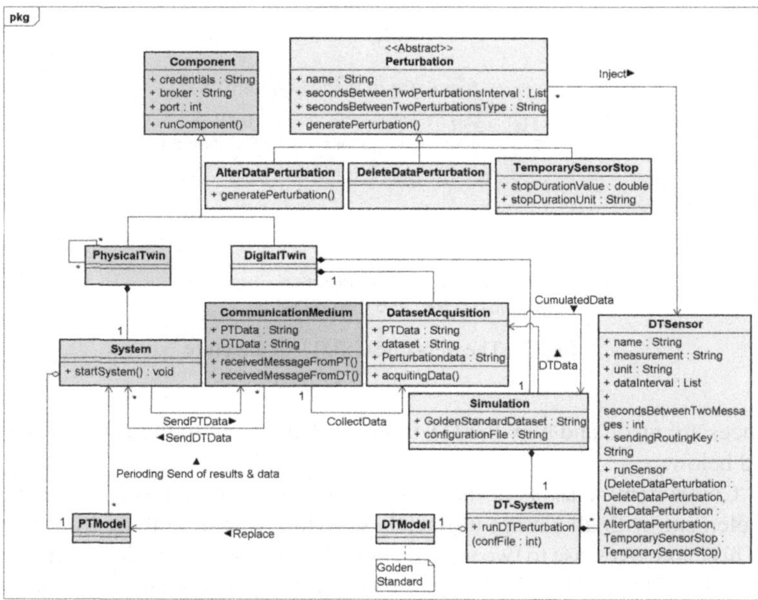

Fig. 3. The Generic CSDT Class Diagram - the colors used here are consistent with the ones in Fig. 2

The data generation in our system is done thanks to a JavaScript Object Notation (JSON) file where the current collected data is viewed and similar data is generated to create the normal data and the different ones are considered the abnormal ones.

Focusing more on the disturbance, the induced anomalies primarily involve manipulating the data. This, of course, is contingent upon the specific environment in which the data is collected, as what might be deemed anomalous in one environment may not be so in another. Another type of anomaly involves the deletion of the data, wherein the data is anticipated by the receiving communication medium but is not transmitted. Lastly, simulating a sensor stop or a sensor halt occurs when the sensor does not generate at all the data for a certain amount of time.

4 Case Study

In this section, a case study is put in place to accomplish several objectives:

- Test the CSDT and how its generation module is effective compared to a specific employed environment,
- Verify the scalability of the CSDT Simulator,
- Find a suitable ML or DL model for the PT and the CSDT.

4.1 Scenario

The use case consists of detecting the room occupancy by collecting data of the environment. By following the workflow presented in [4], the nature of the collected data stream data stream is continuous because the collected data are time-series. The anomaly type is contextual anomaly and we have enough training data. The problem is considered a binary classification problem to the PT (Occupied, not Occupied) but it is a multiclass classification to the DT (Occupied, not Occupied, disturbed) which makes the CSDT model more efficient and with greater knowledge than the PT model and that way the disturbance can be predicted as a result of the generation layer and the prediction model.

4.2 Configuration of the System

As depicted in the Fig. 4, the PT is an IoT system in the form of an Edge/Fog architecture where the Edge is constituted of a Raspberry Pi 3 Model B and three sensors: Grove DHT11 sensor, Grove Light sensor and Grove SGP30 sensor. The Grove Barometer sensor has been added later on to prove that the CSDT's simulator is extensible. The collected data (temperature, humidity, CO2 and Light) are sent to the Fog to train the model along with an external dataset[6] to have enough data in the PT.

The predicted output of the model along with the collected data are displayed and sent as well to the Communication Medium where it is saved in a time-series database InfluxDB.

After all the data have been received, the communication medium forward it to the DT where the data acquisition module makes the preprocessing and combine the three sources of data (DT generated normal and abnormal data, External dataset, collected data) to create on global dataset. The multi-class model is trained on the global dataset. This golden standard model replaces the one available in the PT periodically.

[6] **Occupancy Detection Data Set:** https://archive.ics.uci.edu/dataset/357/occupancy+detection, last access in 21/11/2023.

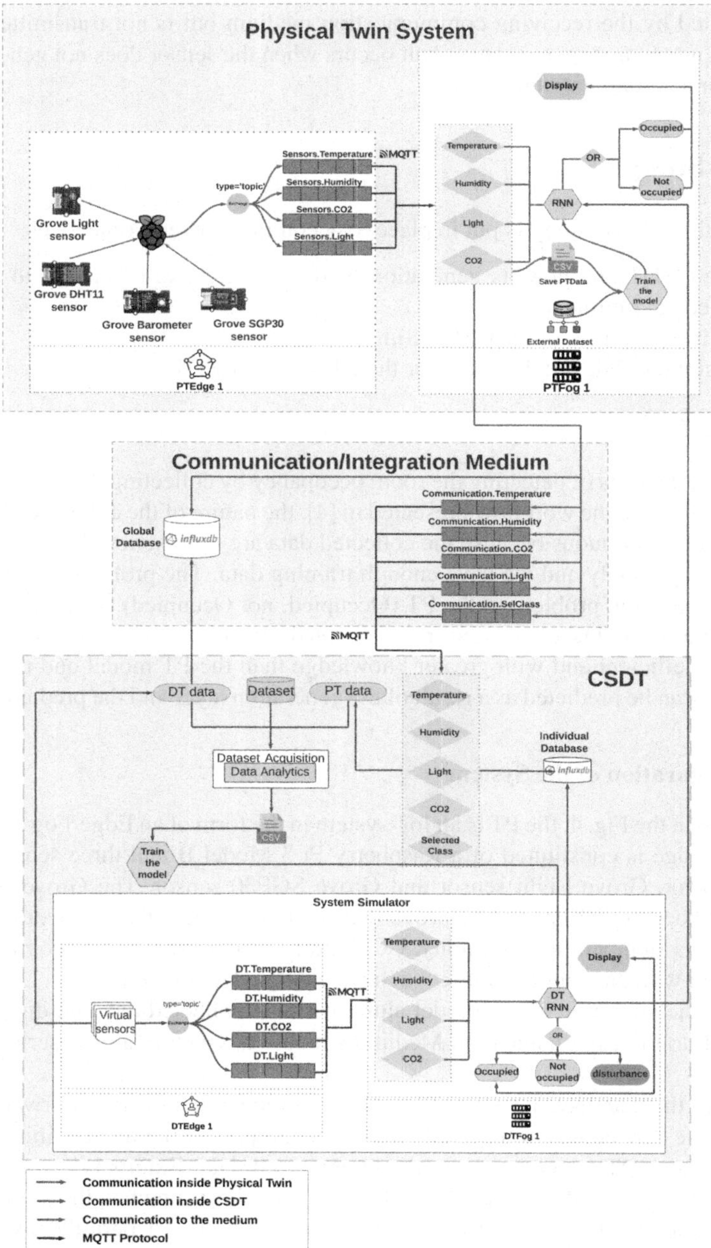

Fig. 4. The use of CSDT on a Case Study.

4.3 Results

The selection of the PT and DT models has been done after training several ML and DL models (Decision Tree, Random Forest, K-Nearest Neighbors (KNN), Naive Bayes, RNN, MultiLayer Perceptron (MLP)). The PT model that gave the best results is a RNN with LSTM layers [15]. The results are present in Table 1: an Accuracy of 99.51%, a Precision of 98.38% and a Recall of 99.81% so it gives an F1_Score of 99.09%, another reason why this model has been chosen is because it takes into consideration timestamps. But the model that outperformed concerning the CSDT are the tree-based models giving 99% scores and the RNN gave 67.03% F1_Score (Table 2). A trade-off must be done, either have correct predictions or taking the timestamps into consideration, in this study, the first choice has been picked, and as mentioned in [7], tree-based models perform better than other models on tabular data.

Table 1. PT Trained models.

	Decision Tree	Random Forest	KNN	Naive Bayes	RNN	MLP
Accuracy	99.14%	99.34%	99.31%	97.03%	99.51%	99.27%
Precision	97.73%	97.85%	97.85%	88.36%	98.38%	97.34%
Recall	98.47%	99.23%	99.13%	99.89%	99.81%	99.46%
F1_Score	98.10%	98.54%	98.48%	93.77%	99.09%	98.39%

Table 2. CSDT Trained models.

	Decision Tree	Random Forest	KNN	Naive Bayes	RNN	MLP
Accuracy	99.27%	99.38%	99.20%	97.84%	67.03%	86.26%
Precision	97.28%	99.39%	99.20%	97.43%	67.03%	77.22%
Recall	99.27%	99.38%	99.20%	97.84%	67.03%	86.46%
F1_Score	99.27%	99.38%	99.20%	97.85%	67.03%	78.39%

The same study was accomplished on a CDT, the faced problems are the lack of abnormal data so the anomalies were not able to be predicted on the contrary of what a CSDT is bringing. So the anomalies could be treated and help the user on the decision-making either by ignoring the anomaly or deleting it or calling for human-help to correct the anomaly.

This case study allowed to verify the accuracy and extensibility of the CSDT as well as injecting data disturbance to stress it and verify the prediction of the selected models that were proven to be suitable for the case study.

5 Conclusion

This paper explores the power and potential of DT as a preventative tool, particularly addressing its efficacy in detecting and mitigating data anomalies, introducing the new framework called CSDT. While DT offers benefits and advantages, it is not a one-size-fits-all solution and requires significant investments in data collection, analytics, computing power, and skilled personnel and they should be seen as a complementary tool to traditional maintenance practices, enhancing decision-making processes. Thanks to the CSDT that goes beyond replicating the physical Twin and can detect and generate perturbations for improved prevention and resilience strategies, aiming to enhance overall system performance and preparedness.

While this contribution brings a lot to a significant issue, namely data insufficiency and the full reliance of the DT on the PT data, it does have its limitations. One of these constraints is related to the data distribution and environmental context and awareness. For example, a temperature of -10 °C within a typical office setting is considered abnormal, whereas the same temperature in a cryogenic chamber or cold storage facility would be considered normal.

Another limitation is that, although the simulator of CSDT is extensible, it can only simulate sensors that collect numerical data. For instance, it is still unable to handle sensors that capture images, videos, frames, and similar non-numeric data.

The future works and prospective endeavors encompass an evolution towards a Systems-of-Systems (SoS) structure rather than working on only one system. This evolution requires and entails the expansion of the CSDT from a component twin to a system twin or a process twin and a transition from a static DT to a dynamic DT that operates in real-time, providing instantaneous data and responses. This will be possible by integrating real-time training mechanisms, allowing continuous learning and improvement in performance over time.

A case study to showcase the power of CSDT could come in hand and incorporate reinforcement learning allowing it to learn and improve its performance over time. Another future work is to analyse the patterns of the sensors to generate the data and predicting the source of anomaly as well as testing different combination of anomaly injections based on Model-Driven Engineering (MDE) which enables the integration of various models representing different facets of the system, this holistic approach enhances anomaly detection accuracy by considering a comprehensive set of factors.

Disclosure of Interests. The authors have no competing interests to declare that are relevant to the content of this article.

References

1. Bonney, M.S., de Angelis, M., Dal Borgo, M., Wagg, D.J.: Contextualisation of information in digital twin processes. Mech. Sys. Signal Process. **184**, 109657 (2023). https://doi.org/10.1016/j.ymssp.2022.109657
2. Eirinakis, P., et al.: Cognitive digital twins for resilience in production: a conceptual framework. Information **13**(1), 33 (2022). https://doi.org/10.3390/info13010033

3. El Mokhtari, K., Panushev, I., McArthur, J.J.: Development of a cognitive digital twin for building management and operations. Front. Built Environ. (2022). https://doi.org/10.3389/fbuil.2022.856873
4. Fahim, M., Sillitti, A.: Anomaly detection, analysis and prediction techniques in IoT environment: a systematic literature review. IEEE Access **7**, 81664–81681 (2019). https://doi.org/10.1109/ACCESS.2019.2921912
5. Flammini, F.: Digital twins as run-time predictive models for the resilience of cyber-physical systems: a conceptual framework. Phil. Trans. R. Soc. A **379**(2207), 20200369 (2021). https://doi.org/10.1098/rsta.2020.0369
6. Grieves, M.: Digital twin: manufacturing excellence through virtual factory replication. White paper **1** (2014)
7. Grinsztajn, L., Oyallon, E., Varoquaux, G.: Why do tree-based models still out-perform deep learning on tabular data? (2022)
8. Hosamo, H.H., Nielsen, H.K., Alnmr, A.N., Svennevig, P.R., Svidt, K.: A review of the digital twin technology for fault detection in buildings. Front. Built Environ. (2022). https://doi.org/10.3389/fbuil.2022.1013196
9. Lamb, K.: Principle-based digital twins: a scoping review. CDBB (2020). https://doi.org/10.17863/CAM.47094
10. Laszka, A., Abbas, W., Vorobeychik, Y., Koutsoukos, X.: Integrating redundancy, diversity, and hardening to improve security of industrial internet of things. Cyber-Phys. Syst. **6**(1), 1–32 (2020). https://doi.org/10.1080/23335777.2019.1624620
11. Mishra, A., Cohen, A., Reichherzer, T.: Detection of data anomalies at the edge of pervasive IoT systems. Computing **103**, 1657–1675 (2021). https://doi.org/10.1007/s00607-021-00927-9
12. Nwogu, C., Lugaresi, G., Anagnostou, A., Matta, A., Taylor, S.J.: Towards a requirement-driven digital twin architecture. Procedia CIRP **107**, 758–763 (2022). https://doi.org/10.1016/j.procir.2022.05.058
13. Riahi, K., Kahn, G., Dafflon, B., Laval, J.: A faulty IoT network: simulating sensors and perturbations. In: Rocha, A., Adeli, H., Dzemyda, G., Moreira, F. (eds.) Information Systems and Technologies: WorldCIST 2022, Volume 3, pp. 87–97. Springer International Publishing, Cham (2022). https://doi.org/10.1007/978-3-031-04829-6_9
14. Saraeian, S., Shirazi, B.: Digital twin-based fault tolerance approach for cyber–physical production system. ISA Trans. **130**, 35–50 (2022). https://doi.org/10.1016/j.isatra.2022.03.007
15. Savaglio, C., Barbuto, V., Awan, F.M., Minerva, R., Crespi, N., Fortino, G.: Opportunistic digital twin: An edge intelligence enabler for smart city. ACM Trans. Sensor Netw. (2023). https://doi.org/10.1145/3616014
16. Sharma, A., Kosasih, E., Zhang, J., Brintrup, A., Calinescu, A.: Digi tal twins: State of the art theory and practice, challenges, and open research questions. J. Ind. Inf. Integr. (2022). https://doi.org/10.1016/j.jii.2022.100383
17. Yasaei, R., Hernandez, F., Faruque, M.A.A.: IoT-CAD: Context-aware adaptive anomaly detection in IoT systems through sensor association. In: Proceedings of ICCAD'20 (2020). https://doi.org/10.1145/3400302.3415672
18. Zheng, X., Lu, J., Kiritsis, D.: The emergence of cognitive digital twin: vision, challenges and opportunities. Int. J. Prod. Res. **60**(24), 7610–7632 (2022). https://doi.org/10.1080/00207543.2021.2014591
19. Zhu, Z., Liu, C., Xu, X.: Visualisation of the digital twin data in manufacturing by using augmented reality. Procedia CIRP **81**, 898–903 (2019). https://doi.org/10.1016/j.procir.2019.03.223

Enhancing Learning Experience Through Micro-learning Activities for Cost Managers in Restaurants Amidst Digital Transformation

Tama Duangnamol[1], Nuttanee Boonthum[1](✉), Orathai Khiawcha-um[1], Tanit Chumsang[2], and Julanit Wangwiwat[3]

[1] College of Arts, Media and Technology, Chiangmai University, Chiang Mai, Thailand
nuttanee.b@kirly.org

[2] Chiang Mai Restaurant and Bistro Association, Chiang Mai, Thailand

[3] Chiang Mai Chamber of Commerce, Chiang Mai, Thailand

Abstract. This study evaluates the effectiveness of microlearning activities in reskilling Cost Managers in the restaurant technology domain. Utilizing the VARK model, microlearning activities were tailored to match diverse learning styles, enhancing engagement and outcomes. Experiential activities like Site-Visits scored highest in satisfaction, knowledge, and learning experience, highlighting the importance of real-world applications. Seminar Q&A sessions also performed well, underscoring the value of interactive learning. While presentations were effective in knowledge dissemination, areas for improvement were identified in Hand-on Workshops and Micro Videos. These formats could benefit from enhanced interactivity and facilitation. The findings suggest that a balanced approach combining interactive and experiential learning can optimize outcomes. Recommendations include integrating varied learning activities, continuous learner feedback for refinement, and further research to optimize designs. This study highlights the value of interactive and experiential learning in enhancing satisfaction, knowledge acquisition, and overall learning experience, offering insights for effective adult education programs.

Keywords: Micro-Learning · Digital Transformation · Micro-Learning Activities · Cost manager · Restaurants

1 Introduction

Chiang Mai, a rising star among Thailand's top tourist destinations according to US News [1], is experiencing a culinary boom. As food plays a crucial role in tourism, as highlighted by Rong-Zhen [2], the city's explosion of restaurants in recent decades has become a major draw for visitors, fueling the local economy. This growth has had a substantial economic impact on the local workforce in Chiang Mai. However, COVID-19 pandemic and rapid technological advancements have disrupted the restaurant and tourism industries, creating a need for new skills among workers, particularly in digital

Published by Springer Nature Switzerland AG 2025
P. Sureephong et al. (Eds.): PLM 2024, IFIP AICT 740, pp. 114–122, 2025.
https://doi.org/10.1007/978-3-031-93319-6_9

literacy [3]. The urgency for reskilling and upskilling is critical, yet many workers find it challenging to attend full-time classes due to their existing job commitments and responsibilities. Furthermore, traditional educational models often do not cater to the diverse learning styles of adult learners, making it difficult to find a one-size-fits-all solution.

The rise of mobile technology offers a promising solution through microlearning. Microlearning involves delivering educational content in small, manageable segments that can be easily accessed via mobile devices. This approach is especially beneficial for adult learners, as it allows them to engage with content in short bursts and integrate learning into their daily routines [4]. Microlearning activities, such as short video-based learning, lectures, Q&A sessions, hands-on workshops, and mini project-based learning, can provide the flexibility and convenience that adult learners need. The proliferation of digital tools and platforms further facilitates the implementation of microlearning, making it a practical and effective approach for continuous education [5].

To address the diverse learning preferences of adults, this research applies the VARK model of learning styles to identify which types of microlearning activities are most suitable for our target group. The VARK model, developed by Neil and Colleen [6], categorizes learners into four primary types: Visual, Auditory, Read/Write, and Kinesthetic. By recognizing these differences, we can tailor the microlearning sessions to better meet the needs of each learner. Visual learners might benefit from infographics and video content, auditory learners from lectures and discussions, read/write learners from text-based materials, and kinesthetic learners from hands-on workshops. This study explores the application of various microlearning activities, allowing participants to choose and attend hybrid sessions (online and onsite) based on their interests. By collecting data on their experiences and satisfaction, we aim to determine the most effective microlearning strategies for reskilling individuals for the position of Cost Manager in the restaurant technology domain.

2 Literature Review

2.1 Microlearning in Adult Education

Microlearning, a method of delivering educational content in small, manageable segments, was coined by Theo [5] in the early 2000s as a response to the increasing use of mobile technology and the internet in education. This approach focuses on bite-sized learning units that can be easily accessed anytime and anywhere, making it particularly suitable for adult learners who balance education with other responsibilities. Microlearning contributes significantly to adult learning by accommodating the unique needs of adult learners. It provides flexibility, allowing adults to engage with educational content in short bursts, which fits well into their busy schedules. Additionally, microlearning supports just-in-time learning, enabling adults to acquire knowledge and skills as they need them, thereby increasing the relevance and immediate applicability of the content.

Several successful cases highlight the effectiveness of microlearning in adult education. In Europe, the German company SAP implemented a microlearning strategy to enhance employee skills, resulting in increased engagement and knowledge retention [7]. In the UK, the Open University has adopted microlearning to provide continuous professional development for teachers, leading to improved teaching practices and student outcomes [8]. In Asia, the Singaporean government's Skills Future initiative uses microlearning to help citizens upskill and reskill, contributing to a more adaptable and skilled workforce [9]. These cases demonstrate that microlearning is well-suited to the needs of adult learners, particularly in enhancing their digital literacy and technical skills. By integrating microlearning into our reskilling project for Cost Managers in the restaurant technology domain, we can provide effective, flexible, and relevant learning experiences that align with the busy lives of adult learners.

2.2 VARK Model

The theory of learning styles suggests that individuals have preferred ways of processing information and learning. This concept emerged in the 1970s, with early proponents such as Howard [10], who introduced the theory of multiple intelligences, and David [11], known for his experiential learning theory. These theories emphasize that recognizing and addressing individual learning preferences can enhance educational outcomes. The VARK model is a specific application of the broader theory of learning styles, providing a practical framework for identifying and catering to different learning preferences.

The VARK model, developed by Neil and Colleen [6], is one of the most widely recognized frameworks for understanding learning preferences. VARK stands for Visual, Auditory, Read/Write, and Kinesthetic, representing four primary types of learners. Visual learners prefer diagrams, charts, and other visual aids; auditory learners benefit from listening to lectures and discussions; read/write learners excel through reading and writing activities; and kinesthetic learners learn best through hands-on experiences and practical activities.

In the context of adult education, the VARK model is particularly relevant to microlearning. By tailoring microlearning activities to match the preferred learning styles of adults, educators can enhance engagement and effectiveness. For instance, visual learners might benefit from video-based content, auditory learners from podcasts, read/write learners from text-based materials, and kinesthetic learners from interactive simulations or practical tasks. Therefore, VARK model provides a structured approach to assess and improve the quality of microlearning activities. By understanding and applying the principles of VARK, we can design more personalized and effective microlearning experiences that cater to the diverse preferences of adult learners, ultimately enhancing their learning outcomes and satisfaction.

2.3 Microlearning in the Learning City Project

The Learning City Project, initiated by Chiang Mai University, aims to address the skill gaps in the local workforce by providing targeted education and training programs. One of the key positions identified as crucial for the restaurant industry in Chiang Mai is the Cost Manager. This role requires a comprehensive understanding of financial management, cost control, restaurant technology management, and digital skills to effectively use digital platforms such as Enterprise resource planning (ERP), Point of sale (POS), and basic workplace platform. Given the high demand for this position, the Learning City Project has focused on reskilling individuals to meet these requirements.

The boot camp organized under this project included a variety of microlearning activities designed to cater to adult learners. These activities comprised short video-based learning modules, interactive lectures, Q&A sessions, experience sharing, hands-on workshops, mini project-based learning tasks, presentations, and site visit. Participants were encouraged to choose and attend hybrid sessions (both online and onsite) based on their interests and learning preferences. These activities were carefully mapped to the VARK model to ensure they addressed the diverse learning styles of the participants:

- **Visual Learners**: Benefited from short video-based learning and presentations that included visual aids like infographics and charts.
- **Auditory Learners**: Gained from lectures, Q&A sessions, and experience sharing, where they could listen to and engage in discussions.
- **Read/Write Learners**: Excelled with activities involving detailed written instructions and textual content, such as mini project-based learning and presentations.
- **Kinesthetic Learners**: Thrived in hands-on workshops and site visits, where they could engage directly with practical tasks and real-world environments.

Understanding which activities are most effective for our target group is crucial for the success of the Learning City Project. By collecting data on participants' experiences and satisfaction, we aim to determine which microlearning strategies are most suitable for reskilling Cost Managers. This will help us refine our approach and ensure that the educational programs are both effective and engaging, ultimately contributing to a more skilled and adaptable workforce in the restaurant industry.

3 Research Methodology

The research methodology for this study was designed to evaluate the effectiveness of microlearning activities in reskilling individuals for the role of Cost Manager in the restaurant technology domain. The methodology consisted of three main phases over a duration of 90 days: registration and online learning, participation in microlearning activities, and a final assessment through a mini project and survey.

Phase 1: Registration and Online Learning
During the first month, interested workers were invited to register for the boot camp. This initial phase included online learning through a Massive Open Online Course (MOOC) platform, which provided foundational knowledge in several key areas: Digital & AI Literacy, Leadership Skills, Problem-Solving, Strategic Thinking, and Business Mindset. Participants engaged with the online content at their own pace, ensuring they acquired the necessary background knowledge before moving on to the more intensive microlearning activities. At the end of this phase, participants were selected and interviewed based on their performance and commitment, resulting in a final group of 18 individuals who proceeded to the next phase (Fig. 1).

Phase 2: Microlearning Activities
In the final two months, the selected participants were involved in various microlearning activities designed to cater to different learning styles as per the VARK model. The activities included:

Fig. 1. Massive Open Online Course (MOOC) platform

- Short Video-based Learning: Providing concise, engaging content to introduce key concepts (Visual learners).
- Online Lectures: Delivered by subject matter experts to deepen understanding (Auditory learners).
- Q&A Sessions: Facilitating interaction and clarification of doubts (Auditory learners).
- Experience Sharing: Allowing participants to share their experiences and learn collaboratively (Read/Write and Auditory learners).
- Hands-on Workshops: Practical sessions to apply learned concepts (Kinesthetic learners).
- Site Visits: Real-world exposure to practical applications in restaurant settings (Kinesthetic learners).
- Mini Project-based Learning: Encouraging critical thinking and application of skills to solve real-world problems (Kinesthetic and Read/Write learners).
- Presentations: Participants presented their project findings, enhancing communication and presentation skills (Visual and Read/Write learners) (Fig. 2).

Learners were free to choose and engage in the activities they found most relevant and interesting without penalty, ensuring that their individual learning preferences and needs were met. To facilitate communication and track progress, the boot camp utilized a LINE Official Account platform. This platform allowed learners to receive news, notifications, and updates, connect with peers, and check their learning progress. To complete the boot camp, participants were required to attend more than 80% of the total activities.

Phase 3: Assessment and Survey

In the final part of the program, participants were required to complete a mini project

Fig. 2. Microlearning Activities

related to cost management at a real restaurant. This project allowed them to apply their acquired skills in a practical setting and culminated in a final presentation where they showcased their findings and solutions. After completing the program, a survey was conducted to evaluate participants' reactions and satisfaction with the microlearning activities. The survey used a Likert scale to measure satisfaction levels and included selfassessment questions to gauge the perceived improvement in skills and knowledge application in their professional roles.

4 Data Collection and Analysis

The data collection involved conducting a self-report survey using a 5-point Likert scale at the end of the program. Each of the microlearning activities was rated by participants in three aspects:

- Satisfaction: How satisfied the participants were with the activity.
- Knowledge Improvement: How much participants felt their knowledge improved as a result of the activity.
- Learning Experience: The overall quality of the learning experience, including engagement and applicability.

In addition to these ratings, participants provided self-assessment feedback on their perceived skill improvement and the application of knowledge in their professional roles. The survey results were analyzed to determine which microlearning activities were most effective and aligned with the learning preferences of the adult learners. This analysis aimed to identify the best practices for implementing microlearning in adult education programs, particularly in the context of reskilling for specific roles such as Cost Manager.

5 Result

See Table 1.

Table 1. The Evaluation of Micro Learning activities

Micro Learning Activity	Satisfactions	Knowledge	Learning Experience
Micro Video	4.29	4.00	4.07
Lecture (Online and Onsite)	4.47	4.27	4.33
Experience Sharing	4.40	4.40	4.27
Hand-on Workshop	4.13	4.20	4.27
Project-based Learning	4.20	4.07	4.00
Presentation	4.33	4.47	4.50
Site-Visit	4.79	4.64	4.71
Seminar Q&A session	4.50	4.50	4.50

6 Discussion

The findings from this research offer insightful perspectives into the effectiveness of various micro learning activities on learners' satisfaction, knowledge acquisition, and learning experience. The discussion will focus on the implications of these findings and possible recommendations for future practice.

1. ***Effectiveness of Site-Visit and Seminar q&A Sessions:***
   - Site-Visit activities consistently scored the highest across all evaluated categories—satisfaction, knowledge, and learning experience. This suggests that experiential learning, where learners are exposed to real-world environments and practical applications, significantly enhances their learning outcomes. The high engagement and contextual learning likely contribute to this effectiveness. [11].
   - Similarly, Seminar Q&A sessions also ranked high, emphasizing the importance of interactive learning environments. These sessions provide opportunities for learners to clarify doubts, engage in discussions, and deepen their understanding of the subject matter through direct interaction with experts. [6].

2. ***High Scores for Presentation Activities:***

   - Presentation activities scored highly in all categories too, indicating they are an effective method for knowledge dissemination and engagement. The active involvement required in preparing and delivering presentations likely contributes to higher retention of information and satisfaction among learners. [5].

3. ***Areas for Improvement: Hand-On Workshop and Micro Video:***

   - The relatively lower scores for **Hand-on Workshops** and **Micro Videos** highlight areas needing improvement. For Hand-on Workshops, despite their practical nature, the effectiveness may be hindered by factors such as insufficient hands-on opportunities or inadequate facilitation. Enhancing the quality and interactivity of these workshops could improve learner outcomes. [8].
   - **Micro Videos**, while convenient and accessible, may lack the depth and engagement provided by more interactive formats. To address this, incorporating interactive elements, such as embedded quizzes or discussion prompts, could make micro videos more effective [7].

3. ***Balanced Learning Approaches:***

   - The findings suggest that a balanced approach incorporating both interactive and experiential learning activities can enhance overall learning outcomes. While traditional lecture formats (both online and onsite) provide a solid foundation of knowledge, combining these with more engaging and practical activities like Site-visits and Q&A sessions can create a more comprehensive and effective learning experience [9].

4. ***Recommendations for Future Practice:***

   - Educators and trainers should conSiter integrating a variety of learning activities to cater to different learning styles and preferences. Emphasizing interactive and experiential learning can lead to higher satisfaction and better knowledge retention.
   - Continuous evaluation and feedback from learners can help in refining and improving the learning activities to meet their needs more effectively.
   - Future research could explore the specific elements within each type of activity that contribute most to their effectiveness, providing more granular insights for optimizing learning designs.

In conclusion, the research underscores the value of interactive and experiential learning activities in enhancing learner satisfaction, knowledge acquisition, and overall learning experience. By leveraging these insights, educational practitioners can design more effective and engaging learning programs that better meet the needs of their learners.

References

1. U.S. News & World Report: Best places to visit in Thailand (2023). https://travel.usnews.com/rankings/best-places-to-visit-in-thailand/

2. Huang, R.-Z.: Brand equity, tourist satisfaction, and travel intentions in a post-COVID-19 pandemic scenario: A case study of Thailand Unpublished manuscript (2023). https://typeset.io/papers/brand-equity-touristsatisfaction-and-travel-intentions-in-a-1cx26mrv
3. Ministry of Digital Economy and Society: Digital skill development report (2021)
4. Shoba, K.N.: Microlearning practices in blended educational contexts: a qualitative study. Educ. Administ.: Theory Pract. **30**(4), 3095–3101 (2024)
5. Hug, T.: Mobile learning as 'microlearning': Conceptual considerations towards enhancements of didactic thinking. Int. J. Mobile Blended Learn. **2**(4), 47–57 (2012)
6. Fleming, N.D., Mills, C.: Not another inventory, rather a catalyst for reflection. To Improve Academy **11**(1), 137–155 (1992)
7. SAP: SAP's microlearning strategy (2021). http://www.news.sap.com/2021/03/microlearning-strategy/
8. Open University: Professional development through microlearning (2020). http://www.open.ac.uk/news
9. SkillsFuture Singapore: SkillsFuture and microlearning (2021). http://www.skillsfuture.gov.sg
10. Gardner, H.: Frames of mind: The theory of multiple intelligences, 2nd ed. Basic Books (1983)
11. Kolb, D.A.: Experiential learning: Experience as the Source of Learning and Development. Prentice Hall (1984)

Lifecycle-Orientated Cost Estimation in the Early Phase of Factory Planning

Leonard Rieke(✉), Luca Philipp, Mehmet Demir, and Peter Nyhuis

Leibniz University Hannover, Hannover, Germany
rieke@ifa.uni-hannover.de

Abstract. As the number of variants increases and customer requirements grow, factory planning becomes more relevant in a dynamic market environment. Development and expansion planning, in particular, significantly impact companies' costs due to their high capital requirements and long lifetimes. Inaccurate estimates for the costs of factories can quickly lead to misinvestments and financial instability for companies. The interactions between factory configuration and the associated costs play a significant role in this context. Costs for planning projects must be estimated using a low-effort and reproducible procedure to avoid these misinvestments. With current methods and tools, this is not or only insufficiently possible. In preliminary work, requirements for cost estimation of factory planning projects have already been identified and prepared for a holistic approach. This article presents a methodological framework for a cost estimation approach. In a three-stage process, influences from an organizational perspective, the factory configuration and the related lifecycle-oriented cost estimation are considered. Considering different scenarios in the context of an outlook enables a final assessment of the opportunities and risks of different factory scenarios in the early planning phase.

Keywords: Factory Planning · Cost Estimate · Early Planning Phase · Life Cycle Costs · Scenario Design · Total Cost of Ownership

1 Introduction and State of the Art

Factory planning projects are highly relevant to a company's success due to their long lifetime and high capital requirements [1, 2]. An increasingly dynamic factory environment, rising customer requirements and ever shorter product lifecycles also lead to more significant challenges for factory planning [3, 4]. Different approaches of factory planning exist in science and practice. Various procedures have been consolidated in VDI Guideline 5200 [5] to systematize the various planning activities. This guideline divides the planning process into seven phases, which lead linearly and sequentially from "the rough to the fine" to a successively detailed planning result [2, 5]. The inter-disciplinary approach of synergetic factory planning has made a decisive contribution to the VDI 5200 guideline [2]. This approach links the planning phases of production planning

Published by Springer Nature Switzerland AG 2025
P. Sureephong et al. (Eds.): PLM 2024, IFIP AICT 740, pp. 123–139, 2025.
https://doi.org/10.1007/978-3-031-93319-6_10

(process view) with the service phases of object planning (spatial view) following the Fee Structure for Architects and Engineers [6]. The aim is to combine the planning and design of logistical and technological processes with the interior and exterior design of the factory from an architectural perspective to meet the interdisciplinary character of factory planning [2].

The high investment requirements of planning projects influence companies' liquidity, particularly in the case of cost-intensive development planning and factory expansions [1, 7]. However, in addition to investments in building and production technology [2, 8], operating costs make up the central part of the profitability of factories and their elements [2, 9, 10]. This is partly due to the long lifetime of factories [11] and the associated high-cost relevance for the operating phase of factories [12–14]. Nevertheless, investments are often used as a one-dimensional basis for decision-making, leading to incorrect cost estimates [15, 16]. Studies show that cost targets for factory planning projects are missed in 70 % of cases [17], whereby an average cost target miss of 37 % can be assumed for major projects [18]. SMEs often cannot build up the necessary resources and expertise in the calculating process, resulting in incorrect cost estimates and serious financial difficulties [19].

These points underline the relevance of an approach for holistic cost estimation of factory planning projects. A systematic literature review has already been published in preliminary work, the findings of which were used to derive the need for further research [20]. First of all, it was determined that current methods and tools do not allow for a holistic cost estimate or only to an insufficient extent. Furthermore, four central requirements were identified and prepared for a holistic approach (Fig. 1).

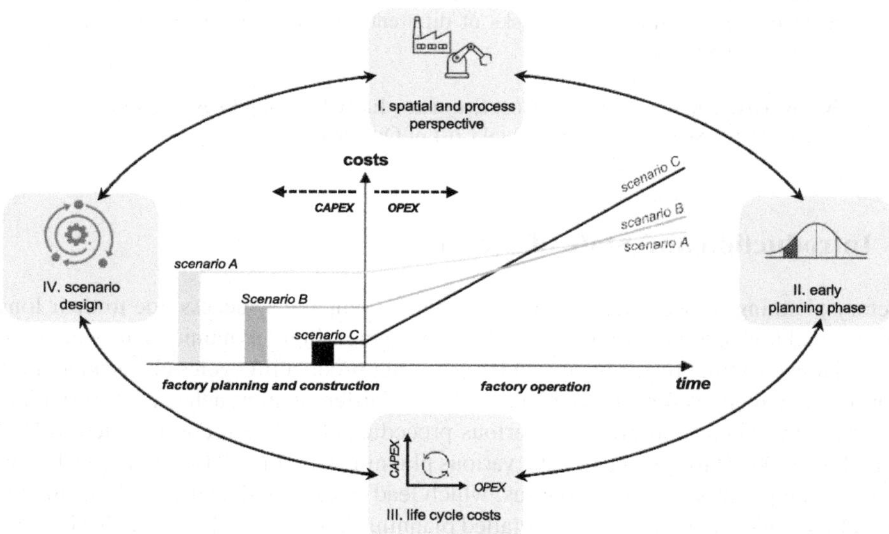

Fig. 1 Requirements for a holistic cost estimate [20]

Accordingly, a holistic approach must consider synergetic planning and the associated costs from a *spatial and process view*. This is due to the strong interactions between the different planning disciplines involved. In addition, the approach must start in the *early phase of factory planning*. This is because costs can only be sufficiently influenced in the early phase; in the subsequent phases, they are already fixed mainly within the developed concept. Furthermore, a *life cycle approach* is necessary due to the long lifetime of factories. Operating and usage costs (OPEX) comprise a large part of the total costs from a spatial and process view and must, therefore, also be included in a cost estimate alongside the investments (CAPEX). The fourth requirement is the *development of scenarios* for possible opportunities and risks. With the help of design influencing factors, e.g., the degree of automation, the factory can be configured individually and different cost scenarios can be compared [20]. In addition, an approach must meet standard requirements such as a high degree of standardization, practical relevance and comparatively low-effort and easy-to-understand applicability. The following step is to develop an approach based on these requirements. The methodological framework for this is explained below.

2 Holistic Approach to Lifecycle-Oriented Cost Estimation

Considering the state of the art and the preliminary work, a methodological framework (Fig. 2) was developed for holistic cost estimation in the early phase of factory planning, which is intended to fulfill the derived requirements. The overall procedure provides a three-stage process, briefly summarized here and detailed in the following chapters according to numbering in Fig. 2.

In the first step of the organizational analysis (Sect. 2.1), information relevant to the factory, such as the production program, sales forecasts and other general frame-work conditions as well as strategic requirements, are extracted. In addition, the focus is on identifying and analyzing design factors that significantly influence the factory design. In the second step, this information is used as a basis for the factory's configuration (Sect. 2.2). The focus is on the impact and requirements of the organizational analysis and the appropriate selection of factory elements. Since, following the philosophy of this article, factories are planned from the "inside out", the initial focus is on the process view. The strategic-organizational input information is used to plan production and logistics concepts and to select the resulting main dimensioning variables in the early planning phase, such as operating resources, personnel and area. In addition to the organizational analysis requirements, the process view's configuration places further requirements on the spatial view, which must be considered in the spatial factory configuration. Following the factory design, the third step of the method is the cost estimation (Sect. 2.3). In this step, the costs of the previously designed factory are determined from a process and spatial perspective and then aggregated in a life cycle model. The individual steps are described in the following chapters.

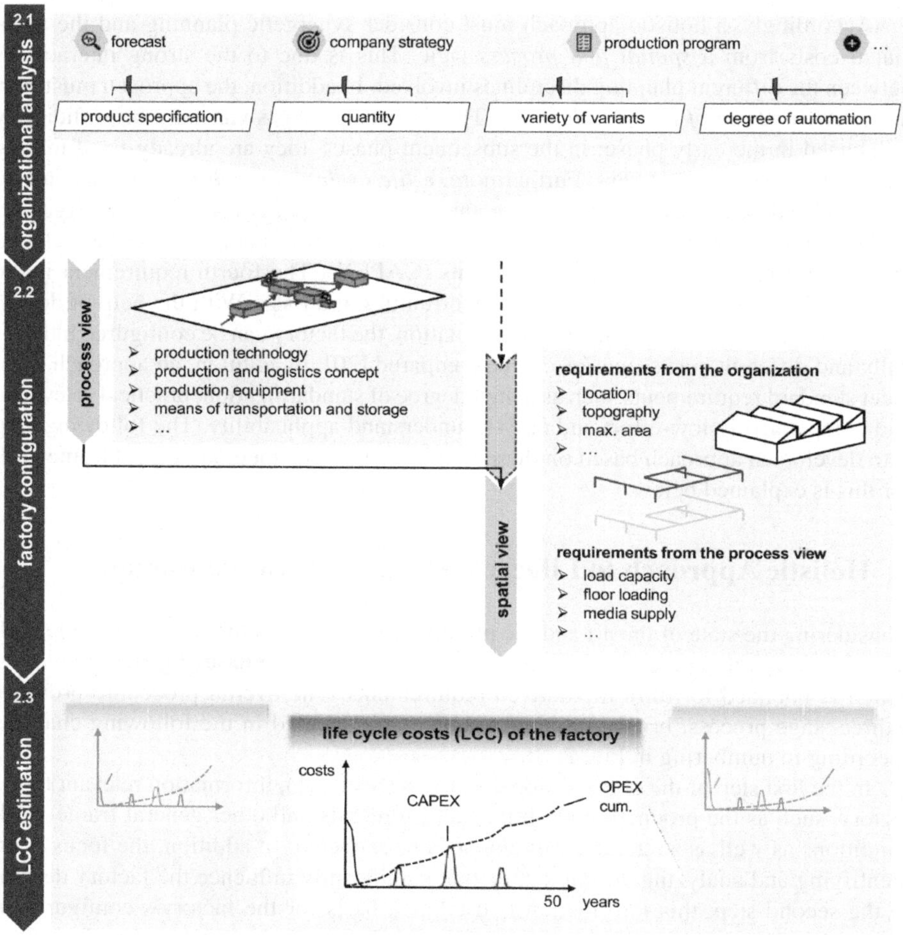

Fig. 2 Holistic approach to cost estimation in the early phase of factory planning

2.1 Organizational Analysis

Factories are roughly dimensioned early on in the planning process. The planned factory's vision, mission and strategic goals are developed as early as the setting of objectives stage, which is the first phase of the factory planning process according to the VDI 5200 [5] or synergetic factory planning [2]. Relevant information, such as the production program, product characteristics and sales forecasts, can be derived from the strategic orientation and the company and environment analysis, with each factory planning project being unique due to the different processes and products [2, 21]. In addition, this information can also be used to identify generally applicable factors influencing the factory configuration. In the context of this work, design influencing factors are factors that affect both the factory configuration and the economic efficiency of the planned factory. Influencing factors can generally be divided into internal and external factors [2,

22]. External factors such as material price increases influence costs but cannot be adequately controlled by factory planning. Conversely, internal factors can be influenced by planning so that they directly impact the factory configuration. For the internal factors, this article particularly emphasizes the *degree of automation*, the *quantity* (number of units), the *variety of variants* and the *product specification*. These influencing factors were identified through research and expert interviews, as they have an exceptionally high influence on the factory's configuration. In the context of this work, the four influencing factors are divided into three qualitative classes: low, medium and high (leads to the triangular shape in Fig. 3). This can impact other factors depending on the level of an influencing factor. The positioning of the influencing factors can result in a positioning corridor, as shown in Fig. 3. To create a uniform understanding and since the influencing factors are partly interrelated, they are explained in more detail below.

The *degree of automation* indicates the proportion of functions performed by automated resources in an object or system [23]. If there is a degree of automation of one, it can, therefore, be considered a fully automated object or system [24]. Conversely, the degree of automation can also be used to describe the manual processes in an object or system, resulting in the proportion of personnel required. The level of automation not only affects key performance indicators such as productivity or personnel requirements but can also be used in the design of production and logistics processes and, above all, in the selection of operating resources (see Sect. 2.2).

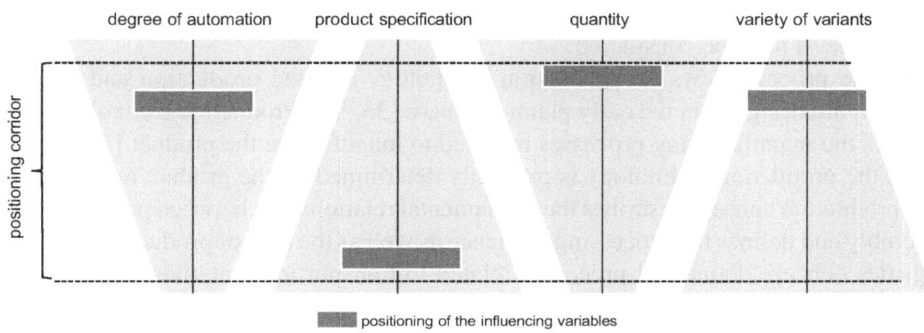

Fig. 3. Design influencing factors, characteristics and interactions

The *quantity* (number of units), as one of the central logistical variables in production control and planning, indicates the number of individual units of a product to be produced in a defined space. The level of automation correlates with the quantity, as comprehensive automation measures are often only economically profitable from a higher number of units [25, 26]. In addition to the pure quantity, the constancy or fluctuation of the quantity over a certain period plays an important role. The *variety of variants* as another central influencing factor reflects the number of possible combinations of different characteristics of a product type [27]. Irrespective of the variety of variants, the lifetime of the variants and the differentiation in the complexity of different variants are also relevant at this point. In contrast to the quantity, fewer variants favor more automated resources [26]. The *product specification* as an influencing factor in this work is essentially determined by the geometric and physical characteristics such as the size and weight of the products. These characteristics and features can be derived from the design data [28]. In addition to the pure size and weight of the products, the complexity and quality requirements must also be considered. This factor can also be used as a significant criterion for selecting and designing the degree of automation [29, 30]. The degree of automation is exceptionally high in production types with medium or large quantities and products with small to medium-sized product specifications [31].

By positioning the design influencing factors, the factory can be configured using the individual elements of the factory (so-called factory elements or factory objects). Factory objects represent the individual elements of a factory, which is too complex and can only be made assessable by breaking it down into individual objects [2, 32]. In the course of the overall objective of cost estimation, these elements are referred to below as cost elements of the factory. The methodical factory configuration of individual factory elements based on the influencing factors is explained below.

2.2 Factory Configuration

The information from the organizational analysis, including the design influencing factors, serves as input information for the factory configuration process step, as shown in Fig. 2. The factory configuration refers to the configuration of the logistical and technological processes from a process perspective and the interior and exterior design configuration from a spatial perspective. In particular, the strong interactions between the two views must be considered.

In the process view, the production technology and the production and logistics concepts are designed in the early planning phase [33, 34]. Production technology summarizes the manufacturing processes required to manufacture the product [2]. Accordingly, the production technology is primarily determined by the product requirements. The production concept describes the fundamental relationships between production and assembly and defines the processing sequence as well as the type of production [32]. The logistics concept defines all processes related to transporting materials, semi-finished products or storing objects [22]. In combination with the influencing factors, these three conceptual planning objects result in the cost elements of production, transportation and storage equipment, which essentially make up the technical design of the factory. In order to derive a reliable configuration of the factory at an early planning stage, the cost elements mentioned above can be divided into superordinate classes based on the design influencing factors. For example, means of transportation are sub-divided into continuous and discontinuous conveyors (this is repeatedly used as a central example in this article for better understanding) [30]. These classes can be assigned characteristics that differ in terms of the associated life cycle costs (see subchapter 2.3). However, the choice of equipment not only influences the factory's life cycle costs but also places requirements on the spatial view. Although the configuration of the spatial view can be started simultaneously with the process view, it is essential to consider the requirements resulting from the process view and the requirements from the organizational analysis. General requirements resulting from the organization are, for example, the location, the topography, or the maximum area available for planning the factory.

In addition, the need for storage space can already be significantly reduced, for example, by defining the order strategy as a make-to-order strategy [2]. In addition to these general requirements, further specific requirements arise from the design of the process view. For example, the loads from the process perspective must be considered when designing the floor or the supporting structure. The selection of an automated conveyor or production equipment can lead to increased floor loads due to high masses or dynamic forces [35], which must be implemented through spatial configuration. Figure 4 shows the exemplary relationship between the design influencing factors and the process-related and spatial configuration of the factory with the help of nine-field matrices, which result from the comparison of the influencing factors. Furthermore, specific characteristics and requirements can be attributed to the fields. Consequently, the combination of a high quantity and low product specification increases the suitability for a higher degree of automation in the process view. A high degree of automation or a high product specification leads to increased requirements in the spatial view, such as increasing floor loads or media supply requirements.

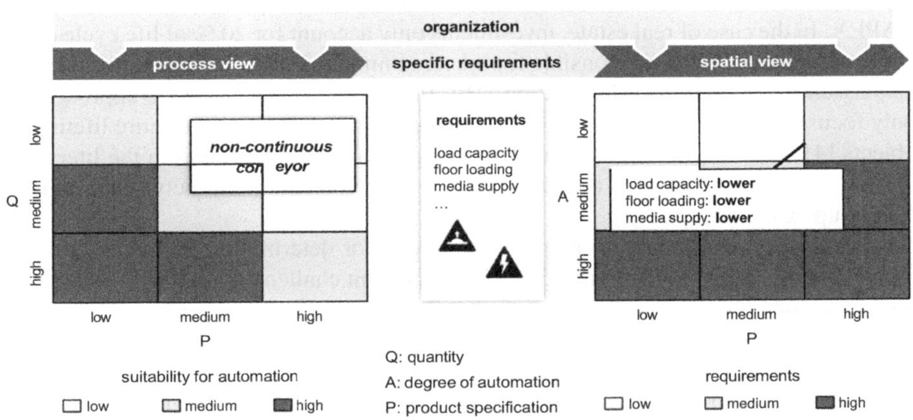

Fig. 4 Process and spatial factory configuration, exemplary interactions

These generally applicable effects can differ from case to case. In addition to the effects on life cycle costs, the requirements for the spatial view, in particular, can be derived from the process view. The fields of the spatial view contain the realization information of these requirements, i.e., the spatial conditions that must be provided for the process view. Figure 4 shows an example of the interaction for the means of transportation element. Unlike continuous conveyors, discontinuous conveyors are generally characterized by greater flexibility and suitability for use with high product specifications, such as heavy products with large dimensions [30]. In contrast to continuous conveyors, which often require complex constructions of the floor or the supporting structure, discontinuous conveyors have lower requirements regarding floor load, load capacity and media supply. This is not least due to the higher average degree of automation of continuous conveyors compared to discontinuous conveyors. Similar correlations also occur when deciding between production equipment, such as automated transfer lines and manual hand tools. Due to their weight and media requirements, automated systems place much higher requirements on the building and the technical building equipment. Slight changes in the requirements can significantly impact the process and associated spatial configurations. This directly impacts the associated costs, which will be discussed in more detail in the next step.

2.3 Lifecycle-Oriented Cost Estimation

To avoid financial risks, the life cycle costs of planned factories must be estimated as accurately as possible. Besides investments, which are often the only basis for decision-making, life cycle costs also consider operating and utilization costs. Capital expenditure (CAPEX) generally refers to the use of capital and the long-term commitment of financial resources to assets [36]. Operating costs (OPEX), which are primarily incurred during factory operation, include costs for energy, materials, personnel or maintenance, for example, and serve to maintain operational capability [37, 38]. Due to the long lifetime of factories (40-50 years), factory operation accounts for the largest share of the life cycle [12, 13], meaning that the share of OPEX in total costs is much higher than the share of CAPEX. In the case of real estate, investments only account for 20 % of life cycle costs, while operating costs are responsible for the remaining 80% [39, 40]. Machine tools can also assume a similar cost distribution [16]. Furthermore, the life cycle approach not only focuses on a specific interval but also considers the costs over the entire lifetime of objects [41]. Although various terms for life cycle costs can be found in the literature, they all share one thing: costs are added up to represent the trade-off between initial and follow-up costs [42].

The early phase of factory planning is crucial for determining life cycle costs. In this phase, the factory planner is faced with significant challenges. On the one hand, the project's scope is determined at this stage [1, 7]. On the other hand, such decisions with enormous implications at this stage are only based on a limited amount of information [43]. In this article, the early phase of factory planning, based on synergetic factory planning, is defined from setting objectives and establishing the project basis up to and including concept planning. Alternatively, a phase 0 of the HOAI [6] or feasibility study [44] can also be considered part of the early planning phase. During this early phase, there is still an opportunity to influence the factory's life cycle costs in the long term [45–47]. The more advanced the planning process, the higher the concept's detail level and the more this possibility decreases [2, 5]. Consequently, costs can only be influenced within the existing concept. Due to the low information availability at an early planning stage, costs cannot be calculated fully, but a reliable cost estimate is required. DIN 276 defines cost estimation as determining costs based on preliminary planning, which supports decision-making in construction planning [48]. According to WIENDAHL, the principle of completeness before accuracy applies to cost estimation [2]. To do justice to this approach, a holistic cost estimate from a process and spatial perspective is required in the early phase. This is the only way to ensure that a reliable statement can be made about the factory's life cycle costs.

The factory configuration (Sect. 2.2) considers the effects and requirements of the organizational analysis (Sect. 2.1). It is determined which selection of factory elements is suitable and what effects the process design, in particular, has on the spatial design. Take continuous conveyors as an example; they are easier to automate but also have higher average requirements in terms of building services and the building. The subsequent cost estimate now evaluates this factory configuration.

A factory cost estimate based on the respective factory's configuration in the early phase is not easily possible with little effort due to its complexity and uncertainty, so various limitations must be made for the process and spatial view. To narrow down the object of consideration from a process perspective, the cost types to be considered are first identified. For this purpose, an evaluation of the Federal Statistical Office is used [49]. This evaluation shows, among other things, the proportion of costs in the manufacturing industry in Germany to the gross production value. As shown in Fig. 5, material, energy, personnel and depreciation costs account for around 75–80% of these operating costs. Accordingly, from a process perspective, the calculation logic focuses on these cost types in addition to the investments, i.e., the acquisition costs of the cost elements.

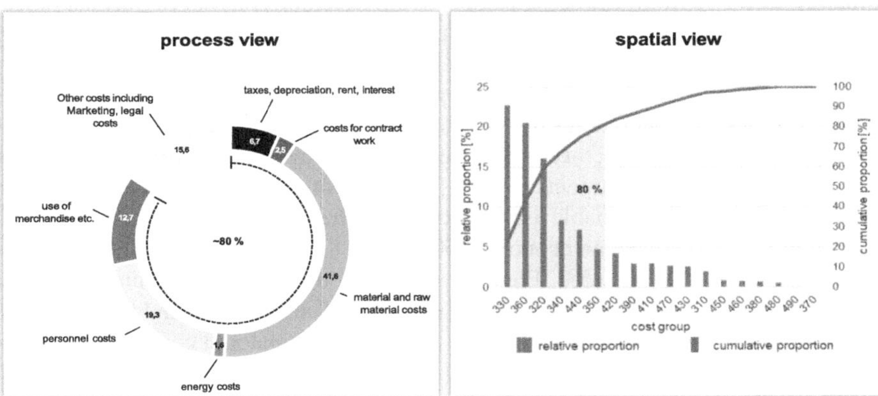

Fig. 5 Systematic limitation of the process and spatial view to relevant cost elements

As described in Sect. 1, the spatial view considers all factory elements from an architectural perspective. This field of consideration is too complex for an initial low-effort cost estimate, so limitations must also be defined. DIN 276 classifies investments from a spatial perspective into eight first-order cost groups, which can be broken down into second and third-order cost groups [48]. The information center for construction costs (BKI) publishes an annual report in which these cost groups are used to present investments made in new buildings [50]. For this purpose, the report provides so-called cost parameters for the first and second-order cost groups based on data collected and statistical evaluations from the previous year. Furthermore, the price development of relevant materials such as steel and wood should be monitored and forecast. Since, according to the information available, cost groups 300 (building – building constructions) and 400 (building – technical building equipment) are particularly relevant in the early planning phase [51], this approach is initially limited to these cost groups. Furthermore, the scope can be specified by analyzing the second level of these cost groups and identifying cost-intensive groups. Here, the second-level cost groups, such as 310, 320, and 330, are analyzed in terms of their percentage share of the total costs of cost groups 300 and 400. The data basis here comes from eight different building types from the BKI, limited to the direct factory area from both a production and logistics perspective. The analysis shows that six of the 18 cost groups in the second level already account for 80% of the total investment costs, so these are considered a suitable focus for further procedure (see Fig. 5).

Once the scope has been successfully narrowed down and the complexity of the procedure reduced, calculation logic for cost estimation can be identified and developed. As can be seen from the delimitation of the process view, in addition to the investments, the personnel, material and energy costs as well as the depreciation should be considered for cost estimation. To model the calculation logic, it is essential to consider the dependency of OPEX on the period under consideration. A life cycle-oriented cost estimate can be ensured only by linking costs and lifetime. The observation period is also essential for estimating CAPEX from a process perspective. A factory consists of various objects with different lifetimes [2]. Therefore, it is possible, that some elements involved in the production process need to be replaced several times during the life cycle of the factory, while other elements do not need to be replaced or only need to be replaced a small number of times. In concrete terms, this means that if the period under consideration exceeds the lifetime of a factory object, replacement investments will be required in addition to the initial investments [33, 52]. A table of the German Federal Ministry can be used to make a reasonable assumption about the lifetime of individual elements. For example, forklifts have an economic lifetime of 8 years, whereas overhead cranes have a lifetime of 15 years [53]. Figure 6 illustrates this using an example life cycle and the associated costs of a factory and its elements. The operating and followup costs increase continuously due to rising energy, personnel and material costs and are shown cumulatively.

In this example, after a high initial investment, subsequent investments are required, for example, in production, transportation or storage equipment. The investments are shown as sudden costs. Suppose the example is based on a machine with a service life of 10 years. In simplified terms, this would result in three further replacement purchases over the life cycle of 40 years in addition to the initial investment, each of which would lead to certain operating cost distributions depending on the design. Instead of considering the entire life cycle, there are often fixed periods in which investments must be amortized.

To holistically determine such life cycle costs, they can first be divided into process and spatial aspects. Figure 7 shows a schematic procedure for estimating life cycle costs. From a process perspective, the costs can be determined by aggregation at the respective factory levels, according to WIENDAHL [2]. At the lowest level, the station, the costs of the operating resources in the individual structural units (production, assembly, logistics) are determined first. By successively aggregating these costs to the factory level above, the aim is to ensure that all costs are systematically recorded. In addition to investments in the cost elements of process and production technology, input data on consumption data can be derived from machine data sheets. Necessary input parameters are, for

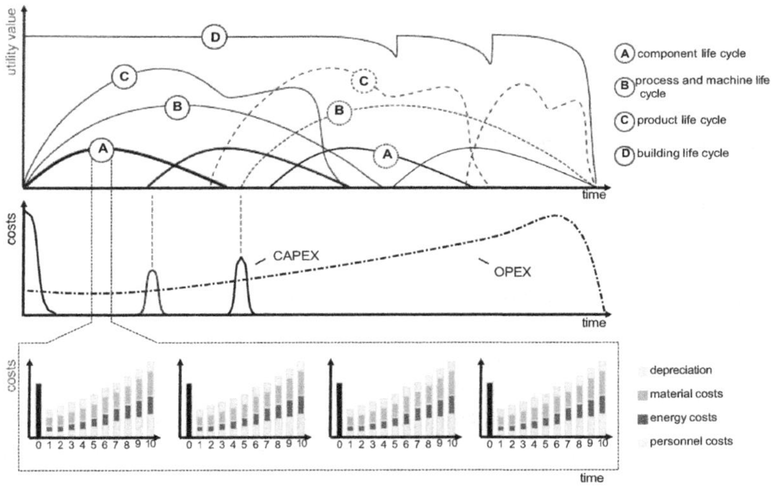

Fig. 6 Exemplary factory life cycle and associated costs of the elements based on [20, 54, 55]

example, the connected load, the drive medium, the consumption and the number of days of use. If information is missing, historical data or expert estimates can be used. Software support, such as an Excel tool with the current prices of the energy sources, can be used to calculate the consumption with the respective costs of the energy sources. Any unconsidered costs due to the simplification of operating costs to 80% are subject to a cost surcharge. In addition, an uncertainty factor can be added (see formula Fig. 7).

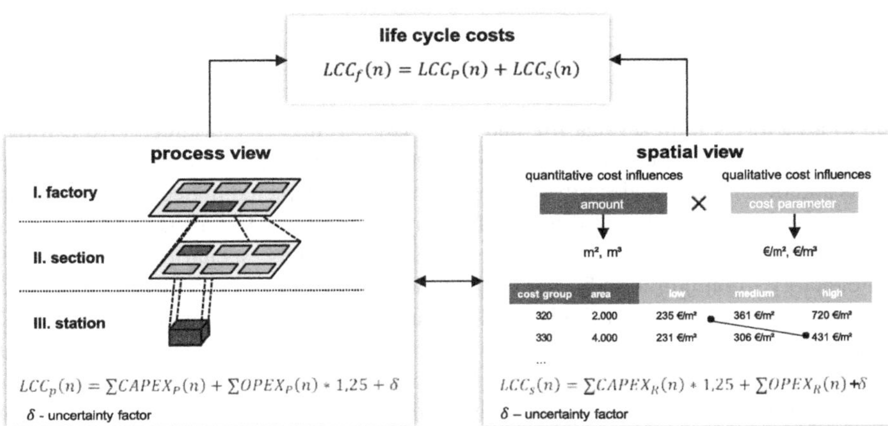

Fig. 7 Schematic calculation logic for cost estimation in the life cycle of a factory

In addition to the process view, a separate calculation logic must be developed for the spatial view. Quantitative and qualitative cost influences can be derived from the literature [56]. Quantity units of m2 and m3 represent the quantitative cost influences and result from the dimensioning of the factory. The explicit allocation of areas can be carried out following DIN 277 [57]. Due to the early stage, an exact determination of the area is impossible [56]. Methods such as the functional area determination or the substitute area method can be used for the directionally reliable determination [2]. Cost parameters represent the qualitative cost influences. Generally, cost parameters can be divided into synthetic and analytical [58]. While synthetic values are derived by combining or aggregating individual or partial values of a more extensive system, analytical values are based on historical data and offer better transparency and completeness [58]. The analytical values described last are published in the BKI's annual report [50], so they can be used as qualitative cost influences for determining investments from a spatial perspective. A procedure consisting of quantitative and qualitative cost influences can also be used to determine the utilization costs from a spatial perspective. For the qualitative cost parameters, for example, data from the fm.benchmarking report [40], which, like the BKI report, is based on historical data, can be used. Systematization from an investment perspective is based on the DIN 276 [48], whereas DIN 18960 is used as the basis for usage costs [48, 59]. For a higher degree of accuracy, depending on the building type, the individual cost group can be used directly instead of applying a flat-rate value for the entire area. Due to the simplification to 80 %, the unconsidered investments are added back with a cost surcharge. It is also possible to add an uncertainty surcharge to the investments, which at 25-30 % represents an average value in practice [60–62].

In summary, it can be stated that a comprehensive cost estimate from a process and spatial perspective is currently not possible or only possible to an insufficient extent. The complexity can initially be reduced by narrowing down the field of consideration. The initial holistic calculation procedure framework can be developed by introducing different schematic calculation logics for the process and the spatial view from the investment and operating cost perspective.

3 Scenario Design

Due to a factory's long service life and high costs, a life cycle-oriented cost estimate must also include different influences and scenarios so that potential financial risks for clients and owners can be described and anticipated. As depicted in Sect. 2.1, the design influencing factors can be used to lay the foundations for different internal scenarios in the early planning phase. It must be emphasized that these scenarios are not limited to the factory's configuration but directly impact the factory's cost structure. Individual design variations in the factory's elements can already significantly impact later life cycle costs – the simplified example of a means of transportation underlines this. Continuous conveyors are characterized by continuous transport and a higher degree of automation than discontinuous conveyors [63, 64]. Due to the higher degree of automation, a continuous conveyor requires a higher investment but lower operating costs over the life cycle [65]. From a process perspective, the increasing degree of automation, using the continuous conveyor as an example, directly impacts the factory's spatial design. Increased demands on the floor level, load capacity and the media supply directly lead to higher characteristic values in the cost estimate. In the case of discontinuous conveyors, higher operating costs can be assumed due to the significantly higher personnel requirements.

Simultaneously, the initial investment is lower due to the lower degree of automation [66].

In addition to internal influences, companies are also exposed to external factors that significantly impact life cycle costs, depending on the probability of occurrence and the extent of damage. External factors include demographic change and the resulting shortage of skilled workers, increasing pressure from sustainability objectives and the energy crisis [2, 14, 67, 68]. As a result of these factors, it is possible that the values assumed for the cost estimate to determine personnel or energy costs may change over the life cycle, increasing the probability of cost-related deviations. However, transparency about rising personnel costs can also lead to an opportunity for companies. Early investment in appropriate automation can provide a remedy here. The risk of sudden cost deviations can be counteracted with statistical methods or scenario analyses. A standard method for scenario planning is the three-point estimate [69]. Internal and external factors can be taken into account here. The modeling considers an optimistic (best case), pessimistic (worst case) and probable value (most likely case) by assuming the effects of the individual factors [70]. This approach can be used for personnel or material costs, for example, to better prepare for long-term risks.

4 Conclusion and Outlook

A long service life and high investments characterize factory planning projects. Misinvestments can have a considerable economic impact on companies. Apart from the investments, the operational and usage expenses specifically form a significant portion of the total costs. Life cycle-oriented cost considerations can provide a remedy here. Especially in the early planning phase, the life cycle costs of a factory can still be sufficiently influenced. Current methods and tools cannot or only insufficiently estimate the life cycle costs of factories methodically. This article, therefore, introduces a methodological framework as a decision support tool that enables systematic life cycle-oriented factory configuration and cost estimation. The method presented is based on the synergetic consideration of the process and spatial view as well as the consideration of investments and operating costs and their interrelationships. The approach is divided into three steps. First, framework conditions, input parameters and design influencing factors are derived in the organizational analysis. The focus is then placed on the requirements for the factory design from a spatial and process perspective as well as the appropriate selection of cost elements. Finally, the cost effects and constellations of the designed factory are evaluated in a schematic life cycle-oriented cost estimate. To reduce complexity, the object of consideration in this method was initially limited to the most cost-intensive elements of the factory. In addition, initial approaches explain how internal and external influencing factors lead to different factory scenarios and cost characteristics and how opportunities and risks can be countered in the early planning phase.

The method presented is initially a methodological approach that must be developed further. Model theory can support a systematic approach at this point. A description model could present the factory transparently in the early planning phase. Developing an impact model could show the relevant interactions between the spatial and process views as well as investments and operating costs. Building on this, an evaluation model could contribute to the systematic monetary assessment of the factory scenarios. At this point, adequate calculation logic for individual cost types and groups in the life cycle must be identified and developed. For a concrete application, the individual process steps must be further detailed and summarized in a process model to use the approach in a low-effort and reproducible manner. Moreover, it can be considered to look closely at the interactions between changeability and automation. This applies particularly under the hypothesis that the ability to change decreases with an increasing degree of automation.

References

1. Grundig, C.-G.: Fabrikplanung: Planungssystematik – Methoden – Anwendungen, 7, aktualisierte, Auflage Carl Hanser Verlag GmbH & Co. KG, München, Germany (2021)
2. Wiendahl, H.-H., Reichardt, J., Nyhuis, P.: Handbuch Fabrikplanung: Konzept, Gestaltung und Umsetzung wandlungsfähiger Produktionsstätten, 3, vollständig überarbeitete, Auflage Hanser, München (2024)
3. Lanza, G., et al.: Wandlungsfähige, menschzentrierte Strukturen in Fabriken und Netzwerken der Industrie 4.0. (acatech Studie). Herbert Utz Verlage, München (2018)
4. Scholz-Reiter, B., Freitag, M.: Autonomous processes in assembly systems. CIRP Ann. **56**(2), 712–729 (2007)
5. Verein Deutscher Ingenieure e. V. Fabrikplanung- Planungsvorgehen. Beuth-Verlag, Düsseldorf, Berlin (2011).
6. HOAI: HOAI – Honorarordnung für Architekten und Ingenieure/official scale of fees for services by architects and engineers, Deutschland (2021).
7. Kettner, H., Schmidt, J., Greim, H.-R.: Leitfaden der systematischen Fabrikplanung: Mit zahlreichen Checklisten. Hanser, München (2010)
8. Dombrowski, U., Hennersdorf, S., Palluck, M.: Die wirtschaftliche Bedeutung der Fabrikplanung am Standort Deutschland. Zeitschrift für wirtschaftlichen Fabrikbetrieb **102**(1–2), 14–18 (2007)
9. Schenk, M. (ed.): Produktion und Logistik Mit Zukunft: Digital Engineering and Operation. Springer, Berlin/Heidelberg (2015)
10. Niemann, J., Westkämper, E.: Investitionskosten versus Betriebskosten – Nicht Preisschild sondern Lebenslaufkosten entscheiden über den Produktnutzen. wt Werkstattstechnik online **94**(3), 43–47 (2004).
11. Pawellek, G.: Ganzheitliche Fabrikplanung: Grundlagen, Vorgehensweise, EDV- Unterstützung, 2, Auflage Springer Vieweg, Berlin (2014)
12. Herrmann, C., Kara, S., Thiede, S.: Dynamic life cycle costing based on lifetime prediction. Int. J. Sustain. Eng. **4**(3), 224–235 (2011)
13. Gebler, M., Cerdas, J.F., Thiede, S., Herrmann, C.: Life cycle assessment of an automotive factory: identifying challenges for the decarbonization of automotive production – A case study. J. Cleaner Prod. **270**, 122330 (2020)
14. Abele, E., Dervisopoulos, M., Kuhrke, B.: Bedeutung und Anwendung von Lebenszyklusanalysen bei Werkzeugmaschinen. In: Schweiger, S. (ed.) Lebenszykluskosten optimieren, pp. 51–80. Gabler, Wiesbaden (2009)
15. Herrmann, C.: Ganzheitliches Life Cycle Management – Nachhaltigkeit und Lebenszyklusorientierung in Unternehmen. Springer, Berlin Heidelberg, Berlin, Heidelberg (2010)
16. Mattes, K., Schröter, M.: Kurzstudie Wirtschaftlichkeitsbewertung: Bewertung der wirtschaftlichen Potenziale von energieeffizienten Anlagen und Maschinen, Kurzstudie. Fraunhofer-Institut für System- und Innovationsforschung ISI (2012).

17. Reinema, C., Pompe, A., Nyhuis, P.: Agiles Projektmanagement: Einsatzpotenziale und Handlungsbedarfe im Rahmen von Fabrikplanungsprojekten. Zeitschrift für wirtschaftlichen Fabrikbetrieb **108**(3), 113–117 (2013). https://doi.org/10.3139/104.110903
18. McKinsey&Company: The art of project leadership: Delivering the world´s largest projetcs (2017).
19. Rautenstrauch, T., Müller, C.: Investitionscontrolling in kleinen und mittleren Unternehmen (KMU). Z Control Manag. **50**(2), 100–105 (2006)
20. Rieke, L., Philipp, L., Nyhuis, P.: Towards a holistic cost estimate of factory planning projects. In: Herberger, D., Hübner, M. (eds.) Proceedings of the Conference on Production Systems and Logistics: CPSL 2023 – 2. Hannover : publishing., pp. 459–793 (2023).
21. Wirth, S., Schenk, M., Müller, E.: FABRIKPLANUNG: Fabrikarten, Fabriktypen und ihre Entwicklungsetappen. Zeitschrift für wirtschaftlichen Fabrikbetrieb **106**(11), 799–802 (2011). https://doi.org/10.1515/zwf-2011-1061107
22. Burggräf, P. (ed.): Fabrikplanung: Handbuch Produktion und Management 4, 2. Auflage, 2021st edn. Springer, Berlin, Berlin (2021)
23. Deutsches Institut für Normung e.V., 2014. DIN IEC 60050-351 – Internationales Elektrotechnisches Wörterbuch – Internationales Elektrotechnisches Wörterbuch – Teil 351: Leittechnik. Beuth Verlag GmbH, Berlin
24. Milberg, J., Reinhart, G.: Produktionssystemplanung. In: Eversheim, W., Schuh, G. (eds.) Produktion und Management 3. Gestaltung von Produktionssystemen. Springer Berlin Heidelberg, Berlin, Heidelberg, pp. 118–244 (1999).
25. Westkämper, E.: Einführung in die Fertigungstechnik, 8, aktualisierte und erweiterte, Auflage Vieweg+Teubner, Wiesbaden (2010)
26. Holtewert, P., Wiendahl, H.-H.: Fertigungs- und Montagesysteme, In: Bauernhansl, T. (ed.), Fabrikbetriebslehre 1. Management in der Produktion, 1. Aufl. 2020 ed. Springer Vieweg, Berlin, Heidelberg, pp. 129–164 (2020).
27. REFA – Verband für Arbeitsstudien und Betriebsorganisation, 1985. Methodenlehre der Planung und Steuerung, [4. Aufl.] ed. Hanser, München.
28. Möhrle, M.: Gestaltung von Fabrikstrukturen für die additive Fertigung. Springer Vieweg, Berlin, Heidelberg (2018)
29. Hagedorn, J., Sell-Le Blanc, F., Fleischer, J.: Handbuch der Wickeltechnik für hocheffiziente Spulen und Motoren: Ein Beitrag zur Energieeffizienz. Springer Vieweg, Berlin, Heidelberg (2016)
30. Hompel, M. ten, Schmidt, T., Dregger, J.: Materialflusssysteme: Förder- und Lagertechnik, 4. Auflage 2018 ed. Springer Berlin Heidelberg, Berlin, Heidelberg (2018).
31. Schmitt, R.: Selbstoptimierende Produktionssysteme. In: Brecher, C. (ed.) Integrative Produktionstechnik für Hochlohnländer, 1. Aufl. ed. Springer, Berlin, Heidelberg, pp. 774–1058 (2011).
32. Heger, C.L.: Bewertung der Wandlungsfähigkeit von Fabrikobjekten. Zugl.: Hannover, Univ., Diss., 2006. PZH, Produktionstechn. Zentrum, Garbsen (2007).
33. Hingst, L., Dér, A., Herrmann, C., Nyhuis, P.: Towards a holistic life cycle costing and assessment of factories: qualitative modeling of interdependencies in factory systems. Sustainability **15**(5), 4478 (2023)
34. Dér, A., Gabrisch, C., Kaluza, A., Cerdas, F., Thiede, S., Herrmann, C.: Integrating environmental impact targets in early phases of production planning for lightweight structures. Procedia CIRP **80**, 168–173 (2019)
35. Helbing, K., Mund, H., Reichel, M.: Handbuch Fabrikprojektierung, 2., aktualisierte und ergänzte Auflage ed. Springer Vieweg, Berlin, Heidelberg (2018).
36. Weber, W., Kabst, R., Baum, M.: Einführung in die Betriebswirtschaftslehre. Springer Fachmedien Wiesbaden, Wiesbaden (2018)

37. Müller, D.: Betriebswirtschaftslehre für Ingenieure, 3, wesentlich überarbeitete, Auflage Springer Gabler, Berlin, Heidelberg (2020)
38. Helmus, F.P.: Anlagenplanung: Von der Anfrage bis zur Abnahme, 1, Auflage, neue, Ausg Wiley-VCH, Weinheim, Online-Ressource (2009)
39. Steinmetz, F., Gürtler, V.: Lebenszyklusmanagement/ PPP an Hochschulen: Ansätze im Hochschulbau. HIS Hannover (2008).
40. Rotermund, T., Weiland, C.: fm.benchmarking Bericht 2022 – Der FM- Kennzahlenvergleich (2022). Accessed 13 January 2024.
41. Verband Deutscher Maschinen- und Anlagenbau e.V. (VDMA): Prognosemodell für die Lebenszykluskosten von Maschinen und Anlagen. Beuth-Verlag, Berlin (2006)
42. Pelzeter, A.: Lebenszykluskosten von Immobilien: Schriften zur Immobilienökonomie, European Business School (2006).
43. Unzeitig, W.: Methodik zur frühen Fabrikplanung bei Unsicherheiten. Dissertation, Graz (2014).
44. Cevirgen, C.: Methodik zur Durchführung typorientierter Vorstudien in der Fabrikplanung. Dissertation, Erstausgabe TEWISS, Garbsen (2023)
45. Gairola, A.: Montagegerechtes Konstruieren ein Beitrag zur Konstruktionsmethodik, Dissertation (1981).
46. Zehbold, C.: Lebenszykluskostenrechnung. Gabler, Wiesbaden (1996)
47. Chou, J.-S.: Web-based CBR system applied to early cost budgeting for pavement maintenance project. Expert Syst. Appl. **36**(2), 2947–2960 (2009)
48. Deutsches Institut für Normung e.V. DIN 276 – Kosten im Bauwesen. Beuth-Verlag, Berlin (2018)
49. Statistisches Bundesamt (Destatis): Produzierendes Gewerbe. Kostenstruktur der Rechtlichen Einheiten des Verarbeitenden Gewerbes (2018).
50. Baukosteninformationszentrum Deutscher Architektenkammern: BKI-Baukosten 2020 Neubau: Statistische Kostenkennwerte. BKI, Stuttgart (2020).
51. Kalusche, W., Herkel, S. (eds.): BKI Baukosten 2018 Neubau. BKI Baukosteninformationszentrum, Stuttgart (2018)
52. Albach, H.: Allgemeine Betriebswirtschaftslehre: Einführung, 3, überarbeitete, Auflage Gabler Verlag, Wiesbaden (2001)
53. Bundesministerium der Finanzen: AfA-Tabelle für die allgemein verwendbaren Anlagegüter AfA-Tabelle "AV"). Bundesministerium der Finanzen (2019).
54. Nielsen, L., Schmidt, C., Blume, S., Schmidt, M., Thiede, S., Nyhuis, P., Herrmann, C.: Towards quantitative factory life cycle evaluation. Procedia CIRP **55**, 266–271 (2016)
55. Wirth, S., Enderlein, H., Petermann, J.: Kompetenznetze der Produktion. In: Vortragsband IBF-Fachtagung "Vernetzt planen und produzieren" (2020)
56. Kalusche, W. (ed.): Handbuch Kostenplanung im Hochbau, 4. komplett überarbeitete, Auflage BKI, Stuttgart (2021)
57. Deutsches Institut für Normung e.V. DIN 277 – Grundflächen und Rauminhalte im Bauwesen. Beuth-Verlag, Berlin (2016).
58. Stoy, C.: Kostenermittlung in frühen Projektphasen – Ein Verfahren auf der Grundlage von Planungs- und Kostenkennwerten. Bautechnik **78**(12), 882–888 (2001)
59. Deutsches Institut für Normung e.V.: DIN 18960 – Nutzungskosten im Hochbau. Beuth-Verlag, Berlin (2020).
60. Manteufel, T., Frechen, F., Rodemann, T., Heinzerling, K., Wagner, K.: Der Bauprozess: Prozessuale und materielle Probleme des zivilen Bauprozesses, 18, umfassend überarbeitete, Auflage Werner Verlag, Hürth (2023)
61. Keldungs, K.-H., Baldringer, S. (eds.): Architektenrecht: Praxishandbuch zu Honorar und Haftung, 7, Auflage Werner Verlag; Wolters Kluwer Deutschland GmbH, Köln, Hürth (2021)

62. Siemon, K.D.: Baukosten bei Neu- und Umbauten. Vieweg+Teubner Verlag, Wiesbaden (2012)
63. Martin, H.: Transport- und Lagerlogistik: Systematik, Planung, Einsatz und Wirtschaftlichkeit, 10. Aufl. 2017 ed. Springer Vieweg, Wiesbaden (2016).
64. Heiserich, O.-E.: Logistik: Eine praxisorientierte Einführung, 4, vollständig überarbeitete und erweiterte, Auflage Gabler Verlag/Springer Fachmedien Wiesbaden, Wiesbaden (2011)
65. Lotter, B., Wiendahl, H.-P. (eds.): Montage in der industriellen Produktion. Springer, Berlin (2006)
66. Wehking, K.-H., et al.: Technisches Handbuch Logistik: Fördertechnik, Materialfluss, Intralogistik. Springer Vieweg, Berlin, Heidelberg (2020).
67. Kraus, W., Bauernhansl, T.: Wie Automatisierung die Zukunft der Produktion verändern wird. Zeitschrift für wirtschaftlichen Fabrikbetrieb **116**(10), 652–656 (2021)
68. Denkena, B., Wichmann, M., Kettelmann, S.: Prozesskettenplanung unter Unsicherheit/Process chain planning under uncertainty – Consideration of internal and external uncertainties during planning of process chains across companies. wt Werkstattstechnik online **113**(04), 146–152 (2023).
69. Fiedler, R.: Controlling von Projekten: Mit konkreten Beispielen aus der Unternehmenspraxis – alle Aspekte der Projektplanung, Projektsteuerung und Projektkontrolle, 7, aktualisierte und überarbeitete Auflage, 2016th edn. Springer Vieweg, Wiesbaden (2016)
70. Klausing, H. (ed.) Kompetenzbasiertes Projektmanagement (PM4): Handbuch für Praxis und Weiterbildung im Projektmanagement, 1. Auflage ed. GPM Deutsche Gesellschaft für Projektmanagement e.V, Nürnberg, Berlin (2019).

Enhancing German Manufacturing Competitiveness: Integrating B2B Platforms into the Industrial Value Chain

Mehmet Demir[1(✉)], Leonard Rieke[1], David Herberger[2], and Peter Nyhuis[1]

[1] Institute of Production Systems and Logistics, Leibniz University Hannover, 30823 Garbsen, Germany
demir@ifa.uni-hannover.de

[2] Institute for Production and Logistics Research GbR, Stuttgart, Germany

Abstract. The manufacturing sector in Germany, a vital component of its economy, faces a transformative phase amidst the digital revolution accelerated by the global pandemic. This paper highlights the integration of Business-to-Business (B2B) platforms into the industrial value chain, underlining their potential, complexities and implementation strategies. Evolving market dynamics and the surge in customized demands have reshaped traditional strategies, emphasizing the need for B2B platform adoption. Integrating B2B platforms promises streamlined processes, cost reductions and innovation opportunities. Yet, complexities arise from regulatory uncertainties, cybersecurity concerns and the absence of standardized practices. The current literature does not cover comprehensive methods that address how B2B platforms work within industrial settings. To close this gap, this paper proposes a three-phase methodology – 'Observation', 'Understanding' and 'Implementation' - designed to systematically study, comprehend and implement B2B platforms. The approach includes stakeholder analysis, risk identification, governance framework development and a maturity model to guide companies through platform integration. This comprehensive framework aims to provide tailored recommendations for companies navigating the complexities of B2B platform adoption. By addressing existing research gaps, this methodology strives to assist the manufacturing industry in effectively leveraging B2B platforms to enhance competitiveness in an increasingly digitalized landscape.

Keywords: B2B Platforms · Governance · Industrial Value Chain

1 Introduction

The manufacturing sector constitutes a cornerstone of the German economy. This sector underlines its high economic significance, accounting for approximately 24% of the German Gross Domestic Product (GDP) in 2022 [1]. This high economic relevance emphasizes this sector's central role in the future German economy. Although small and medium-sized enterprises (SMEs), with almost 2.5 million companies, make up the majority of the corporate landscape in Germany (99.4%) and shape the economic

Published by Springer Nature Switzerland AG 2025
P. Sureephong et al. (Eds.): PLM 2024, IFIP AICT 740, pp. 140–151, 2025.
https://doi.org/10.1007/978-3-031-93319-6_11

landscape of Germany, they only account for a third of the total turnover of all companies [2]. Advancing digitalization opens up new opportunities for SMEs to increase their sales and make their business processes more efficient. To avoid losing touch in this area, the Center for European Economic Research advocates the intensified presentation of digitalization potentials to SMEs, clarifying these through best practice examples [2].

For decades, successful companies have been able to orient themselves on Porter's value chain concept [3, 4]. However, advancing developments in production technology have led to a shift from a traditional seller's market to a buyer's market [5]. In this saturated market, buyers today have an abundance of choices [5, 6]. In order to gain market shares in such a market environment, companies strive to develop customized products to meet the needs of their customers better [6]. This concept's once decisive success factors, which aimed at controlling activities from purchasing to sales and emphasized cost efficiency, quality or market leadership, are increasingly losing relevance [4, 7]. This is attributed to the constantly changing market and the increasing diversity of individualized customer requirements [4, 7].

Amid the digital transformation accelerated by the global COVID-19 pandemic and the increasing demand for real-time data, the business models of traditional companies and industries are losing relevance [7, 8]. In contrast, platform-based business models such as those of Amazon or Meta are experiencing significant growth and are changing traditional market structures [7, 9]. These business-to-consumer (B2C) platforms represent a successful model of digitalization. They provide companies with access to digitalization and opportunities to increase efficiency and open up new purchasing and sales channels and innovative approaches to business models [8, 10]. Transferring platform-based value creation from the end consumer segment to the industrial sector opens up promising opportunities to improve manufacturing companies' competitiveness further [11, 12]. In this context, digital platforms serve as an essential tool [7, 13, 14]. B2C platforms targeting end consumers account for 27% of Europe's global e-commerce value, whereas B2B platforms designed to facilitate business transactions represent only 4% [15]. This indicates that B2B platforms are still in their early stages of development in Europe. In this context, there is the potential for companies in Germany to become leading users by implementing innovative industrial platforms.

2 Integration of B2B Platforms into the Industrial Value Chain

A value chain comprises a sequence of interlinked activities that generate an end product of greater value than the individual components considered independently [16]. In contrast, the concept of platforms is characterized by a multidimensional and interdependent structure [4]. It differs significantly from the traditional value chain concept regarding stakeholders, success factors and potential strategies [4]. As a result, competition is no longer limited to individual companies but extends to complex value chains [7].

A digital platform operates as a complex construct of rules, infrastructure and stakeholders [17, 18]. Standardized components and fundamental principles characterize the structural nature of these platforms. These technological systems connect different stakeholders to enable the exchange of services or information [19]. A platform's business model consists of four essential components: the platform operator, architecture, suppliers and demanders [18]. The platform operator is the company that provides the platform and controls the interface between suppliers and demanders [18]. Moreover, the platform operator is responsible for designing the technical framework (platform architecture) and establishing the governance within the platform [18]. The platform operators gain significant advantages from the robust user loyalty to their infrastructure [20]. The governance establishes the framework for the actions of components, sets clear rules and determines which stakeholders are allowed to interact with each other [4, 18]. The platform architecture enables interactions between platform components [17].

2.1 The Advantages of an Integration of B2B Platforms into the Industrial Value Chain

As the center of value creation, the factory is experiencing significant development and is evolving from an isolated production site to a central point within vertical and horizontal value chains [7]. In this sense, it no longer solely functions as a site for pure manufacturing but as a hub where diverse phases and processes of value creation are interconnected [7]. The vertical value chain comprises all activities required to manufacture, market and distribute a product or service [3]. It comprises various stages such as product and technology development, marketing, sales and customer service. The horizontal value chain describes the cooperation between companies, suppliers and customers to bundle core competencies [21]. B2B platforms can, for example, enable manufacturing companies to improve production quality and reduce costs by efficiently utilizing shared resources. A key advantage of these platforms is that time-consuming processes, such as sales, can be accelerated and streamlined through B2B platforms [8]. This is accomplished by platforms that create intermediary connections like networks, marketplaces or by providing a smart infrastructure to facilitate the exchange of data between machines and devices [8]. Moreover, using digital platforms allows companies to automate communication processes with suppliers and customers [19, 22]. Figure 1 demonstrates the streamlining of interaction processes with a platform as an intermediary, reducing the contacts from 16 (without platform) to 8 (with platform).

Currently, leading internet platforms like "Spanflug" [24] and "Part Factory" [25] are providing components without demanders knowing the exact supplier. These platforms allocate potential suppliers from a pool who can apply for the contract [7]. Platform users, for instance, can benefit from the experiences of others, either through their reviews or indirectly through statistical insights [8].

Platforms provide the opportunity to redesign and adjust business models to take full advantage of the networked economy [19]. Firstly, digital platforms enable companies to gain strategic advantages by developing innovative business models, opening up new market segments and entering into partnerships and collaborations with other companies [18, 19]. Secondly, these platforms offer operational advantages through the opportunity to optimize processes in various business areas, including the reduction of contacts [18, 19]. Digital platforms enable efficiency gains and cost savings throughout the entire supply chain, from procurement through production to distribution [19]. Furthermore, they enable the standardization of business activities, leading to additional cost savings [22].

Moreover, digital platforms demonstrate high scalability [8, 18]. This is based on positive network effects, where the benefits grow stronger as the number of users increases

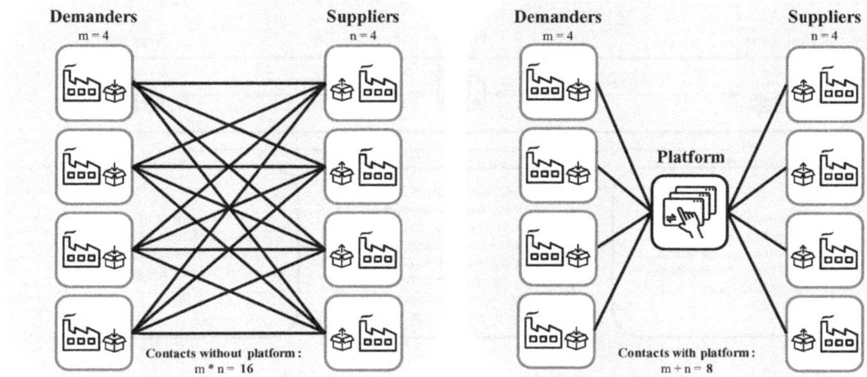

Fig. 1. Enhancing operational efficiency via platform-driven Interaction processes (adapted from TOPOROWSKI [23])

[4]. In addition to the positive network effects on digital platforms, there are other types of network effects. Direct network effects are generated when users attract other users, while indirect network effects result from the increasing variety of additional products and services [26]. In contrast, negative network effects arise when the value of a good decreases as the number of users increases, often due to uncontrolled growth and infrastructure quality issues [26]. From a demander's perspective, having multiple suppliers means a wider range of goods and services to choose from. It also leads to more transparent pricing due to competing offers, which ultimately results in more affordable prices [8]. For suppliers, having more buyers on a platform means increased reach, a larger volume of transactions and the potential for higher revenue [8]. The overall scalability is facilitated by the standardization of processes on platforms [27]. The extensive data within digital platforms enables the integration of Artificial Intelligence (AI) into data-driven business models, facilitating tailored, instantly available services [12]. This involves, for example, training data to develop more precise advertising campaigns [8]. This enables companies to tailor their strategies more precisely to their customers' individual preferences and needs. The success of a platform-based business model can be measured not only by the revenue generated but also by key figures such as click rates, dwell time or abandonment rates [12]. The effects of platforms for suppliers and demanders are summarized in Fig. 2.

2.2 The Complexities of an Integration of B2B Platforms

Despite the promising potential of B2B platforms, companies face several challenges when implementing such platforms, making integration into existing structures more complex. There is a lack of comprehensive understanding regarding the dynamics and potentials of platforms, particularly concerning the transformation of traditional physical value creation into digital platforms.

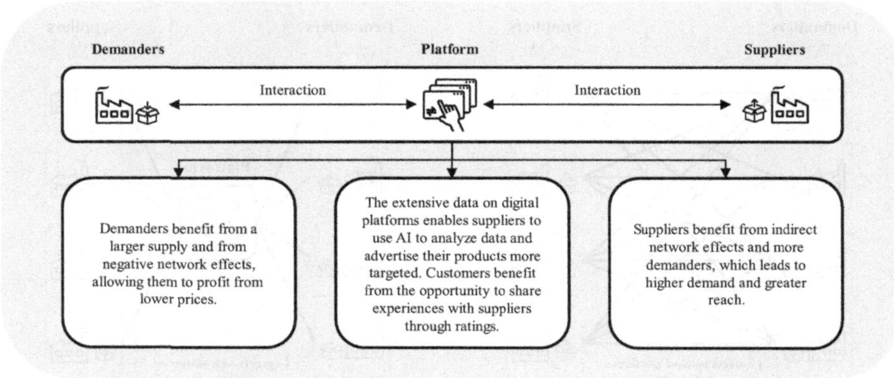

Fig. 2. Effects of B2B platforms for suppliers and demanders

While companies are facilitated in their access to digital platforms through various basic software and open-source solutions, there is a missing foundation for offering scientific concepts and methods to businesses [28]. These developments offer cost-effective opportunities to integrate companies with limited resources into the digital transformation [4]. This enables broader participation and strengthens the competitive- ness of companies that previously did not have the resources to compete in an increas ingly digital market. For companies to fully leverage the potential of B2B platforms, they require support based on scientific insights regarding the dynamics, success factors and risks of digital platforms.

A central need for research exists due to the lack of a clear regulatory framework for the involved stakeholders on B2B platforms [29]. While platform openness and per ceived independence can be identified as crucial success factors [12], the current chal lenge with an open platform lies in its flexibility, exposing the company to an unknown and unacceptable risk and leading businesses to refrain from adopting such platforms [10]. In addition, introducing a platform requires considerable organizational effort, which is directly linked to costs for companies [29]. In order to establish digital plat forms, companies must first minimize risks concerning investments and define security mechanisms to maintain competitiveness and legal foundations [30]. A direct barrier to introducing digital platforms can also be linked to deficits in data security as sensitive company data is exchanged on these platforms [29]. This barrier is highlighted by a study that revealed that 46% of companies surveyed in Germany fell victim to cyberat tacks in the past 12 months [31].

In manufacturing, value creation using platforms is still in its early stages but will play a more critical role in the future [7]. Currently, a primary issue lies in the absence of standards, serving as a major impediment for business models based on platforms [32]. The limited literature stemming from fundamental research amplifies this obsta cle, highlighting the need for theoretical groundwork in research [8, 30]. In order to respond actively to the opportunities presented by platforms, a foundation and a concept are required that enable the transfer of physical value creation to B2B platforms.

In summary, companies face various challenges when implementing such platforms, including the need for specific knowledge to use them effectively [19]. It can be stated that the identification of factors responsible for functional competition among B2B platforms has received limited scientific attention. This paper, therefore, presents a methodology for overcoming the identified challenges.

The first step is an in-depth examination of the integration of B2B platforms into the industrial value chain. The benefits of this integration are highlighted to emphasize the potential of an integration. The need for research is then explained by outlining the current integration challenges. After presenting the current state of research, a three-stage methodology for effectively integrating B2B platforms is introduced.

3 Framework and Current State of Research

In order to capture all potential interactions in industrial digital platforms, a holistic and interdisciplinary view of the factory as the center of value creation is required. A *Factory* is defined as a "Place where value is created by the manufacture of industrial goods based on division of labour while utilizing production factors" [33]. Five key requirements that a methodological approach to researching B2B platforms in an industrial context should fulfill can be derived from the existing literature.

To fully comprehend the impact of a B2B platform in an industrial context, it is necessary to consider all factory processes, including technical, logistical, personnel and organizational aspects (requirement 1). The scalability must also be considered, encompassing the network effect and the platform's adaptability to changes such as user growth, technological adjustments, new stakeholder configurations or market shifts (requirement 2). From identified cause-and-effect relationships on digital platforms, targeted recommendations for action need to be derived (requirement 3). These recommendations will assist companies in utilizing digital platforms within the manufacturing industry. Moreover, the approach must allow an assessment of the impact of integrating a digital platform into a company's current value chain. This assessment helps to estimate potential disruptions, compatibility issues and integration obstacles (requirement 4). Ultimately, the approach must establish rules with defined responsibilities and security standards to ensure the platform's efficient management and long-term stability (requirement 5).

A systematic literature review was conducted to evaluate the current state of research and determine the extent to which the defined requirements have been met. The systematic literature review follows the approach of WEBSTER AND WATSON [34]. According to them, a high-quality literature review should focus on the concepts to better synthesize the literature [34]. Therefore, a concept-based literature research was conducted, with a focus on the predefined requirements.

Table 1 shows numerous approaches to researching the dynamics and integration of different platforms in the industrial environment in the current literature. These approaches already offer initial recommendations for action and highlight barriers andt opportunities. In order to assess their relevance for the development of a holistic approach, the previously defined requirements are evaluated in this context.

The holistic view of the factory system plays a subordinate role in the current literature. However, it has been recognized that a comprehensive approach is required

Table 1. Current state of research

● Requirements fully met
◑ Requirements partially fulfilled
○ Requirements not met

Nr.	Requirements	[35]	[10]	[36]	[37]	[38]	[27]	[30]	[39]	[40]	[41]
1	Holistic view of the factory	○	○	○	○	○	○	○	○	●	◑
2	Scalability	◑	○	◑	○	○	◑	◑	○	◑	○
3	Recommendations for action	●	◑	◑	◑	●	◑	◑	●	●	●
4	Assessment of the integrability	○	○	○	○	○	○	○	○	◑	◑
5	Effective governance	○	◑	◑	◑	●	◑	◑	○	○	○

when implementing a platform to cover all relevant application areas of a company and exploit synergy effects [40]. Various research papers, in particular [27, 30, 35, 36], emphasize the scalability of digital platforms, which results from the low marginal costs of platforms and network effects. Despite the recognized potential of platforms, there is a lack of established approaches or models that support companies in optimal use of network effects. The research gap in this area suggests that developing strategies and frame-works to maximize these effects is crucial to supporting companies in effectively using digital platforms. Some studies strongly focus on a specific industry, such as the automotive sector [10, 35]. As a result, the transferability of their results remains unclear. BRAUN et al. [38] presented a six-step governance model that enables the development of platform governance even without comprehensive prior information. This approach integrates best practices from similar platforms and thus provides an initial guideline for a systematic approach to defining platform governance. However, it is noticeable that prior research has focused less on the governance structure and more on the procedural aspects [37].

Although the significance of efficient governance has been highlighted in various studies [27, 30, 35–38], a comprehensive concept has not yet been developed. Previous approaches neglect a detailed analysis of the individual initial situation of companies and the derivation of specific recommendations for action. This is the starting point for developing a method to close this research gap and empower companies to overcome the previously identified problems. In summary, it can be stated that none of the approaches presented in Table 1 can completely fulfill the requirements. Even combining the approaches does not lead to comprehensive fulfillment of all requirements. Consequently, there is a need for research into the development of a new methodology that provides recommendations for design and action based on the specific maturity level of a company.

4 Methodological Approach

The devised methodology is structured into three primary phases – 'Observation', 'Understanding' and 'Implementation' – which are necessary due to the complexity of the subject (see Fig. 3).

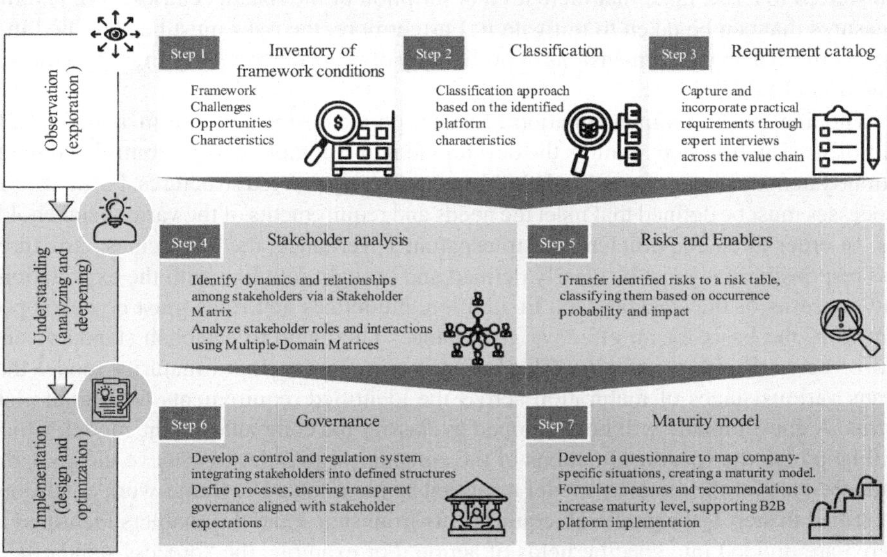

Fig. 3. Methodology of the approach

The initial phase, 'Observation', focuses on exploring the subject area. First, the frame-work, challenges, opportunities and characteristics of B2B platforms within the industrial context must be identified. Subsequently, a classification of B2B platforms must be established based on the identified characteristics (step 1). Different classes of industrial B2B platforms, such as transactional or data platforms, must be evaluated using predefined aspects like user type, industry focus and functionality to attribute specific features to the identified platform classes (step 2). In the final step of the 'Observation' phase, a survey must be conducted to transfer practical requirements into a catalog that shapes further activities. Industry-independent requirements must be captured and incorporated through expert interviews with industry professionals in various positions across the value chain, such as sales or suppliers (step 3).

The second phase, 'Understanding', focuses on analyzing dynamics and deepening knowledge about B2B platforms. Initially, the yet unknown dynamics of digital B2B platforms must be identified by analyzing the relationships among the involved stake-holders. A stakeholder analysis must be performed to analyze stakeholders (e.g., customers, operators, or providers). The results will be documented in a Stakeholder Matrix, encompassing the qualitative assessment of each stakeholder's conflict potential, influence, expectations and concerns. Subsequently, the roles of the stakeholders must be further outlined. The interactions among these actors will be analyzed through systematic comparisons using Multiple-Domain Matrices. A quantitative assessment will be conducted by capturing the intensity of these interactions. The insights gained from these analyses will ultimately be utilized to derive design recommendations for B2B platforms (step 4). The factors that facilitate the platform's integration and possible obstacles and risks must be identified in a further step. Any identified risks should be transferred to a risk table that includes a description of the risk, its causes, type and the measures that can be taken to mitigate it. Furthermore, the risks must be classified in a risk portfolio using the dimensions of probability of occurrence (low or high) and impact (low or high) (step 5).

The final phase, 'Implementation', focuses on the design and optimization of B2B platforms. The first step requires the development of a comprehensive control and regulation system that integrates all stakeholders into clearly defined structures. Furthermore, processes must be defined that meet the needs and requirements of the various stakeholders. In order to enable efficient and transparent governance, the hierarchies, structures and responsibilities must be clearly defined and brought into line with the expectations and concerns of the stakeholders. In addition, guidelines and rules must be developed that form the basis for an effective governance. They help to establish standards and utilize network effects (step 6). The last step aims to develop a maturity model that spans various stages of maturation across the identified requirements for digital platforms. A questionnaire will be developed to classify the companies in the model, which will help map the specific situations of the companies. In order to achieve the goal, the requirements for the maturity model must first be identified. The frame-work conditions identified in step 1, the practical requirements from step 3 and the enablers identified in step 5 are divided into specific fields of action. For example, the "process standardization, automation and digitalization" field of action serves as a reference point. In this context, suitable evaluation criteria and maturity levels are defined to structure the fields of action clearly and targeted. The maturity model aims to present the individual development levels of companies in detail to derive individual needs for action and potentials that contribute to increasing the level of maturity in the context of B2B platform integration. This questionnaire will consist of specific questions that will assist companies in conducting a subjective and comparable self-reflection. Afterward, measures (design and action recommendations) will be developed to improve the level of maturity in respective fields of action. These measures will be transferred into a catalog of measures to support companies implementing B2B platforms (step 7).

5 Summary

The paper explores the integration of Business-to-Business (B2B) platforms into the German manufacturing sector. It highlights the significant role of small and medium-sized enterprises (SMEs) in Germany's economic landscape and emphasizes the necessity for these companies leveraging digitalization to enhance sales and operational efficiency. The shift from traditional value chains to buyer's markets, intensified by the global digital transformation and the rise of platform-based business models, under-scores the need for industrial companies to adapt. The advantages of integrating B2B platforms into the industrial value chain are discussed, focusing on increased production efficiency, cost reduction, streamlined communication and strategic business model innovations. Despite these benefits, challenges such as lacking regulatory frame-works, cybersecurity concerns and the absence of standardized approaches for platform integration pose obstacles to widespread adoption. The paper identifies a methodology to overcome these challenges and presents a roadmap for observing, understanding and implementing B2B platforms in the industrial context, aiming to empower companies and drive the evolution of German manufacturing.

References

1. Statistisches Bundesamt (Destatis): Volkswirtschaftliche Gesamtrechnungen: Bruttoinlandsprodukt (2023). https://www.destatis.de/DE/Themen/Wirtschaft/Volkswirtschaftliche-Gesamtrechnungen-Inlandsprodukt/_Grafik/_Interaktiv/bip-wirtschaftsstruktur.html. Accessed 22 Jan 2024
2. Zentrum für Europäische Wirtschaftsforschung GmbH: Digitalisierung im Mittelstand: Status Quo, aktuelle Entwicklungen und Herausforderungen, Mannheim (2016)
3. Porter, M.E.: Competitive Advantage: Creating and Sustaining Superior Performance. FreePress, New York (1985)
4. Wallenhorst, D.: Perspektivenwechsel in der Digitalisierung: Interdisziplinäre Ansätze für die digitale Epoche. Springer Fachmedien Wiesbaden, Wiesbaden (2023)
5. Maune, G.,: Möglichkeiten des Komplexitätsmanagements für Automobilhersteller auf Basis durchgehender IT-gestützter Systeme (2002)
6. Wiendahl, H.-P., Gerst, D., Keunecke, L. (eds.): Variantenbeherrschung in der Montage. Springer, Berlin Heidelberg (2004)
7. Wiendahl, H.-P., Reichardt, J., Nyhuis, P.: Handbuch Fabrikplanung: Konzept, Gestaltung und Umsetzung wandlungsfähiger Produktionsstätten, 3, überarbeitete und erweiterte, Auflage Hanser, Carl, München (2023)
8. Koenen, J., Falck, O.: Industrielle Digitalwirtschaft–B2B-Plattformen. Studie im Auftrag des Bundesverbands der Deutschen Industrie eV, ifo Institut, München (2020)
9. Evans, P.C., Gawer, A.: The rise of the platform enterprise: A global survey (2016)
10. Svahn, F., Mathiassen, L., Lindgren, R.: Embracing digital innovation in incumbent firms. MIS Q. **41**(1), 239–254 (2017)
11. Hein, A., Weking, J., Schreieck, M., Wiesche, M., Böhm, M., Krcmar, H.: Value cocreation practices in business-to-business platform ecosystems. Electron. Markets **29**(3), 503–518 (2019)
12. Haller, S., Wissing, C.: Normstrategien für das Service Business Development zur Berücksichtigung digitaler Plattformen in der Wertschöpfung. In: Bruhn, M., Hadwich, K. (eds.) Service Business Development, pp. 167–188. Springer Fachmedien Wiesbaden, Wiesbaden (2018)
13. Li, Y., Fu, K., Gong, X., Xiang, Z., Zhang, J., Liao, C.: Research on value co-creation mechanism of platform enterprises in digital innovation ecosystem: A case study on Haier HOPE platform in China. Front. Psychol. (2022)
14. Brambring, F.: Produktion am Standort Deutschland: Ergebnisse der Untersuchung im Rahmenkonzept "Forschung für die Produktion von morgen" sowie der Fördermaßnahme "Intelligente Vernetzung in der Produktion - Ein Beitrag zum Zukunftsprojekt Industrie 4.0" 2013, Ausg. 2013 ed. FIR an der RWTH Aachen, Aachen (2013)
15. Berger, R.: B2B Marketplaces are blossoming. New opportunities for B2B players to capitalize on a growing business (2018)

16. Gawer, A. (ed.): Platforms, Markets and Innovation. Edward Elgar, Cheltenham (2009)
17. van Alstyne, M.W., Parker, G.G., Choudary, S.P.: Pipelines, platforms, and the new rules of strategy. Harv. Bus. Rev. **94**(4), 54–62 (2016)
18. Meier, P.: Digitale plattformen als innovationstreiber. Innovationsumgebungen gestalten: Impulse für Start-ups und etablierte Unternehmen im globalen Wettbewerb, pp. 207–217 (2018)
19. Lundborg, M., Gull, I.: Digitale Plattformen als Chance für den Mittelstand. Relevanz, Anwendungen, Transfer, Eine Erhebung der Mittelstand-Digital Begleitforschung, WIK, Bad Honnef (2019)
20. Nachtwey, O., Schaupp, S.: Ungleicher Gabentausch – User-Interaktionen und Wertschöpfung auf digitalen Plattformen. Köln Z Soziol **74**(S1), 59–80 (2022)
21. Kaufmann, T., Forstner, L.: Horizontale Integration der Wertschöpfungskette in der Halbleiterindustrie. In: Vogel-Heuser, B., Bauernhansl, T., ten Hompel, M. (eds.) Handbuch Industrie 4.0 Bd.1, pp. 95–103. Springer Berlin Heidelberg, Berlin, Heidelberg (2017). https://doi.org/10.1007/978-3-662-45279-0_31
22. Dr. Isabel Gull, Martin Lundborg: Digitale Plattformen als Chance für den Mittelstand. Begleitforschung Mittelstand-Digital WIK GmbH (2019)
23. Toporowski, W.: Der Baligh-Richartz-Effekt. Kontaktkostenreduktion durch die Einschaltung von Handelsbetrieben. Wirtschaftswissenschaftliches Studium **28**, 81–84 (1999)
24. Spanflug Technologies GmbH: https://spanflug.de. Accessed 29 Dec 2023
25. Part Factory: https://www.partfactory.com/en. Accessed 29 Dec 2023
26. Jaekel, M.: Die Macht der digitalen Plattformen. Springer, Wiesbaden (2017)
27. Haucap, J., Kehder, C., Loebert, I.: B2B-Plattformen in Nordrhein-Westfalen: Potenziale, Hemmnisse und Handlungsoptionen (2020)
28. Bundesministerium frü Bildung and Forschung (BMBF): Zukunftsbild "Industrie 4.0". Bonn (2013).
29. Bertenrath, R., Fritsch, M., Thorsten, L., Karl, L., Agnes, M., Thomas, S.: Digital-Atlas Deutschland, 1–338 (2018)
30. Lerch, C., et al.: Die volkswirtschaftliche Bedeutung von digitalen B2B-Plattformen im Verarbeitenden Gewerbe (2019)
31. Hiscox: Anteil der Unternehmen, die in den letzten 12 Monaten eine Cyber-Attacke erlebt haben, in ausgewählten Ländern im Jahr (2022). https://de.statista.com/statistik/daten/studie/1230157/umfrage/unternehmen-die-in-denletzten-12-monaten-eine-cyber-attacke-erlebt-haben/. Accessed 6 Dec 2023
32. Brossardt, B.: Plattformen - Infrastruktur der Digitalisierung: Studie Stand: Januar 2019. Eine vbw /bayme vbm Studie, erstellt von der Insitut der deutschen Wirtschaft Köln Consult GmbH (2019)
33. Verein Deutscher Ingenieure (VDI). VDI 5200: VDI-Fachbereich Fabrikplanung und -betrieb; Fabrikplanung; Planungsvorgehen; VDI-Gesellschaft Produktion und Logistik
34. Webster, J., Richard, T.W.: Analyzing the past to prepare for the future: writing a literature review. MIS Q. **26**(2), xiii–xxiii
35. Weiß, N., Schreieck, M., Brandt, L.S., Wiesche, M., Krcmar, H.: Digitale Plattformen in der Automobilbranche – Herausforderungen und Handlungsempfehlungen. HMD Praxis der Wirtschaftsinformatik **55**(2), 349–365 (2018)
36. Gawer, A.: Bridging differing perspectives on technological platforms: toward an integrative framework. Res. Policy **43**(7), 1239–1249 (2014)
37. Brüning, A., Gluchowski, P., Kaiser, A.: Data Governance Einordnung, Konzepte und aktuelle Herausforderungen. Chemnitz Economic Papers 015, Chemnitz (2017). http://hdl.handle.net/10419/170675

38. Braun, H.-M., Rottenegger, D., Braunreuther, S.: Prozessmodell für die Gestaltung einer Plattform-Governance für digitale mehrseitige Plattformen. In: Weber, B. (ed.) Data Governance: Nachhaltige Geschäftsmodelle und Technologien im europäischen Rechtsrahmen, pp. 329–348. Springer, Berlin Heidelberg, Berlin, Heidelberg (2023)
39. Oberländer, A.M., Häckel, B., Übelhör, J.: IIoT-basierte Geschäftsmodellinnovation im Industrie-Kontext – Archetypen und praktische Einblicke. In: Meinhardt, S., Wortmann, F. (eds.) IoT – Best Practices. EH, pp. 23–35. Springer, Wiesbaden (2021). https://doi.org/10.1007/978-3-658-32439-1_2
40. Cevirgen, C., Bischoff, L.-M., Heinen, T., Nyhuis, P.: I4PE-Plattform zur Unterstützung der unternehmensweiten Digitalisierung von KMU. J. Product. Syst. Logist. (2020)
41. Lanza, G., Nyhuis, P., Ansari, S.M., Kuprat, T., Liebrecht, C.: Befähigungs- und Einführungsstrategien für Industrie 4.0: Vorstellung eines reifegradbasierten Ansatzes zur Implementierung von Industrie 4.0. Zeitschrift für wirtschaftlichen Fabrikbetrieb **111**(1–2), 76–79 (2016). https://doi.org/10.3139/104.111462

Managing Heterogeneous Historical Data Within the Digital Twin Paradigm

Oghenemarho Orukele[1](✉), Arnaud Polette[1], Aldo Gonzalez-Lorenzo[2], Jean-Luc Mari[2], and Jean-Philippe Pernot[1]

[1] École Nationale Supérieure d'Arts Et Métiers, Aix-en-Provence, France
oghenemarho.orukele@ensam.eu
[2] Aix Marseille Univ, CNRS, Marseille, LIS, France

Abstract. The rise of the fourth industrial revolution has propelled research and development in Digital Twins (DTs) in the manufacturing sector. DTs establish a crucial link between physical systems and their digital replicas, enabling real-time monitoring, process optimisation, and predictive analysis. This paper introduces a historical data management framework for DTs in manufacturing, addressing existing research gaps. Our framework includes a data classification scheme based on change frequency, distinguishing static, dynamic, and periodic data. We propose an architecture involving registration, data transformation for unified schema creation, utilisation of queues for asynchronous data storage, and a tailored database selection process. Implementation in a UR5e 6-axis collaborative robot's digital twin demonstrates the practicality and efficiency of the proposed framework in the aspect of the latency and throughput of the digital twin. The proposed architecture is benchmarked against existing architectures, and the results are discussed.

Keywords: Digital twin data · Historical data management · Industry 4.0

1 Introduction

As the fourth industrial revolution continues in the manufacturing sector, the Digital Twin (DT) has emerged as a critical concept in this revolution. The Digital Twin seeks to create a high-fidelity digital representation of a physical system and establish a bidirectional connection between the digital representation and the physical system [1]. This interconnectivity facilitates seamless communication between the physical and digital entities, enabling multiple use cases, including real-time process monitoring, process simulation or optimisation, process control, asset management, production planning, performance prediction and other manufacturing applications [2, 3]. The proliferation of data within the manufacturing process, combined with the advancements in data processing and analytics technology, has increased the importance of historical data in the manufacturing sector. This historical data is ideally available to the digital twin services (DTS) and utilised in applications ranging from predictive analysis to process optimisation, historical process simulation, machine learning, and other data analytics applications.

Published by Springer Nature Switzerland AG 2025
P. Sureephong et al. (Eds.): PLM 2024, IFIP AICT 740, pp. 152–165, 2025.
https://doi.org/10.1007/978-3-031-93319-6_12

This paper presents a comprehensive historical data management solution tailored for digital twins within the manufacturing domain. Our framework is designed to accommodate the heterogeneous nature of manufacturing data. The paper is organised as follows: Sect. 2 delves into the state-of-the-art of historical data management practices and their utilisation within the digital twin paradigm. In Sect. 3, we introduce our proposed historical data management framework. Section 4 provides insights into the implementation of our framework, detailing practical aspects of its deployment. Finally, Sect. 5 discusses the implementation and outlines future perspectives for this research endeavour.

2 Literature Review

With the expansion of data acquisition and processing technologies in the manufacturing and industrial sectors, there is now a high level of availability and accessibility of production data for a variety of applications. One of these applications is the digital twin; the digital twin uses the data to create a high-fidelity digital replication of the physical entity with real-time data from the system [4, 5]. This data is usually utilised in the real-time applications of the digital twin and rarely re-exploited afterwards. However, these data have historical value, and their management is essential to the digital twin concept.

The significance of data management within digital twin frameworks is illustrated in various proposed architectures [6, 7]. Talkhestani et al. explored the potential integration of historical data management within their digital twin architecture [8]. Within a broader architectural context, F. Tao et al. proposed the utilisation of distributed nodes to store the data from both the physical entity and the digital twin services [9]. Lee Jay et al. introduced a framework for managing historical data within digital twins in the context of cyber-physical manufacturing systems, suggesting the capture of system snapshots upon "drastic" change instead of constant storage [10]. Similarly, another architecture proposed for digital twins in cyber-physical systems outlines a data management platform tasked with data storage, acquisition, and management [11]. Despite recognising the necessity for historical data management within digital twins, these frameworks lack explicit methodologies or architectures for implementing historical data management systems within digital twins.

2.1 Historical Data Management in Digital Twins

As part of the literature review, an analysis is conducted on the research in digital twins to examine the state of historical data management solutions. The main objective is to evaluate if historical data management is considered in the digital twin. Specifically, how the digital twin manages historical data within the proposed approaches. A systematic literature review methodology was followed, and literature studies were critically appraised and analysed, with the results synthesised quantitatively. The search was conducted with the SCOPUS database; since the objective of the review is to examine the management of historical data in digital twins within the industrial sector, the search was executed on a combination of the following keywords: "digital twin" OR "virtual twin" AND "Production" OR "manufacturing" OR "industrial".

Table 1. Analysis of selected articles on historical data management in digital twins.

Historical Data Management								
Authors	Reference	Publication Type	Field of Application	Data Storage	Data Transformation	Level of Abstraction	Data Write Strategy	Data Storage Technology
F. Yepeng et al. 2021	[12]	J	M	x		H		RDB
L. Fang et al. 2021	[13]	J	M					
Í. A. Fonseca et al. 2022	[14]	J	MR	x	x	L	S	FS (JSON)
K. Wang et al. 2021	[15]	J	PI					
W. Luo et al. 2019	[16]	J	M	x		H		
A. Hyre et al. 2022	[17]	J	M					
G. Steindl et al. 2022	[18]	J	E	x		H		
H. Jiang et al. 2021	[19]	J	M	x		H		RDB, C
M. Moretti et al. 2021	[20]	J	M					
S. Gopalakrishnan et al. 2020	[21]	J	I					
J. Conde et al. 2022	[22]	J	IM	x	x	L	S	USDB
L. Yi et al. 2021	[2]	J	M					
J. Conde et al. 2022	[23]	J	I	x		L		FS (CSV)
Y. Cai et al. 2020	[24]	J	M					
G. Schroeder et al. 2022	[25]	C	P					
C. Liu et al. 2022	[26]	J	M	x		H		
W. Liu et al. 2023	[27]	J	M					
S. Martinez et al. 2021	[28]	J	M					
G. A. Moslott et al. 2021	[29]	J	MR					
D. K. Bong et al. 2022	[30]	J	M					
G. Barbieri et al. 2021	[31]	J	M					
A. I. Vodyaho et al. 2022	[32]	J	M	x		H		

(*continued*)

Table 1. (*continued*)

Historical Data Management								
Authors	Reference	Publication Type	Field of Application	Data Storage	Data Transformation	Level of Abstraction	Data Write Strategy	Data Storage Technology
L. Zhao et al. 2021	[33]	J	M					

Legend:
Publication Type: J = Journal Articles, C = Conference Article
Field of Application: M = Manufacturing, P = Process, IM = Infrastructure Management, E = Energy, I = Industrial, MR = Maritime
Level of Abstraction: H = high level, L = low level
Data Write Strategy: S = Synchronous, AS = Asynchronous
Data Storage Technology: RDB = Relational Database, USDB = Unstructured Database, FS = File Storage, C = Cache
*The empty columns indicate there is no implementation of the category in the work

The literature search resulted in 297 articles. These articles were filtered in two phases. The initial phase entailed reading the titles and abstracts to select articles within the review's objective, i.e., digital twins in the industrial sector. The second phase involved reading the contents of the selected articles to analyse the implemented digital twin. The output of this analysis is 23 papers that have been reviewed, and the findings are presented in Table 1 and are discussed in the subsequent sections. The literature review consists of journal and conference papers on digital twins written between 2016 and 2022. The industrial application of the research papers includes manufacturing, process industry, energy sector, infrastructure management and industrial application. The 'historical data management' column discusses the features of the implemented data management solutions in the digital twins. Specifically, it scrutinises the facets of implemented data management solutions, focusing on key attributes such as the presence of historical data storage, the level of abstraction characterising the historical data management system, the involvement of data transformation processes, and the technology facilitating the realisation of historical data management.

Data Storage. Among the reviewed literature, emphasis on data storage within digital twins is observed in 40% of the examined articles. This prevalence underscores the significance attributed to historical data management in digital twins. However, more than the mention of historical data is needed. Thus, it becomes imperative to interrogate the intricacies of the proposed historical data management systems to understand the nuances of their implementation within the digital twin frameworks.

Level of Abstraction. This criterion investigates the level of detail given for the historical data management system. Notably, a significant proportion of surveyed works predominantly present these systems at a high level, lacking substantive implementation details [12, 16, 18, 19, 23]. Conversely, a minority of the papers have a low level of abstraction; the low-level abstraction presents the details of the historical data management in the digital twins, then discusses the implementation technology and how the data is treated before being stored.

Data Transformation. F. Tao et al. underscored the centrality of data transformation within the historical data management framework for digital twins [9]. Data transformation is a process of transforming different data formats (e.g. numbers, text, pictures, etc.) into a unified form. This is significant for digital twins, given the heterogeneous nature of data in the industrial sector. For example, Fonseca et al. transformed data to adhere to ISO 8601 standards for their proposed digital twin of a marine vessel [14]. In contrast, Conde et al. proposed an application-specific data transformation approach for standardising diverse data sources within a car park digital twin [22]. The data transformation enables the digital twin services to have a standardised interface for interacting with the digital twin's historical data.

Data Writing Strategy. The data writing strategy criterion revolves around the mechanisms and strategy employed in writing the data to the data storage system. The two strategies are synchronous and asynchronous data write. In the synchronous strategy, the data is written to the database in a blocking manner, such that the write operation must be completed and acknowledged before the application can proceed. In contrast, in the asynchronous strategy, the data is written into the database without waiting for confirmation or acknowledgement from the database system.

Data Storage Technology. The technology used to store historical data from digital twins is essential to digital twins' historical data management systems. The reviewed literature showcases a spectrum of storage technologies ranging from structured and unstructured databases to in-memory caches and computer file storage systems employing formats such as CSV or JSON. The selection of storage technology directly impacts the performance of digital twin operations, necessitating meticulous consideration during implementation.

2.2 Research Gaps

The existing literature on historical data management in digital twins reveals several notable research gaps. Firstly, while there is recognition of the importance of historical data management, there is a lack of detailed implementation frameworks delineating the steps involved in deploying such solutions in digital twins. Secondly, there is a dearth of standardised approaches for data transformation, which is essential for harmonising heterogeneous data formats within digital twin environments. The absence of standardised methodologies impedes interoperability and consistency across digital twin applications. Lastly, while optimising data storage technologies is crucial for digital twin performance, there is limited empirical evidence evaluating the performance impact of different storage technologies in real-world scenarios.

In light of these identified research gaps, this paper endeavours to proffer a novel architecture for historical data management within digital twin frameworks alongside a methodology for data transformation explicitly tailored for digital twin contexts. Furthermore, the paper aims to present empirical data by performing performance benchmarking of the proposed architecture compared to existing historical data management solutions in the literature.

3 Historical Data Management Solution

As established in the literature review, historical data management in digital twins has significant importance and interest. This section discusses the historical data management architecture we propose for digital twins.

3.1 Digital Twin Historical Data Classification

Given the heterogeneous nature of data within digital twins, it becomes imperative to categorise the various types of data inherent in these systems. Such classification facilitates more refined modelling of a historical data management architecture and ensures their suitability within the manufacturing context. In our paper, we propose a classification scheme based on the frequency of data change, which we contend is particularly pertinent within digital twins. By considering the frequency of data change, the data in digital twins can be categorised into three distinct classes: static data, dynamic data, and periodic data; the basis for this classification stems from our interest in capturing the evolution of data over time. These classifications of digital twin data are illustrated in Fig. 1.

Static Data. This category of data encompasses information from the physical entity that remains unchanged over the life cycle of the digital twin. This data type is similar to those found in the Asset Administration Shell (AAS) in various industry 4.0 frameworks [34]. Examples of static data include equipment dimension, specification, name, material properties, and fixed system configurations. Due to its stability and persistence, static data is the bedrock for creating digital twins.

Dynamic Data. In contrast, dynamic data represents continuously evolving parameters within the digital twin. This class of data embodies real-time sensor readings of both the physical entity and environmental parameters. Characterised by frequent changes, the dynamic data drives the real-time nature of the digital twin. Additionally, the dynamic data is at the core of the historical data generated in the digital twin.

Periodic Data. Periodic data exhibits a hybrid nature, showcasing periodic alteration while maintaining a degree of stability over the lifetime of the digital twin. This category encompasses configuration settings that may change in inconsistent intervals which are not as frequent as the dynamic data, examples of such data change include, intermittent operational parameters, modification of CAD files during the production process, and predictive maintenance schedules based on analysis of historical data in the digital twins.

The different classifications of data require data storage solutions suitable for the nature of the data. A tabular database is adequate for storing static and periodic data because of its stable nature, while a time-series database is suitable for dynamic data.

3.2 Proposed Historical Data Management Architecture

We propose an architecture for historical data management in digital twins based on the data classification. The proposed historical data management architecture involves a registration process, the use of data transformation to create a unified data schema for the data acquired from the physical entity, database selection based on the nature of the data, and a method for linking related data.

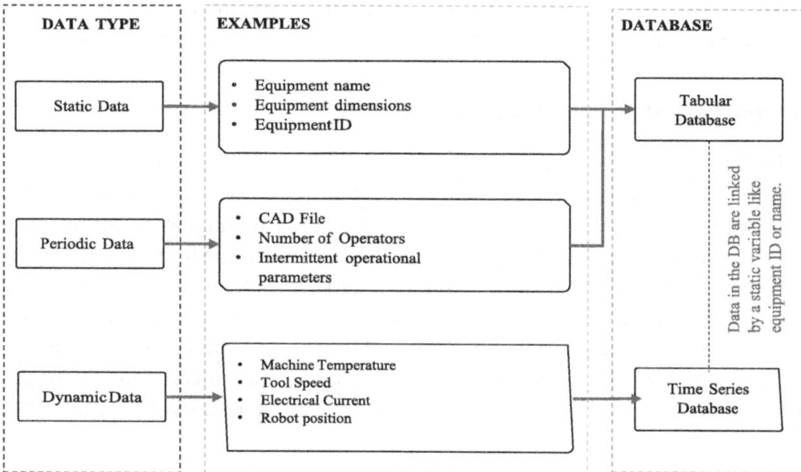

Fig. 1. Overview of classification of digital twin data.

Data Registration, Transformation and Linking Process. This process focuses on generating unique IDs for the different parts of the digital twin model. After the unique IDs have been generated, metadata of the attributes are stored in a reference table, this table allows for retrieving of the data at run time of the digital twin, and it also allows adding extra context to the time-series data. After the registration process, the next phase is the data transformation. The data transformation process links attribute metadata and attribute values into a unified schema. The unified schema aims to create a data block that is enriched with the contextual information of the data while preserving the original value of the data. The contextual information includes the class of the data, the ID of the attribute that owns the data, attribute-specific metadata, the class of the data (static, dynamic, or periodic), and lastly the timestamp when the data was obtained from the physical entity which can enable the construction of an accurate timeline of the physical attribute data. This data schema is described by the set in Eq. 1.

$$D = \{I, C, V, M, T\} \tag{1}$$

where:

D is the transformed data.

I is the ID of the attribute of the physical entity the data was acquired from.

C is the class of the data (static, dynamic, or periodic).

V is the value of the data obtained from the physical twin attribute.

M is attribute-specific meta-data.

T is the timestamp when the data was obtained.

This data transformation creates a standardised interface for storing the historical data of the physical entity within the digital twin framework.

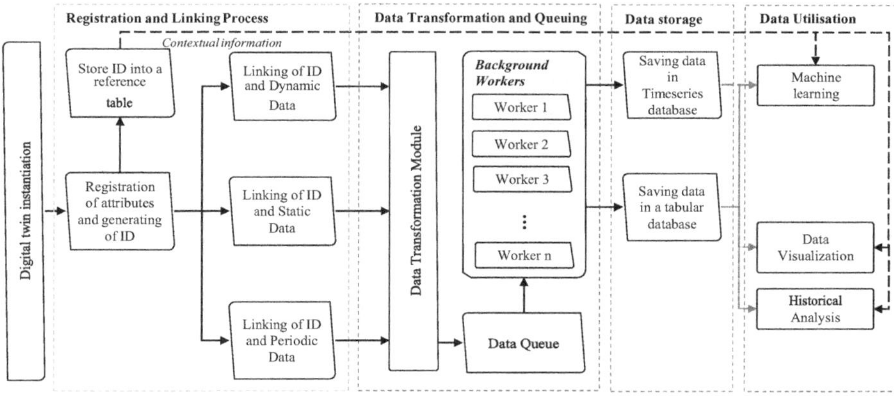

Fig. 2. Digital twin historical data management framework workflow.

Data Queuing. Data writing is a type of I/O operation which is blocking. This blocking produces a bottleneck which affects the ability of the digital twin to deliver real-time performance which makes synchronous data writing unsuitable for digital twin applications. To eliminate this blocking, we propose the use of a queue, to enable asynchronous storage of the data into the database. Since the data has been transformed, and enriched with contextual data such as the ID and the Timestamp, the order of entry into the database is no longer of importance, as the timeline can be reconstructed with the contextual data. This enables the insertion of the transformed data packets into the queue as they arrive and the utilization of background workers to store them in the database. Naturally, there is the potential for a backlog to build up in the queue, particularly if the arrival rate of data packets into the queue is greater than the departure rate of data packets out of the queue. This backlog has to be dealt with when shutting down the digital twin to avoid data loss.

Data Storage. This phase of the architecture determines the data storage mechanism. It comprises a tabular database for static and periodic data, a time-series database for dynamic data, and a module for determining which database to use. The tabular database is suitable for static and periodic data because of the low frequency of change and the ability to represent the fields of the transformed schema as columns, while on the other hand, the time-series database is optimised for storing high-frequency data, which is exactly what the dynamic data is. The decision of which database to write the data to is determined by the data class information in the schema of the transformed data object.

The proposed architecture enables digital twin services or other data analytic services to access the historical data of the digital twin from the relevant database and to utilise the reference table to obtain contextual information about the data. The proposed architecture workflow is illustrated in Fig. 2.

4 Implementation and Results

This section presents an implementation of the proposed digital twin historical data management. We implemented this architecture on the digital twin of a UR5e 6-axis collaborative robot with integrated sensors. The digital twin is developed to obtain real-time joint position and temperature information from the six joints of the robot, as illustrated in Fig. 3. The digital twin of the robot is modelled using a digital twin data structure (DTDS) [35].

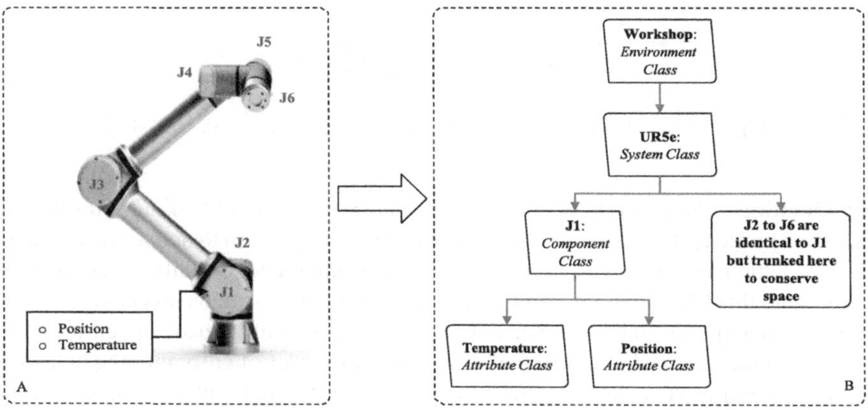

Fig. 3. (A) UR5e Robot with labelled joints and joint attributes, (B) UR5e digital twin data structure.

4.1 Implementation

The digital twin of the UR5e robot is modelled using the DTDS with the Python programming language (Fig. 3). During the instantiation process, unique IDs are generated for the joint attributes and this attribute information is stored in the reference table to allow for contextualising the data in the time-series database; an example of the reference table for the base and shoulder joint attributes of the UR5e robot is presented in Table 2.

The data within the DTDS is transformed into the proposed schema by associating the attribute ID with its corresponding value. This transformed data is then placed in a thread-safe queue, facilitating the utilisation of a background thread for synchronous data writing. This workflow is illustrated in Fig. 4. In this implementation, SQLite is used as the tabular database [36], while InfluxDB is used as the time-series database [37].

A benchmarking test is conducted to evaluate the performance of the proposed digital twin data management architecture. This test involves tracking the real-time position and temperature of the UR5e robot's joints while executing continuous movement over a five-minute duration. The architecture is benchmarked against those proposed by Fonseca

Table 2. Example of a reference table for UR5e base and shoulder joint attributes.

Attribute Name	Attribute ID	Parent Name	Metadata
Base Joint Position	A-0xx-xxx	UR5e Base Joint	$M1$
Shoulder Joint Position	A-1xx-xxx	UR5e Shoulder Joint	$M2$

et al. [14], which writes the digital twin data to a JSON file and stores it in the computer file storage system, and Jiang Haifan et al., which writes all the digital twin data to an SQL database without any specified data transformation [19]. These approaches are recreated with Python and the relevant database storage systems as modules, based on the description presented in the papers and are used within the DTDS.

4.2 Results

The various architectures undergo benchmarking based on two performance parameters typically used in I/O intensive applications; the throughput, measured in requests per second, and latency, measured in milliseconds. Throughput signifies the volume of data requests the digital twin can handle within a specified timeframe. This metric considers the duration taken by the digital twin historical data management system to process and store data before delivering the requested information to the caller.

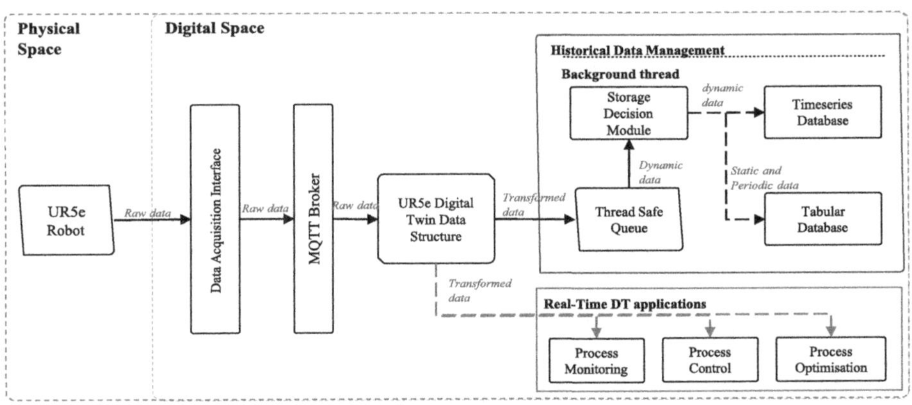

Fig. 4. Data flow for real-time data utilisation and historical data storage in the digital twin implementation.

Conversely, latency represents the time taken for data to be returned to the caller. This metric is important as the digital twin aims to provide real-time data, necessitating minimal latency. A third metric, the "time to save a packet" is used to evaluate the performance of the proposed architecture for writing data into the database, this metric directly affects the time taken to clear the data storage queue backlog. These experiments are conducted at three different data acquisition frequencies: 60 Hz, 120 Hz, and 500 Hz (which is the maximum data acquisition frequency of the UR5e robot) using a Dell laptop with 32 GB of RAM and an Intel i9-11950H 2.60 GHz processor; the average value of the benchmark parameters for five runs at a given frequency is recorded. The "time to save a packet" benchmark is conducted at the 500 Hz data acquisition frequency; it is further expanded by examining the effect of the number of worker threads on the metric. The results of these experiments are presented in Figs. 5 and 6 and are subsequently discussed in Sect. 5.

5 Discussion and Conclusion

5.1 Discussion

The benchmarking results provide valuable insights into the performance of the different historical data management architectures at varying data acquisition frequencies. The throughput metric indicates the capacity of the digital twin historical data management system to handle data requests within a given time frame. The proposed historical data management architecture demonstrates consistently high throughput across all frequencies, outperforming Fonseca et al. and Jiang Haifan et al. architectures at all data acquisition frequencies. The proposed architecture displayed a consistent level of throughput that surpasses the maximum data acquisition frequency; indicating that architecture can be utilised in applications with data acquisition frequencies up to approximately 2500 Hz. Latency measures the time taken for data to be returned to the caller, and lower latency signifies quicker response times, which is crucial for real-time applications.

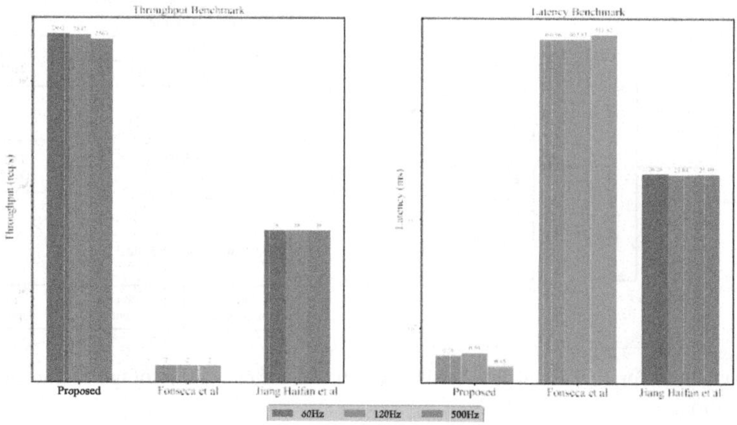

Fig. 5. Digital Twin data management benchmark results. (best seen in colour).

The proposed architecture exhibits significantly lower latency compared to the other architectures in all the data acquisition frequencies, indicating that it is suitable for use in low-latency digital twins that require historical data management. In the proposed architecture, "time to save a packet" varies with the number of worker threads for writing data into the database as illustrated in Fig. 6. The results indicate that using more than one thread results in a faster saving time. However, from the experiments, the optimal number of threads to use with the proposed architecture is 4. These results suggest that the proposed architecture is suitable for utilisation in digital twins and that it compares sufficiently with the other architectures. The consistent performance of the historical data management system is due to the asynchronous nature of the data writing. The proposed architecture improves the real-time response of the digital twin while ensuring that the management and storage of the historical data are not discarded.

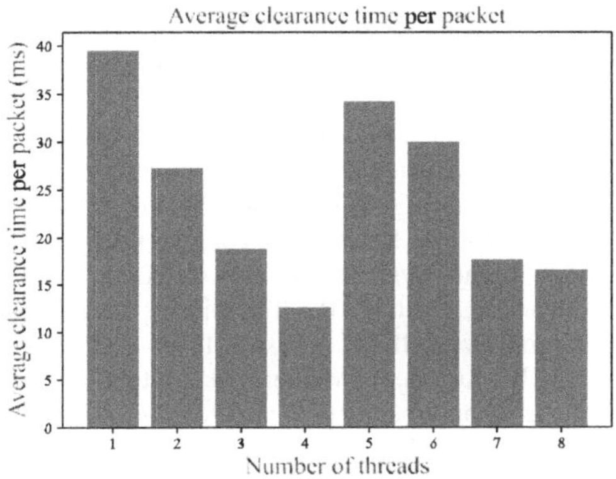

Fig. 6. Average time to save a data packet in a data storage queue at 500 Hz data acquisition vs. the number of threads.

5.2 Conclusion and Future Work

In conclusion, this paper presents a comprehensive historical data management solution tailored for digital twins within the manufacturing domain. Through a literature review and analysis of existing research, we identified notable gaps in implementing historical data management systems in the digital twin paradigm. Subsequently, we proposed a novel architecture that addresses these gaps, incorporating data classification, registration, transformation, and tailored database selection processes. Implementing the proposed architecture on a digital twin of a UR5e robot demonstrated promising results in terms of throughput and latency, showcasing its effectiveness in handling real-time data acquisition and management tasks. While the proposed architecture exhibited competitive performance across different data acquisition frequencies, further experiments must be conducted to determine the cause of this limitation by testing different time series databases. Moving forward, strategies to enhance the performance of the proposed architecture will be explored, to address the potential scalability issues with handling real-time data streams. Implement the architecture on other digital twin models to validate the architecture. There is also a question about data security and privacy that needs to be addressed.

Acknowledgements. This work has been supported by the French ANR PRC grant COHERENCE4D (ANR-20-CE10-0002).

Disclosure of Interests. The authors have no competing interests to declare that are relevant to the content of this article.

References

1. Grieves, M., Vickers, J.: Digital Twin: Mitigating Unpredictable, Undesirable Emergent Behavior in Complex Systems. Transdisciplinary Perspectives on Complex Systems, 85–113 (2016)
2. Yi, L., Glatt, M., Ehmsen, S., Duan, W., Aurich, J.C.: Process monitoring of economic and environmental performance of a material extrusion printer using an augmented reality-based digital twin. Additive Manufacturing (48), 102388 (2021)
3. Liu, M., Fang, S., Dong, H., Xu, C.: Review of digital twin about concepts, technologies, and industrial applications. J. Manuf. Syst. **58**, 346–361 (2020)
4. Autiosalo, J., Vepsalainen, J., Viitala, R., Tammi, K.: A feature-based framework for structuring industrial digital twins. IEEE Access **8**, 1193–1208 (2020)
5. Hribernik, K., Cabri, G., Mandreoli, F., Mentzas, G.: Autonomous, context-aware, adaptive digital twins—state of the art and roadmap. Computers in Industry (133), 103508 (2021)
6. Abburu, S., Berre, A.J., Jacoby, M., Roman, D., Stojanovic, L., Stojanovic, N.: COGNITWIN – hybrid and cognitive digital twins for the process industry. 2020 IEEE International Conference on Engineering, Technology and Innovation (ICE/ITMC) (2020)
7. Al-Ali, A.R., et al.: Digital Twin Conceptual Model within the Context of Internet of Things. Future Internet. (12), 163 (2020)
8. Ashtari Talkhestani, B., et al.: An architecture of an intelligent digital twin in a cyber-physical production system. at - Automatisierungstechnik. (67), 762–782 (2019)
9. Tao, F., Zhang, M., Nee, A.Y.C.: Five-Dimension digital twin modeling and its key technologies. Digital Twin Driven Smart Manufacturing, 63–81 (2019)
10. Lee, J., Bagheri, B., Kao, H.-A.: A cyber-physical systems architecture for industry 4.0-based manufacturing systems. Manufacturing Letters (3), 18–23 (2015)
11. Josifovska, K., Yigitbas, E., Engels, G.: Reference framework for digital twins within cyber-physical systems. 2019 IEEE/ACM 5th International Workshop on Software Engineering for Smart Cyber-Physical Systems (SEsCPS) (2019)
12. Fan, Y., et al.: A digital-twin visualized architecture for flexible manufacturing system. J. Manuf. Syst. **60**, 176–201 (2021)
13. Fang, L., Liu, Q., Zhang, D.: A digital twin-oriented lightweight approach for 3D assemblies. Machines (9), 231 (2021)
14. Fonseca, Í.A., Gaspar, H.M., de Mello, P.C., Sasaki, H.A.U.: A standards-based digital twin of an experiment with a scale model ship. Computer-Aided Design (145), 103191 (2022)
15. Wang, K., Liu, D., Liu, Z., Wang, Q., Tan, J.: An assembly precision analysis method based on a general part digital twin model. Robotics and Computer-Integrated Manufacturing (68), 102089 (2021)
16. Luo, W., Hu, T., Zhang, C., Wei, Y.: Digital twin for CNC machine tool: modeling and using strategy. J. Ambient. Intell. Humaniz. Comput. **10**, 1129–1140 (2018)
17. Hyre, A., Harris, G., Osho, J., Pantelidakis, M., Mykoniatis, K., Liu, J.: Digital twins: representation, replication, reality, and relational (4Rs). Manufacturing Letters. **31**, 20–23 (2022)

18. Steindl, G., Stagl, M., Kasper, L., Kastner, W., Hofmann, R.: Generic digital twin architecture for industrial energy systems. Applied Sciences. (10), 8903 (2020)
19. Jiang, H., Qin, S., Fu, J., Zhang, J., Ding, G.: How to model and implement connections between physical and virtual models for digital twin application. J. Manuf. Syst. **58**, 36–51 (2021)
20. Moretti, M., Rossi, A., Senin, N.: In-process monitoring of part geometry in fused filament fabrication using computer vision and digital twins. Additive Manufacturing (37), 101609 (2020)
21. Gopalakrishnan, S., Hartman, N.W., Sangid, M.D.: Integrating materials model-based definitions into design, manufacturing, and sustainment: a digital twin demonstration of incorporating residual stresses in the lifecycle analysis of a turbine disk. J. Comp. Info. Sci. Eng. (21), (2020)
22. Conde, J., Munoz-Arcentales, A., Alonso, A., Lopez-Pernas, S., Salvachua, J.: Modeling digital twin data and architecture: a building guide with FIWARE as enabling technology. IEEE Internet Comput. **26**, 7–14 (2021)
23. Zhidchenko, V., Startcev, E., Handroos, H.: Reference architecture for running computationally intensive physics-based digital twins of heavy equipment in a heterogeneous execution environment. IEEE Access. **10**, 54164–54184 (2022)
24. Cai, Y., Wang, Y., Burnett, M.: Using augmented reality to build digital twin for reconfigurable additive manufacturing system. J. Manuf. Sys. (56) (2020)
25. Schroeder, G.N., Steinmetz, C., Pereira, C.E., Espindola, D.B.: Digital twin data modeling with AutomationML and a communication methodology for data exchange. IFAC-PapersOnLine. **49**, 12–17 (2016)
26. Liu, C., et al.: Digital twin-enabled collaborative data management for metal additive manufacturing systems. J. Manufact. Sys. (62), (2020)
27. Liu, W., et al.: A 5M synchronization mechanism for digital twin shop-floor. Chinese J. Mecha. Eng. (36) (2023)
28. Martinez, S., et al.: A digital twin demonstrator to enable flexible manufacturing with robotics: a process supervision case study. Production & Manufacturing Research. **9**, 140–156 (2021)
29. Moslått, G.-A., Padovani, D., Hansen, M.R.: A digital twin for lift planning with offshore heave compensated cranes. J. Offshore Mecha. Arctic Eng. (143) (2020)
30. Bong Kim, D., Shao, G., Jo, G.: A digital twin implementation architecture for wire + arc additive manufacturing based on ISO 23247. Manufacturing Letters. **34**, 1–5 (2022)
31. Barbieri, G., et al.: A virtual commissioning based methodology to integrate digital twins into manufacturing systems. Prod. Eng. Res. Devel. **15**, 397–412 (2021)
32. Vodyaho, A.I., Zhukova, N.A., Shichkina, Y.A., Anaam, F., Abbas, S.: About One Approach to Using Dynamic Models to Build Digital Twins. Designs (6), 25 (2022)
33. Zhao, L., Fang, Y., Lou, P., Yan, J., Xiao, A.: Cutting parameter optimization for reducing carbon emissions using digital twin. Int. J. Precis. Eng. Manuf. **22**, 933–949 (2021)
34. Abdel-Aty, T.A., Negri, E., Galparoli, S.: Asset administration shell in manufacturing: applications and relationship with digital twin. IFAC-PapersOnLine. **55**, 2533–2538 (2022)
35. Orukele, O., Polette, A., Lorenzo, A.G., Mari, J.-L., Pernot, J.-P.: A Data Structure for Developing Data-Driven Digital Twins. IFIP advances in information and communication technology, 25–35 (2024)
36. SQLite: SQLite Home Page, https://www.sqlite.org/
37. InfluxData: InfluxDB | Real-time insights at any scale, https://www.influxdata.com/home

Enabling Ontology-Based Knowledge Linking in Smart Product Innovation Processes

Damun Mollahassani(✉), Martin Becker, and Jens C. Göbel

Institute of Virtual Product Engineering (VPE), Rheinland-Pfälzische Technische Universität Kaiserslautern-Landau (RPTU), Gottlieb Daimler Str. 44, 67663 Kaiserslautern, Germany
damun.mollahassani@mv.rptu.de

Abstract. In the state of art innovation processes different domain-specific products or subsystems are integrated into single systems, resulting in the engineering of multidisciplinary smart products. Through the integration of different disciplines and stakeholders, value chains are more and more transformed into complex value-creation networks for the development of these new product. These interdisciplinary innovation processes need to be supported regarding collaboration and especially knowledge-based collaboration by considering existing expert knowledge of different engineering domains. Knowledge sharing can tremendously impact the progress and success of innovation processes and help to overcome these upcoming challenges. A knowledge base and a supporting knowledge network are required to enable knowledge sharing across value creation networks. This paper introduces a smart product domain ontology that will support knowledge networking within an existing knowledge base. The knowledge base uses a multi-level tagging approach to create a network of different knowledge models, which needs to be supported by an automatic ontology-based networking. Based on these requirements, an analysis of various existing smart products and ontologies is conducted to develop a new ontology in the Protégé ontology-editor that considers and integrates previously defined ontologies for microelectronics. The main focus of this approach is the support of knowledge sharing by enabling enhanced knowledge networking. The ontology will be validated by using it as an application ontology with the validation use case automotive wireless smartphone charging.

Keywords: Ontology · Smart Products · Knowledge Management · Knowledge Base

1 Introduction

The transformation of traditional mechatronic products towards interdisciplinary smart products increases the complexity and number of involved stakeholders during the development phase [1]. Product-related and service-oriented aspects are combined for the servitization of smart products [2]. Simultaneously, the increasing complexity requires a holistic collaboration of all involved stakeholders. At the same time, the value chain

Published by Springer Nature Switzerland AG 2025
P. Sureephong et al. (Eds.): PLM 2024, IFIP AICT 740, pp. 166–179, 2025.
https://doi.org/10.1007/978-3-031-93319-6_13

oriented development undergoes a transformation into a development in value creation networks [3]. An early collaboration in a value creation network can help to handle the complexity of the process and accelerate the development phase [4]. Especially in the innovation phase, regular and comprehensive collaboration can mitigate the effects of missing expertise and competencies a single stakeholder may have. Regarding the collaboration in the early phases of the development, they still have not reached their maximum and have deficits in regard to quality, tempo and especially in volume [5, 6]. The sharing of knowledge in such early phases can be vastly beneficial for the ongoing innovation development. To guarantee the required willingness and volume of collaboration a supporting method and environment are needed [7]. A knowledge base can be used for the management of the available knowledge but need additional capabilities to network it. The network of knowledge can create synergies between different stakeholders and create new knowledge, due to the combination of different knowledge states [8]. Networking of knowledge related to smart products becomes more complex, through the aspect of the heterogeneity of the related knowledge, which is caused by the interdisciplinarity of smart products and their different components [9]. An ontology can be used for the networking of interdisciplinary smart products knowledge.

The approach in this paper presents an ontology-based networking within an innovation knowledge base, to enable the knowledge-based collaboration. Specific characteristics and especially the service aspects of smart products are considered, for consistent and complementary networking through an ontology. For an individual usage in the context of smart products, the approach can be customized for specific use-cases. Thereby a domain ontology will be enriched to an application ontology. The goal of the approach is to create access to knowledge and set it in relations between different stakeholders, within an innovation value creation network using an enhanced knowledge base.

2 State of the Art

Smart Products are characterized by the combination of a physical product and an internet-based service. Thus, smart product innovation processes require the combination of different disciplines based on specific domain knowledge [10]. The innovation phase requires a large amount of indirect knowledge, which can influence innovation development [11]. Indirect knowledge in this context is defined as knowledge of and about subsystems that are originally not considered but influence the innovation. The innovation phase is a very early stage in the development lifecycle, where collaboration can be done without any restrictions or risk of competence loss. Additionally one of the highest impacts on the innovation process can be reached in the following phases [12]. An example of such a smart product is the innovation of an automotive wireless charger, which will be used later as a validation use case of the ontology. The wireless charger allows the wireless charging of a mobile device while driving. The concept of wireless charging combines different disciplines and domains from automotive, microelectronics and seems to be a simple innovation, but in case it is very complex due to the effect of an integration into an another system and interaction with it. The corresponding stakeholders all have expertise and knowledge about the innovative idea, but from their point of view exclusively. To successfully develop an innovation, the available and shared knowledge needs to be set in relation to each other.

A multitude of tools capable of setting and managing such relations as well as facilitating the networking are available, e.g.: glossaries, thesauri, meta-data or ontologies [13]. The concept of knowledge networking is essential for the successful development of innovation and the transfer into a real product. Often, only new external knowledge allows common development, as it is used in the concept of open innovations [14]. Regarding the development within a value creation network, different stakeholders can integrate their knowledge or share it with other relevant stakeholders. At the same time, as a result of the collaboration, different stakeholders can create and use new knowledge [15]. This newly created and the previously existing knowledge needs to be networked, by creating relations between the different knowledge aspects. Various concepts of networking approaches can be found in literature, such as the concept of thesaurus, which can be used for networking using tags and a harmonization of the nomenclature with a common dictionary [16]. These networking concepts can be used in knowledge bases for correct knowledge management and provision. This basic networking with a thesaurus is often limited to the nomenclature and has no focus on the content itself. An example of a thesaurus being used in this way can be found in [16], which is a knowledge management approach that uses a tag-based thesaurus for the harmonization of the nomenclature. The following knowledge management approach is later used for an instantiation and validation of the ontology approach of this paper. More specific and holistic networking is needed to guarantee the required knowledge-based collaboration. Therefore, ontologies can be used for a common and holistic networking of all involved knowledge models within a value creation network. Another future use case of such ontologies could be a structure for generative artificial intelligence. Which can help to enhance knowledge networking and support stakeholders while searching for additional unknown but related knowledge to them.

Aspects of smart products are considered in different ontologies approaches, which allows the knowledge networking of different subsystems. The approach of [9] focuses the context of smart home, which are often realized through the combination of different smart products in a combined ecosystem. Even though the aspect of internet-based services is present as well, it is not a general smart product ontology. The basis of this ontology is the Semantic-Sensor-Network ontology (SSNO) from the World Wide Web Consortium [17]. Different smart products that are used within the smart home context are modeled separately. The whole ontology is modeled with a minimum number of classes through a controller peripheral model. Aspects of physical components are summarized in one class, which makes the usage for smart products difficult because of the required differentiation of the components. The primary focus is the usage and the control of the different smart products in one ecosystem, therefore the Semantic-Sensor-Network ontology utilization as a smart product ontology is not sufficient. Nevertheless, the usage of the Semantic-Sensor-Network ontology is useful and can be used as a foundation for the approach to be developed.

Smart products need to be recycled, therefore the approach of [18] was developed and supports decision making processes regarding the recyclability. Therefore, three types of information are differentiated: components, architecture and usage. This is based on the Semantic-Data-Management approach [19]. Through the networking of these different knowledge classes, the decision making will be supported. Thus, smart products are an independent subclass of the class products within the ontology and focus on the communication capabilities. Further classes are material, usage or component that are all networked together. However, the focus is mainly on recyclability, but aspects of the architecture from the different classes can be used later as an inspiration. Smart product service systems are considered by the ontology presented in [20], which enables networking of the smart product, services and service enablers related to knowledge. The ontology is structured using the subdivisions products, service requirements, integration solutions and sensor properties. Through the usage of a collaborative platform, the ontology can create a common data model of related data which can be used by different domain engineers. Services are again considered by the Semantic- Sensor-Network ontology from the World Wide Web Consortium, which supports the aspect of a late re-usage of this ontology. Existing modules of the approach were connected to the Semantic-Sensor-Network ontology for holistic networking. This results in a main focus on sensors, which is addressed by this approach. Through the focus on the sensors, other related aspects of smart products are not taken into account, which makes it unsuitable as a networking tool for smart products knowledge.

Microelectronics are a main component that are used in smart products and can, for example, create the required capabilities. An ontology that focuses this aspect is the GENIAL! Basic Ontology, which was developed for the innovation roadmapping for microelectronics [20]. The reusability of this ontology in this approach need to be considered regarding the original use in the development of microelectronics with a focus on innovation. As a top-level ontology the Basic-Formal-Ontology [21] is used in this approach and is supported by the ontology units of measure [22], which itself uses ISO 26262 for orientation. Context, function, component, software part, hardware part or element are among other parts of the classification and are hierarchized to each other by the relation *"has_part"*. With the ontology units of measure, the results of the innovation development can be analyzed and networked together. The hierarchy is refined by the specific class relation such as "execution" for the software part. Aspects of smart products are not directly considered by the GENIAL! Basic Ontology, which makes directly reusing it impossible.

However, an adaption of the structure and used components of [20], which is similar to a smart product and a combination with the results of [18] can create a baseline for the development of the ontology, that helps with the networking of smart products innovation knowledge within a knowledge base.

3 Methodical Approach for Developing an Adapted Smart Product Ontology

The approach from Uschold and King [23] is used to develop the ontology, due to it being well-known and established. For an easier and faster development of the ontology, the approach of [23] was extended and adapted, with regard to smart products and is shown in Fig. 1. The different tasks are divided into: indentify of purpose, ontology capture, integration of existing ontologies, ontology coding and evaluation. The initial purpose of the ontology is to enable sharing and enriching of existing knowledge models within an innovation value creation network. The knowledge base in [24] is used during an innovation value creation network development and offers a knowledge base exploration space that includes different knowledge models from different stakeholders. Stakeholders in this case are different engineers with innovation focus, thereby they have knowledge and responsibility in the context of smart products and their respective components. Currently, the networking within the knowledge base is done by a three-step tagging process that includes one thesaurus and two content-based networking steps. The networking allows the stakeholders to perform searches easily with varying scopes to enhance and accelerate the knowledge acquisition process. In the current iteration all networking steps have to be performed manually. To enhance the continuous networking the utilized ontology must allow for a networking of the tags and therefore, the knowledge. The smart product ontology is defined as a domain ontology for the purposes of re-usage it in different use cases. With the usage of a specific product, it becomes an application ontology and can be used for specific smart product knowledge networking. Therefore, the final purpose of the ontology is to support the three tagging steps and especially to enable a condensed and automatic semantic networking of the different knowledge models.

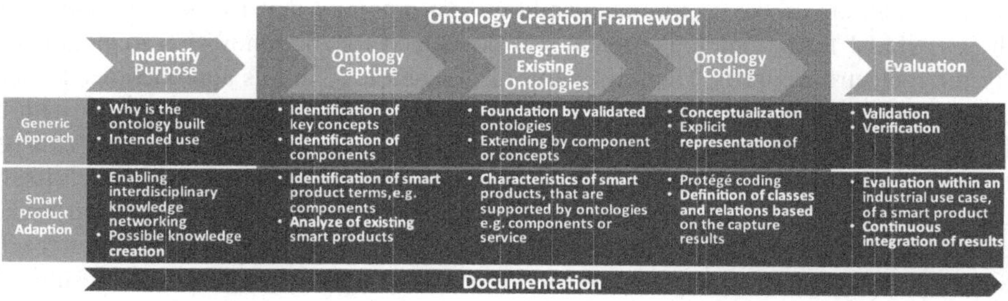

Fig. 1. Smart Product Ontology Development Process based on [23]

A foundation is required to conceptualize the ontology, which can be used later for the implementation purposes. Thereby key concepts and relations of smart products need to be analyzed. Additionally, research about core components within smart products must be done. Smart products are hierarchically structured in different and independent modules. To better characterize the ontology and subsequent subclasses, existing smart products were analyzed with a focus on the components. This procedure allows the realization of the task of the ontology capture, to define key concepts, components and relationships. Existing smart products, such as a smartphone, smart watch, Smart TV, Smart Fridge and a smart TV-Stick were analyzed with a focus on their components like sensors, software and inor output devices. The result was a matrix, which classified the different components into the modules: sensor, physical component, networking component, input device, actuator, energy source, case, processing unit, service, data storage, and software. An excerpt of this matrix is given in Fig. 2. These different modules are compatible with the previously in [2] defined components but do not describe it in a complementary way. As an example of this, a smart products physical component is represented by a case and input device. Electronic components are represented by energy sources and actuators. Within the networking components, the wireless networking is established via interfaces and a connection protocol. Additional examples for smart components are sensors, software, storage and user interfaces. For a refined description, the aspects of the internet and services must be considered and integrated as well [1]. Furthermore, the function of the smart product must be implemented.

	Smartphone	Apple Watch	Samsung Watch	Smart TV	Smart Fridge	Smart TV Stick
Sensor Components	x	x	x	x	x	x
Acceleration			x			
Temperature	x	x	x		x	
...						
Connectivity Components	x	x	x	x	x	x
Wi-Fi-Antenna	x	x	x	x	x	x
GPS Antenna	x	x	x			
...						
Service	x	x	x	x	x	x
Digital Assistant	x	x		x	x	x
Navigation	x	x	x			
...						

Fig. 2. Component Matric of Smart Products

Quality is another aspect that goes hand in hand with the functions and their satisfactory execution in the context of smart products. These functions are described by different stakeholders through requirements. These different components need to be represented hierarchically in the ontology and be cross-linking with their respective relations.

The next required step is the integration of existing ontologies of services and the different components. By using pre-existing and established ontologies, user acceptance can be increased, redundancy and errors can be avoided and effort can be reduced. As described in the previous chapter, the GENIAL! Basic Ontology is classified through the detailed description of the different components resulting from the decomposition into hardware and software components. This can be adapted into the smart products ontology. The GENIAL! Basic Ontology also includes the Basic-Formal-Ontology and the Ontology for units and measure, thus these ontologies will be integrated too. For the aspect of services, the Web Ontology Language for Services (OWL-S) [25], which describes services in semantic networks, will be merged. Within the OWL-S, the services are presented by a profile, supported by a basis and described by a model. An overview of the integrated ontologies and their scope is given in Fig. 3.

Coding of the smart product ontology is done by using the ontology editor Protégé [26]. Protégé facilitates the easy integration of existing ontologies and an usage of plugins for visualization. Additionally, classes or properties can be described multilingual. As mentioned, the OWL-S is merged completely. From the GENIAL! Basic Ontology the classes of element are reused, that includes the hardware and software, the actuator, context, quality, unit, requirements, role and sensor. These classes are also part of the Basic-Formal-Ontology [21] and the ontology units of measure [22]. The smart product ontology has therefore the same basis as the GENIAL! Basic Ontology and is built up from the classes entity, continuant, element, element of the system level and element of the non-system level. For the smart product ontology these classes will be extended and customized. Therefore, in Fig. 4 is a complete overview of the smart product ontology is shown, which has two focuses highlighted, that describe this customization and extension. Focus 1 describes the smart components of a smart product, which is part of the extension to the GENIAL! Basic Ontology. Smart products are mentioned as smart systems within the ontology, for larger coverage of use cases and to allow for the combination of different devices into one system. In focus 1 smart components and connections components are added and the physical component is divided into a mechanical and electrical component. This subdivision is required due to the previous component research on smart products. Software is a fundamental component and is realized through the software component, which is part of the smart component, that includes, the sensors, storage, or user interfaces. These different components are part of the required components for the realization of a smart product and are based on the component matrix, as shown in Fig. 2. Services are a further enabler and are fully included by the OWL-S and represent a required class that is offered by the smart system. The aspect of quality, which also considers units, properties, and functions, is imported by the GENIAL! Basic Ontology [20] and originally from the Basic-Formal-Ontology [21] and the ontology units of measure [22].These classes are required for services, as they allow the smart product to analyze data and generate information from it. This information can then be used to offer internet-based and use-case-specified services. Functions especially need to be refined regarding product and element functions, as is represented in focus 2 in Fig. 4. Product functions describe parts that are realized by a smart product such as, monitoring or optimization. Element functions focus on different functions from the non-system level, such as controlling or measuring. Another aspect that needs to be considered is the added subclass context. As shown in focus 2, context allows aspects like roles, stakeholders, standards, the environment, or the internet to be focused. Stakeholders and especially their different roles become an important class due to the different required disciplines for the realization and running of a smart product. In focus 2, the roles are subdivided into more concrete roles for an easy integration of different knowledge views of the different roles. By the direct combination of the subclass's role and stakeholder, each stakeholder must represent at least one role within the value creation network. Each of these stakeholders will include stakeholder-related requirements for the innovation. The stakeholders are then included in the context, which will allow them to define the smart product holistically. Overall, the smart product ontology includes 69 different classes, 29 different object-related relations and 3 data-related relations, an excerpt of this is shown in Fig. 4.

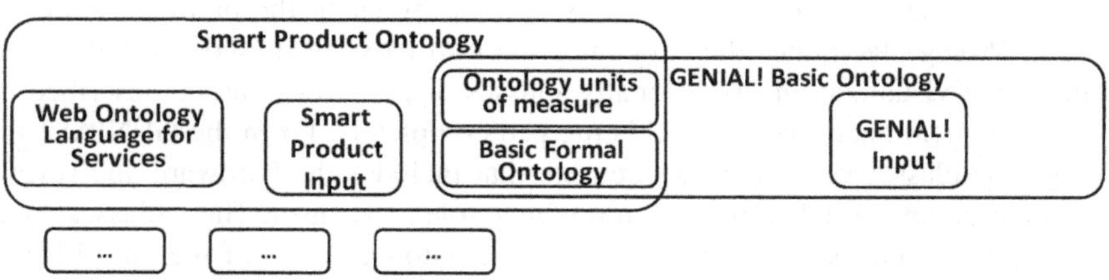

Fig. 3. Overview of Integrated Ontologies

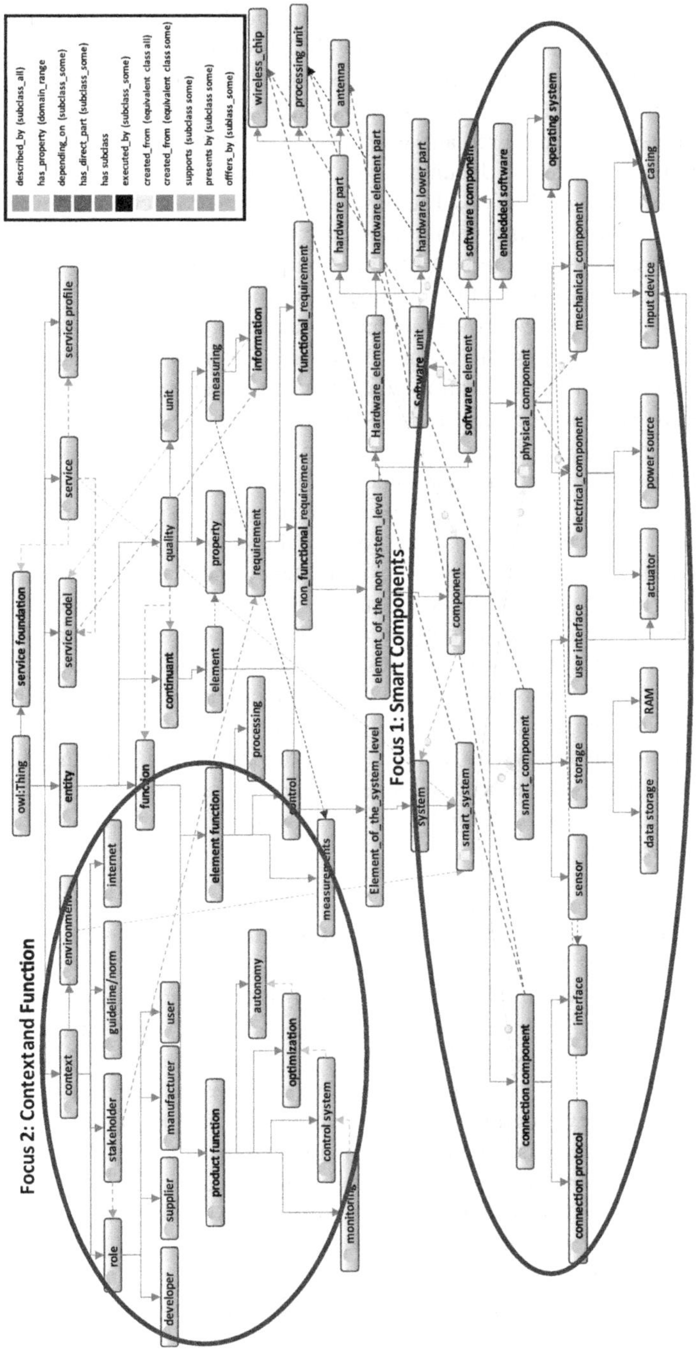

Fig. 4. Overview of the adapted Smart Product Ontology

As the last part of the Uschold and King approach, the smart product ontology needs to be evaluated from the theoretical view. With the formalized research and analyze of existing smart products, the required concepts and relations could be modeled and guarantee the correct scope. For a detailed validation of the ontology, a validation utilizing an industrial use case is required and will be done in the following chapter.

A continuous to-do is the documentation of all tasks and results, from the indentify of the purpose to the evaluation. The developed smart product ontology was initially developed for the networking within a specific knowledge base, but can now be used widely in different use case and for different purposes. The so-called potentials of the ontology can be separate into different categories. Knowledge linking is the first purpose of the ontology. During the innovation phase it can support the combination of different existing knowledge models and ensure the creation of new knowledge. This purpose is not only limited on the innovation phase but can be adapted into different lifecycle phases as well. Stakeholder networking is another aspect that can be done by using the ontology. Regarding the innovation phase within value creation networks, the possibility of the involvement of unknown stakeholders is given. By networking the knowledge, the stakeholders are networked as well, which is an important aspect during the innovation phase. Another important potential that focuses the innovation phase is the early safeguards regarding completeness. Ensuring consistency and completeness is an important feature in development and at the same time, an early ensuring is required.

As previously described the ontology will be validated in the next chapter by the wireless charger innovation, which represents a single function. By using the ontology on more complex products with different sub-functions, a divation of the system is required. Through the divation the ontology can be used for each layer, this step can be supported by the existing system architecture of the innovation. The knowledge of each of these different subsystem need to be networked seperately by a different application ontology. These different ontology can then be organized and manged by an overall domain ontology.

4 Instantiation of the Smart Product Ontology

There are three steps required for the validation of the smart product ontology and to provide a reliable usage during the later knowledge networking. In the first step, the consistency of the ontology will be checked. Afterwards the customization and as the last step the possibility of an integration in a suitable environment must be proven. The smart product ontology is defined as a domain ontology and for a validation of it, it needs to be enriched by a use case and become an application ontology. By this transfer, the correct usage can be analyzed and the ontology validated. Criteria about the correctness, usability and applicability must be highlighted too. To guarantee correctness, the HermiT OWL reasoner is used to identify possible inconsistencies [27], which represents the first step of the validation.

For the second step of the validation, a wireless charger application ontology will be specified, using the smart product domain ontology as a basis. The new application ontology then support the knowledge collaboration and networking within the knowledge base exploration space [24]. Initial requirements must be defined to characterize the innovation. Different required and already-known components and restrictions can then be consolidated in a matrix, as was done in the previous section. Based on the initial requirements, the application ontology will be created. Dynamic usage of the ontology within the knowledge base and continued extension of the ontology are provided to satisfy upcoming conditions. Through the common development within a value creation network, different stakeholders are involved. In the case of the wireless charger, these could be stakeholders from the disciplines mechanics, electrics, and software development, which makes the dynamic adjustment so important. These different stakeholders define each subsystem developed by them, as well as components and necessary restrictions. Due to the usage of the wireless charger within another smart product, in this case a car, not only the concept and properties of the smart product ontology must be adjusted, but the environment and interaction with other systems as well. After the definition of all pre-known components and restrictions, the ontology must be extended to different classes and subclasses. The subclass wireless charger must be added under the smart system, which is then a subclass of the class element. Thus, the wireless charger represents all previously defined properties and requirements of a smart product. Due to this integration, the smart, physical and connectivity components must be adapted and expanded. In Fig. 5, an excerpt from this extension is shown, where wireless charger-specific components were added to the class of physical components. All subclasses in Figs. 5 and 6 highlighted in red, are added through the use case wireless charger. On the mechanical side, there are classes like heat sink or shielding added. They are needed as they allow other knowledge models concerning cooling can be networked. At the electrical component site, e.g., the coil module is added, which is responsible for the charging procedure. Other important components, are the specific temperature sensor, and the antenna board which were added to the smart components and the connectivity components respectively. Not only were the component classes extended within the ontology, but the classes regarding the quality, context and service as well. The quality must be extended concerning the functional requirements, towards safety a correct energy transfer. As an example, the class Celsius must be added to the class of units, as it is used in a metric environment and is required for temperature control. Within the context the environment of the wireless charger will be detailed, as shown in Fig. 6. An important restriction is added by one of the stakeholders, which considers the keyless-go system and is another smart product, that can have influence or be influenced by the wireless charger and thus knowledge of the keyless-go system must also be considered and networked. Recharge is also added and is a subclass of service. For the standards, the subclass Qi-standard is added due to the previously detected and set energy transfer standard. Due to the addition of the different selected components and the extension in regard to environment and quality, the smart product ontology is customized to a wireless charger ontology. To check the accuracy of the ontology, it will be integrated into a knowledge base and support the knowledge networking of different models. The knowledge base exploration space offers a tag-based networking of knowledge models, such as Computer Aided Design (CAD),Systems Modeling Language (SysML) or a Business Process Modeling Notation (BPMN) models. These models are imported by the different involved stakeholders and represent a short overview of the wireless charger innovation model. Thereby the models need to be networked by tags, for a condensed and better structured search. For the last step of the validation a correct networking, within the knowledge base is required and will be done automatically by the ontology. The tagging of the different models can be realized by the owner of the innovation and ontology or by any of the involved stakeholders.

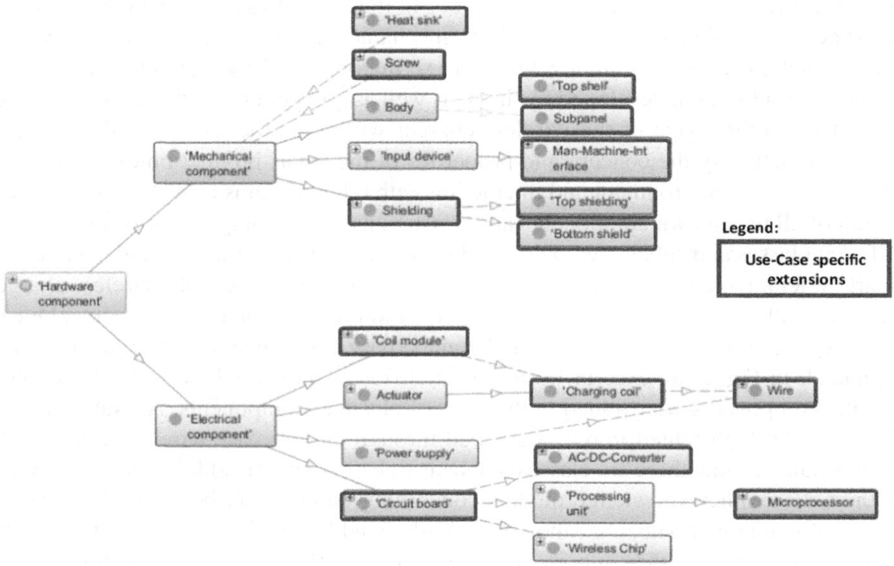

Fig. 5. Physical Components and Parts of the Wireless Charger

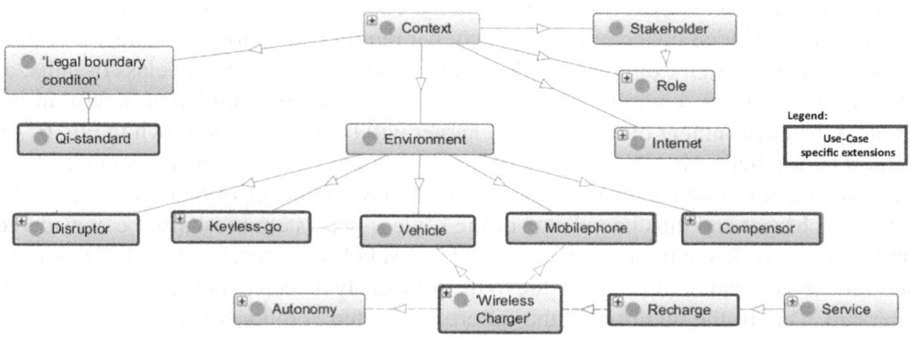

Fig. 6. Context, Service and Function of the Wireless Charger

Tagging by stakeholders can be faulty, therefore, the ontology will support the value creation network. Around twelve different tags exist and are used for networking within the knowledge base with a description and their matching synonyms. These synonyms are added to the tags by the different stakeholders. For the validation, the tagging of the temperature sensor will be analyzed and used for the ontology. All tags can be set with the relation has_part, has_influence and their inverse to the different knowledge models and groups. First, we have the tags: sensor, hardware, safety, software and coil, which is also a synonym of the tag inductor. The used subclasses within the ontology must be matched and networked with the used tags in the knowledge base as shown in Fig. 7. A special matching is the match between safety and the Qi-standard, which is based on the given safety regulations of the standard. Through the Owlready2-package [28] and sqlite3-module [29], the ontology and database can be connected. The twelve existing and used tags can now be retrieved and connected with the new subclasses of the ontology. As shown in Fig. 4, the ontology also uses different relations between the subclasses. The superclasses of these relations match with the relations of the knowledge base. By running the ontology, new tags can now be added to the temperature sensor model in Fig. 6. The tags sensor, antenna board, overheat protection, burn protection, charging and heat, which was added as a synonym for temperature, is now related to the temperature sensor. Some of these added tags were not part of the previously defined tags of the knowledge base, so they will be added automatically and are available for usage for all other stakeholders. By using the ontology iteratively, it was possible to enrich the available tags from twelve to fifty-five.

Class of Ontology	Tag of the Wireless Charger Model
AC-DC-Converter	AC-DC-Converter
Cabel	Cabel
Hardware Element	Hardware
Primary Coil	Coil
Stakeholder	Personal Group
Qi-Standard	Safety
Sensor	Sensor
Software Element	Software
Temperature Sensor	Temperature Sensor

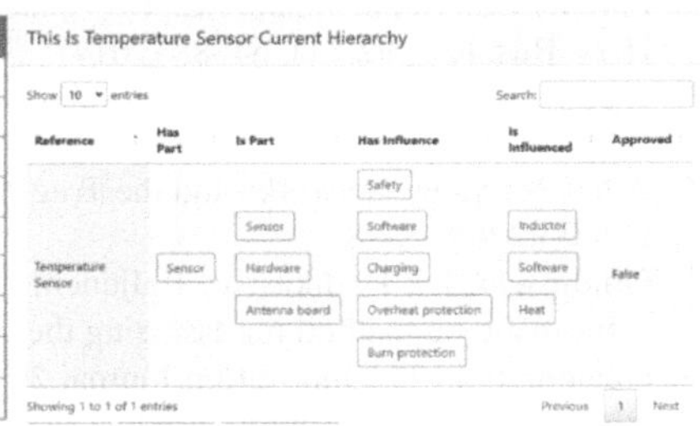

Fig. 7. Overview of the Matching Tags and Integration of the Ontology

5 Conclusion

Networking of heterogeneous knowledge is always complex, due to the difficult relationships between the referencing systems. Considering future smart products, there is an increasing challenge of how different disciplines and heterogeneous knowledge can be combined in one product. Thus, collaboration is a core competence for the ongoing and future innovation processes. During innovation processes knowledge sharing could accelerate the innovation development, especially in value creation networks with various stakeholders. For a successful knowledge sharing combined with collaboration, networking of various knowledge models are required. With the smart product ontology as a domain ontology, this networking can be done for each different smart product, especially during the innovation phase. By transferring the smart product ontology with a use case to an application ontology, it can be adapted to every application and be used individually. During the usage in the knowledge base exploration space, it has shown successfully that knowledge from different disciplines can be integrated and support a common development. Additionally, the networking allows for an easier knowledge search for the involved stakeholders, which in return results in an enhanced and better collaboration during the innovation. For an easier validation and continuous improvement, the smart product ontology must undergo more application transfers, and this feedback must be transferred back into the domain ontology. Another aspect that will be considered in future work is an automatic extension and simultaneous supplementation of the domain or application specific ontology classes based on artificial intelligence. An empirical validation is also planned, to solidify the results and to improve the applicability.

References

1. Abramovici, M., Göbel, J.C., Savarino, P., et al.: Towards smart product lifecycle management with an integrated reconfiguration management. In: Ríos, J., Bernard, A., Bouras, A., et al. (eds.) Product Lifecycle Management and the Industry of the Future, vol 517, pp. 489–498. Springer International Publishing, Cham (2017). https://doi.org/10.1007/978-3-319-72905-3_43
2. Porter, M.E., Heppelmann, J.E.: How smart, connected products are transforming competition. Harv. Bus. Rev. **92**(11), 64–88 (2014)
3. Ricciotti, F.: From value chain to value network: a systematic literature review. Manag Rev Q **70**, 191–212 (2020). https://doi.org/10.1007/s11301-019-00164-7
4. Allee, V.: Value Networks and the True Nature of Collaboration, Tam-pa, FL (2015). ISBN: 978-0-929652-52-8
5. Grudinschi, D., Hallikas, J., Kaljunen, L., et al.: Creating value in networks: a value network mapping method for assessing the current and potential value networks in cross-sector collaboration. The Innovation Journal **20**(2) (2015)
6. Horvath, L.: Collaboration: the key to value creation in supply chain management. Supply Chain Manag. **6**, 205–207 (2001). https://doi.org/10.1108/EUM0000000006039
7. Mollahassani, D., Gries, J., Forte, S., et al.: Supporting collaborative in-novation processes in smart product value creation networks. Procedia CIRP **109**, 349–355 (2022). https://doi.org/10.1016/j.procir.2022.05.261
8. Xu, L., Zeng, D.: When does the diverse partnership of R&D alliances promote new product development? The contingent effect of the knowledge base. Technol. Soc. **65**, 101590 (2021). https://doi.org/10.1016/j.techsoc.2021.101590
9. Berat Sezer, O., Can, S.Z., Dogdu, E.: Development of a smart home ontology and the implementation of a semantic sensor network simulator: An Internet of Things approach. In: 2015 International Conference on Collabo-ration Technologies and Systems (CTS), pp. 12–18. IEEE (2015). https://doi.org/10.1109/CTS.2015.7210389
10. Tomiyama, T., Lutters, E., Stark, R., et al.: Development capabilities for smart products. CIRP Ann. **68**, 727–750 (2019). https://doi.org/10.1016/j.cirp.2019.05.010
11. Bergman, J., Jantunen, A., Saksa, J.-M.: Enabling open innovation process through interactive methods: Scenarios and group decision support system. Int. J. Innov. Manage. **13**, 139–156 (2009). https://doi.org/10.1142/S1363919609002224
12. Geurts, A., Geerdink, T., Sprenkeling, M.: Accelerated innovation in crises: the role of collaboration in the development of alternative ventilators during the COVID-19 pandemic. Technol. Soc. **68**, 101923 (2022). https://doi.org/10.1016/j.techsoc.2022.101923
13. Uschold, M., Gruninger, M.: Ontologies and semantics for seamless connectivity. SIGMOD Rec **33**, 58–64 (2004). https://doi.org/10.1145/1041410.1041420

14. Chesbrough, H.W.: The Era of Open Innovation. MIT Sloan Management Review, Cambridge (2003)
15. Büchel, B., Raub, S.: Building knowledge-creating value networks. European Manage. J. **20**, 587–596 (2002). https://doi.org/10.1016/S0263-2373(02)00110-X
16. Aitchison, J., Clarke, S.D.: The thesaurus: a historical viewpoint, with a look to the future. Cataloging & Classification Quarterly **37**, 5–21 (2004). https://doi.org/10.1300/J104v37n03_02
17. W3C Semantic Sensor Network Incubator Group (2009–2011) Semantic Sensor Network Ontology. https://www.w3.org/2005/Incubator/ssn/ssnx/ssn. Accessed 05 Sep 2023
18. Jiang, H., Yi, J., Zhou, K., et al.: A decision-making methodology for the cloud-based recycling service of smart products: a robot vacuum cleaner case study. Int. J. Comp. Integr. Manuf. **32**, 58–71 (2019). https://doi.org/10.1080/0951192X.2018.1535195
19. Abramovici, M., Göbel, J.C., Dang, H.B.: Semantic data management for the development and continuous reconfiguration of smart products and systems. CIRP Ann. **65**, 185–188 (2016). https://doi.org/10.1016/j.cirp.2016.04.051
20. Wawrzik, F.: Knowledge representation in engineering 4.0, Technische Universität Kaiserslautern (2022). https://doi.org/10.26204/KLUEDO/6888
21. Ruttenberg, A.: BFO Basic Formal Ontology (2020). https://basic-formal-ontology.org/. Accessed 12 Jan 2024
22. Rijgersberg, H.: OM - Ontology of units of Measure (2023). https://github.com/HajoRijgersberg/OM. Accessed 12 Jan 2024
23. Uschold, M., King, M.: Towards a methodology for building ontologies. In: IJCAI International Joint Conference on Artificial Intelligence (ed) Workshop on Basic Ontological Issues in Knowledge Sharing (1995)
24. Mollahassani, D., Eickhoff, T., Juresa, Y., et al.: Knowledge collaboration approach in smart product innovation networks. Procedia CIRP **119**, 662–668 (2023). https://doi.org/10.1016/j.procir.2023.02.158
25. W3C World Wide Web Continuum: OWL-S: Semantic Markup for Web Services (2004). https://www.w3.org/submissions/OWL-S/. Accessed 12 Sep 2023
26. BMIR Stanford Center for Biomedical Informatics Research (2020) Protégé. https://protege.stanford.edu/. Accessed 17 Jan 2024
27. Knowledge Representation and Reasoning Group: HermiT OWL Reasoner (2023). http://www.hermit-reasoner.com/. Accessed 19 Jan 2024
28. Lamy, J.-B.: Owlready2 Documentation (2023). https://owlready2.readthedocs.io/_/downloads/en/latest/pdf/. Accessed 19 Jan 2024
29. Python Software Foundation sqlite3: DB-API 2.0 interface for SQLite data-bases. https://docs.python.org/3/library/sqlite3.html. Accessed 19 Jan 2024

Integrating Assembly Sequence Planning into Product Lifecycle Management

Rafael Parzeller[1,2(✉)], Tizian Dagner[1], Oliver Lohse[1], and Detlef Gerhard[2]

[1] Siemens AG, Munich, Germany
{rafael.parzeller,tizian.dagner,oliver.lohse}@siemens.com
[2] Ruhr-University Bochum, Bochum, Germany
detlef.gerhard@rub.de

Abstract. This paper presents a concept for the integration of Assembly Sequence Planning into Product Lifecycle Management. The use of Assembly Sequence Planning in the product lifecycle is described to show how it can enable Design for X. The presented concept covers the main phases of the product lifecycle and highlights the resulting advantages for the Product Lifecycle Management. Two new Assembly Sequence Planning methods based on Reinforcement Learning using a physics-based simulation environment are presented and validated.

Keywords: Assembly Sequence Planning · Artificial Intelligence · Reinforcement Learning · Design for X

1 Introduction

The concept of Product Lifecycle Management (PLM) involves designing and managing the entire life cycle of a product. It is important to consider all aspects of the product life cycle during product development [19]. In order to consider all aspects of the product life cycle, product development requires an increased automation through the use of 3D-CAD models. Assembly Sequence Planning (ASP) is a process that can derive the assembly sequence of a product based on the 3D-CAD model, through use of which ASP can create a link between assembly planning and product design, resulting in reduced manufacturing costs and increased automation throughout the product lifecycle [21]. The outcome of ASP, a simulation of the product assembly process, can be utilized in other stages of the product life cycle, including product usage and replacement. For instance, the ASP result can serve as a guide for disassembly during maintenance and recycling.

Although Industry 4.0 has led to an increase in technology in the manufacturing industry, some manufacturing processes are still implemented by humans. Furthermore, the demand for customized products is on the rise, resulting in an increasing number of variants. To meet these requirements, operators must be equipped with new technologies that enable them to complete their tasks more quickly and accurately. Extended reality (XR) technologies can be used in the manufacturing industry to provide operators with

Published by Springer Nature Switzerland AG 2025
P. Sureephong et al. (Eds.): PLM 2024, IFIP AICT 740, pp. 180–189, 2025.
https://doi.org/10.1007/978-3-031-93319-6_14

an improved perception of instructions in a visual and interactive way. Mixed reality (MR), which is part of the XR spectrum, can help reduce manual errors and rework in complex assembly tasks. To accomplish this, MR necessitates the visualization of the simulation of the assembly process, which can be generated using an ASP method [7].

This paper presents a concept for integrating ASP into PLM to enable Design for X (DfX) and places ASP in the context of MR technology. It discusses possible applications based on ASP principles and presents two ASP methods based on Reinforcement Learning (RL), which are trained to disassemble an industrial assembly. The ASP methods presented differ in their assembly component selection approach for disassembly by the RL agent. One method iterates over individual components at time intervals, while the other method enables the RL agent to iterate over individual components independently with the help of an additional action. During validation, the two ASP methods are compared with each other in terms of their ability to disassemble an assembly over time.

This paper is structured as follows: Sect. 2 provides an overview of the current state of research in ASP and the fundamentals of RL. Section 3 presents a concept for the application of ASP in the PLM. Additionally, potential use cases of ASP are analysed across the phases of the product lifecycle. Section 4 presents two ASP methods based on RL. Section 5 shows an implementation and validation of the presented ASP approaches based on an industrial assembly. Section 6 concludes with a summary and outlook.

2 Related Works

2.1 Assembly Sequence Planning

One of the core tasks of the product assembly planning process within the production phase of the product lifecycle is ASP, whose objective is to utilize the assembly design to generate implementable assembly sequences [14, 20]. The planning of the assembly sequence is a crucial phase of the production cycle, as it accounts for up to 70% of the total manufacturing costs and time [2]. ASP is classified as a NP-hard combinatorial problem due to the considerable expansion of the search space as the number of components increases [4, 9]. Depending on the complexity of the assembly, only 0.4% of the possible assembly sequence combinations are viable [6]. To address this problem, several soft computing techniques have been proposed in the literature, including Genetic Algorithms (GA), Ant-Colony Optimization (ACO), and Neural Networks (NN) [3].

Current research uses the principle of Assembly by Disassembly and assembly path planning (APP) methods to solve ASP. Assembly by Disassembly is a methodology that assumes the already assembled product as the initial start for ASP. Paths are then searched for the individual components of the assembly in order to disassemble them. Once all components have been removed, a disassembly sequence is identified, which is then reversed to produce the assembly sequence. APP is used in ASP to calculate collision-free paths for the parts during the disassembly process in Assembly by Disassembly. By using a physics simulator that supports collision detection via signed distance fields (SDF) instead of collision proxy through meshes, tasks such as unscrewing a nut from the thread of a screw can be performed [17]. Furthermore, the assembly paths produced by APP and the resulting assembly sequence can be utilised to perform assembly in the physical world with the assistance of a robot [16]. SDF for collision detection is supported, for example, by the physics-based simulators Mujoco [18] and Nvidia Isaac Gym [10].

2.2 Reinforcement Learning

Reinforcement learning (RL) is a sub-area of machine learning, which enables the automation of decision-making through a learned strategy for the execution of actions to achieve an objective. The five basic elements of RL are an agent, an environment, actions, states and rewards [15].

The agent aims to achieve a goal in the environment referring, to a certain state that the agent can assume in the environment. Through action based interactions over time, the agent improves and approaches the target state. All actions are collected in an action space. A policy is used to decide which actions the agent performs depending on the agent's current state. The policy improves over time so that the selection results in actions that are the most likely to lead to the target state. The objective of an RL agent is described by a reward function. After performing an action, the agent receives rewards indicating whether the action was a good decision. The rewards are then passed to the agent after each time step to update its policy. In general the objective of the agent is to maximize its reward. RL is used in various areas, including video games, self driving cars and autonomous robotics [1].

3 Concept of Automated Assembly Sequence Planning in the Product Lifecycle Management

Based on the work described above, a concept for the use of ASP in the PLM is presented and possible applications in the individual main phases of the product life cycle are described.

To achieve the possible applications described in the following, a functionally designed 3D-CAD model of an assembly is necessary. This assembly is transferred as output from the design phase to the ASP method to derive feasible assembly sequences from the 3D-CAD model. ASP is used in the product development phase of the product life cycle to enable the designer to create a DfX compliant design (see Fig. 1) The ASP method's output provides feedback to the designer on whether a realizable assembly sequence has been found. This feedback is crucial in determining whether the assembly has been suitably designed for assembly and disassembly. The method may output none, one, or multiple applicable assembly sequences, the requirement to be considered as such being that all components can be removed without collision.

During the manufacturing phase of production, ASP facilitates assembly preparation and execution. The assembly sequence is determined in a 3D simulation environment, which allows for a digital view of the assembly process. This replaces the previously used 2D assembly instructions, resulting in time savings that can be used to derive the 2D assembly instructions. Viewing the assembly in a 3D simulation can help the assembler to become familiar with the process quicklier due to a more intuitive view. What is more, the use of mixed reality (MR) headsets, in combination with the 3D simulation data from the

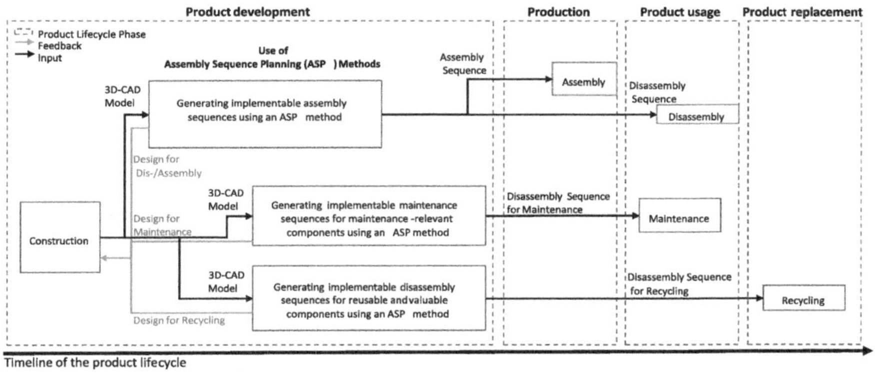

Fig. 1. Concept for the integration of Assembly Sequence Planning into the product lifecycle.

ASP, provides an almost realistic view of the assembly process [12]. The 3D simulation of the assembly sequence can be projected into the real world, allowing employees to carry out assembly steps digitally using the 3D-CAD model and the assembly sequence generated by ASP. MR headsets eliminate the need for controllers, as the hands and eyes serve as the control system, which allows the assembler to follow the 3D simulation without having to stop working [7].

The use of ASP in product development enables design for assembly, benefiting both human and robot-based assembly. The 3D-based ASP generates assembly paths that can be used to create robot assembly plans [16]. Figure 2 displays the use of ASP for the assembly and disassembly phase. First, the designer finalises the design of his assembly. This is then transferred to the ASP method, which attempts to derive an assembly sequence using the 3D-CAD model. The simulation of the process of deriving the assembly sequence is passed to the designer together with the information as to whether an assembly sequence has been found. As ASP derives the assembly sequence via Assembly by Disassembly, the designer receives feedback on the assembly and disassembly capability of his assembly (design for assembly and design for disassembly). If an assembly sequence is found by the ASP method, this can be used as a guide for assembly and disassembly. To make this process as real as possible, mixed reality glasses can be used to visualize the assembly sequence during the actual assembly.

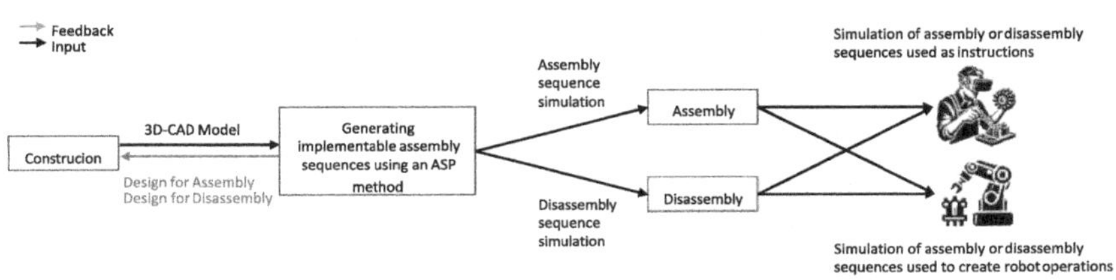

Fig. 2. Use of Assembly Sequence Planning for the assembly and disassembly process.

Furthermore, the simulation of the assembly sequence can be used to generate robot operations that can be used for the automatic assembly and disassembly of a product.

To maintain and recycle a product using ASP, the classic ASP objective must be modified. The previous application options aimed to assemble all components into an assembly or check if the assembly can be assembled. However, maintenance or recycling aims to quickly reach the component that requires maintenance or recycling. Therefore, it is necessary to have an assembly sequence that includes the required component as early as possible, so that once the sequence has been generated successfully, it results in the corresponding component's maintenance or recycling sequence. In the case that components are recycled, the sequence is the same as the one used for assembly in the production phase, which also applies to disassembly, as described above for assembly in the production phase. The product utilization and replacement phase benefits from the generation of the maintenance sequence and recycling sequence. The resulting simulation of the maintenance or recycling process saves time by eliminating the need to create disassembly instructions in 2D. While disassembly instructions may not always be created, employees possess the knowledge of how to maintain the product. The 3D simulation allows maintenance technicians to better prepare for maintenance work and become less reliant on the knowledge of experienced colleagues, while MR headsets equally enable technicians to view the 3D simulation generated by an ASP method during maintenance without having to take their eyes off the component being serviced [7]. This also applies to recycling during the product replacement phase.

4 Method for Automated Assembly Sequence Planning Using Reinforcement Learning

This paper presents two new Assembly by Disassembly approaches for automated ASP by employing RL. The input required is a 3D-CAD assembly that does not have any components with overlapping solids. The ASP methods produce disassembly sequences that correspond in reverse to the assembly sequences. A finite-horizon Partially Observable Markov Decision Process (POMDP) is defined to utilize RL. The two presented ASP methods POMDP descriptions differ only their action space. Method 2 includes an additional action that modifies the selected component of the RL agent in a time independent manner, as described below.

4.1 Partially Observable Markov Decision Process for Assembly Sequence Planning

A finite-horizon Partially Observable Markov Decision Process (POMDP) is described in order to utilize RL in an dynamic environment. A POMDP is characterized by a 7-Tuple [S,A, Ω,T,O,R, γ]. S describes a set of partially observable states S = $[s_1,...,s_n]$. A set of actions A = $[a_1,...,a_n]$ is described by A. Ω defines a set of observations Ω = $[o_1,...,o_n]$. T describes the transition from state st to the next state st+1 after taking action a by a set of conditional probabilities T(S_{t+1}|S_t, a). The observation probability O(o|S_t, a is determined by the observation function O. O(o|S_t, a) is the probability of obtaining a specific observation o after taking an action a in the current state st. The reward function R calculates the reward, which the agent receives after taking an action a in its current state st. The weight of future rewards compared to immediate rewards is balanced by the discount factor γ, which is an value between 0 and 1 [5]. An observation o = $[p_x, p_y, p_z, o_x, o_y, o_z, d]$ of the observation space Ω of the POMDP describes the position, orientation and status of assembly of each component of the product. The status d of the assembly contains the information whether the component is disassembled by the method or not. The defined POMDP for method 1 contains a discrete action space A, which includes 12 actions. Each action of method 1 is described by Eq. 1. Each action applies a force f or a torque q in a single axis x,y or z.

$$a = [f_x, f_y, f_z, q_x, q_y, q_z] | f_x, f_y, f_z, q_x, q_y, q_z \in \{-1, 0, 1\} \tag{1}$$

In comparison to method 1, method 2 includes a 13th action. Method 2 applies a force f or a torque q in a single axis (x, y, or z) for each action, except for action 13 which sets the value of b to 1, triggering a change in the selected component of the RL agent. The discrete actions of method 2 are described by Eq. 2.

$$a = [f_x, f_y, f_z, q_x, q_y, q_z] | f_x, f_y, f_z, q_x, q_y, q_z \in \{-1, 0, 1\} \tag{2}$$

If the RL agent selects the 13th action, the current selected component is reset to the starting position, and the next component is randomly chosen from the list of all components in the assembly that have not yet been removed. In contrast, method 1's RL agent has a restricted number of 512 time steps to disassemble a single component. If the component is not removed within the allotted time, its position is reset to the initial position, and the next component is randomly selected from the list of all components in the assembly that have not yet been removed.

The observation space Ω of the POMDP contains observations o=$[p_{x1}, p_{y1}, p_{z1}, ..., p_{xn}, p_{yn}, p_{zn}, d_1, ..., d_n]$ which are described by the positions $px1$, $py1$, $pz1$ of each component of the assembly and the information whether each of the component is disassembled or not, denoted by d. The RL method's reward function penalises the agent for each time step taken to disassemble the assembly via Assembly by Disassembly by a value of -0.1. The agent receives a reward of $+100$ if it successfully removes a component from the assembly. A component is considered removed as soon as its centre of mass touches a removal zone that is located near the assembly.

5 Implementation and Validation

For the validation of the proposed methods for automated ASP using RL, a 3D-CAD model of an industrial assembly is used, where the assembly is based on a data set that has been adapted to remove intersections between the solids of individual components in the initial state of the original assembly [8]. The assembly that has been adapted consists of nine components (see Fig. 3).

The objective of the presented RL method is to determine a feasible assembly sequence in the shortest possible time. For the implementation of the RL method a Deep Q-Network (DQN) algorithm is used [11]. The assembly sequences are determined within the open source physics-based simulator Mujoco [18]. Collisions between

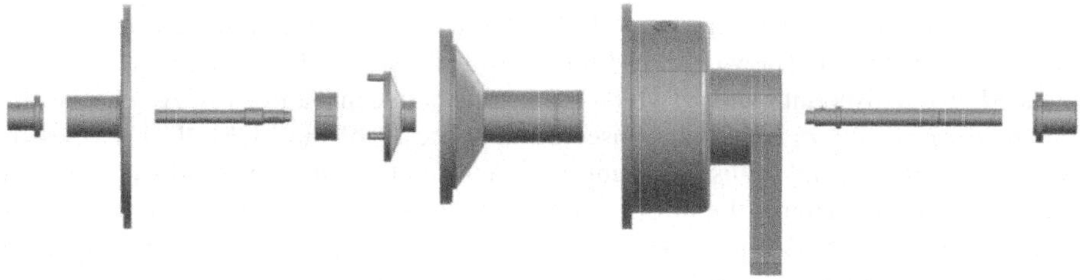

Fig. 3. Exploded view of the assembly used for validation.

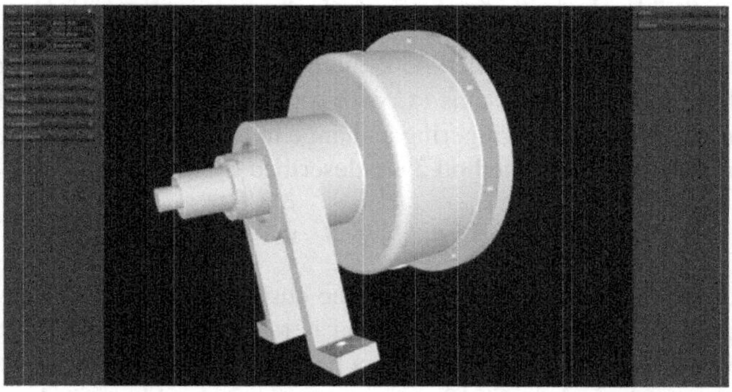

Fig. 4. Validation Assembly within the Mujoco Simulator.

the solids are detected by SDF [13]. Figure 4 displays the example assembly within the Mujoco simulation environment.

For the learning process of the ASP methods, a general-purpose system (13th Gen Intel® Core™ i7-13850HX CPU @ 5,3 GHz) and an internal GPU (NVIDIA RTX3500 Ada Generation Laptop GPU) are used. The validation shows that both methods are able to successfully disassemble the validation assembly. Both methods where trained for over 30000 time steps. Method 1 completely disassembled the assembly 4 times within the training time (see red crosses in Fig. 5).

Method 2 removed the assembly 27 times during the training period, which is 6.75 times more than method 1 (see Fig. 6). On average, the RL agent using method 1 required 7594 steps to disassemble the assembly completely. This is 7.5 times more than the average of 1009 steps required by the agent using method 2. The cumulative reward shows that method 2 performs well after successfully disassembling the assembly one time. Method 2 collected a cumulative reward of 728 on average until the complete disassembly of an assembly. This is 23 times higher than the average value of method 1, which had an average cumulative reward of 31.5.

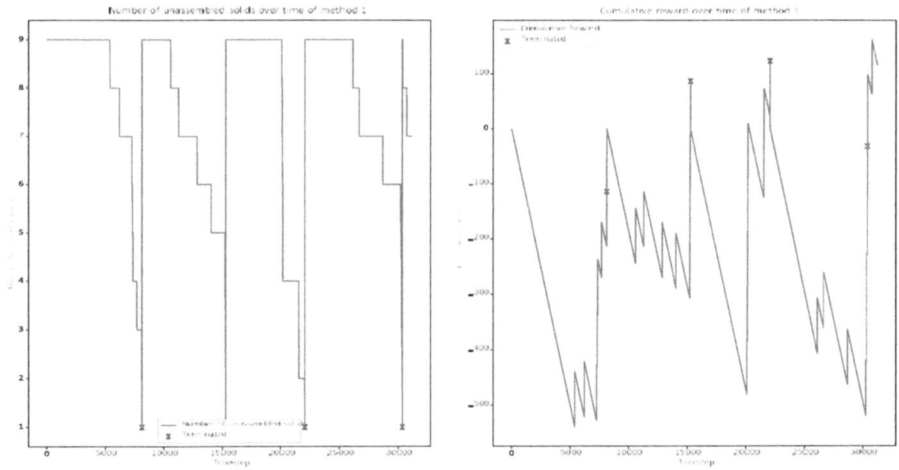

Fig. 5. Validation results of Assembly Sequence Planning Method 1. The diagrams show the number of unassembled solids and the cumulative reward over the time.

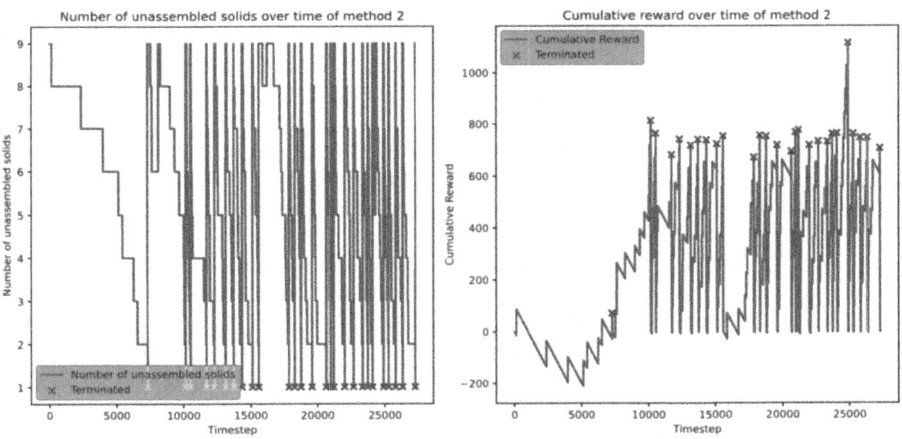

Fig. 6. Validation results of Assembly Sequence Planning Method 2. The diagrams show the number of unassembled solids and the cumulative reward over the time.

6 Conclusion and Outlook

The presented concept demonstrates that ASP has significant potential for the enhancement of the automation and networking of individual phases in PLM. ASP can be used in the product development phase after the design phase. It enables forward- looking planning of the PLM phases, including assembly, maintenance, disassembly, and recycling, in the sense of Design for X. An industrial assembly was utilized to validate two new approaches of RL-based ASP in a physics-based environment. The validation of ASP method 2 has shown that an RL agent can learn to switch between the components of an assembly in order to successfully disassemble the assembly. ASP method 1, which was unable to perform time-independent component changes, was outperformed by ASP method 2, which disassembled the assembly 7.5 times more frequently during the training period. Future research should explore the combination of ASP with robotics to train robots to disassemble complex assemblies. The use of robots is feasible in many of the applications mentioned. In assembly, a robot can be used to assemble components. In maintenance, it can replace a component. In disassembly, it can dismantle the assembly, and in recycling, it can separate components or remove valuable components. Further Research should investigate whether the combination of ASP and mixed reality headsets positively influences the learning ability of operators in assembly or maintenance.

Disclosure of Interests. The authors have no competing interests to declare that are relevant to the content of this article.

References

1. AlMahamid, F., Grolinger, K.: Reinforcement learning algorithms: An overview and classification **4**, 1–7 (2021). https://doi.org/10.1109/CCECE53047.2021.9569056
2. Champatiray, C., Bahubalendruni, M.V.A.R., Mahapatra, R.N., Mishra, D.: Optimal assembly sequence planning with tool uncertainties. Proceedings of the Institution of Mechanical Engineers, Part B: Journal of Engineering Manufacture (2023). https://doi.org/10.1177/09544054231189012
3. Deepak, B., Bala Murali, G., Bahubalendruni, M.R., Biswal, B.B.: Assembly sequence planning using soft computing methods: a review. Proceedings of the Institution of Mechanical Engineers, Part E: Journal of Process Mechanical Engineering **233**(3), 653–683 (2019). https://doi.org/10.1177/0954408918764459
4. yong fa, Q., zhi gang, X.: Assembly process planning using a multi-objective optimization method. In: 2007 International Conference on Mechatronics and Automation, pp. 593–598. IEEE (2007). https://doi.org/10.1109/ICMA.2007.4303610
5. François-Lavet, V., Henderson, P., Islam, R., Bellemare, M.G., Pineau, J.: An introduction to deep reinforcement learning. Foundations and Trends in Machine Learning **11**(3–4), 219–354 (2018). https://doi.org/10.1561/2200000071
6. Ghandi, S., Masehian, E.: A breakout local search (bls) method for solving the assembly sequence planning problem. Eng. Appl. Artif. Intell. **39**, 245–266 (2015). https://doi.org/10.1016/j.engappai.2014.12.009
7. Kolla, S., Kumar, A.A., Plapper, P.: Mixed reality for industry 4.0 in manufacturing systems. In: Zaman, U.K.u., Siadat, A., Baqai, A.A., Naveed, K., Kumar, A.A. (eds.) Handbook of Manufacturing Systems and Design, pp. 123–138. CRC Press, Boca Raton (2023). https://doi.org/10.1201/9781003327523-10
8. Lupinetti, K., Giannini, F., Monti, M., Pernot, J.P.: Content-based multi-criteria similarity assessment of cad assembly models. Comput. Ind. **112**, 103111 (2019). https://doi.org/10.1016/j.compind.2019.07.001
9. Lv, H., Lu, C.: A discrete particle swarm optimization algorithm for assembly sequence planning. In: 2009 8th International Conference on Reliability, Maintainability and Safety, pp. 1119–1122. IEEE (2009). https://doi.org/10.1109/ICRMS.2009.5270057
10. Makoviychuk, V., et al.: Isaac gym: High performance gpu-based physics simulation for robot learning. http://arxiv.org/pdf/2108.10470v2

11. Mnih, V., et al.: Playing atari with deep reinforcement learning. http://arxiv.org/pdf/1312.5602v1
12. Poh, E., Liong, K., Lee, J.: Mixed reality for mechanical design and assembly planning. https://doi.org/10.48550/arXiv.2209.01252
13. Pujol, E., Chica, A.: Adaptive approximation of signed distance fields through piecewise continuous interpolation. Comput. Graph. **114**, 337–346 (2023). https://doi.org/10.1016/j.cag.2023.06.020
14. Rashid, M.F.F., Hutabarat, W., Tiwari, A.: A review on assembly sequence planning and assembly line balancing optimisation using soft computing approaches. The Int. J. Adv. Manufact. Technol. **59**(1–4), 335–349 (2012). https://doi.org/10.1007/s00170-011-3499-8
15. Sutton, R.S., Bach, F., Barto, A.G.: Reinforcement Learning: An Introduction. Adaptive Computation and Machine Learning series, 2nd edn. MIT Press Ltd, Massachusetts (2018)
16. Tian, Y., et al.: Asap: Automated sequence planning for complex robotic assembly with physical feasibility. https://doi.org/10.48550/arXiv.2309.16909
17. Tian, Y., et al.: Assemble them all: Physics-based planning for generalizable assembly by disassembly. ACM Transactions on Graphics **41**(6), 1–11 (2022). https://doi.org/10.1145/3550454.3555525, http://arxiv.org/pdf/2211.03977v1
18. Todorov, E., Erez, T., Tassa, Y.: Mujoco: A physics engine for model-based control. In: 2012 IEEE/RSJ International Conference on Intelligent Robots and Systems, pp. 5026–5033. IEEE (2012). https://doi.org/10.1109/IROS.2012.6386109
19. Vajna, S., et al.: Übergreifende informationsverarbeitung im produktlebenszyklus. In: Vajna, S., Weber, C., Zeman, K., Hehenberger, P., Gerhard, D., Wartzack, S. (eds.) CAx für Ingenieure, pp. 515–547. Springer Berlin Heidelberg, Berlin, Heidelberg (2018). https://doi.org/10.1007/978-3-662-54624-6_12
20. Xing, Y., Chen, G., Lai, X., Jin, S., Zhou, J.: Assembly sequence planning of automobile body components based on liaison graph. Assem. Autom. **27**(2), 157–164 (2007). https://doi.org/10.1108/01445150710733423
21. Yang, H., Chen, J., Wang, C., Cui, J., Wei, W.: Intelligent planning of product assembly sequences based on spatio-temporal semantic knowledge. Assembly Automation **40**(5) (2020). https://doi.org/10.1108/AA-11-2018-0196

Discrete Event Simulation Based Approach for Cyber Physical Logistics System Development

Rachid Benmoussa(✉)

Cadi Ayyad University, UCA, ENSA Marrakech, LARTID Laboratory, BD Abdelkarim Khattabi, BP 575, Marrakesh, Morocco
r.benmoussa@uca.ac.ma

Abstract. Setting up a CPS is a complex process, especially for SMEs with limited financial means and a technological legacy that is difficult to replace. Besides, the concepts and approaches for developing a CPS, which are too generic and not yet mature, have not been adapted to specific sectors of activity such as logistics. To fulfill this gap, we propose in this paper a new approach, centered on DES as an analysis technique, to build a CPLS (Cyber Physical Logistics System). In this context, we addressed the following research questions: How to transpose the concepts of the CPPS to logistics to build the CPLS? Which maturity model should be adopted for a CPLS progressive development? An academic case study was conducted to demonstrate the approach's applicability. It proves also that the CPPS current model maturity is not suitable for CPLS progressive development. Hence the need for a new model as perspective.

Keywords: CPS · CPPS · CPLS · Discrete Event Simulation · Logistics · Maturity

1 Introduction

Industry 4.0 and Cyber Physical Production System (CPPS) are two closely related concepts that share similarities but have key differences. Industry 4.0 provides a holistic view of the future of the manufacturing industry, encompassing not only CPS but also other technologies and approaches such as artificial intelligence, robotics, augmented reality and blockchain. Whereas CPS is a specific technology within the underlying Industry 4.0 infrastructure. It is a System that connects the physical and digital components of production, enabling real-time communication between machines, products and people. It uses sensors and actuators to collect data about machines and production processes, then uses data analysis algorithms to optimize and improve production.

Setting up a CPS is a complex process, especially for SMEs that have limited financial means and a technological legacy that is difficult to replace. These SMEs require a gradual deployment process over time that helps them overcome their technological limits. Based on a continuous, bearable and risk-free reconfiguration logic, it aims

Published by Springer Nature Switzerland AG 2025
P. Sureephong et al. (Eds.): PLM 2024, IFIP AICT 740, pp. 190–207, 2025.
https://doi.org/10.1007/978-3-031-93319-6_15

to increase SMEs competitiveness through digitalization and gradually move towards industry 4.0. However, the concepts of CPS remain too generic and are not tailored to specific industry sectors, including logistics.

Discrete event simulation (DES) is one of the tools classically used for decision support in the context of the supply chain. DES models the behavior of a complex system as a series of well-defined and ordered events [6]. Although it requires the system model to be known, DES sometimes remains the only suitable technique for logistics systems because it allows the analysis of decoupled processes by a common performance criterion such as stock [3].

This paper thus aims to develop a design approach for what we refer to as CPLS (Cyber Physical Logistics System), specific to the logistics sector and centered on discrete event simulation as an analysis technique, placing the system to be developed in a maturity cycle that allows its progressive evolution towards excellence. To address this issue, two main research questions will be investigated: How to transpose the concepts of the CPPS to logistics to build the CPLS? Which maturity model should be adopted for a CPLS progressive development?

This paper is divided into 5 sections. Following the introduction, the second section presents the CPS/CPPS paradigm, providing its definition and its architecture. The third section introduces the generic CPPS metamodel. The fourth section develops the proposed approach for building a CPLS. The fifth section presents an academic case study to prove the applicability of the proposed approach. Finally, the conclusion discusses the findings and presents future perspectives.

2 CPPS Paradigm

2.1 Basic Concepts

The first definition of cyber-physical systems (CPS) can be traced back to 2006 [14], during a workshop held by the American National Science Foundation (NSF). Throughout their development, increasingly synthetic definitions have been proposed, such as those of [2, 15, 18, 19], who define CPS as "cooperative systems, with decentralized control, resulting from the fusion between the real and virtual world, with behaviors that are autonomous and dependent on the context in which they are located, that can be constituted as systems of systems with other CPS and conduct deep collaboration with humans". To achieve this, the software embedded in the CPS uses sensors and actuators, connects with each other and with human operators by communicating via interfaces, and has the ability to store and process data from the sensors or the network [20]. This interconnection of systems, as asserted by [10], stems from the fact that a CPS encompasses control, computation and communication devices [13].

The model proposed by [17] focuses on the control function commonly implemented in CPS. In particular, it makes it possible to specify and clearly illustrate the real paradigm shift proposed by the concept of CPS. Regardless of the system considered (Fig. 1.a), classical control systems allow (usually in closed loop) the system to achieve the desired performance indicators by absorbing disturbances from the external environment. A basic CPS (Fig. 1.b) uses its cyber component to assess and potentially adapt the control system, in particular through simulation techniques. It is this evaluation loop, ideally carried out in real-time and automatically, that differentiates a CPS from a traditional system.

The concept of CPS is, therefore, very broad and encompasses an extremely broad class of systems and potential application areas [5]. Consequently, many fields of research are also affected by this notion. This is probably a huge opportunity, as it gives the possibility to create a cohesive ecosystem in many application domains such as autonomous vehicles [9], health devices, power grid management [1].

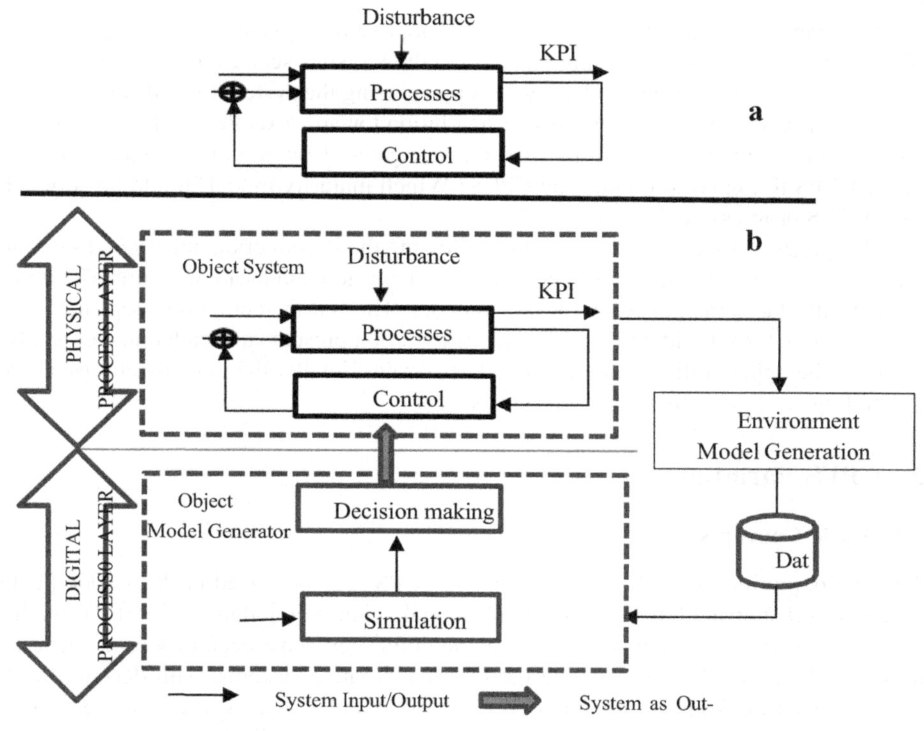

Fig. 1. Basic architecture of a CPS, adapted from [5]

2.2 From CPS to CPPS

CPPS is an adaptation of the CPS concepts to production systems. [5] define CPPS as follows "CPPS are constituted; of physical elements with their control systems, offering and consuming services along a value chain, integrated with digital systems with means of online analysis and reconfiguration of these elements. These elements, more or less autonomous, connect and cooperate with each other and with humans depending on the situation, via information systems, to improve actions and decision-making processes in real-time, react to hazards, and facilitate the integration of industrial system changes over time".

2.3 CPPS Architecture and Maturity

Presentation. Figure 2 illustrates a relatively generic modeling of such a CPPS, inspired by the representation of [17]. In this representation, the physical part and the cyber part are clearly decoupled, and their functionalities and objectives are largely differentiated. On one side, the physical part focuses on real-time control of activities, while on the other side, the cybernetic part engages in dynamic system evaluation. The lever of action of this cyber part is more extensive on an industrial system than it can be on a classic CPS [5]. Indeed, a reconfiguration decided by the cyber part can be proposed both to the control system and to the physical elements.

Within the physical part, the main emphasis is obviously on the level of the elements that must be controlled in fine. Simultaneously, these elements contain assets (machinery, transport, etc.), products (manufacturers, drugs, building materials, etc.), and human beings (operator, user, etc.) considered, at this level, as contributing elements to the operational management of the physical part. These operators have different roles and impacts on the physical part depending on the type of application, sometimes carrying out activities within the system, making decisions, or even generating disturbances or on the contrary preserving the situation or the security of the CPPS. Data from these elements is transmitted to a control system through sensors or computer inputs, for example. The control system then issues a series of commands to the physical elements to achieve the desired objectives for the CPPS. Typically, interaction with human supervisors is established. Through this interaction and based on various performance indicators observed in the physical elements, a supervisor can either enforce or modify decisions through the control system.

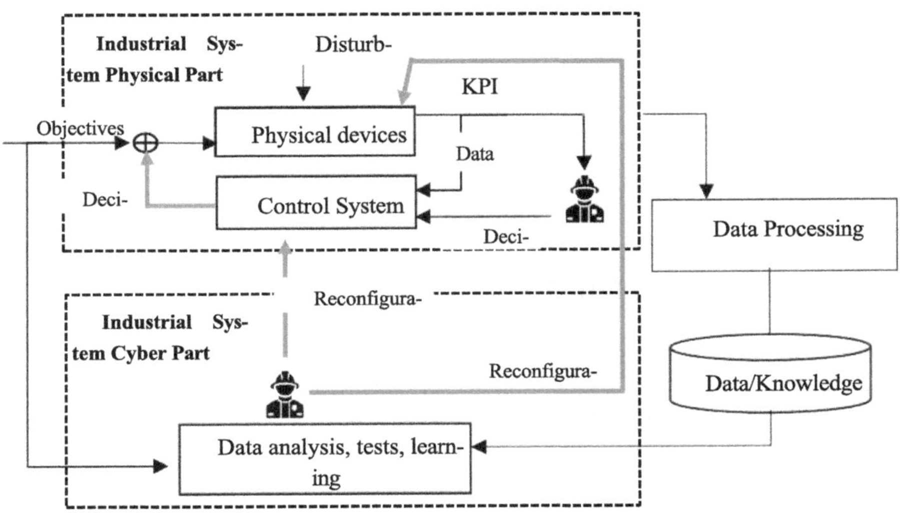

Fig. 2. CPPS modeling, adapted from [17]

CPPS Loop Maturity. There are four main scenarios for human interaction with CPPS (Figure 3): complete, automation, tool and manual scenarios [4, 7]. In a complete scenario, humans only have a supervisory role in the CPPS, while the CPPS can make all the necessary decisions without any human intervention. In an automation scenario, most of the decisions are taken by CPPS, while human beings are guided by CPPS to perform activities. Conversely, in a tool scenario, the majority of decisions are made by humans, while CPPS are guided by humans. In a manual scenario, all decisions are made by humans, while CPPS only provides data to humans. By grouping human roles in CPPS, different types of humans can be identified, e.g., operator [21], worker [11], manager [8], supervisor [16] and decision-maker [16].

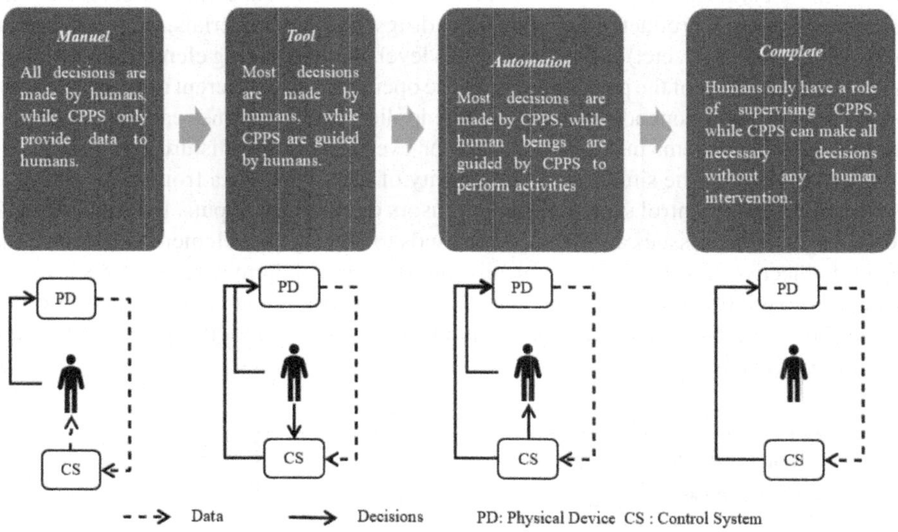

Fig. 3. CPPS loop maturity

3 CPPS Proposed Meta-Model

In this section, we synthesize the knowledge acquired through CPPS literature in the meta-model of Figure 4. This meta-model was inspired by the work of [22] which we tried to complete, simplify and clarify by focusing on other bibliographical references. Due to page limitations, the detailed description of the metamodel has been excluded. The reader could consult the references.

4 Proposed Approach

This section first introduces a general approach in accordance with the earlier metamodel for CPPS design and deployment. Following that, it outlines a specific adaptation of this approach to the logistics sector to set up a CPLS, based on Discrete Event Simulation (DES) as an analysis technique.

4.1 General Approach

The proposed approach (Figure 5) consists of designing separately the two complementary views, functional and technical. The functional (conceptual) view aims to design the CPPS specifications independently of any technological choice. While the technical view translates the functional view into a technological solution implemented through specific technical choices. Thus, CPPS technical view development follows a classic systems design, production and deployment cycle, including generic phases such as testing, training, operating and audit. Note that a functional view can be broken down into several technical views depending on the technological choices made at each stage of the functional view.

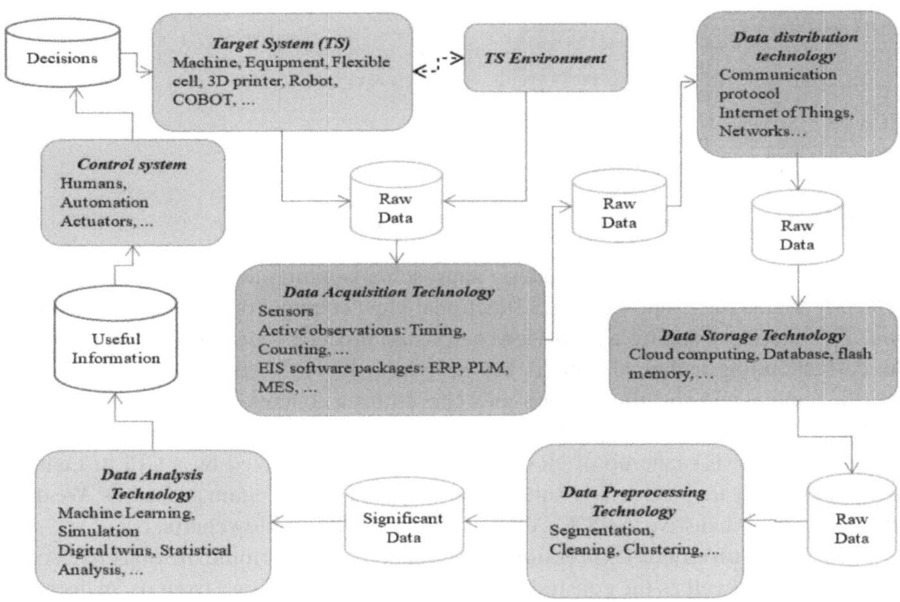

Fig. 4. Proposed Meta-model

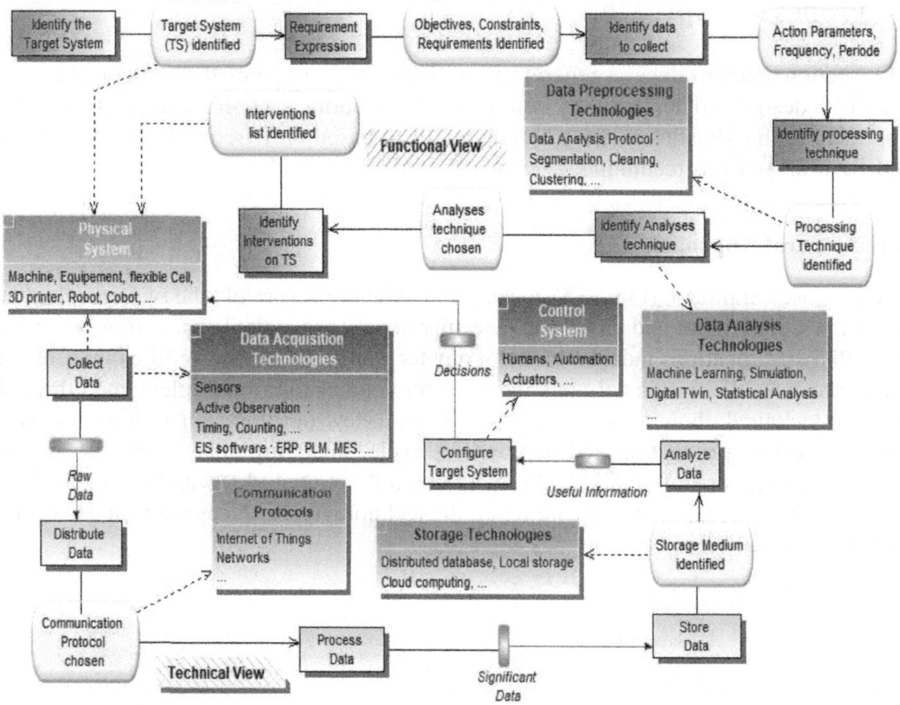

Fig. 5. Proposed Approach

4.2 Simulation Approach for Logistics: The Functional View

This section presents the CPPS generic approach adaptation to a specific approach for CPLS. It details the proposed CPLS functional view centered on discrete event simulation. To concretely deploy a such view, a six-step process is designed (Table 1). Since the corresponding technical view is not generic, it has to be designed according to a specific context and specific technologies chosen for a concrete supply chain.

Step 1: Identify TS. This step aims to identify the Target System which, in the case of logistics, is made up of all the processes to be digitized by a CPLS. Lean/Six Sigma tools [12] are particularly suitable for mapping supply chain processes. We quote without being exhaustive, SIPOC, VSM, UML and BPMN Flowcharts, …

Step 2: Requirement Expression. This step aims to determine the objectives of the end-customer as well as the constraints that the TS has to undergo from its environment to achieve these objectives. The critical to quality (CTQ) and Kano diagram [12] are particularly interesting for capturing the voice of the customer.

Step 3: Identify Raw Data. Raw data can be categorized into four groups:

- Action parameters (AP): parameters internal to the TS that act directly on its performance. They represent the levers under the control of the TS.

Table 1. Design Process of DES based CPLS functional view

Steps	Label	Tools
Step1	Identify Target System (TS)	Flowchart, SIPOC, VSM, …
Step2	Requirement Expression	Objectives: CTQ, Voice of the customer Constraints: Operational, Financial,…
Step3	Identify Raw Data	Action Parameters (AP), Environmental Disturbance Evaluation parameters (EP) from: the voice of cus- tomers, constraints
Step4	Identify Processing Techniques	Statistical processing of data
Step5	Identify Analysis Technique	Discrete Event Simulation (DES)
Step6	Identify Potential Decisions	***Decisions to Reconfigure (Di):*** The target system (TS) The decision-making process

- Environmental disturbances: parameters external to the TS that positively or negatively influence its performance. They are out of control and undergone by the TS from its environment.
- The evaluation parameters (EP): parameters which make it possible to evaluate the TS performance.
- Functional constraints: operational, financial, juridical parameters, etc. that the TS must respect during its operability.

Step 4: Identify Processing Techniques. The use of discrete event simulation in general requires the statistical processing of data. Professional software like Minitab can be used for this purpose.

Step 5: Identify Analysis technique. The core of the proposed approach relies on the use of discrete event simulation as an analysis technique.

Step 6: Identify Potential Decisions. Knowing the types of TS interventions is the key issue. This knowledge enables the anticipation of the different configuration possibilities, including potential automation, whether at the TS level or within the decision-making process itself.

5 Case Study

This academic case study focuses on a "Manufacture to stock, Distribution to order" supply chain. Its purpose is to demonstrate the feasibility and effectiveness of the proposed approach.

5.1 Design of the CPLS Functional View

Step1: Identify the Target System (TS) and its environment. The company manufactures for stock and distributes products to order. The production process consists of 3 manufacturing stations. Between two successive operations, the products are transferred using forklifts. The distribution process is initiated by customer order. In cases of stock shortages, orders are held until the stock is replenished. The preparation of the orders is carried out by operators. The distribution is ensured by trucks. Its duration is proportional to the outward and return distance to the customer. The TS environment is composed of the market and its requirements as well as the operational and financial capacity limits of the supply chain.

Step 2: Requirement Expression. Customers expect to receive orders ideally just in time without any delay in delivery compared to the time just needed to process their orders. This requirement is expressed by the theoretical delay (Dth) as key performance indicator: Dth = Reception order time – just necessary delay of processing – Order time The supply chain is subject to certain operational and financial constraints and capacity limits:

- Number of all operators $< = 6$: Pertains the limits of Recruitment/Reassignment of production and distribution operators.
- Number of Resources $< = 7$: Represents investment limits in transport and handling equipment.
- Truck occupancy rate $> = 70\%$: Reflects the minimum utilization rate of trucks in distribution activities.
- Forklift occupancy rate $> = 70\%$: Refers to the minimum utilization rate of forklifts in handling activities.

Step 3: Identify the raw data to collect. The raw data are the action parameters, the external disturbances, the evaluation parameters and the constraints. These data can be classified into two categories:

- Type 1 (T1): Represents random data that changes from one occurrence to another, such as the execution times of products at workstations.
- Type 2 (T2): Signifies deterministic data unchanged for the duration of the simulation, such as the number of operators at the workstations.

Step 4: Identify the processing method. Statistical data processing is unavoidable if the chosen analysis technique is discrete event simulation.

Step 5: Identify the method of analysis. Discrete event simulation is chosen to technically perform the data analysis. This choice constitutes the basis of the proposed approach.

Step 6: Identify Decisions (Interventions on the TS). This section presents the list of potential decisions to reconfigure the TS:

D0: Whether or not to reconfigure the target system in its current state (not capable) or when facing a change in its environment (not resilient).

Decisions to reconfigure the target system in the case that the latter is not capable or not resilient:

D1: Recruitment, overtime, reassignment of human resources in charge of production in the 3 stations.

D2: Recruitment, overtime, reassignment of human resources responsible for order preparation.

D3: Increase or reduction in the number of forklifts dedicated to product transfer between stations.

D4: Modification of the batch rule when transferring products between stations by forklifts.

D5: Increase or reduce the number of trucks dedicated to order delivery. D6: Modify the policy for launching production orders.

Decisions to reconfigure the decision-making process itself:

D7: Reconfiguring the decision-making process for multidimensional performance involving conflicting evaluation parameters.

5.2 Design and Deployment of the Technical View

In this case study, as a reminder, the objective is to develop a CPLS for a «manufacture to stock, distribution to order» supply chain, based on "discrete event simulation" as analysis technique. To translate the functional view (section 5.1) into the technical view of this CPLS, ARENA simulator and its complementary tools have been chosen to deploy our proposed six-step approach presented in Table 2.

Table 2. CPLS technical view of the case study

Steps	Label	Tools
Step1	Collect data	• Action Parameters (AP): Active observations • Evaluation Parameters (EP): Enterprise Resource Planning (ERP)
Step2	Distribute data	Local entry of collected series
Step3	Process data	Input Analyzer
Step4	Store data	Local disk storage
Step5	Analyze data	• Implementation of the decision-making process: Design, Test, Validation of the simulation program with ARENA • Analyzes: A0 (Capability, Resilience); A1 & A2 (AP influence on EP); A3 (Interaction between AP); A4 (Optimal Scenario); A5 (Correlation between EPs)
Step6	Reconfigure TS	Continuous Improvement (Data Type 1); Investment (Data Type2)

Step 1: Collect data. The action parameters and environmental disturbances are recorded manually using the Active Observations Lean Manufacturing tool [12].

Step 2: Distribute data. For this case study, local input of the collected series onto a computer has been chosen.

Step 3: Process data. To determine the significant data, i.e., the distributions of probability that best represent the series of raw data collected, "Input Analyzer" has been chosen in this case study. [6].

Step 4: Store data. Local hard drive storage has been chosen in this case study.

Step 5: Analyze data. To analyze the raw data collected and processed using the simulation technique, one must first set up the corresponding decision-making process. This involves designing, testing, and validating the simulation program built with the ARENA simulator, according to the classic approach proposed by [6]. The target system must then be monitored by simulation for a sufficient period of time to obtain significant results. We use in this case study a duration of 7 days (1 day = 24 h). Finally, it is a question of collecting the "Useful information" i.e., the evaluation parameters resulting from the simulation program execution. From this information, various analyses had been conducted. These analyses are of six types. A0 measures the overall characteristics of the target system: capability, resilience to a change in the environment. A1 identifies the most influential action parameters on the Dth (Causality between a given parameter and Dth). A2 identifies the most influential action parameters on the Dth (Comparison of the averages of Dth (Threshold 95% and 99%) for 2 values of a given parameter). A3 identifies the interaction between 2 given action parameters. A4 identifies the best scenario involving the action parameters that are actually influential to minimize the Dth. A5 identifies the correlation between 2 given evaluation parameters.

Step 6: Reconfigure the TS. In the current system, the implementation of decisions resulting from various analyses conducted through the simulator falls under the responsibility of the plant manager. Some configurations (type 1 data) require Lean continuous improvement actions [12] while others (type 2 data) entail a financial investment.

6 Conclusion

The main contribution of this paper lies in the approach we have proposed to develop CPLS. Comprising a six-step functional view and a six-step technical view, it places discrete event simulation at the center of data analysis. The academic case study conducted in the context of a "manufacture to stock, distribution to order" supply chain, demonstrated its applicability and allowed the definition of additional concepts specific to CPLS compared to those provided by CPPS.

In alignment with the maturity model presented in section 2.3, the CPLS developed in the case study has a "Manual" maturity level. Indeed, all decisions are made and implemented by the plant manager to reconfigure the target system based on useful data (scenarios) provided by the simulator. Given the nature of the decisions made in the logistics context in general and the case study in particular (section 5.1 step 6), the developed CPLS cannot attain a maturity level higher than the "Manual" level. Table 3 nevertheless presents for each type of data, a minimal scenario and an ideal one from a technological point of view. However, both scenarios still fall under the "Manual" maturity level, as per the model presented in section 2.3.2. This underscores that the current maturity model is unsuitable for making the CPLS evolve in small steps according to the different dimensions of the Figure 4 metamodel.

As a prospective direction, the natural progression of this work would entail proposing a more precise maturity model that supports gradual changes within companies. Additionally, evolving the proposed approach to incorporate such a maturity model could be a significant step forward.

Table 3. Case study CPLS maturity

Data	Maturity	Collecting	Distribution	Storage	Processing	Analysis	Configuration
Type1	Initial	Active Observation	Manual Entry	On paper	Human	Human	Factory manager
	Ideal	Sensors, EIS	IOT	On disk	Computer program	Computer program	Actuators
Type2	Initial	Observation	Manual Entry	On paper	-	Human	Factory manager
	Ideal	Sensors	IOT	On disk	-	Computer program	Factory manager

Appendix A

Figure 6.

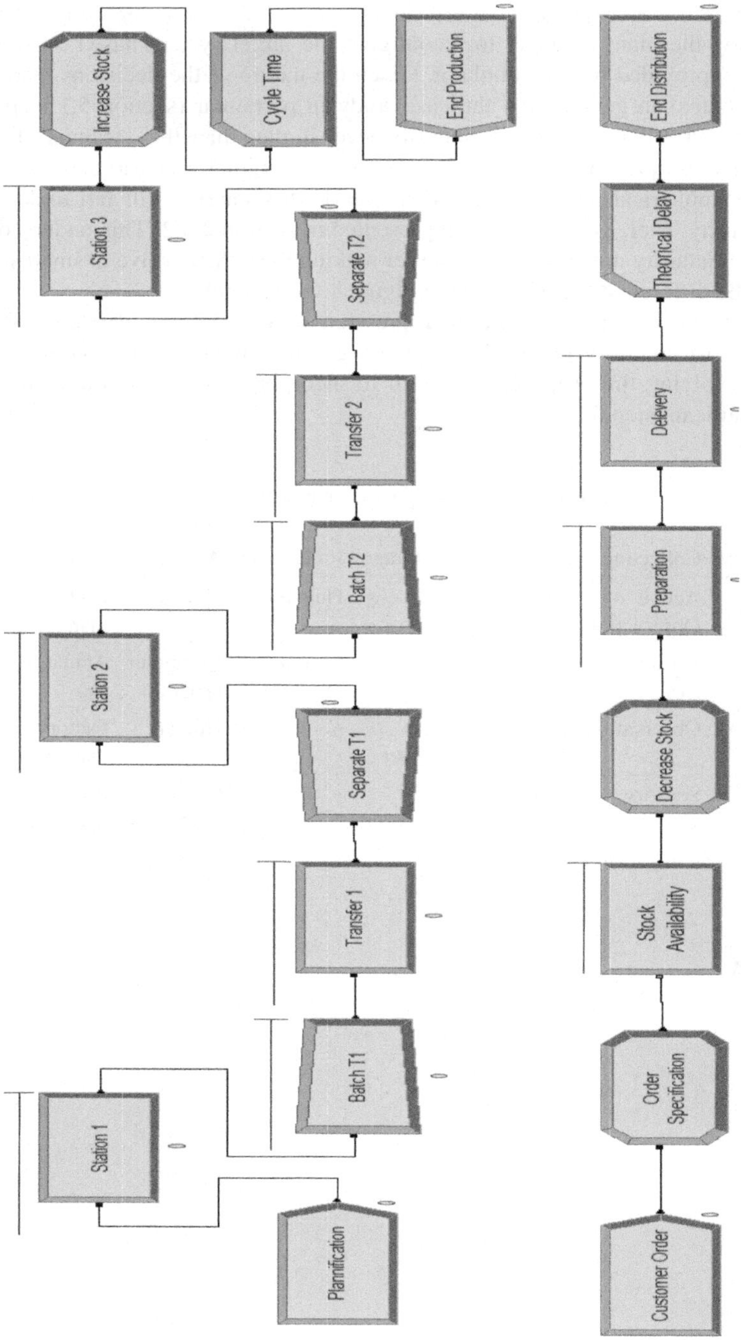

Fig. 6. Target System ARENA Flowchart

Appendix B

Table 4.

Table 4. Case Study Raw Data

Parameter Designation	Nature	Workstation	Source	Type
Time Between Arrivals of Manufacturing Orders	Action	Planification	Production	T1
Execution time Station 1	Action	Station 1	Production	T1
Number Operator Station 1 (Ostat1)	Action	Station 1	Production	T2
Transfer Batch Size Transfer 1	Action	Transfer 1	Production	T2
Number of forklifts Transfer 1	Action	Transfer 1	Production	T1
Duration Transfer 1	Action	Transfer 1	Production	T1
Operator Number Station 2 (Ostat2)	Action	Station 2	Production	T2
Execution time station 2	Action	Station 2	Production	T1
Transfer Batch Size Transfer 2	Action	Transfer 2	Production	T2
Number of forklifts Transfer 2	Action	Transfer 2	Production	T2
Duration Transfer 2	Action	Transfer 2	Production	T1
Number Operator Station 3 (Ostat3)	Action	Station 3	Production	T2
Execution time Station 3	Action	Station 3	Production	T1
Time Between Arrivals of customer orders Quantities ordered Distances from customers	Disturbance	Customer Order	Environment	T1
Number of Preparation Operators (Pop)	Action	Preparation	Distribution	T2
Execution Time Order preparation (Operation)	Action	Preparation	Distribution	T1
Number of trucks	Action	Delivery	Distribution	T1
Order delivery time Average truck speed	Disturbance	Delivery	Environment	T1
Theoretical delay	Evaluation	-	Customer	T1
Occupancy rate of transport resources Occupancy rate of handling resources Number of operators Number of transport and handling resources	Constraints	Company	Strategy	T2

Appendix C

Table 5 and Table 6.

Table 5. Analyzes carried out in the case study

Analysis	Objectif	Mode	Technical Tools	Example
A0	Measure the overall characteristics of the target system: capability, resilience to a change in the environment, etc. for a given scenario	**Capability:** a. Record Dth (Theoretical Delay) for the given scenario.	Use the block: Statistics to generate the data file (.dat)	***Example:*** Initial Scenario (Table 05). ***Result:***
		b. Calculate the capability of the Target System: Cp = (Ts-Ti) /6s. Ts: upper limit. Ti: lower limit. S: Standard deviation	Use the tool: Output Analyzer//Analyze/Confident Interval on Standard Deviation	• Cp= (Ts-Ti)/6s = 4 / 6*11.7 = 0,057 ***Conclusion:*** • TS not capable • Requires reconfiguration
		Resilience: • Introduce environmental change into the simulation program • Repeat a and b	-	
A1	Identification of the most influential action parameters on the Dth - Causality between a given parameter and Dth for a given scenario (the other parameters being unchanged). For decisions (D1 … D6).	Draw the curve Dth = F(Capacity)	Use the tool: Process Analyzer	***Example:*** Analyze the causality between the "Truck Capacity" and "Dth" for the initial scenario ***Result:*** beyond 4 trucks, there is no longer any impact on Dth
A2	Identification of the most influential action parameters on the Dth - Comparison of the averages of Dth (Threshold 95% and 99%) for 2 values of a given parameter for a given scenario (the other parameters being unchanged). For decisions (D1 … D6).	• Record Dth (Theoretical order delay) for the given scenario with the parameter at value1 level. • Record Dth (Theoretical delay of orders) for the given scenario with the parameter at value2 level. • Compare the averages of the two series of data obtained by simulation	Use the block: Statistics to generate the data files (.dat) then use the tool: Output Analyzer/Compare Means	***Example:*** Compare the averages (Threshold 95% and 99%) for the Truck Capacity 2/3 then 4/5 scenarios ***Result:*** H0 rejected for Truck 2/3 (Threshold 95% and 99%) H0 rejected for Truck 4/5 (Threshold 95%) H0 accepted for Truck 4/5 (Threshold 99%)
A3	Identification of the interaction between 2 given action parameters. For decisions (D1 … D6).	• Carry out the experimental plan (4 experiments: 2 parameters at 2 levels) • Record Dth for each experiment • Optionally draw the interaction diagram • Calculate the level of interaction	Use the tool Process Analyzer to automate experiments and a spreadsheet to calculate the level of interaction	***Example:*** Study the interaction between the "Truck Capacity" and "Batch Size" action parameters Truck capacity (2/3); Batch Size (5/6) ***Result:*** Very weak interaction (2%)
A4	Identify the best scenario involving the action parameters that are actually influential in their areas of influence to minimize the Dth taking into account the constraints defined in the "Expression of needs" phase. For decisions (D1 … D6).	• Define the interval and step for each parameter (Min, Max, Step) • Define the constraints to be taken care of • Search for the best scenario in the value ranges defined for each parameter	Use the tool OptQuest	***Example:*** Factors: capacity (Truck, Pop, Ostat1, Ostat2, Ostat3, forklifts), Batch Size, Opreparation. For each parameter, here is the (Min, Max, Step). Truck(1,4,1), Pop(1,2,1), Ostat1(1,2,1), Ostat2(1,2,1), Ostat3(1,2,1), forklifts(2,4,1),
				Batch Size(4,10,1), Opreparation(0.05,0.3,0.05) ***Constraints :*** [Pop] + [Ostat1] + [Ostat2] + [Ostat3] <=6 [Truck]+ [Forklift]<=7 [Truck.Utilization] >=70% [Forklift.Utilization]>=60% ***Result:*** Optimal scenario: Truck (2), Pop (2), Ostat1(1), Ostat2(2), Ostat3(1), Forklift(2), Batch Size(5), Opreparation(0.05)
A5	Identification of the correlation between 2 given evaluation parameters (D7)	• Carry out the experimental plan involving the influential parameters (A1, A2) on their value intervals (A4) • Calculate the degree of correlation	Use the tool Process Analyzer to automate the experimental design and a spreadsheet to calculate the degree of correlation	***Example***: Analyze the correlation between Average Stock and Dth. ***Result:*** There is no correlation

Table 6. Initial Scenario

Poste	Planification	Station 1		Transfer 1			Station 2		Transfer 2			Station 3		Customer Orders			Preparation		Delivery	
Parameter	TBA	NBOp	Delay	Batch	NB forklifts	Delay	NBOp	Delay	Batch	NB forklifts	Delay	NBOp	Delay	TBA	Quantity	Distance	NBOp	Durée	NB trucks	Speed
Value	10	1	N(10,2)	5	1	N(45,5)	1	N(10,2)	5	1	N(30,5)	1	N(10,2)	1h	T(1,10,15)	T(5,100,400)	1	6/p	2	70km/h

Abbreviations: Delay in minutes; N: Normal law; T: Triangular law; p: product

References

1. Al-Hammouri, A.T.: A comprehensive co-simulation platform for cyber-physical systems. Comput. Commun. **36**(1), 8–19 (2012)
2. Baheti, R., Gill, H.: Cyber-physical systems. The Impact of Control Technology **12**, 161–166 (2011)
3. Benmoussa, R. : De la simulation a évènement discret au machine Learning pour la chaine logistique. 13ème Conférence internationale de Modélisation, Optimisation et Simulation (MOSIM2020), 12-14 NOV 2020, Agadir, Maroc. HAL-03177538 (2020)
4. Cardin, O.: Classification of cyber-physical production systems applications: Proposition of an analysis framework. Comput. Ind. **104**, 11–21 (2019)
5. Cardin, O., Derigent, W., Trentesaux, D.: Digitalization and Control of Industrial Cyber-Physical Systems, Concepts, Technologies and Applications. ISTE Ltd and John Wiley & Sons (2022)
6. David Kelton, W., Sadowski, P., Zupick, B.: Simulation with Arena, sixth Edition. Mc Graw Hill Eduction (2015)
7. Dworschak, B., Zaiser, H.: Competences for cyber-physical systems in manufacturing – first findings and scenarios. Procedia CIRP, 8th International Conference on Digital Enterprise Technology - DET 2014 Disruptive Innovation in Manufacturing Engineering towards the 4th Industrial Revolution 25, 345–350 (2014). https://doi.org/10.1016/j.procir.2014.10.048
8. Fantini, P., Pinzone, M., Taisch, M.: Placing the operator at the centre of Industry 4.0 design: Modelling and assessing human activities within cyber-physical systems. Computers & Industrial Engineering (2018). https://doi.org/10.1016/j.cie.2018.01.025
9. Fink, J., Ribeiro, A., Kumar, V.: Robust control for mobility and wireless communication in cyber–physical systems with application to robot teams. Proc. IEEE **100**(1), 164–178 (2012)
10. Gengarle, M.V., Bensalem, S., McDermid, J., Sangiovanni–Vincentelli, A., Törngre, M.: Characteristics, capabilities, potential applications of cyber–physical systems: A preliminary analysis. Deliverable D2.1, CyPhERS FP7 Project (2013)
11. Gräßler, I., Wiechel, D., Roesmann, D.: Integrating human factors in the model-based development of cyber-physical production systems. 31st CIRP Design Conference 2021 (CIRP Design 2021) **100**, 518–523 (2021). https://doi.org/10.1016/j.procir.2021.05.113
12. Hennion, R., Makhlouf, A.: LES FICHES OUTILS DU LEAN SIX SIGMA, Eyrolles (2016)
13. Klimeš, J.: Using formal concept analysis for control in cyber-physical systems. Procedia Engineering **69**, 1518–1522 (2014)
14. Lee, E.A.: Cyber-physical systems – Are computing foundations adequate? Position Paper for NSF Workshop on Cyber-Physical Systems. Austin, Texas (2006). http://ptolemy.eecs.berkeley.edu/publications/papers/06/CPSPositionPaper/
15. Monostori, L.: Cyber-physical production systems: Roots, expectations and R&D challenges. Procedia CIRP **17**, 9–13 (2014)
16. Müller, R., Narciss, S., Urbas, L.: Interfacing cyber-physical production systems with human decision makers. In: Song, H., Rawat, D.B., Jeschke, S., Brecher, C. (eds.) Cyber-Physical Systems, pp. 145–160. Academic Press (2017). https://doi.org/10.1016/B978-012-803801-7.00010-9
17. Putnik, G.D., Ferreira, L., Lopes, N., Putnik, Z.: What is a cyber-physical system: definitions and models spectrum. FME Transactions **47**(4), 663–674 (2019)
18. Rajkumar, R., Lee, I., Sha, L., Stankovic, J.: Cyber-physical systems: The next computing revolution. 2010 47th ACM/IEEE Design Automation Conference (DAC), pp. 731–736 (2010)
19. Schuh, G., Potente, T., Varandani, R., Hausberg, C., Fränken, B.: Collaboration moves productivity to the next level. Procedia CIRP **17**, 3–8 (2014)

20. Strang, D., Anderl, R.: Assembly process driven component data model in cyberphysical production systems. Proceedings of the World Congress on Engineering and Computer Science (2014). [Online]. Available at: http://www.iaeng.org/publication/WCECS2014/WCECS2014_pp947-952.pdf
21. Zolotová, I., Papcun, P., Kajáti, E., Miškuf, M., Mocnej, J.: Smart and cognitive solutions for Operator 4.0: Laboratory H-CPPS case studies. Computers & Industrial Engineering (2018). https://doi.org/10.1016/j.cie.2018.10.032
22. Wu, X.: Transformation of existing production systems with integrated enterprise information systems (EISs) into cyber physical production systems (CPPSs) in the context of Industry 4.0 (2022)

Universal, Bi-directional Real Time Communication Between Real and Digital Twin in an MBSE Environment

Silvan Lack(✉), René Klopfer, and Daniel Schmid

ZHAW Zurich University of Applied Sciences, Winterthur, Switzerland
{silvan.lack,rene.klopfer,daniel.schmid}@zhaw.ch

Abstract. The paper presents a solution for establishing a universal, bi-directional real-time communication between the Digital and Real Twins in mechanical engineering. The exemplary set-up consists of a three-axis XYZ cross table, which was developed together with the Digital Twin Framework in a Model-based Systems Engineering (MBSE) environment, applying the RFLP V-model method and using Dassault Systèmes' 3DExperience platform. The Real Twin is controlled by a standard industrial programmable logic controller (PLC). The communication is implemented using Modelica, Named Pipes, and the OPC UA protocol.

The paper describes the design and implementation of the communication interface, the connections, and the communication modes. It also describes the performance, precisely the expected turnaround time (expectation value) for one axis value, which is 10 ms for the chosen reference communication loop. Furthermore, the paper describes the potential applications of the solution in industry and education and discusses future challenges and opportunities. The scientific project constitutes the evidence that it is feasible to establish a complete Digital Twin Framework in a complex MBSE environment and yet achieve high performance. It bridges the gap between the Real and Digital Twin in mechanical engineering.

Keywords: Digital Twin Framework · Model-Based Systems Engineering · Named Pipes · OPC UA · Real-time communication · PLC · XYZ cross-table · Axes system · 3DExperience · Dymola · Modelica

1 Introduction

1.1 Bridging the Gap Between Real and Digital Twin in Mechanical Engineering

Digital Twins have been known [1] and described [2], and their values have been investigated and highlighted [3] accordingly. A Digital Twin Framework consists of the Real Twin (physical entity), the Digital Twin (virtual model) and the bi-directional communication. It enables users to monitor, understand, and predict complex systems better, among many other benefits, such as detecting and diagnosing problems [4].

Published by Springer Nature Switzerland AG 2025
P. Sureephong et al. (Eds.): PLM 2024, IFIP AICT 740, pp. 208–221, 2025.
https://doi.org/10.1007/978-3-031-93319-6_16

Despite the significant development history of the Digital Twin, the bi-directional communication of the Real and Digital Twin still poses a notable challenge. The problem is accentuated when the requirements are demanding, e.g. the communication has to occur in real-time.

Therefore, the investigations and development had to verify or falsify the following hypothesis: "It is possible to establish real-time bi-directional communication between the Digital Twin in a Model-Based System Engineering (MBSE) development environment and the Real Twin, controlled by a standard, commercially available industrial Programmable Logic Controller (PLC), with a universal approach." (Fig. 1)

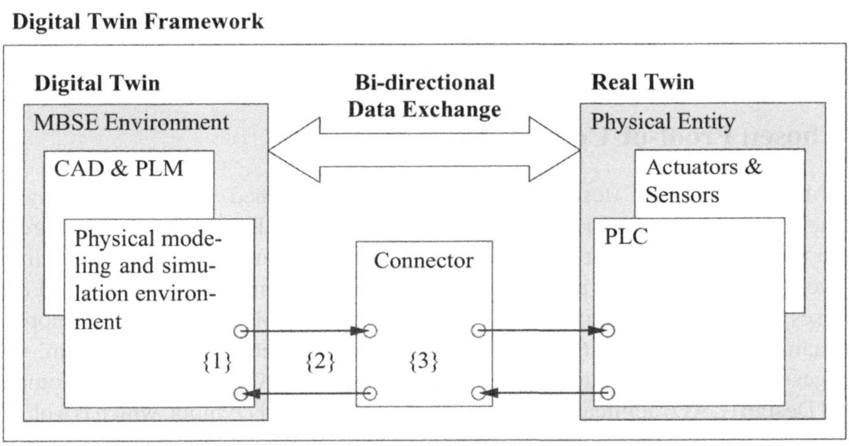

Fig. 1. Digital Twin Framework illustrating the hypothesis to be verified or falsified.

The aim of exploring the hypothesis is to advance the communication capabilities of industry-oriented Digital Twin Frameworks and to apply the MBSE approach to product development. The insights generated shall be applied in subsequent commercial projects and as part of the mechanical engineering curriculum at the ZHAW. The universal approach, meaning vendor-neutral or readily adaptable to other vendors, and real-time communication are the two outstanding requirements for communication with standard industrial PLC systems [5, 6], such as Siemens, Rockwell Automation, and B&R [7].

1.2 Key Development Areas

Enabling a vendor-neutral approach for a bi-directional, real-time communication between the Digital and the Real Twin requires developments in three key areas.

The first area involves the conceptualisation of a multi purpose data concept and interface within the physical modeling and simulation environment {1} to enable bidirectional control of the kinematic model. The simulation includes kinematic connections for the assembly and computes physical behavior, such as friction. It typically encompasses a user interface displaying the virtual assembly in a three-dimensional view or as simplified blocks. The simulation environment may be Dymola/Modelica, Simulink/Simscape, or any other simulation environment supporting the integration of custom program code in languages like C, C++, or C#, which can be called from the model.

The second aspect entails defining and implementing a method for real-time bi-directional communication {2} between the simulation environment and external sources.

The third area involves the development of a Connector {3}, which serves as the communication master responsible for managing data exchange between the Digital and Real Twin. The data to be exchanged must be defined by the developer to align with the relevant movements and PLC variables (and within the PLC to the physical motors). In vendor-locked systems, such as Siemens [8], this process also requires definition but is usually easier within the vendor's environment.

2 Chosen Proof-of-Concept Environment

A proof-of-concept (PoC) test environment (Fig. 2) was defined with the selected systems that lack a Digital Twin interface to a market-typical PLC (Real Twin) and that are used within the academic context of the university. On the Digital Twin side, this primarily involves the 3DExperience platform from Dassault Systèmes, which is a cloud-or on-premises environment that integrates various software tools for product development. The main tools in use are: the Product Lifecycle Management (PLM) system, which manages product data throughout the whole lifecycle; CATIA, which is a Computer-Aided Design (CAD) application for mechanical design; and Dymola, which is a physical modelling and simulation environment based on the Modelica language.

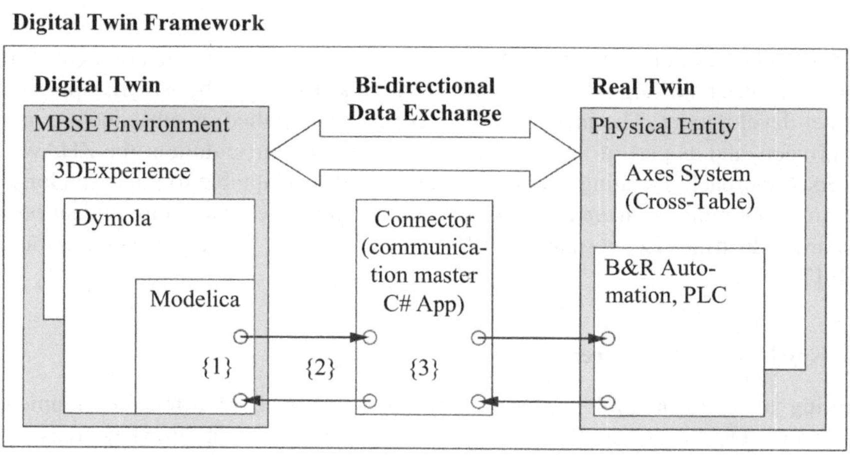

Fig. 2. Proof-of-Concept environment configuration and vendors

Dymola uses the Modelica [9] language, which is essentially a standard due to its application in various commercially available simulation environments [10]. A variety of applications within 3DExperience have been used to develop the Digital Twin Framework according to the Model-Based Systems Engineering (MBSE) method to include requirement (R) and function (F) definitions and the related logical (L) and physical (P) design (RFLP [11, 12]).

To enable communication between the model and external sources, an implementation must be created in Modelica {1}. This implementation provides the model with a function block with data sources, sinks, and typically a command number and trigger. The function block in Modelica is the interface for the communication to/from the kinematic model and calls C-code functions that are compiled with the simulation.

A communication master application (Connector, described below) that runs on Microsoft Windows{2} communicates with the C-code functions via Named Pipes [13], also known as First-In-First-Out (FIFO), a method of inter-process communication that enables communication between processes on a computer.

The Connector {3} or communication master application has been chosen to be a C# application running on Microsoft Windows, implemented with a commercial Software Development Kit (SDK) for OPC UA [14], a standard for communication with PLCs. It masters the data exchange between the Digital and the Real Twin, or to be more exact, their controllers.

While the Digital Twin's controller is implemented in Modelica, the Real Twin's controller has been chosen to be a programmable logical controller (PLC) from B&R Industrial Automation [15]. It controls a motor-driven three-axis system (XYZ crosstable) with two main axes, x and y, defining the horizontal plane and a short z-axis perpendicular to the x-y-plane.

This concept for the chosen environment can easily be adapted to other modelling systems that use the Modelica language as well as other physical modelling and simulation environments that allow intercommunication between processes with a similar approach. (Further effort may be required to adapt to another system.) On the side of the Real Twin, compatibility with other PLC vendors is given due to the widespread use of the OPC UA protocol. Finally, other hardware configurations of mechatronic systems other than a cross-table can be implemented with a similar approach.

3 Solution

3.1 Component Interaction Within the Digital Twin

The Digital Twin in this solution is built with apps from the 3DExperience platform. For a better understanding of the solution, the workflow of the Digital Twin in this environment is shown in Fig. 3. Within 3DExperience, Dymola uses kinematic information from the CAD assembly that was built with CATIA, to create a corresponding multi-body-system. In an optimal simulation, multiple aspects from the CAD assembly are taken into account, such as mass and inertia calculated from the geometry and the assigned materials, as well as collision handling [16]. In Dymola, the environment for model creation, testing, simulation and post-processing [17], the kinematic model is translated to a Modelica model. From there, the model can be expanded upon using the graphical user interface and a large library to implement further logical behaviour with drag & drop operations or directly by editing the Modelica code. In this solution, a graphical function block was created so that it can be integrated easily into an existing graphically modelled logic. Furthermore, it was necessary to extend the function block with Modelica code for the invocation of the external C-functions, in which the Named Pipes transactions for external communication were implemented.

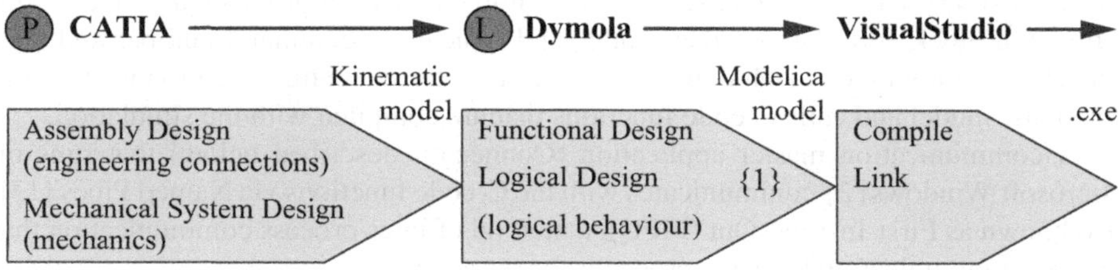

Fig. 3. Workflow from CAD to the Simulation Model

Unlike other programming languages, the Modelica code is not based on instructions but on systems of equations, which are translated into C-program code and connected to e.g. a numerical input. The Modelica translated program code is then compiled and linked together with the C-functions for the Named Pipe transactions.

3DExperience does not come with a C-compiler. Therefore, VisualStudio's compiler was chosen to build and link the final combined simulation, which is then executed within the simulation environment of 3DExperience.

3.2 Nomenclature

For further describing the solutions for elements of the Digital Twin and the Real Twin, The concept needs some major definitions to be aligned with the RFLP approach (see Fig. 4)

- **Virtual Logical Controller (VLC)** – The VLC represents the actual control system in the virtual environment. It combines the related functionalities with the simulation of the physical laws. In the RFLP approach, the VLC is part of the product's logical design (L).
- **Virtual Assembly (VA)** – The VA represents the actual physical product in the virtual environment and can be described best as the dynamic simulation of the product, including the CAD model with material specifications and other physical properties. The VA is part of the product's physical design (P) in the RFLP approach.
- **Real Logical Controller (RLC)** – The RLC is the control system of the axis system and is based on a PLC.
- **Real Assembly (RA)** – The RA is the actual physical product in the real world.
- **Digital Twin (DT)** – In this test environment, the DT is the combination of the VLC and the VA and the "sibling" of the Real Twin.
- **Real Twin (RT)** – The RT is a fully functional product comprising the RLC and the RA.
- **Connector** – The Connector enables bi-directional real-time communication between the Real and the Digital Twin.

- **Digital Twin Framework (DTF)** – The DTF is established with the three required elements DT, RT, and Connector to fulfil the related definition of a DTF.

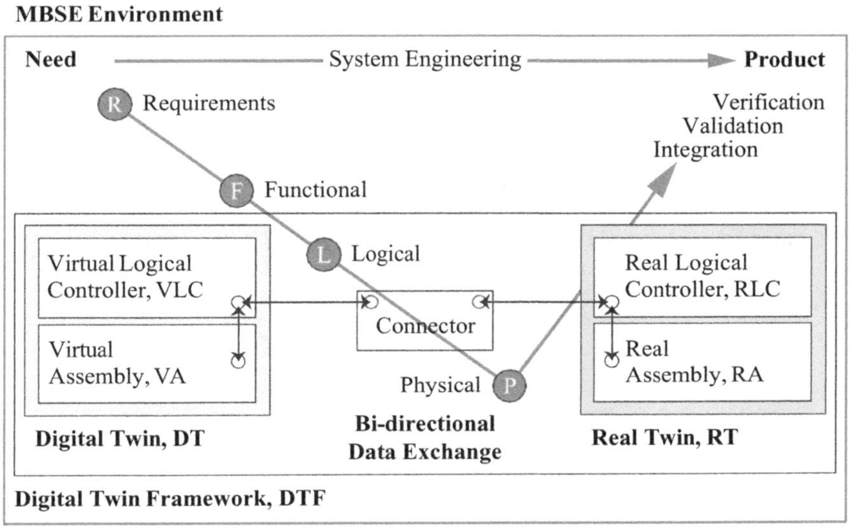

Fig. 4. MBSE environment (RFLP) and elements of the Digital Twin Framework, DTF

3.3 Interface Implementation in Modelica {1}

The implementation of the interface starts within the Modelica model, as shown in Fig. 3 {1}. To facilitate further uses, the interface is designed as part of a Dymola library and can be used within the graphically programmed logic in Dymola or the *Functional and Logical Design*-Application respectively.

The Virtual Assembly (VA) contains the internally generated kinematics model. The generation method is left out of focus here, since this is a process that is already established and functional.

The Virtual Logical Controller (VLC) (Fig. 5, left hand side) contains the logic for the Mode of Operation (see 3.6) as well as the interface.

The interface in Modelica is a function block defined with a general-purpose set of boolean, real, and integer inputs and outputs, as well as a command number and a trigger input. Figure 6 shows the input part, which currently uses only the x-, y-and z-axes for the tool position. The interface, as shown on the right-hand side with its Modelica code, calls a function described in 3.5 which is provided to the VisualStudio compiler in separate C-code for compilation.,

3.4 Communication from Modelica to the Outside and Vice Versa {2}

As explained in Sect. 3.3, the function call in Modelica invokes a C-function that is written in C-code outside of Modelica. Despite initialisation routines, a read and a write

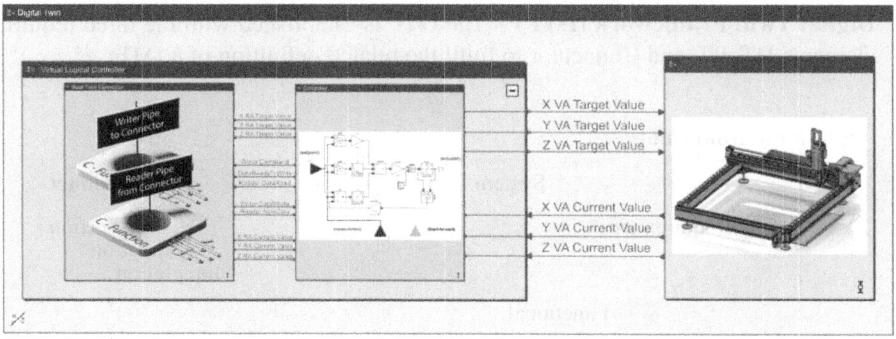

Fig. 5. Functional and logical design view of the Digital Twin in 3DExperience

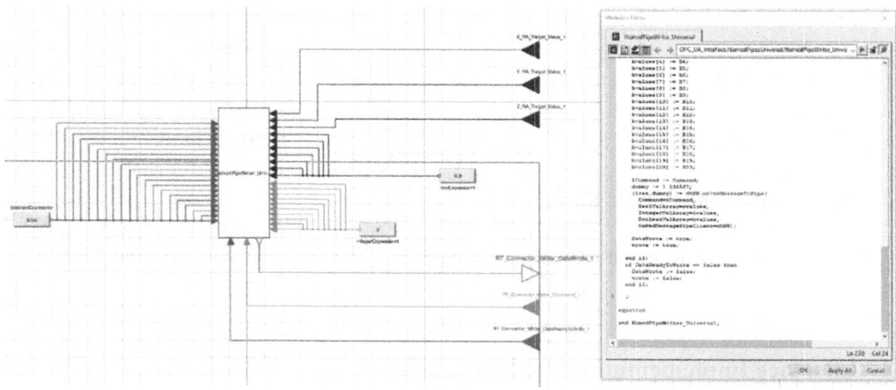

Fig. 6. Named Pipe write function block in Modelica including external function call

functionality is provided (Fig. 7). This will supply a Named Pipe for write operations (Writer Pipe) from the Digital Twin to the Connector. Read operations use a separate Named Pipe (Reader Pipe) to communicate from the Connector to the Digital Twin. Writer and Reader Pipe are of the same type.

3.5 Real Twin Connector with OPC UA Interface {3}

The Connector (or Real Twin Connector) is a C# program that runs as a separate application (Fig. 8). It enables the communication between the compiled simulation (the Modelica model with additional C-code functions) and an OPC UA server on the PLC. The OPC UA implementation in the Connector is based on a software development kit (SDK) from Unified Automation GmbH, which provides a high-level Application Programming Interface (API) for OPC UA client applications. The Connector processes the commands from a Named Pipe (Writer Pipe) and returns its answer to the Reader Pipe. A command table must be provided in order to pair the OPC nodes [18] with inputs/outputs from/to the Modelica model. With a command identifier sent by Modelica on the Writer

```
int writeMessageToPipe(int command, double *rene, double rvalues[],
  size_t anzRvalues, int ivalues[], size_t anzIvalues, int bvalues[],
  size_t anzBvalues, void* externalObject)
{
        int nbCharsToWrite;
        char strBuffer[READBUFSIZE];

        collectMessage(strBuffer, command, rvalues, anzRvalues, ivalues,
           anzIvalues, bvalues, anzBvalues);
        nbCharsToWrite= (int) strlen(strBuffer);
        NamedMessagePipeStruct* namedMessagePipeStruct = (
           NamedMessagePipeStruct*)externalObject;

        namedMessagePipeStruct->fSuccess = WriteFile(
                namedMessagePipeStruct->hPipe,
                strBuffer,
                nbCharsToWrite,
                NULL,
                NULL);

        if (! namedMessagePipeStruct->fSuccess)
        {
                return -1;
        }
        return nbCharsToWrite;
```

Fig. 7. WriteMessageToPipe C-function, Writer Pipe

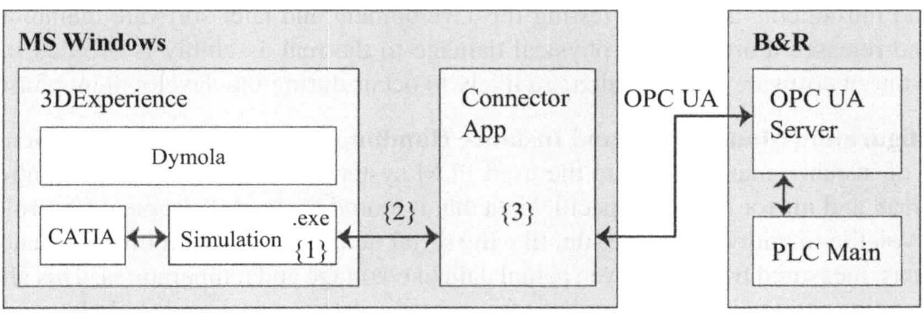

Fig. 8. Process overview

Pipe and the looked up OPC UA node identifier for that command, the Connector reads the values of the PLC variables through the OPC UA server and writes the results back to the Reader Pipe. It is also possible to write values to OPC UA nodes.

3.6 Bi-directional Real-Time Communication

Modes of Operation. The command source for any movement is usually a program that runs either on the Real Logical Controller (RLC) of the Real Twin or on the Virtual Logical Controller (VLC). With a mode of operation-switch, the VLC controls which command source is active for the Digital Twin's movements. Modes of operation are:

1. Sending real data to the Digital Twin (Passthrough). The Virtual Assembly (VA) follows the movement of the Real Assembly (RA), while the master movements come from the Real Controller (RLC) or, to be more specific, the B&R Automation PLC main program. The purpose of this mode could be debugging of the Real Twin's movement, while the Real Assembly (RA) is not accessible, not visible or located in a remote facility. In a more sophisticated version of this set-up, the Real Assembly (RA) and the Virtual Assembly (VA) movements can be tracked independently and compared to discover physical issues and consider automated routines at system startup.

2. Sending planned movements to the Real Twin. The Real Assembly (RA) and Virtual Assembly (VA) execute the same movement, which is initiated by the Virtual Logical Controller (VLC). This mode is intended for testing either the kinematic model or the quality of the real assembly by capturing and comparing the planned and real movements.

3. In a third mode of operation, the Digital Twin uses real data, which can include physical (human-made) manipulations of the Real Assembly (RA) kinematics. The Virtual Logical Controller (VLC) could (with further investigations) also be replaced by a virtual instance of the Real Logical Controller (RLC). Most PLC vendors have a simulation environment for running the program being developed, which is then used as the source for the motion. This is mainly used to develop PLC software without having a real machine, with the advantage of significantly shortening the project timeline and introducing automated testing for development and later software maintenance and releases. Furthermore, physical damage to the real assembly is avoided in the event of software errors, which are likely to occur during the development phase.

Configuration Management and Instance Handling. As an option to circumvent the lack of instance management in the used PLM system, the Virtual Logical Controller can read and mirror instance-specific data that is stored in the Real Logical Controller's non-volatile memory during set-up, like the serial number, maximum travel distance of the axis, measured friction or even actual data like voltage and temperatures. This allows for running the Digital Twin not only from master data (as designed) but also on built instances (as built) and their health state (as maintained).

Real-Time Communication. Real-time systems have different meanings and requirements in different domains. A common classification of real-time systems distinguishes between hard, soft, and firm real-time systems [19].

The shown set-up is considered a soft real-time or near real-time system (slightly slower than real-time [20]), as it delivers fast and hardly humanly traceable responses. For an in-depth understanding of the response times, Sect. 3.7 provides detailed results. It's noteworthy that minor time variations and lateness do not hinder the user experience, the value of the Digital Twin or a task that is carried out with it, such as detecting anomalies, optimising performance, predicting outcomes, and recommending solutions.

3.7 Achieved Performance

The Digital Twin tracks a motion from the Real Twin quickly but with a certain delay. This delay depends on various factors, such as how many axes are controlled or whether other data is transmitted simultaneously. Moreover, the cycle time is not always constant, which can result in a jerky appearance in the graphical 3D view.

The experienced end-to-end delay between the Real and the Digital Twin was measured to be about 150 ms, by analysing a recorded video that framed both the Real Assembly (RA) in real and the Virtual Assembly (VA) on a computer monitor.

The typical performance of the communication only, starting from a request message from Modelica until the received response for an axis position (also called turnaround time) of the Real Twin is shown in Fig. 9. This test was performed with two computers of different performance. Both have the 3DExperience platform running incl. CAD.

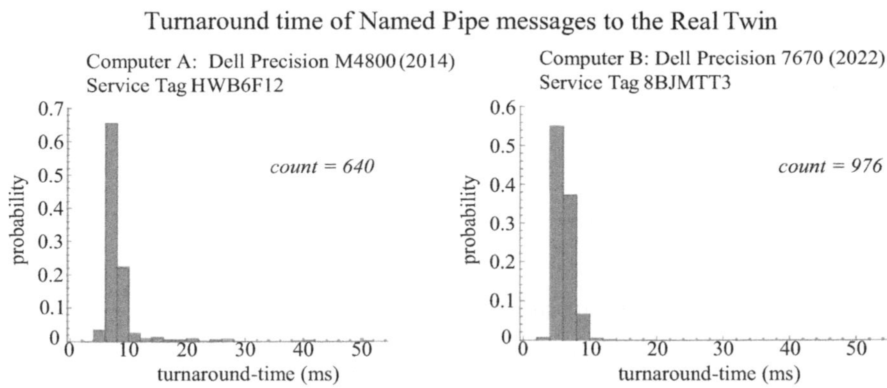

Fig. 9. Communication performance for a single value

The turnaround time for one axis value on computer B can generally be expected to be lower than 10 ms. On the outdated computer A, a generally higher response time and a certain number of outliers with much higher response times are expected.

However, periods of much higher response times were found on both systems (Fig. 10), which are inconclusive at this stage but have been shown by process monitoring to be related to background processes, including malware detection of the operating system. To further reduce the problem, the system could be manually optimised by disabling unnecessary background processes, using a dedicated central processing unit (CPU) core [21], or using special software for real-time optimisation [22]. A final illustration of the reaction time's typical composition can be found in Fig. 11.

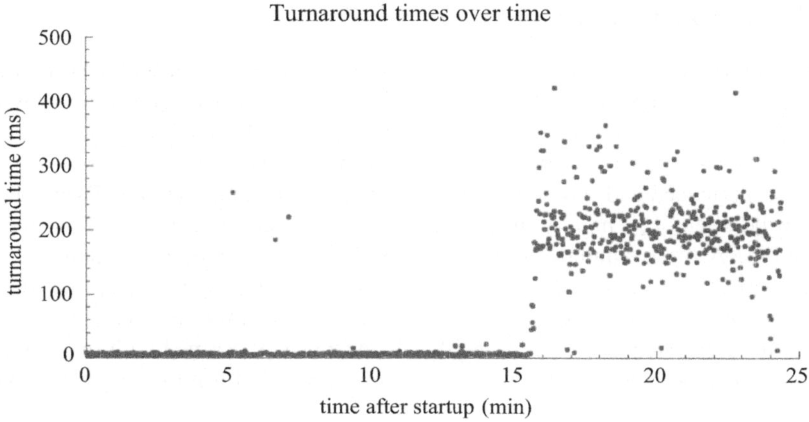

Fig. 10. Phase of higher turnaround times due to background processes

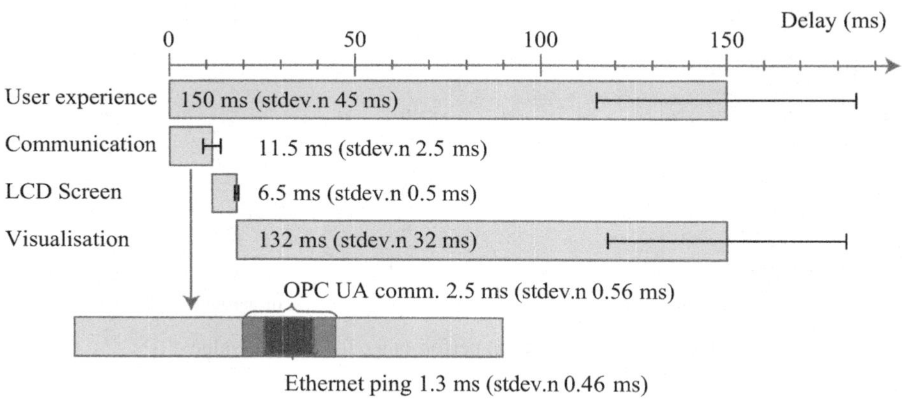

Fig. 11. Analysis of delay sources with three position values being transmitted (x, y, z)

4 Conclusions, Applications and Outlook

4.1 Conclusions

The findings demonstrate that the universal, bi-directional real-time communication between Real and Digital Twin in an MBSE environment is possible. Therefore, the hypothesis in Sect. 1.1 Bridging the Gap Between Real and Digital Twin in Mechanical Engineering", page 1, is verified accordingly, based on the presented axes system (cross-table). Figure 12 shows the Real Twin in front (the three-axes system and the commercially available industrial PLC) and the Digital Twin in the MBSE environment on a screen in the background.

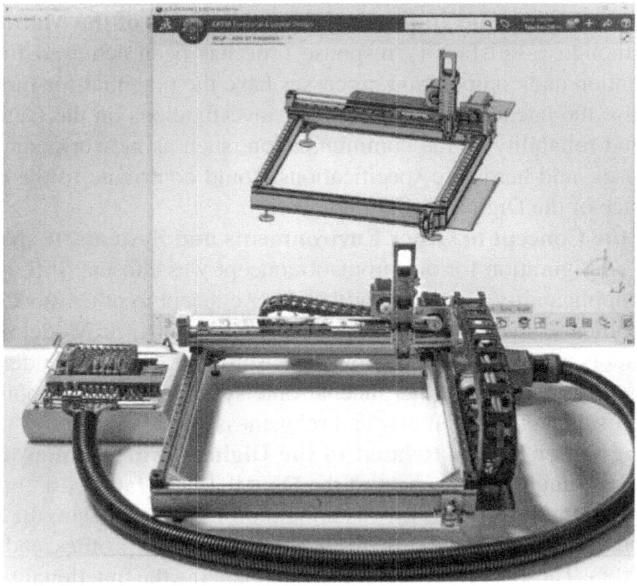

Fig. 12. Digital and Real Twin

4.2 Applications

The elaborated case of the presented Digital Twin Framework based on the axes system (cross-table) is applied in multiple contexts within the ZHAW. The axes system is part of the mechanical engineering bachelor curriculum; specifically, it is a fundamental element of the specialisation module Smart Products and Production (SPP [23]), which spans two semesters. In the Advanced Digital Engineering module (ADE [24]), the students are given more in-depth instruction on the relevant aspects of the Digital Twin Framework.

Varying elements of the Digital Twin Framework are content of the continuing education, e.g. for Certificate of Advanced Studies (CAS). One example is the course Product Innovation and Leadership for Engineers [25] as part of the Master of Advanced Studies (MAS) Industry 4.0 [26].

The Digital Twin Framework of the axes system (XYZ cross-table) is a demonstrator to attract research and development projects. For example, related government-funded research projects shall be promoted (e.g. Innosuisse [27]). It is also conceivable to incorporate alternatives to the software tool used. For example, Dymola can be substituted by tools such as OpenModelica [28], SimulationX [29], Modelon Impact [30], Wolfram System Modeler [31], or MapleSim [32].

4.3 Outlook

This paper demonstrated the feasibility and performance of a universal, bi-directional real-time communication between the Digital and Real Twin in an MBSE environment. However, some limitations and open questions still need to be addressed in future work. Some of them are:

Reducing the Delays and Improving the Smoothness of the Motion of the Digital Twin. Although a satisfactory response time has been achieved for our system, the communication and computation processes have the potential for further optimisation to minimise the latency and jitter. Further investigations on the factors that affect the stability and reliability of the communication, such as network congestion, background processes, and hardware specifications, could contribute to the understanding and performance of the Digital Twin Framework.

Applying the Concept to Other Environments and Systems. A specific software and hardware configuration for our proof-of-concept was chosen. Still, it is of interest to explore the applicability and adaptability of our concept to other modelling and simulation environments, such as OpenModelica, SimulationX, or Modelon Impact, and other PLC vendors, such as Siemens, Rockwell Automation, or Schneider Electric. The target is to test our concept on other mechatronic systems, such as robots, vehicles, or machines, and evaluate the scalability and robustness of our solution.

Addressing Other Topics Related to the Digital Twin Framework. The paper focused on the communication aspect of the Digital Twin Framework, but other topics deserve further attention, such as the instantiation of the Real Twin, the integration of master data and ERP systems, the storage and analysis of load profiles, and the automatic detection of errors. These topics are relevant for enhancing the functionality and usability of the Digital Twin Framework and providing more value and benefits to the users and stakeholders.

The paper shall inspire and motivate other researchers and practitioners to advance the state-of-the-art of the Digital Twin Framework and bridge the gap between the Real and Digital Twin in mechanical engineering. The results and conclusions are convincing that this is a promising and exciting field of research and innovation with great potential for improving the design, development, operation, and maintenance of mechatronic systems.

References

1. Grieves, M.: Digital Twin: Manufacturing Excellence through Virtual Factory Replication (2015)
2. Crespi, N., Drobot, A.T., Minerva, R. (eds.) The Digital Twin. Springer International Publishing, Cham (2023). https://doi.org/10.1007/978-3-031-21343-4
3. Rasheed, A., San, O., Kvamsdal, T.: Digital Twin: Values, Challenges and Enablers (2019)
4. Qi, Q., et al.: Enabling technologies and tools for digital twin. J. Manuf. Syst. **58**, 3–21 (2021). https://doi.org/10.1016/j.jmsy.2019.10.001
5. Brecher, C., Weck, M.: Machine Tools Production Systems 3: Mechatronic Systems, Control and Automation. in Lecture Notes in Production Engineering. Springer Fachmedien, Wiesbaden (2022). https://doi.org/10.1007/978-3-658-34622-5
6. Bolton, W.: Programmable Logic Controllers. Newnes (2015)
7. Steve: PLC Manufacturers: The Latest PLC Brands, Rankings & Revenues. Ladder Logic World. Accessed: 08 Dec. 2023. https://ladderlogicworld.com/plcmanufacturers/
8. 'DigitalTwin@Education Module 150-002'. https://www.automation.siemens.com/sce-static/learning-training-documents/tia-portal/digital-twin/sce-150-002-mcd-tia-com-digital-twin-at-education-development-tia-hsd-0919-de.pdf
9. Modelica. Accessed: 22 Jan. 2024. https://modelica.org/

10. 'Tools'. Accessed: 08 Dec. 2023. https://openmodelica.org/useresresources/tools/
11. Hofmann, A., Menager, N., Belhaj, I., Mikelsons, L.: Integrated Engineering based on Modelica, p. 901 (2015). https://doi.org/10.3384/ecp15118893
12. Mejía-Gutiérrez, R., Carvajal-Arango, R.: Design Verification through virtual prototyping techniques based on systems engineering. Res. Eng. Design **28**(4), 477–494 (2017). https://doi.org/10.1007/s00163-016-0247-y
13. Ince, D.I.: Named Pipes. In: Ince, D. (ed.) A Dictionary of the Internet. Oxford University Press (2019). Accessed: 13 Dec. 2023. https://www.oxfordreference.com/display/. https://doi.org/10.1093/acref/9780191884276.001.0001/acref-9780191884276-e-2215
14. Mahnke, W., Leitner, S.-H., Damm, M.: OPC Unified Architecture. Springer, Berlin, Heidelberg (2009). https://doi.org/10.1007/978-3-540-68899-0
15. Industrial automation | B&R Industrial Automation. B&R: Perfection in Automation. Accessed: 08 Dec. 2023. https://www.br-automation.com/
16. Elmqvist, H.: Modelica Methodology for Systems Engineering (2014)
17. Dymola: Dassault Systèmes. Accessed: 06 Feb. 2024. https://www.3ds.com/products/catia/dymola
18. Devices - 4.2.1 Information Modelling in OPC UA. Accessed: 21 Feb. 2024. https://reference.opcfoundation.org/DI/v102/docs/4.2.1
19. Echtzeitsystem: Wikipedia (2023). Accessed: 10 Jan. 2024. https://de.wikipedia.org/w/index.php?title=Echtzeitsystem&oldid=236555618
20. NEAR REAL-TIME definition and meaning | Collins English Dictionary. Accessed: 07 Dec. 2023. https://www.collinsdictionary.com/dictionary/english/nearreal-time
21. Real Time under Windows: Kithara Software. Accessed: 23 Feb. 2024. https://kithara.com/en/solutions/real-time-for-windows
22. Windows Real-time, Windows Realtime-acontis. Accessed: 23 Feb. 2024. https://www.acontis.com/en/windows-realtime.html
23. Modul im Detail - SPP1. Accessed: 25 Aug. 2023. https://eventoweb.zhaw.ch/Evt_Pages/Brn_ModulDetailAZ.aspx?node=2901247e-aa27-4f84-a5d6-d6b33b234dbd&IDAnlass=1746265&clearcache=true&date=662248224000000000
24. Modul im Detail - ADE. Accessed: 28 Feb 2024. https://eventoweb.zhaw.ch/Evt_Pages/Brn_ModulDetailAZ.aspx?node=c594e3e5-cd9a-4204-9a61-de1e43ccb7b0&IDAnlass=1558276
25. CAS Product Innovation and Leadership for Engineers: ZHAW School of Engineering. Accessed: 12 Feb. 2024. https://www.zhaw.ch/de/engineering/weiterbildung/detail/kurs/cas-product-innovation-and-leadership-for-engineers/
26. MAS Industrie 4.0: ZHAW School of Engineering. Accessed: 12 Feb. 2024. https://www.zhaw.ch/de/engineering/weiterbildung/detail/kurs/mas-industrie-4-0/
27. Innosuisse is the Swiss Innovation Agency. Accessed: 12 Feb. 2024. https://www.innosuisse.ch/inno/en/home.html
28. OpenModelica, Introduction. Accessed: 12 Feb. 2024. https://openmodelica.org/
29. SimulationX. Accessed: 12 Feb. 2024. https://ch.mathworks.com/products/connections/product_detail/simulationx.html
30. Modelon Impact Platform. Modelon. Accessed: 12 Feb. 2024. https://modelon.com/modelon-impact/
31. Wolfram System Modeler: Modellieren, simulieren & analysieren. Accessed: 12 Feb. 2024. http://www.wolfram.com/system-modeler
32. MapleSim-Advanced System-Level Modeling & Simulation-Maplesoft. Accessed: 12 Feb. 2024. https://de.maplesoft.com/products/maplesim/

Embracing Lifecycle Modularity and Postponement in Urban Architecture

Vighneshkumar Rana(✉) and Vishal Singh

Indian Institute of Science Bengaluru, Bengaluru, Karnataka 560012, India
{vighneshkuma,singhv}@iisc.ac.in

Abstract. The research explores how lifecycle modularity and delayed differentiation can change the way we build and design spaces. Lifecycle modularity means incorporating modular elements at every stage of a building's life, making it more adaptable and sustainable. Through a review of existing studies, the research identifies the main challenges and opportunities of modularity, especially during the use phase. Emerging trends and technologies, like adaptable architecture and smart building systems, provide promising solutions to these challenges. Additionally, the study examines delayed differentiation, which involves postponing product customization until later in the supply chain. It explores how this concept can be applied to the built environment with innovative designs such as plug-in and plug-out modular architecture. These designs allow for greater customization, enabling users to tailor their spaces to their specific needs. The research includes examples of space product innovations, showcasing the potential of lifecycle modularity and delayed differentiation. These innovations offer flexible and adaptable alternatives to traditional construction methods, from prefabricated housing to mobile modular units. Furthermore, the study highlights the role of startups and innovative companies in driving these changes, using advancements in technology and design to create more responsive and sustainable built environments.

Keywords: Built Environment · Modular Construction · Lifecycle Modularity · Delayed Differentiation · Platform Methods · Urban trends

1 Introduction

The concept of modularity has gained significant traction in the Architectural, Engineering, and Construction (AEC) sector, promising transformative implications for how buildings and infrastructure projects are designed, constructed, and managed. Modularity involves breaking down a system into smaller, discrete components or modules that can be independently created, modified, replaced, or exchanged within a larger system. This approach offers many benefits and opportunities for innovation within the AEC industry. In the built environment context, modularity is brought to life through innovative approaches like modular and offsite construction [1]. Modular construction, also known as offsite construction, marks a significant shift in the building industry. This method involves creating building components in a factory and then assembling

Published by Springer Nature Switzerland AG 2025
P. Sureephong et al. (Eds.): PLM 2024, IFIP AICT 740, pp. 222–234, 2025.
https://doi.org/10.1007/978-3-031-93319-6_17

them on-site, a process that is technology-driven and production-centric [2]. Modular construction offers numerous advantages over traditional methods, including enhanced flexibility, increased efficiency, reduced waste, improved cost-effectiveness, and elevated safety and quality standards [3].

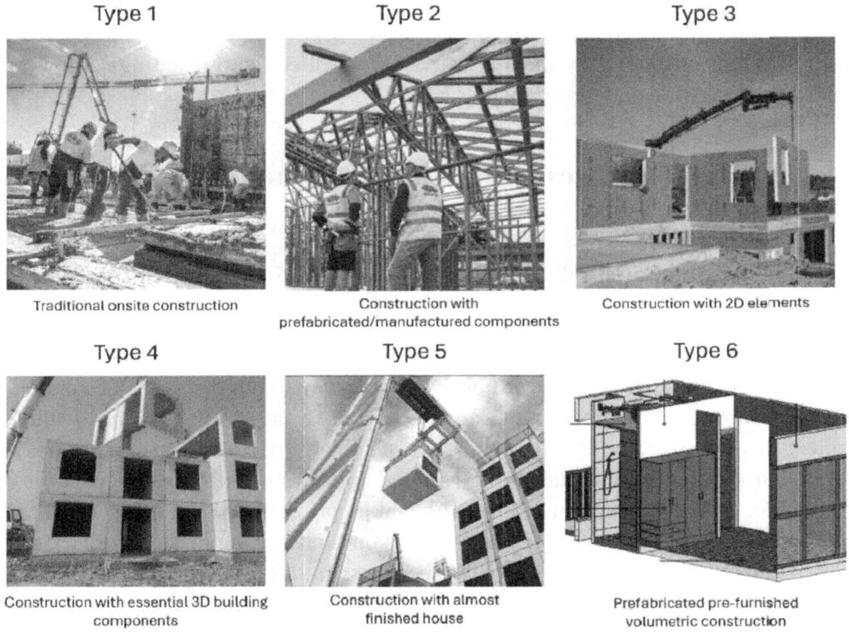

Fig. 1. Classification of Construction Methods Based on Modularity in Building Processes

When discussing the spectrum of construction methodologies, particularly in modularity and prefabrication, it's evident that practices range broadly from traditional site-centric methods to advanced prefabricated modular systems, as shown in Fig. 1. These methods are categorized into a continuum from Type 1 to Type 6 modular construction, each representing a distinct approach from offsite fabrication to onsite assembly. Construction methodologies range from traditional on-site construction (Type 1), which is labour-intensive and time-consuming, to advanced modular systems. Type 2 utilizes prefabricated components made in a factory and assembled on-site, speeding up construction and reducing labour. Type 3 involves 2D elements like prefabricated walls requiring on-site assembly. Type 4 uses 3D basic components to form the building's shell, with additional parts added on-site. Type 5, known as volumetric construction, involves nearly complete modules constructed in a factory and assembled using cranes. The most advanced Type 6 or Prefabricated Prefinished Volumetric Construction (**PPVC**) delivers fully finished and furnished modules to the site, minimizing on-site work. Each type offers a unique blend of efficiency, speed, and customization, allowing for a diverse range of construction projects tailored to specific needs and contexts [4, 5].

Construction methodologies shown in Figure 1 show a clear progression in modularity from Type 1, which has minimal modularity, to Type 6, which has the highest level of modularity. As we move from Type 1 to Type 6, the modularity increases and the need for manual labour on-site decreases. Type 6, the most modular, minimizes on-site work by using fully finished and furnished modules. This increase in modularity in construction phases offers significant benefits in terms of time, cost, and quality. Higher modularity streamlines construction, speeds up project completion [4], reduces labour costs, and enhances quality by shifting work to controlled factory settings, allowing for precise manufacturing and better quality control [5].

2 Analyzing Challenges in the Built Environment: A Literature Review

2.1 Unlocking the Potential: Embracing Lifecycle Modularity in the Built Environment

The discussion on modularity in construction highlights its benefits for the built environment. Lifecycle modularity goes further by emphasizing the importance of maintaining modularity throughout a building's entire lifecycle, from design to end-of-life. This approach ensures adaptability, efficiency, and sustainability at every stage. Studies show that modularity enhances maintainability, upgradability, reusability, and recyclability [7]. For example, modular design in products like inkjet printers allows for easy updates and adaptability to new technology [6]. This section emphasizes the need for lifecycle modularity in construction to fully unlock these benefits, highlighting the interconnectedness of different building phases and the importance of a holistic approach.

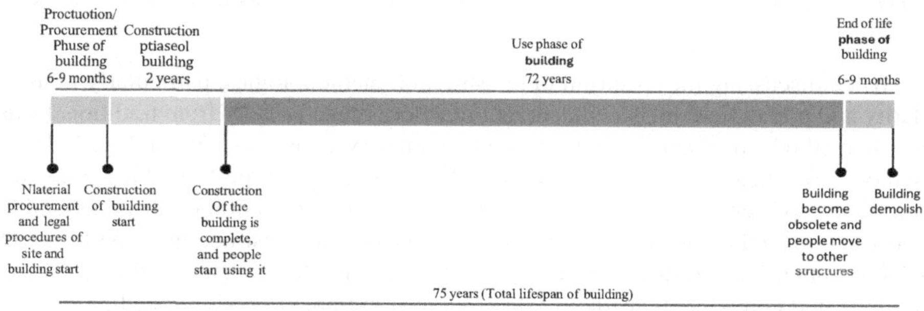

Fig. 2. Lifecycle Mapping of Building Lifespan

The lifecycle of a building has four main phases: production/manufacturing, construction, use, and end-of-life. [8] The production phase includes raw material supply, transportation, and manufacturing. The construction phase involves transporting materials and building the structure. The use phase covers the building's usage, maintenance, repairs, replacements, and energy and water use. Finally, the end-of-life phase involves deconstruction, demolition, waste processing, and disposal. The use phase is the longest, while the other phases are shorter. By adopting lifecycle modularity in building design and construction, we can achieve greater adaptability, efficiency, and sustainability throughout the building's entire life in all four stages. This ensures buildings remain resilient and functional as needs change. When examining a building's lifecycle with a focus on modularity, a clear difference emerges compared to portable products like cars or smartphones, which consistently use modularity across all phases. Modularity is unevenly applied in buildings, especially during the use phase, which makes up about 95% of a building's lifecycle, as shown in Figure 2. Unlike the production and construction phases, the use phase lacks modularity, presenting a significant challenge. It is essential to implement consistent modularity throughout all stages of a building's life to adapt to changing needs and technologies.

Table 1. Modularity level in the lifecycle phase of a building

	Modularity level in Lifecycle phases			
Types of Construction	**Manufacturing/Production**	**Construction**	**Use and maintenance**	**End of life**
Type 1 – Traditional onsite construction				
Type 2 – Construction with manufactured components				
Type 3 – Construction with 2D components				
Type 4 – Construction with 3D components				
Type 5 – Construction with almost finished structure				
Type 6 – PPVC				
Ideal Scenario				

In technology and automotive industries, modularity is integrated throughout the product lifecycle, allowing easy adaptation, upgrades, and repairs by swapping components. This contrasts sharply with the built environment, especially during the use phase of buildings. While production, construction, and end-of-life phases may embrace some modularity, the use phase, characterized by the structural permanence of buildings, often lacks such flexibility. This discrepancy is due to buildings being inherently tied to their locations, which limits their modular adaptability [9]. This location-product coupling makes buildings more static, embedded in complex geographical, social, and economic contexts that resist modular changes. Consequently, buildings are seen as fixed structures rather than dynamic products capable of evolving. Overcoming this challenge requires innovative design strategies that balance structural permanence with modular benefits, allowing buildings to adapt to changing needs and technologies while respecting their anchored relationship to specific locations [9].

Table 1 highlights a key issue in the built environment: the lack of modularity during the use phase of a building's lifecycle compared to its initial production and construction phases. This gap shows the challenge of integrating modularity throughout a building's life but also points to an opportunity for innovation. By incorporating modular principles into the use phase, buildings can become more adaptable and flexible. Despite the challenges due to buildings' fixed locations and their integration with surrounding areas, new trends and technologies offer solutions. Concepts like adaptable architecture, modular interior design, and smart building systems can increase a building's flexibility, allowing it to better respond to changing needs, technological advancements, and environmental factors. These innovations aim to create a more dynamic and responsive built environment. By improving modularity during the use phase, buildings can have longer lifespans and be more sustainable, efficient, and tailored to their occupants' needs. This approach not only addresses current limitations but also paves the way for future developments. Buildings designed with adaptability in mind can seamlessly integrate changes, ensuring their relevance and utility across generations. This emphasizes the importance of considering the entire building lifecycle, from conception to use and beyond.

Problem Statement: "A noticeable disparity exists in the level of modularity across lifecycle phases, particularly during the use and maintenance phase."

2.2 Strategic Flexibility: Harnessing Delayed Differentiation for Built Environment Resilience

This section will delve into the concept of delayed differentiation within the built environment context. Delayed differentiation, as defined in the literature, is a strategic approach where product differentiation is postponed until later stages in the supply chain [10]. This strategy offers numerous benefits, including cost savings, reduced inventory requirements, improved operational processes, increased product variety, and opportunities for innovation. By delaying the final customisation or specification of products until closer to the point of delivery or installation, organisations can better respond to changing market demands, optimise resource allocation, and enhance overall efficiency within the built environment sector. For example, McDonald's implements the concept of delayed differentiation by anticipating the fluctuating demand for various types of French fries. They achieve this by prepping large quantities of peeled, cut, and fried potatoes in advance. When an order is placed, they customize the fries to meet the customer's specific preferences.

This research endeavours to explore the concept of delayed differentiation within the different types of modular construction sectors, aiming to shed light on its potentially transformative impact. Illustrated in Fig. 3 is a comprehensive process flow timeline depicting various stages involved in modular construction, ranging from material procurement to building handover. However, upon closer examination, it becomes evident that current practices in modular construction often lack diversity in product offerings. Standardized configurations and structures, because of the mass production techniques, prevail, thereby limiting consumer choice and constraining opportunities for customization and innovation at the individual level. This conventional approach typically results in fixed sizes and layouts, thereby restricting options for room configurations and quality standards. By delving deeper into these challenges, this research seeks to uncover the potential benefits and implications of implementing delayed differentiation strategies in the modular construction sector, ultimately paving the way for greater flexibility, innovation, and consumer satisfaction.

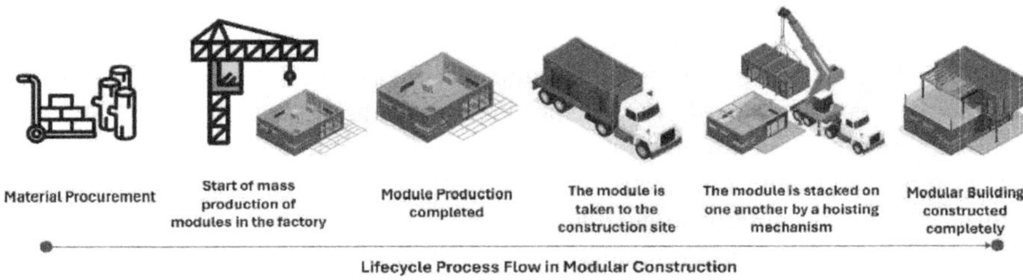

Fig. 3. Lifecycle Process Flow in Modular Construction

Figure 4 depicts the mapping of the point of differentiation in the modular construction process timeline. This "point of differentiation" refers to the juncture at which variability within the process initiates, creating distinct product families. A closer examination reveals that in the modular construction of built environment differentiation occurs even before the commencement of material procurement. During the design discussion stage, decisions regarding the standardisation of the building, overall appearance and complete design of the building and decisions are finalized. Only after these crucial design details are established does the process proceed to material procurement. It's essential to recognize that these processes vary for each modular building, with each modular building considered as a distinct product. As illustrated in Fig. 4, this differentiation process creates various products by the end of construction.

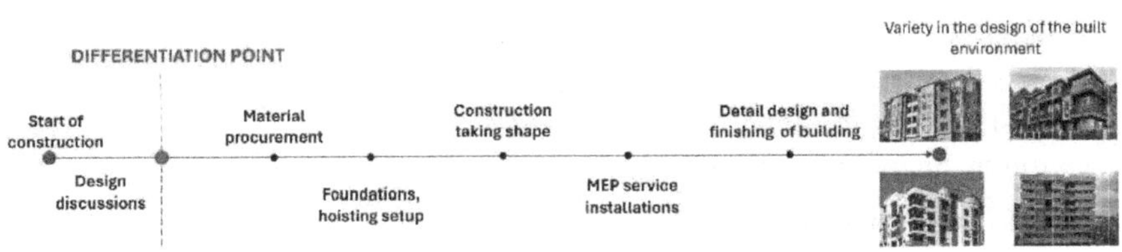

Fig. 4. Differentiation Point Mapping in Modular Construction Timeline

The literature on delayed differentiation underscores two fundamental principles crucial to its implementation. Firstly, it emphasizes that the product and production processes should be structured in a way that situates the point of differentiation at the end of the supply chain process. This ensures that customization and variability can be efficiently accommodated, maximizing advantages such as product variety, customization options, operational cost efficiency, and streamlined processes. Secondly, it advocates for concentrating the differentiating elements of products into a limited number of distinct chunks or components. This principle aims to optimize the benefits derived from delayed differentiation, facilitating smoother processes and enhanced product diversity.

However, as illustrated in Fig. 4, a deviation from these principles is evident in the modular construction industry. Here, the point of differentiation is situated at the initial stage of the construction processes, contradicting the first principle of delayed differentiation. Moreover, achieving product variety in the built environment entails a multitude of chunks or elements after the initial differentiation point, violating the second principle. In construction, chunks refer to groups of building elements utilized in the construction process. In ideal scenarios, delayed differentiation entails concentrating variability into a few distinct elements that differentiate products from each other. However, in the modular construction of the built environment, these differentiating elements or chunks are typically larger and more numerous. Thus, to achieve product variety in modular construction, many differentiating elements must be incorporated at the onset, contravening the principles of delayed differentiation.

Problem Statement: "The conventional construction process flow is in conflict with both fundamental principles of delayed differentiation."

The current research aims to address two key problem statements: the lack of lifecycle modularity and the need to postpone product differentiation in urban architecture. By focusing on these challenges, the goal is to foster urban architecture that is more adaptable, flexible, and resource efficient.

3 Research Methodology and Approach

This research is grounded in transformative and reflective methodologies, drawing insights from a comprehensive academic literature review and future projections in the construction-built environment discourse. Illustrated below is the research flowchart, which delineates the sequential steps undertaken in this study. 1. Problem Identification: Assessing Issues with Lifecycle Modularity and Postponed Differentiation 2. Literature Review: Exploring Solutions to Address Identified Challenges 3. Application in Built Environment: Aligning Solutions with Industry Context 4. Implementation Challenges: Identifying Potential Hurdles in Solution Deployment 5. Recommendations and Future Implementations

The following hypotheses guide the research:

Hypothesis 1: Introducing modularity in the use phase of the built environment's lifecycle enhances adaptability, innovation, and flexibility within the construction sector.

Hypothesis 2: Implementing delayed differentiation principles offers opportunities for customer-centric approaches in the built environment, leading to increased product variety and selection.

4 Solution Approach

4.1 Exploring Volumetric Space Innovations: A Diverse Array of Options

Exploring spatial configurations from literature highlights a shift from traditional fixed layouts to innovative, adaptable solutions. Traditional spaces are built onsite and are permanent, while prefabricated ones are created offsite, allowing for faster construction and better quality control. Prefabricated/Standardized spaces focus on uniformity, often using modular parts like shipping containers. Modular units, on the other hand, enable flexible layouts with interchangeable parts. Mobile spaces cater to needs for temporary, movable settings, while temporary spaces are designed for short-term use with quick setup and low environmental impact. Internal spaces, such as pods, maximize the utility within larger areas for specific functions. Reconfigurable spaces can be adapted in layout and function as needs change. System-integrated spaces view individual areas as parts of a bigger system, enhancing overall functionality and sustainability. This spectrum of spatial products showcases the industry's shifi towards more adaptable, efficient, and user-centric spatial solutions. Through the lens of these diverse spatial product categories, the discourse highlights a transformative shift from the static, one-size-fits-all model of traditional spaces to a dynamic, customizable, and user-centric approach. This evolution not only expands the realm of possibilities in spatial design but also underscores the importance of adaptability, innovation, and sustainability in the built environment.

The concepts presented diverge significantly from the traditional onsite space paradigm, advocating for a shift towards more modular and flexible spatial solutions. Table 2 illustrates how the degree of modularity varies across the lifecycle of various volumetric space solutions, highlighting a trend towards flexible and adaptable spaces that can evolve over time to meet changing needs. When examining the lifecycle phases of volumetric space solutions, it becomes evident that certain types, such as mobile, temporary, internal, and reconfigurable spaces, exhibit a high degree of modularity in the use phase of their lifecycle. This adaptability is crucial for modern urban development, allowing spaces to be repurposed or redesigned with minimal effort and cost. The aim is to leverage the attributes, characteristics, and features of these innovative spaces to re-envision and refine traditionally onsite spaces, thereby introducing modularity into the use phase of buildings. These volumetric space solutions innovations are not just theoretical concepts; they are practical, visible examples that have been successfully implemented in various contexts, such as temporary accommodations, food trucks, mobile health units, and caravan vans. Despite their current niche applications, there is a significant opportunity to broaden the scope of these solutions, integrating them with traditional built environments to foster innovation in the design, planning, and utilization of both residential and commercial spaces. Rana and Singh discussed the product-oriented perspective of volumetric space innovations, emphasizing their potential opportunities when integrated with traditional residential and commercial spaces [12].

By reimagining the integration of volumetric space innovation solutions with traditional construction methods, there is potential to revolutionize the built environment. This approach not only challenges conventional design and utilization paradigms but also promotes a more dynamic, flexible, and sustainable approach to urban development. The key to this transformation lies in recognizing the benefits of modularity— such as adaptability, efficiency, and reduced environmental impact—and applying these principles to redesign and rethink the use of space in a way that better aligns with the evolving needs of society.

Table 2. Modularity Levels Across Lifecycle Phases of Volumetric Space Product Innovations

	Modularity level in Lifecycle phases			
Volumetric Space product Innovation around us	**Manufacturing/Production**	**Construction**	**Use and maintenance**	**End of life**
Prefabricated space	■	■		■
Standard space	■	■		■
Space in parts (Kit house)	■	■		■
Mobile space	■	■	■	■
Temporary space	■	■	■	■
Internal space	■	■	■	■
Reconfigurable space	■	■	■	■
Space as a part of the system	■	■		■

4.2 Examples of Startups Who Are Working on Volumetric Space Product Innovations

The exploration of modularity in the built environment is not just a theoretical exercise; it's a burgeoning field of innovation driven by startups and forward-thinking companies worldwide. These entities are pushing the boundaries of what's possible, demonstrating the practical application and market viability of modular and flexible spatial solutions. This movement is not confined to residential spaces but extends into the commercial realm, where adaptability and efficiency are increasingly prized.

Prefabrication and Modular Spaces: Companies like PopUp House, Cube Haus, and Blokable are revolutionizing construction through prefabricated components that make building processes quicker, cheaper, and greener. Nexii, BuildNext, and Factory_OS advance modular construction, significantly cutting down on construction time and waste.

Houses in Parts and Mobile Homes: Innovators such as Wikkelhouse offer sustainable, customizable building blocks, while Bobvila and The Popomo cater to the demand for mobile, flexible housing solutions.

Internal Spaces and Reconfigurability: BOOQED, Mikomax, and Outfit are transforming interior spaces with modular, reconfigurable designs that adapt to changing needs, simplifying renovations and reconfigurations.

Commercial Innovations: Framery, SmartBox, and Hushoffice respond to the commercial sector's needs for modular internal spaces, providing products like soundproof pods and flexible meeting spaces that cater to the demand for adaptable and modular commercial environments.

These startups and companies are not just creating products; they are pioneering a shift towards a more flexible, sustainable, and efficient use of space. Their success and innovation underline a significant trend in the built environment: the move away from one-size-fits-all, permanent structures to dynamic, adaptable, and sustainable solutions that can evolve with the needs of their users. This trend is not only about embracing new technologies but also about changing our mindset towards how spaces can be created, utilized, and transformed.

5 Discussions

5.1 The Implication of Space Product Innovations on Lifecycle Modularity and Delayed Differentiation

The exploration of solutions within the context of lifecycle modularity and delayed differentiation in the built environment necessitates a closer examination of volumetric space product innovations mentioned in the previous section. These innovative configurations present a compelling blueprint for addressing the complexities of the use phase in building lifecycles, emphasizing the importance of adaptability, flexibility, and resource efficiency. The concept of plug-in plug-out modular architecture, as presented by scholars like Pasquale [3] and Rana [9], illustrates a groundbreaking solution that exemplifies modularity during the use phase of a building's lifecycle. This architectural concept entails a fixed frame structure capable of seamlessly integrating multi-purpose modules, allowing for easy addition, removal, or exchange of components to accommodate diverse functions and adapt to evolving needs. This architectural approach involves the design of buildings with a core structure that allows volumetric spaces— such as individual rooms or modules—to be attached or detached as needed. This innovative design enables buildings to adapt over time to various requirements, facilitating a dynamic evolution of space that can respond to changing needs, preferences, or functions.

Figure 5 illustrates the lifecycle phase mapping of plugin plug-out modular architecture, showcasing its flexibility to incorporate a fixed frame structure capable of hosting various multipurpose modules tailored to individual requirements. These modules may vary in size, quantity, cost, quality, and configuration, offering customizable options to accommodate diverse needs and preferences. For example, certain modules may cater to different societal classes or individual preferences, allowing for customization according to varying requirements. The key feature of this approach is the ability to introduce differentiation at the end of the lifecycle stage, as depicted in Fig. 5.

In the plugin plug-out modular architecture, customers are empowered to customize their living space according to their preferences and needs. For instance, upon acquiring a slot within the fixed frame structure, customers have the flexibility to choose from various module options, such as 2-bedroom or 3-bedroom layouts, within the allocated floor area, such as 900 sqft. This concept emphasizes customer-centricity by enabling them to tailor their homes to their unique requirements throughout the lifecycle of the frame, particularly during the use phase. Additionally, if customers find their modules outdated, they can easily replace them with newer options featuring different configurations, higher quality, or enhanced features, thereby ensuring adaptability and longevity. By integrating delayed differentiation, this approach prioritizes customer choice and satisfaction,

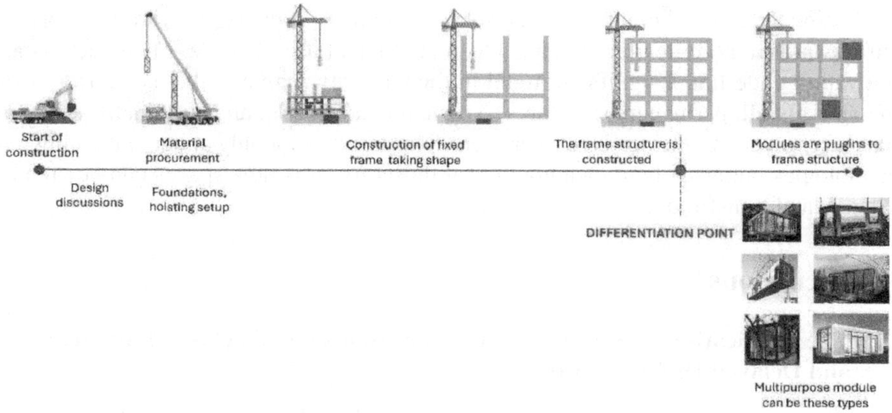

Fig. 5. Lifecycle Mapping of Plugin Plug-Out Modular Architecture: Identifying Differentiation Point

fostering a more personalized and customer-oriented built environment ecosystem. The framework employs a balance of standardization in the frame structure and customization in module selection, providing customers with a tailored yet standardized building experience.

5.2 Platform Paradigm: Revolutionizing Building Development with Platform Methods

The frame design within the plugin plug-out modular architecture embodies standardization, ensuring uniformity and consistency in structural elements. Conversely, the module design offers customization, allowing customers to personalize their living spaces according to their preferences. This approach aligns with the concept of delayed differentiation, wherein standardization precedes the introduction of variety in the product. This strategy is akin to the platform method, which advocates for standardizing processes and products before the point of differentiation. By establishing a standardized platform capable of accommodating various modules, the system facilitates seamless integration of diverse product options while maintaining operational efficiency and quality standards.

In the plugin plug-out architecture, the frame construction occurs prior to the point of differentiation, allowing for the implementation of standardized practices in frame design, as shown in Fig. 6. This standardization encompasses various aspects, including the design of frame components, dimensions and sizes of slots or vacant spaces, connection designs, locking mechanisms, and interfaces between the frame and modules. Standardizing these elements ensures consistency and compatibility across different modules and facilitates seamless integration within the framework. Moreover, by establishing standardized protocols for frame construction, manufacturers can streamline production processes, optimize resource utilization, and enhance overall efficiency. This approach not only simplifies the assembly and installation of modules but also contributes to the scalability and versatility of plug-in plug-out architecture in accommodating diverse spatial configurations and user preferences.

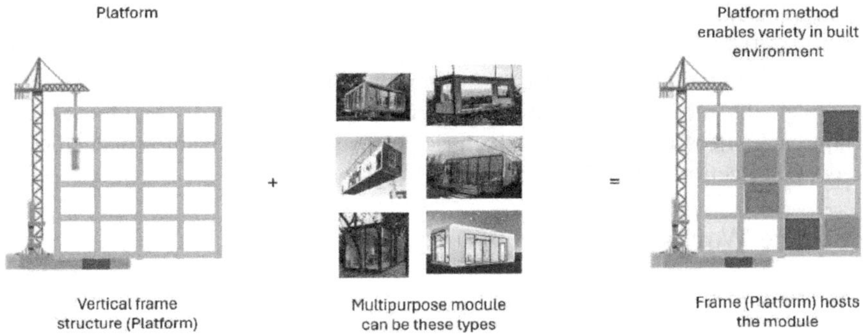

Fig. 6. Vertical Frame Structure Hosting Multipurpose Modules

5.3 Implication in Urban Evolution: Meeting the Demands of Growing Populations

The built environment encompasses the physical spaces where human activities unfold, from individual buildings to entire cities and infrastructural systems. It plays a pivotal role in shaping our daily lives, offering not just shelter but also a sense of identity and belonging. With urbanization on the rise and populations migrating to cities, there's a pressing need for innovative solutions to address housing shortages and meet diverse urban living needs effectively. Traditional approaches to construction and urban planning must evolve to adapt to changing social dynamics, technological advancements, and lifestyle preferences. Moreover, the rapid pace of urbanization underscores the urgency for innovative approaches to address the diverse and fluctuating housing needs of urban populations. It's essential to reimagine traditional solutions in light of our increasingly technologically advanced and interconnected world, ensuring they resonate with evolving lifestyles, social dynamics, and expectations. [3]

In this context, the emergence of startups and innovative space solutions has highlighted the potential of alternative spatial concepts. These include features like mobility, reconfigurability, temporariness, and modularity, which challenge conventional notions of built spaces and housing. These attributes align with emerging urban lifestyle trends characterized by fluidity and adaptability. However, despite their promise, these concepts may have initially struggled to gain traction due to a lack of support from construction and market ecosystems. With the advent of Industry 4.0 technologies, such as digitalization and automation, the landscape is now more conducive to their successful implementation. This shift paves the way for future product innovations in the built environment, where modular and personalized spaces can meet the evolving needs of urban populations more effectively.

6 Conclusion

The current research delves into the lifecycle perspective of product innovations within the built environment, focusing specifically on enabling modularity during the use phase. Through a comprehensive review, the study explores various space product innovations that introduce modularity in this phase, thereby enhancing adaptability, innovation, and flexibility in built environments. Additionally, the research delves into the principles of delayed differentiation, exemplified by the introduction of plug-in plug-out architecture. By strategically moving the point of differentiation to the end of the lifecycle stage, this approach offers a more customer-centric perspective, resulting in expanded product variety and selection options.

7 Future Works

The ongoing research is going on to tackle the structural challenges associated with plug-in plug-out modular architecture while also assessing its feasibility and utility in the real estate market. By examining the practical application of this architectural concept within the context of the real-world market, the study aims to determine its potential for success. The underlying hypothesis suggests that if the plug-in plug-out modular architecture proves successful in the real estate market, it is likely to gain widespread traction and adoption globally. This entails evaluating its structural integrity, cost-effectiveness, adaptability to varying market demands, and overall market appeal.

References

1. Di Pasquale, J., Innella, F., Bai, Y.: Structural concept and solution for hybrid modular buildings with removable modules. J. Arch. Eng. **26**(3), 04020032 (2020)
2. Ferdous, W., Bar, Y., Tuan, D.N., Manalo, A., Mendis, P.: New advancement, challenges and opportunities of multi-storey modular buildings-a state-of-the-art review. J. Eng. Struct. **183**, 883–893 (2019)
3. Di Pasquale, J.: PhD Thesis: Hybrid Modular Architecture: A Strategic Framework of Building Innovation or EmergingHousingBehaviours inUrban Contexts (2018).ABC-Department of Architecture, Built Environment and Construction Engineering
4. Bertram, N., Fuchs, S., Mischke, J., Palter, R., Strube, G., Woetzel, J.: Modular construction: From projects to products. McKinsey & Company: Capital Projects & Infrastructure, 1–34 (2019)
5. Lawson, M., Ogden, R., Goodier, C.: Design in Modular Construction. CRC Press (2014)
6. Ishii, K.: Modularity: A key concept in product life-cycle engineering. Handbook of Life-cycle Engineering, pp. 511–530 (1998)
7. Umeda, Y., Fukushige, S., Tonoike, K., Kondoh, S.: Product modularity for life cycle design. CIRP Ann. **57**(1), 13–16 (2008)
8. Nwodo, M.N., Anumba, C.J.: A review of life cycle assessment of buildings using a systematic approach. Build. Environ. **162**, 106290 (2019)
9. Rana, V., Singh, V.: How may modular construction transform the built environment ecosystem? In: Noël, F., Nyffenegger, F., Rivest, L., Bouras, A. (eds.) Product Lifecycle Management. PLM in Transition Times: The Place of Humans and Transformative Technologies: 19th IFIP WG 5.1 International Conference, PLM 2022, Grenoble, France, July 10–13, 2022, Revised Selected Papers, pp. 506–515. Springer Nature Switzerland, Cham (2023). https://doi.org/10.1007/978-3-031-25182-5_49
10. Eppinger, S. D., Ulrich, K.: Product design and development (1995)
11. Jose, A., Tollenaere, M.: Modular and platform methods for product family design: literature analysis. J. Intell. Manuf. **16**(3), 371–390 (2005)
12. Rana, V., Singh, V.: Analyzing the dimensional aspects of 3D volumetric spaces: a product-oriented perspective. Proc. Des. Soc. **4**, 2433–2442 (2024)

A Value Driven Digital Twin for Academic Purposes

Mansur Asranov[1,2](✉), Francesco Serio[1], Makhkam Kayumov[1,2], and Paolo Chiabert[1,2]

[1] Politecnico di Torino, Corso Duca degli Abruzzi 24, 10129 Torino, Italy
{francesco.serio,paolo.chiabert}@polito.it,
makhkam.kayumov@polito.uz

[2] Turin Polytechnic University in Tashkent, Kichik Khalka Yuli 17, 100095 Tashkent, Uzbekistan
mansur.asranov@polito.it

Abstract. The competitive challenge proposed by Industry 5.0 requires the development of high skills in the technological and organizational fields. To provide these skills it is useful to develop devices that can be used to expe-rience the situations that future technicians will face in their professional career. Furthermore, it is possible to design products and processes integrating techno-logical and organizational tools. This is the experience presented in this paper: the creation of IoT devices (the most recent is the IoT Useless Box) using IT tools, PLM design and Lean organization. The result is a Digital Twin, based on open-source authoring tools and low-cost components, that fully supports bidirectional communication between the physical object and its digital counterpart.The experience, fully described in terms of technological solution, project man-agement, cost and benefit analysis, has been carried out within university courses to prepare our students and future technicians to operate effectively in Industry 5.0 ecosystems.

Keywords: Digital Twin · Industry 4.0 and 5.0 · Internet of Things

1 Introduction

Nowadays it is possible to state that the nine enabling technologies commonly as-sociated with the industry 4.0 umbrella are quite mature and they are making room for the Industry 5.0 paradigm.

The main objective of I5.0 is to increase the sustainability and resilience of the industry sector. It promotes five main actions: enabling and accelerating industrial innovation for global sustainability; exploiting digital technologies improving human-machine interactions; adopting a human-centric approach empowering workers' role; introducing new skills for more effective organizational structures; promoting a greener economy and increasing competitiveness of the European industry [1].

Published by Springer Nature Switzerland AG 2025
P. Sureephong et al. (Eds.): PLM 2024, IFIP AICT 740, pp. 235–248, 2025.
https://doi.org/10.1007/978-3-031-93319-6_18

In order to support the previous actions, I5.0 identifies six new enabling technologies [2]: Digital Twin (DT) and simulation, Data management, Artificial intelligence, Energy management, Human-machine interfaces, Smart materials.

Finally, I5.0 should identify a methodology for the implementation of the six enabling technologies that support the five actions while pursuing its two main goals.

Authors believe that an appropriate methodology could be the Lean approach, originally developed by Japanese automotive companies. The Lean focus on waste removal, value increase, continuous improvement and workforce training, perfectly matches the I5.0 main concepts: Human centricity, Resilience and Sustainability.

According to such idea, this paper investigates technological, methodological and gnoseological aspects in the development, according to Lean principles, of a DTed device.

The more recent device is the Useless box, originally conceived by Marvin Minsky at Bell Laboratories in the early 1950s, but enlarged with the application of sensors and actuators and embedded into an IoT infrastructure.

The development of the IoT Useless box allows for several investigations on the awareness of I4.0/I5.0 underlying concepts, the set of required competences, the difficulties in the development of DT and, finally, the analysis of end-user interaction with such enlarged IoT objects.

The paper illustrates the principles that guided the development of this IoT device, the experiments that were conducted in the academic field and the details for the construction and use of the IoT device.

The IoT Useless Box creates and transfer digital data representing physical variables (temperature, humidity, light intensity, current, voltage etc.) and mechanical behavior from the physical object to the virtual model. On the other hand, operating the digital model it is possible to modify actuators in the physical model thus completing a bidirectional communication within the DT system.

With this ecosystem authors would like to create a solution for teaching purposes that will help students in understanding how to build, interact, simulate and evaluate DT system.

2 State-of-the-Art

2.1 Digital Twin Paradigm

The six enabling technologies identified by I5.0 proposal, derive from the nine enabling technologies of I4.0 strongly related to Human centricity, Resilience and Sustainability:

1. DT and simulation.
2. Technologies for system interoperability and data management (storage, transmission, and analysis).
3. Artificial intelligence.
4. Technologies for energy efficiency, renewability, and storage.
5. Human-machine interfaces promoting cooperation instead of replacement.
6. Smart material and bio-inspired technologies.

Among these technologies, this paper focuses on the DT concept because of its relevant impact on manufacturing systems, enterprise competitiveness and workforce competences.

Exploring the literature, the most cited definitions of the DT concept is: "an integrated multiphysics, multiscale, probabilistic simulation of an as-built vehicle or system that uses the best available physical models, sensor updates, fleet history, etc., to mirror the life of its corresponding flying twin" [3].

In addition, considering some other definitions of DT [4–10], it is possible to summarize the essential sub-systems of a DT in three main components:

- A physical system that collects real-time data by means of sensors and receive instructions.
- A digital representation of the physical system.
- A network to connect physical and virtual environments.

Focusing on the last point and considering other definitions of the DT concept [11–17], it is possible to state that the most frequent implementation of DT nowadays is unidirectional: data flow from the physical system to the virtual environment while a reverted communication is usually performed by a human in charge of operating the feedback. Bidirectional DT models, where the digital environment can even send data and control the behavior of the physical system, are less common in industrial applications.

2.2 The Lean Thinking and the Value Stream Mapping Method

The term Lean was introduced for the first time by Krafcik in 1988 [18] and then widely used in "The machine that changed the world" [19] to express a new way of managing production, which opposes the methodologies of mass production, focused primarily on producing what is perceived as true value by the customer.

Therefore, the Lean approach describes the value stream mapping (VSM) of a production process and preserves operations that increase the value of the product while eliminating or reducing activities that do not increase the value of the product.

Maskell et al. [20] expanded the scope of the VSM tool by defining waste and value according to accounting criteria, thus providing a Lean-oriented corporate accounting system that promotes Lean management of the entire enterprise and not just its production system.

The VSM tool is used to capture the actual situation in an "As-Is" scenario and propose a general improvement, based on the other lean thinking tools, in a single or multiple "To-Be" scenario. Such design ability of VSM recalls many of the digital twin definitions cited in the previous paragraph. However, surprisingly, in the literature there are no documents that simultaneously contain the words "Digital twin" and "Value Stream Mapping" in their title.

In the same way, there are no results for documents that in the title mixes the words "Digital twin" and, one or the numerous Lean buzzwords: kanban, 5s, six sigma etc.

Taking into account all these considerations the focus of discussion in this paper is about the usage of digital twin and VSM as a combined teaching tool.

3 VSM and DT as Combined Teaching Tool

Having acknowledged the absence in the scientific literature of documents that can guide the development of a digital twin according to Lean principles, research was started at the Politecnico di Torino to experiment with the integration of these two tools. The experience proposed in a study course at the Politecnico di Torino consisted in training students in the use of VSM and DT tools in order to acquire the necessary skills to operate in an I5.0 environment.

The experiment consisted in the creation of a physical device equipped with Internet of things (IoT) components and the creation of its DT to support teaching activi-ties in the area of digital technologies.

In the product life cycle (Fig. 1), which starts from the Beginning of Life phase and ends with the End Of Life phase [21], the VSM analysis was applied only to the de-sign phase whose duration was easily controllable.

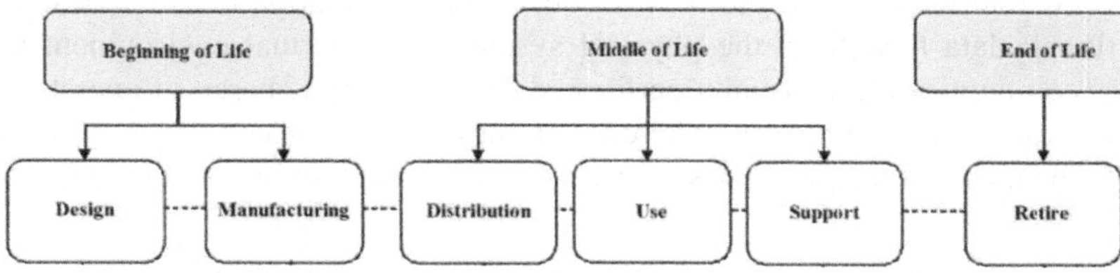

Fig. 1. Product Lifecycle periods [21]

The experimentation was conducted in two consecutive academic years with different classes of students. In this way it was possible to identify an As-Is scenario, based on the experience of the first year, and predict a To-Be scenario implemented in the following academic year.

This way, to act implicitly, an iterative teaching process spawned that started with the application of the method in a university course. This triggered the whole process giving the first design and production phase of the IoT tool and the first VSM analysis on the design phase, that even generated the first "as is" and "to be" scenarios.

During the second academic year, new students followed the instructions of the "to be" scenario from the first year to manufacture an advanced version of the IoT tool and its DT, meaning that it should be equipped with at least one additional functioning feature. At the end of the second year, the students will create another "as is" scenario proposing the respective "to be" scenario.

The third year, new students will start from the "to be" scenario of the second year to create the third version of the IoT tool with its DT and the third couple of "as is" and "to be" scenarios. In this way the university course will iteratively produce an improved version of the IoT tool teaching at the same time the digital twin concept and VSM analysis.

3.1 The Potential Identified Processes to Apply VSM Analysis

The application of the VSM tool requires the identification of the value associated with the IoT device. There are different values corresponding to different processes involving the IoT device.

The first identifiable process is the teaching activity carried out with the IoT device. According to this definition, value consists in increasing students' knowledge of different educational topics.

Based on communication theory [22], the *Value* (Eq. 1) consists of the difference between the student's final (*CF*) and initial (*CI*) behavior:

$$\mathit{Value} = C_F - C_I \tag{1}$$

Conversely, the Waste of the process (Eq. 2) is the difference between the best achievable final behavior (*max{CFj}*) and the average final behavior of the class (*avg(CFj)*) or that of the individual student (*CFj*):

$$\mathit{Waste} = max\{C_{F_j}\} - avg(C_{F_j}) \quad j = 1, \ldots, n \tag{2}$$

Obviously, the measurement of value requires the definition of the concepts that are presented in the lectures (Eq. 3) and the respective communication units (*Uk*):

$$\mathit{Value} = C_F - C_I = \{U_1, U_2, \ldots, U_k, \ldots U_{l-1}, U_l\} \tag{3}$$

Each communication unit has its own efficiency η which depends on the identification of the concepts (ηc), their exposure (ηe) by the teacher as well as their acquisition (ηa) and understanding (ηu) by the student. These efficiencies contribute to the waste (Eq. 4) of the overall communication process (ηo):

$$\eta_o = \eta_c \cdot \eta_e \cdot \eta_a \cdot \eta_u \tag{4}$$

On the other hand, the study of efficiencies allows the improvement of the communication process through the modification of the course material, the exposition of the topics, and the proposed examples.

A second possible scenario is the IoT device manufacturing process. According to the well-known New Product Development Process, we could apply the VSM tool to one of the phases of the NPDP, for example the design phase or the production phase, as well as to the entire manufacturing of physical and virtual objects. In this scenario, the application of the VSM tool would be similar to what is reposted in the books that explain the Lean approach. The value will be the level of satisfaction of teachers (customers) related to each of the physical and digital characteristics of the IoT device and its Digital twin. Similarly, it will be possible to identify the classic 7 wastes by identifying useless, redundant or duplicate activities, as well as errors made during production or even time wasted waiting for the arrival of materials or the completion of work. The third possible scenario is to apply the VSM vision on the product, therefore on the functioning of the useless box. The added value would be the correct functioning of the device, i.e. the accuracy of the measured data, the correct execution of the commands, the exact correspondence between the behavior of the physical device and that of its virtual twin.

On the other hand, waste could be represented by all the times that the data flow system does not work correctly and by the inefficiencies related to the correctness of the measurements.

In order to correctly identify added value and waste, it will be necessary to map the data flows exchanged bidirectionally between the physical and virtual models and verify their syntactic correctness as well as the actual semantics associated with them. Verifying their correctness. Furthermore, the direction of the data flows must be, from the physical environment to the digital one and vice versa. Then a human investigation of the entire system would be necessary to understand whether everything is working correctly. The added value would be the number of times the system works correctly, and all the data is consistent. Waste, on the other hand, could be represented by all the times that the data flow system does not work correctly and by the inefficiencies linked to the correctness of the measurements.

All these three scenarios are educationally interesting because they can simulate many industrial scenarios, with different processes and characteristics. Students can therefore acquire skills on DT and VSM. As mentioned before, it was decided to apply the VSM analysis only to the design phase because its duration was easily schedulable during the academic course.

3.2 The First "As-Is" Scenario of the IoT Box Design Phase

This paragraph reports the first outcome of the teaching method proposed comprising the design phase and production of the IoT tool (see Fig. 2), widely reported in [23], based on the application of Ulrich & Eppinger's (U&E) NPD model [24] and to the related digital environment. This experiment was conducted with the first class of students who designed the IoT device with a Product Lifecycle Management approach.

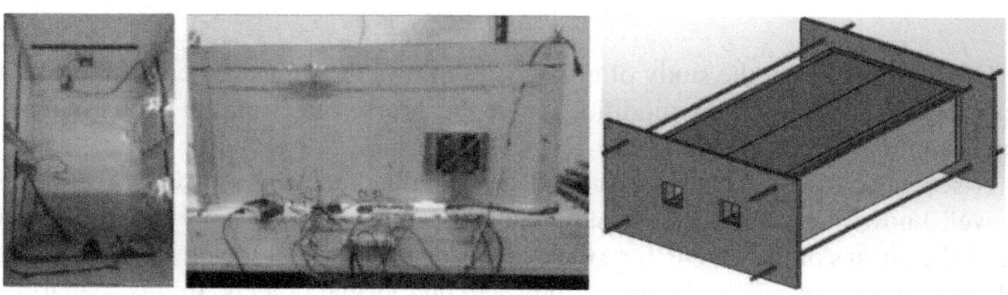

Fig. 2. "As Is" scenario of the IoT box

In the planning phase of the U&E model, the general requirements of the IoT solution were defined: the choice of components widely available on the market, artisanal and low-cost production and assembly, the use of flexible open-source solutions with minimal maintenance required. In the U&E concept phase, the students examined the technical requirements necessary to guarantee openness, simplicity, scalability and replicability of the system.

In the U&E system level design phase, the general architecture of the system was defined, including structure/body, thermal and environmental subsystem, sensors, actuators and logical subsystem. The U&E detailed design phase explained each sub-system from a technical and technological point of view, while the U&E testing and refinement phase guaranteed the correct functioning of the entire solution created.

At the end of the academic year was introduced to the students the VSM tool in order to create the first version of the "as is" scenario on the design phase. On the contrary to the original approach of applying VSM, that requires to physically draw all the actions in the value stream, it was created a chronological list of all the main actions performed during the academic year, which represented the "as is" scenario, with the related solutions or improvements to be applied in the future and so creating the first "To Be" scenario of the iterative process proposed.

The time span of this process was divided into stages in order to monitor the activity of each student, track the progress of the project, validate the results achieved and intercept non-conformities. Unfortunately, in the first academic year it was not possible to create the related digital twin part because the course did not cover the theoretical part behind.

3.3 The First "To Be" Scenario of the IoT Box Design Phase

The "To Be" scenario created by the students of the first academic year proposed the following improvements for each phase of the U&E model:

- In Planning Phase, it is possible to extend the requirements by introducing readiness of the architecture of the IoT tool
- In the concept development phase will be strategically convenient to introduce a boost of replicability of the IoT tool by following the same framework of technical requirements.
- The system development phase can be supported by a new general arrangement of components, which will foster a better integration and harmony of the entire architecture.

4 The Proposed DT Framework of Useless IoT Box

This section presents the IoT tool that was created in the second academic year based on the "to be" scenario reported before, following the same time constraints and milestones but with the addition of another technical feature that improves the IoT tool functioning. In particular, it was decided to create a Useless IoT Box adding a new mechanical system (roof, servomotor, and toggle switch), while also upgrading the type and number of sensors to capture more possible variables, therefore allowing for simulation of more cases, as well as extended visualization and interaction with the introduction of the digital twin concept.

4.1 Description of Physical Twin

This section explains the general architecture of the Useless IoT Box as the outcome of the U&E model steps followed by the students of the second academic year that took even into account the "to be" from the first year. Originally, the idea Useless IoT Box evolved from the Useless Box, also well-known as Useless Machine. The conceptual idea behind Useless Box is pioneered by Marvin Minsky [25], the core function of which is to turn itself off by controlling its own switch. Instead, Useless IoT Box comprises the original "Useless Machine" functionality, while being also extended with IoT capabilities.

Figure 3 describes the general architecture of Useless IoT Box, highlighting the details of each subsystem and component:

- Thermal system – represents the two packages of assembly of Peltier thermoelectric cells, heatsink and fans (thermoelectric assembly). One such assembly generates warm air flow (Pw), while the other generates cold air flow (Pc).
- Sensors – set of digital and analog sensors are used to measure physical variables, such as temperature (T), humidity (H), light intensity (L) of inner and outer environment of the box, voltage (V) and current (C) of the thermoelectric assemblies (Pw, Pc), and the angle of bending (F).

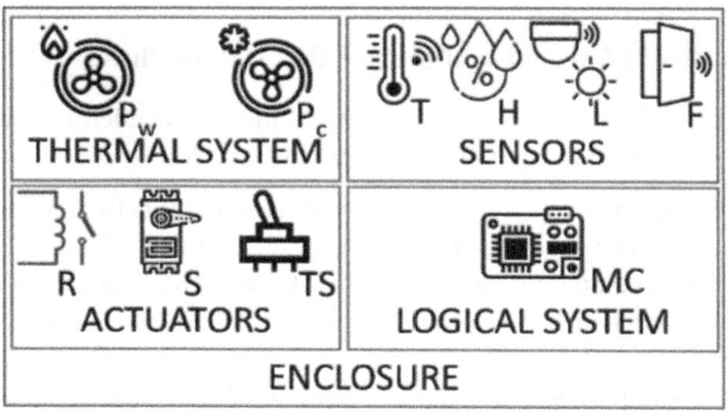

Fig. 3. General architecture of Useless IoT Box

- Actuators – this system comprises relay switches (R) for controlling the thermoelectric assemblies (Pw, Pc), servomotor (S) for opening the roof of the Enclosure and moving the toggle switch handle (TS), toggle switch (TS) which initiates the original "Useless machine" function, as introduced in the beginning of this section.
- Logical system – this system is represented by the set of microcomputers/microcontrollers (MC) and supports logical operations to run combined capabilities - the original "Useless machine" functionality, IoT and systems integration.
- Enclosure – wooden box structure which creates robust facility for hosting the previous subsystems.

While Fig. 3 is focused on the description of general architecture, Fig. 4 illustrates the interaction scheme of the components from different subsystems, previously introduced.

- Enclosure – the wooden box structure which has a roof, consisting of static and moving parts.

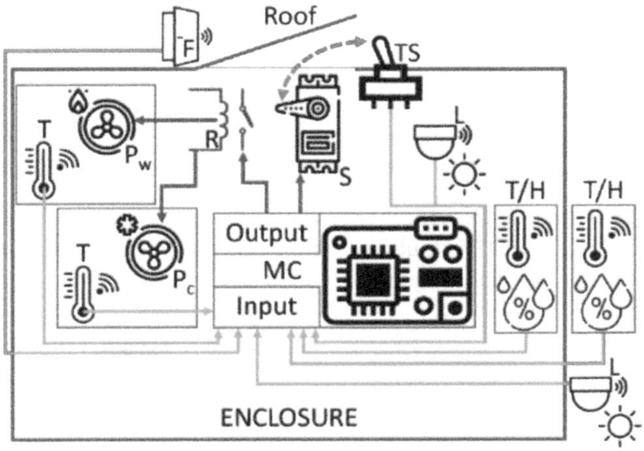

Fig. 4. Interaction of scheme of components of Useless IoT box.

- Sensors – contact Temperature (T) sensors measure temperature state of the thermoelectric assemblies, combined non-contact Temperature/Humidity (T/H) sensors measure inner/outer temperature and humidity of the Enclosure. Similarly, light sensors (L) provide measurements for inner and outer light intensity. Current and voltage sensors (C and V) measure corresponding variables of the thermoelectric assemblies (Pw, Pc). Moreover, flex sensor (F) is responsible for measuring opening angle, being installed between static and moving parts of the roof. All sensors are powered by the Logical system components.
- Logical system – the system consisting of microcomputer (Raspberry Pi – for supporting digital sensors) and microcontroller (Arduino Uno – for supporting analog sensors). This system drives data acquisition and control of actuators, uses readings from the sensors as an Input, allows for application of various logical processing of the Input data to eventually actuate the components of Actuators system - servomotor (S) and thermoelectric assembly (Pw, Pc). On top of that, this system plays the role of gateway (Raspberry Pi) for providing communication between the Useless IoT Box and its Digital Twin, supporting data provision services for state and control data exchange. Logical operations and data provision services are enabled using NodeRed software and MQTT protocol.
- Actuators – using output signals coming from the Logical system, relay switches control thermoelectric assemblies. Instead, toggle switch (TS) is controlled by human – human moves the toggle switch handle to its "on" position, and corresponding signal is sent to the Input of the Logical system, which in turn activates the algorithm of original "Useless Machine" functionality. That algorithm generates the desired Output signal, using which the servomotor (S) is activated to move the handle of the toggle switch (TS) to its original "off" position. While being activat-ed, the servomotor (S) mechanically opens the roof, therefore triggering the flex sensor (F).

4.2 Description of the DT

This section describes the design and development of the Useless IoT Box DT that was introduced for the first time in the second academic year and for this reason, as said before, it extends the U&E development phases to the creation of the DT. The VSM analysis is not reported because will be performed at the end of this academic year.

Description of DT and its Development Pipeline. This subsection describes a DT of the Useless IoT Box. The DT is built to support extensive visualization of the behavior Useless IoT Box and interaction with it.

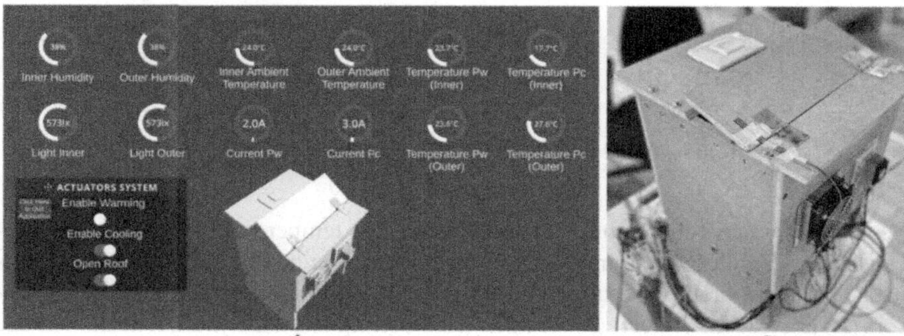

A B

Fig. 5. DT of the Useless IoT Box (A), and its physical counterpart (B)

Figure 5 demonstrates the DT of the Useless IoT Box (A), along with its physical counterpart (B). As can be seen from the Fig. 5A, 3D model of the Useless IoT Box is located at center of the 3D environment, surrounded by the environmental state gauges on top (temperature, humidity, light intensity, current, etc.) and movable window (on the left of the 3D model) for Actuators System (roof control, thermoelectric assembly control). The mechanical behavior of the roof and gauges readings are visualized using the data collected from the previously introduced gateway (Raspberry Pi), therefore representing physical-to-digital communication. While the control buttons in the movable window "ACTUATORS SYSTEM", once interacted, send corresponding signals to control the roof and thermoelectric assemblies. This in turn results in reverse communication, therefore enabling digital-to-physical data flow.

The pipeline of building of the Digital Twin has involved the following sequence of activities:

- CAD modeling – the 3D model has been developed using SolidWorks CAD software. The CAD model implies the architecture of static (the box structure) and moving parts (the roof).
- 3D scene development – the CAD model has been imported into Unity game development engine. Moreover, additional User Interface (UI) elements (gauges, buttons) have been added to support data visualization and control.

- The DT logic development – the logic of DT has been developed within the Unity environment to enable data acquisition and processing from physical counterpart; the correlation of the collected data with behavior of CAD model (the roof) and UI elements; interaction of user with the Digital Twin and replication of such interaction through consequent reverse communication with physical counterpart (roof and thermoelectric assemblies' control).

Process Flow of the DT. The current subsection is devoted to describing communication between the DT and its physical counterpart, giving extensive details about the enabling protocols, data flows and procedures on both ends.

Kritzinger et al. [26] highlighted that crucial feature of a DT is the integration of data flows between an existing physical object and a digital object in both directions. Therefore, DT in focus implements the process flow as depicted in Fig. 6. The communication between physical and digital counterparts is decoupled using MQTT broker. On one hand, the previously introduced Raspberry Pi (MC) of the Logical System, playing the role of gateway, transmits and receives the data from MQTT bro-ker. While the DT in 3D Application Development Environment (Unity) communi-cates with the MQTT broker on the other hand.

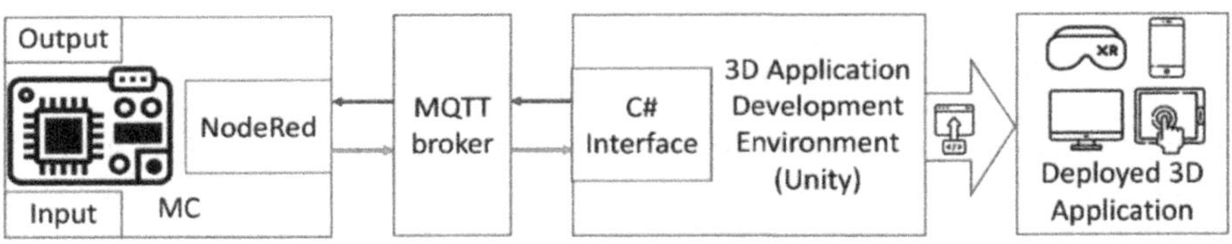

Fig. 6. Process flow of the DT (Colour figure online)

In Fig. 6, the red arrows represent digital-to-physical data flow, while green arrows represent physical-to-digital data flow.

- Physical-to-digital data flow: the NodeRed software, running on Raspberry Pi, implements the logic, configured using NodeRed flows, which publishes the data collected from the sensors, with specific topics to the MQTT broker. While specifically designed C# interface of the 3D Application Development Environment (Unity) subscribes to the previous topics and uses the received data to correlate the behavior of the moving elements of the 3D model (roof) and UI elements (gauges).
- Digital-to-physical data flow – this flow, involving a human, is initiated by the UI elements of the DT – buttons. When a user interacts with the buttons of movable window "ACTUATORS SYSTEM" in the DT space (as described in Sect. 4.2), such interaction initiates the corresponding events of controlling the components of Actuators system (servomotor (S) and relay switches (R)), publishing control messages with specific topics to the MQTT broker. Correspondingly, specific logic running on NodeRed software in Raspberry Pi, subscribes to these control messages, and implements the control of the actuators (servomotor (S) and relay switches (R)) accordingly.

The above-described process flow allows for extensive visualization and monitoring of the Useless IoT Box, while also enables the control of the Actuators system directly from its DT. Although, the previously described bidirectional data flow involves a human to initiate digital-to-physical data flow, it can still be automated. In the current version of the DT, sophisticated automation logic, such as sensors data-driven automatic control can be implemented using NodeRed software running on Raspberry Pi microcomputer.

Finally, the complete design of the DT in 3D Application Development Environment (Unity) can be deployed on different platforms, such as extended reality HMDs (Meta Quest, Microsoft Hololens, HTC Vive and others), mobile devices (smartphone or tablet), and PCs (desktop or laptop), as depicted on the right side of the Fig. 6.

To conclude this section about the DT, it's worth noting that, the following enhancements have been applied:

- The detailed design phase has been extended to the DT design, focusing on defining the pipeline to build the DT, as well as bidirectional communications between physical and digital counterparts and human-system interaction logic.
- Testing and refinement phase involved testing the whole cyber-physical system, focusing on testing the physical Useless IoT Box, its DT, the communications between the two and human-system interactions.

4.3 Comparing First and Second Version of the IoT Tool

The main advantages of the second IoT tool version with respect to the first one are represented by the increased number of sensors for measuring a lot of variables (temperature, humidity, light, roof opening angle, current, etc.), the presence of mechanical system (servomotor, toggle switch to control the roof) additionally to the actuation system (relay and thermal modules/fans) and, thanks to these last additional parts, the opportunity to simulate more cases. Moreover, the most valuable point is represented by the DT of the whole physical system, which enables 3D visualization and interaction in virtual space on the contrary of the first IoT tool. The general mission of the Useless IoT Box is to help professors and students in learning the DT paradigm (the I5.0 enabler) and Value Stream Mapping (the Lean Principle), to drive DT towards the value generation (which is a building block of I5.0).

Using the same approach, hopefully, the interested parties will build their own Value-driven DT in their prospective companies or startups, stimulating development of society.

5 Discussion and Conclusion

The challenge proposed by Industry 4.0 has rapidly evolved into the new I5.0 proposal, combining technological innovation with organizational innovation and shifting interest towards a human-centric vision of the production process.

To support the development of the technological and organizational knowledge necessary to operate effectively in these contexts, it is useful to develop educational devices that reproduce, in a simplified, protected and also fun environment, the situations that will be encountered in professional activity.

In this article the authors propose their experience in experimenting with devices that integrate technological (IoT and DT), design (PLM) and organizational (VSM) aspects.

The most recent result is the IoT Useless box, which updates the original idea of an object that has no other function than to preserve its own state, with IoT sensors and actuators that integrate it and make it available in an IT infrastructure.

This object was conceived, designed and created in its physical and digital version by the students of the Politecnico di Torino with the support of teachers, technicians and doctoral students.

The IoT Useless Box allowed us to experiment with the integration of different development tools and produced an object that will be useful in future teaching activities. Moreover, the Useless IoT Box could be even used as a sandbox for additional learning activities related to Cybersecurity, as well as Energy Management within the Industry 4.0/5.0 context.

Obviously, the result is not perfect, and further developments are possible and desirable. The current DT setup allows for system monitoring and control. However, creating sophisticated decision simulations still requires configuring the desired automation algorithms in NodeRed software. On the other hand, the system still needs to be evaluated from the point of view of cognitive load and user experience.

To address these limitations, in future developments, the DT will be extended to support the application of complex decision-making and control algorithms directly in the DT space. Furthermore, in-depth studies will be conducted to evaluate the overall cognitive load and user experience, involving various target groups, such as professors, students and industry professionals.

References

1. Directorate-General for Research and Innovation European Commission, Breque, M., De Nul, L., Petridis, A.: Industry 5.0: Towards a Sustainable, Human-Centric and Resilient European Industry. Publications Office, Luxembourg (2021)
2. Directorate-General for Research and Innovation European Commission, Müller, J.: Enabling Technologies for Industry 5.0—Results of a Workshop with Europe's Technology Leaders. Publications Office, Luxembourg (2020)
3. Glaessgen, E.H., Stargel, D.S.: The digital twin paradigm for future NASA and U.S. air force vehicles. In: The 53rd AIAA/ASME/ASCE/AHS/ASC structures, structural dynamics and materials conference, Honolulu, Hawaii, USA (2012)
4. Hou, H.C., Lai, J.H.K., Wu, H., Wang, T.: Digital twin application in heritage facilities management: systematic literature review and future development directions. Eng. Construct. Architect. Manage. **31**(8), 3193–3221 (2023)
5. Liu, S., Bao, J., Zheng, P.: A review of digital twin-driven machining: From digitization to intellectualization. J. Manuf. Syst. **67**, 361–378 (2023)
6. Psarommatis, F., May, G.: A literature review and design methodology for digital twins in the era of zero defect manufacturing. Int. J. Prod. Res. **61**, 5723–5743 (2023)
7. Tariq, U., et al.: A state-of-the-art digital factory integrating digital twin for laser additive and subtractive manufacturing processes. Rapid Prototyping J. **29**, 2061–2097 (2023)
8. Yin, Y., Zheng, P., Li, C., Wang, L.: A state-of-the-art survey on Augmented Reality-assisted Digital Twin for futuristic human-centric industry transformation. Robot. Comput.-Integr. Manuf. **81**, 102515 (2023). https://doi.org/10.1016/j.rcim.2022.102515

9. Zhang, H., Liu, Q., Chen, X., et al.: A digital twin-based approach for designing and multi-objective optimization of hollow glass production line. IEEE Access **5**, 26901–26911 (2017)
10. Tao, F., Zhang, M.: Digital twin shop-floor: a new shopfloor paradigm towards smart manufacturing. IEEE Access **5**, 20418–20427 (2018)
11. Böttjer, T., et al.: A review of unit level digital twin applications in the manufacturing industry. CIRP J. Manuf. Sci. Technol. **45**, 162–189 (2023)
12. Lee, H., Yang, H.: Digital twinning and optimization of manufacturing process flows. J. Manuf. Sci. Eng. (2023)
13. Ojstersek, R., Javernik, A., Buchmeister, B.: Optimizing smart manufacturing systems using digital twin. Adv. Product. Eng. Manage. **18**, 475–485 (2023)
14. West, T.D., Blackburn M.: Is digital thread/digital twin affordable? A systemic assessment of the cost of DoD's latest Manhattan project. In: Complex adaptive systems conference with theme: engineering cyber physical systems, pp. 47–56. Chicago, Illinois, USA (2017)
15. Söderberg, R., Wärmefjord, K., Carlson, J.S., Lindkvist, L.: Toward a digital twin for real-time geometry assurance in individualized production. CIRP Ann. **66**(1), 137 (2017)
16. Madni, A., Madni, C., Lucero, S.: Leveraging digital twin technology in model-based systems engineering. Systems **7**(1), 7 (2019)
17. Gehrmann, C., Gunnarsson, M.: A digital twin based industrial automation and control system security architecture. IEEE Trans. Industr. Inf. **16**(1), 669–680 (2020)
18. Krafcik. J.F.: Triumph of the Lean Production System. MIT International Motor Vehicle Program, Sloan Management Review, pp. 41–52 (1988)
19. Womack, J.P., Jones, D.T., Roos, D.T.: The Machine that Changed the World. Harper Perennial, New York (1991)
20. Maskell, B.H., Baggaley, B., Grasso, T.: L. Practical Lean Accounting: A Proven System for Measuring and Managing the Lean Enterprise. CRC Press (2011)
21. Grieves, M.: Product Lifecycle Management: Driving the Next Generation of Lean Thinking: Driving the Next Generation of Lean Thinking. McGraw Hill Professional (2005)
22. Jakobson, R.: Linguistics and Poetics. In: Sebeok, T. (ed.), Style in Language. MA: M.I.T. Press, Cambridge (1960)
23. Serio, F., Awouda, A., Asranov, M., Chiabert, P.: Development of IoT solutions according to the PLM approach. In: Danjou, C., Harik, R., Nyffenegger, F., Rivest, L., Bouras, A. (eds.) Product Lifecycle Management. Leveraging Digital Twins, Circular Economy, and Knowledge Management for Sustainable Innovation. PLM 2023. IFIP Advances in Information and Communication Technology, vol 702. Springer, Cham. (2024)
24. Ulrich, K., Eppinger, S.: Product Design and Development, 6th edn. McGraw-Hill Education (2016)
25. Pesta, A.: Looking for something useful to do with your time? Don't Try This. Wall Street Journal 1 (2013). https://www.wsj.com/articles/SB10001424127887323628804578348572687608806. Last accessed 1 Feb 2024/
26. Kritzinger, W., Matthias Karner, M., Traar, G., Henjes, J., Sihn, W.: Digital Twin in manufacturing: a categorical literature review and classification. IFAC-PapersOnLine **51**(11), 1016–1022 (2018)

Framework Decision Support System Software Comparison in the Agri-Food Industry: A Preliminary Study

Romeo Bandinelli[1](✉), Virginia Fani[1], and Ilaria Bucci[1,2]

[1] University of Florence, Florence, Italy
romeo.bandinelli@unifi.it
[2] Politecnico di Milano, Milan, Italy

Abstract. The paper focuses on developing a logical model for selecting Decision Support System (DSS) software in the agricultural sector. This model aims to address the challenges identified in existing literature, such as the lack of evaluation methodologies and information, which often leads to a mismatch between expectations and outcomes. By creating a logical framework that assesses company-specific needs and know-how, the study proposes a tailored software selection tool. This approach is validated through a comprehensive analysis involving literature review, surveys of software developers, and the creation of a feature database based on ISO/IEC 25010:2011 standards and software comparisons. The model utilizes the Analytic Hierarchy Process (AHP) methodology with Superdecision software for the practical comparison of DSS options. The effectiveness of the developed model is tested through seven case studies on vineyards primarily in Tuscany, offering insights into future developments and potential improvements in the decision-making process for agricultural enterprises seeking to adopt DSS technologies.

Keywords: Agri-food · Decision Support System · DSS · Framework

1 Introduction

The agricultural sector stands as a cornerstone of global food security, economic development, and sustainability. This industry, characterized by its vast diversity and complexity, faces an array of challenges including climate change, resource scarcity, and the need for sustainable practices. To navigate these challenges, the adoption of technology, specifically Decision Support Systems (DSS), has emerged as a pivotal strategy for enhancing decision-making processes, optimizing resource use, and improving overall productivity [1].

Decision Support Systems in agriculture are sophisticated software tools designed to aid in the complex decision-making processes inherent in farm management. These systems integrate data from various sources, including weather forecasts, soil conditions, crop models, and market trends, to provide actionable insights. The objective is to enable

P. Sureephong et al. (Eds.): PLM 2024, IFIP AICT 740, pp. 249–256, 2025.
https://doi.org/10.1007/978-3-031-93319-6_19

farmers and agricultural managers to make informed decisions that optimize production, ensure sustainability, and enhance profitability.

However, the adoption and effective use of DSS in agriculture are hampered by several factors. A primary concern is the selection of appropriate DSS software that aligns with the specific needs, resources, and capabilities of an agricultural enterprise. The market is flooded with a plethora of DSS tools, each boasting different features, capabilities, and levels of complexity. This abundance of choices, while beneficial, also poses a significant challenge for decision-makers in identifying the most suitable DSS tool for their operations.

The literature reveals a notable gap in comprehensive methodologies for the evaluation and selection of DSS software in agriculture. This gap highlights the need for a systematic approach that considers the unique characteristics of agricultural operations, the specific requirements of decision-makers, and the technical capabilities of DSS tools. Moreover, there is a pressing need for an evaluation framework that is both robust and flexible, capable of adapting to the dynamic nature of the agricultural sector and the evolving landscape of DSS technologies.

In response to these challenges, this research proposes a logical model for the selection of DSS software in the agricultural sector. This model aims to provide a structured approach to software selection, incorporating a comprehensive set of criteria derived from industry needs, technical standards, and user preferences. The model leverages the Analytic Hierarchy Process (AHP) methodology, a decision-making tool that facilitates the evaluation of alternatives based on multiple criteria. This approach is designed to offer agricultural decision-makers a clear, systematic, and objective means of assessing and comparing different DSS tools, ultimately guiding them toward the most appropriate software selection.

Starting from this background, this paper aims to answer to the following Research Questions (RQs):

RQ1: What are the key factors that influence the selection of DSS software in the agricultural sector?

This question seeks to identify the critical criteria that agricultural decision-makers consider when selecting DSS tools, including technical features, user requirements, and contextual factors such as the specific challenges and opportunities within the agricultural industry.

RQ2: What are the implications of the proposed model for decision-making in the agricultural sector?

This question aims to assess the practical impact of the proposed model on the decision-making processes of agricultural enterprises, including its potential to enhance efficiency, optimize resource use, and improve overall productivity through the informed selection of DSS software.

By addressing these questions, the research aims to contribute to the body of knowledge on technology adoption in agriculture, offering insights into the effective selection and use of DSS tools. This study not only addresses a critical gap in the literature but also provides a practical tool for enhancing the decision-making capabilities of agricultural enterprises, ultimately supporting the sector's sustainability and growth.

The paper is structured as follows. Section 2 reports a theoretical background on Decision Support Systems and Agriculture. Section 3 describe the proposed framework, while in Sect. 4 are reported the main findings. Last, in Sect. 5 ends with the conclusions and further development.

2 State of the Art

As previously mentioned, many DSS are now on the market, each with multiple functions and fields of application. However, all share a high technological level and thus require a correct evaluation, a level of specialization, knowledge, and extensive experience in their sector of use. Experience and knowledge are easily found in all fields of DSS application, but specialization in terms of education level is one of the main limits in their diffusion and application, consequently affecting the correct choice of these systems in the agricultural field. David C. Rose, in [2] identifies age and education level as relevant factors in the use of DSS following interviews with farmers and consultants in Wales and England. Moreover, in his work, Rose identifies key factors in software use. Performance: Must ensure results perceived by the customer and be continuously verified. Ease of Use: Efficient and easily understandable visual presentation with minimal text. Colleague Recommendation: Software selection based on word of mouth among colleagues. Trust: Based on the algorithms used and on which experience the DSS was developed. Cost: Increase in the use of DSS if public subsidies are present or if an economic advantage in using the software is demonstrated. Habit: Age and aversion to technology influence the use of these tools. Customer Adaptability: Information and customization for the customer.

This initial analysis already highlights the main comparison factors for a functional choice, including cost evaluation, user interface customization, and the presence of the DSS in similar crops in the same territorial area. The effect of the level of education on the use of decision support systems is also highlighted.

From the results obtained from these articles, the importance of understanding for a functional choice the level of technological knowledge of the company under examination emerged. To deepen the topic, literature and web research was conducted on evaluation models for the technological maturity level of primary sector companies.

In [3] some limits of current support systems are highlighted. For the purposes of this work, as analyzed by [3], one of the limits of current software in the agricultural field is being developed for a specific task and for a specific culture, making it difficult to compare software because the specifications are inherent to the particular case study. But as also mentioned by the author, a reference is given to us by ISO/IEC 25010:2011. In [4] a comparison between many DSS for irrigation is carried out. Of particular interest for this work is the table extracted from the article in which the comparison characteristics are identified, in particular: Cost, Release year, Where it was developed and tested, Developing company, Whether the version is public, Literature references. An interesting aspect to consider in choosing software emerges from [5] in which the importance of using a precise and specific taxonomy is emphasized. This problem analyzed in the case of weed management is solved by developing a DSS in collaboration with users and sector consultants.

In [6] the importance of integration with other software and the presence of online training and tutorials is emphasized. These characteristics are fundamental regardless of the field of application and will therefore be among the criteria analyzed in the model developed in this work. As already highlighted, a characteristic that emerges in other documents, such as that in [7] is the importance of having customization of output data and the level of detail based.

3 Framework Implementation

Starting from the analysis conducted in Sect. 2, some relevant characteristics for a correct evaluation of software emerged. Below is a list with the respective sources: [3–5, 8–12]. This analysis answered RQ1 and identified the key factors that influence the selection of DSS software in the agricultural sector. The first step was to start from the listed characteristics to create groups that contained them according to the ISO/IEC 25010:2011 standard. As a result, as reported in Table 1, a classification of the functionality of the software has been defined. Starting from that, the goal of this section is to address RQ2, specifically presenting the logical and mathematical pathway for identifying the most suitable DSS. Subsequently, the identified characteristics (and others created in accordance with the definitions of the standard) were defined in such a way as to have a YES/NO answer or a numerical value or a numerical value corresponding to the group of belonging, obtained through the division of the value scale into groups. The following figure shows the characteristics and their classification.

Hereafter, the categories, and the entries within them will be described, along with the respective evaluation that will be used for comparing software.

Functional Suitability encompasses features describing the adequacy of requested functions, especially if they match the end user's goals. *Functional Completeness*: Assessed as the percentage of satisfaction of required objectives, calculated as the ratio of requested functions to those fulfilled by the software. *Functional Correctness*: Describes the match between required data accuracy and the system's output accuracy. A "YES" is assigned if a match exists, otherwise "NO". *Forecasting Horizon Adequacy*: If the forecasting horizon of data matches the main feature requested by the user (e.g., weather forecast horizon, disease development forecast horizon), a "YES" is given if it's less than or equal to the system's output data horizon, otherwise "NO".

Performance And Efficiency describes software features in terms of required performance and efficiency: *Sufficient Input Data Available*: Evaluates if the company's devices (agrometeorological stations, sensors, etc.) or collected data are sufficient as input for the software. A "YES" is assigned if the match is verified, otherwise "NO". *% Satisfaction of Required Outputs*: Expressed as the ratio of required outputs to those provided by the software.

Compatibility describe compatibility with existing company systems and commercial ones. *Interoperability with Other Systems*: Assesses how integrable the software is with other systems, evaluated as the ratio of systems it integrates with to total systems analyzed. *Interoperability with Company Systems*: Evaluates integration with existing company management or traceability systems, assigning a value of 1 if integrable with all systems, 0.5 if only one (if both are present), 0 if no systems are present or if not integrable.

Table 1. Software categories and characteristics.

Categories	Characteristics Functional
suitability	Functional Completeness
	Functional Correctness
	Forecasting Horizon Adequacy
Performance and efficiency	Sufficient Available Input Data
	% Satisfaction of Requested Outputs
Compatibility	Interoperability with Other Systems
	Interoperability with Legacy Systems
Usability	Training Courses
	Customizable User Interface
	24/7 Assistance
	Input Error Recognition
Flexibility	Application Fields
Reliability and security	Years of Market Experience
	Data Security System
	User Operations Log
Maintenance and integration	Integration with Brand Devices
	Assistance and Warranty Package
Risk mitigation	Economic Impact Assessment
	Environmental Impact Assessment
Trust	Cost Range
	Software Distribution
	Conducted Case Studies
	End-User Involvement in Development

Usability considers features related to software usage, focusing on support, customization, and training due to the inability to test usage directly. *Training Courses*: If the DSS manufacturer provides specific on-site training and an online course, a value of 2 is assigned; 1 for on-site only; 0.5 for a standard online training; otherwise, a value of 0. *24/7 Support*: If the company offers a 24/7 support and problem-solving service to all its clients, a "YES" is evaluated; otherwise "NO". *Input Error Recognition*: If the DSS has a system for evaluating potential input data errors due to device malfunction or human error, a "YES" is assigned; otherwise "NO". *Customizable User Interface*: If the company allows users to customize their Dashboard autonomously and continuously, a value of 2 is assigned; 1 if customization is only available upon request to the manufacturer; otherwise 0.

Reliability and Security category includes parameters describing data security and operation history control, as well as system reliability updates to address development or usage defects. *Years of Market Experience*: To assess this feature, 4 ranges with corresponding scores are created and expressed in a table.

Maintenance and Integration is expressed through the presence of support service and warranty extension, while compatibility through the assessment of integration with devices from multiple brands. *Integration with Devices from Multiple Brands*: To assess this feature, 4 ranges with corresponding scores are created and expressed in a table. *Support and Warranty Package*: If there's an option to extend the warranty with related support, a value of 1 is assigned; 0.5 for one of the two options; otherwise 0.

Risk Mitigation category evaluates if the software includes functions for assessing the economic and environmental impact of its use: *Economic Impact Assessment*: If the software has a model for assessing the economic impact in absolute or differential terms, a "YES" is evaluated; otherwise "NO". *Environmental Impact Assessment*: If the software can evaluate savings in raw material and water consumption, a value of 1 is assigned; if it assesses the global impact of the production cycle in terms of CO_2, a value of 2 is given; otherwise 0.

Trust category evaluates the software's quality and spread through different factors. *Cost Range*: Assesses the cost through two values, one representing the annual subscription cost and the other the installation cost of a central unit and soil moisture sensor, based on the basic package that guarantees the required functions. *Software Spread*: Evaluated through two values, one expressing the software's bibliographic spread and the other its number of clients.

Flexibility contains a single feature describing the system's flexibility, specifically in how many application fields the software can be used. *Application Fields*: Evaluated as the ratio of the number of application fields in which it is usable to the total number of considered application fields.

Once the characteristics for comparison have been identified, some of these require the definition of the evaluation set, and all must be included within a logical evaluation process that concludes with the comparison and definition of the best alternative. In this paragraph, we will analyze the logical path that leads to the definition of the parameters required by the company, which will be the reference for the final evaluation. The need for an analysis of the company and its requirements is one of the fundamental steps of this process, which aims to overcome the problems identified in the literature, such as a lack of internal know-how in the company or a misalignment between expectations/requests and software functionality.

The first step to take is the evaluation of the company's objectives and, in particular, it is important to define the characteristics of the required software based on the needs or problems that it intends to address. For this purpose, the "House of Quality" is used, a tool employed during the product development phase but also useful for translating the functionalities (for us, the objectives) of a product into its necessary characteristics. Since the aim of this phase is merely the translation of objectives into software characteristics, not all the typical steps of the quality house (such as competitors' evaluation, interactions, and relationships) will be carried out.

The second step of the process is the evaluation of the software available in the company (management systems, Windows and Mac devices, sensors, weather stations, etc.) and an assessment of the available connectivity (2G, 3G, 4G, 5G).

The third step of the process is the analysis of internal capabilities, and in particular, technological capabilities. Since the development of the following model has as its first target the Italian companies (and specifically the small-medium enterprise), we will refer to the use of the SELFIE4.0 tool, but any other evaluation of the company's technological maturity remains valid for this phase.

The next step is to define more detailed characteristics required of the software, which is achieved by completing the table developed for this work (see Table 1).

The final step to perform is a qualitative assessment of the match between the available capabilities and the capabilities required for the use of the software.

4 Findings

This work provides a comprehensive methodology for the analysis and selection of DSS software that was not previously available in the literature. The definition of the comparison characteristics starts from ISO/IEC 25010:2011 and from the bibliographic research of comparison parameters used in the analysis of some software functions. To compare the software based on the defined characteristics, a database was created using information available in the literature, webography, and through interviews with the software houses themselves. The next step was to better understand the problems encountered in the introduction and use of these software in order to find a selection methodology that would solve these issues, which was articulated in the definition of a logical path. The logical path, with the contribution of tools such as the "house of quality" for defining objectives and "SELFIE 4.0" for quantifying the technological level of the company, aims to identify problems, such as lack of know-how and misalignment of software features-business objectives, before a detailed analysis of the DSS. This step is essential to make the subsequent analysis more effective and/or to avoid carrying out superfluous and inappropriate analyses. Before carrying out the analysis to quantify the correspondence between business needs and the defined characteristics of the software (included in the previously developed database), a questionnaire was developed to be completed in the company.

For the comparison and selection of the best alternative, the AHP methodology was identified, whose weights are defined based on the priorities described by the company in the aforementioned questionnaire.

Since the developed model is unique in its intended use, it cannot be compared with similar systems present in the literature, but from the analysis of the results carried out in Sect. 5.4.10, it seems to well represent the correspondence between the type of company and the strengths of the software.

5 Conclusions

DSS technology is already used in many fields, but another area of application, close to agriculture, could be in the management of parks and forests. This application could be used not only in the public sector by state, regional, and municipal entities but also by private companies, particularly in both the development and management phases of green areas. Especially for the private sector, the technology would be more interesting if it included systems for assessing the environmental impact in terms of CO_2. There are already case studies and pilot projects in this direction. Starting from the results of this work, it would be interesting to test the model on different types of companies in terms of crops and sizes. The model created and the database developed already allow for the analysis of a wide range of crops. The technology under analysis is characterized by significant development, and therefore future revisions of the database will be necessary to carry out analyses on the AS IS state of the software. Conducting case studies will also be relevant to identify repetitive cases of solution and thus be able to define standard scenarios that could be useful for companies not willing to undertake a detailed analysis path like the one developed but are simply interested in a result of business similarity.

References

1. d'Avolio, E., Pinna, C., Bandinelli, R., Rinaldi, R., Terzi, S.: When product development meets luxury: a case study analysis in fashion, food and furniture companies. In: Lecture Notes in Electrical Engineering, vol. 525, pp. 15–24 (2019)
2. Rose, D.C., et al.: Decision support tools for agriculture: towards effective design and delivery. Agric. Syst. **149**, 165–174 (2016)
3. Zhai, Z., Martínez, J.F., Beltran, V., Martínez, N.L.: Decision support systems for agriculture 4.0: survey and challenges. Comput. Electron. Agric. **170**, 105256 (2020)
4. Ara, I., Turner, L., Harrison, M.T., Monjardino, M., deVoil, P., Rodriguez, D.: Application, adoption and opportunities for improving decision support systems in irrigated agriculture: a review. Agric. Water Manag. **257**, 107161 (2021)
5. Colas, F., et al.: Co-development of a decision support system for integrated weed management: contribution from future users. Eur. J. Agron. **114**, 126010 (2020)
6. Giupponi, C.: Decision support systems for implementing the european water framework directive: the MULINO approach. Environ Model Softw. **22**(2), 248–258 (2007)
7. Thar, S.P., Ramilan, T., Farquharson, R.J., Chen, D.: Identifying potential for decision support tools through farm systems typology analysis coupled with participatory research: a case for smallholder farmers in Myanmar. Agriculture (Switzerland) **11**(6), 516 (2021)
8. Zhai, Z., Martínez, J.F., Beltran, V., Martínez, N.L.: Decision support systems for agriculture 4.0: survey and challenges. Comput. Electron. Agric.put. Electron. Agric. **170**, 105256 (2020)
9. Gutiérrez, F., et al.: A review of visualisations in agricultural decision support systems: an HCI perspective». Comput. Electron. Agric. **163**, 104844 (2019)
10. Bhatnagar, V., Chandra, R., Sunda, S.: State of the Art and gap analysis of precision agriculture: a case study of indian farmers. Int. J. Agric. Environ. Inf. Syst. **10**, 72–92 (2019)
11. Zhang, J., et al.: Challenges and opportunities in precision irrigation decision support systems for center pivots. Environ. Res. Lett. **16**(5), 053003 (2021)
12. Gagliardi, G., Cosma, A.I.M., Marasco, F.: A decision support system for sustainable agriculture: the case study of coconut oil extraction process. Agronomy **12**(1), 177 (2022)

Twin Transition Waste Electric and Electronic Equipment Roadmapping Tool to Support Improvement of Environmental Sustainability Performance of Reverse E-waste Value Chains

Daniele Perossa(✉), Federica Acerbi, Paolo Rosa, and Sergio Terzi

Politecnico di Milano, 20133 Milan, Italy
daniele.perossa@polimi.it

Abstract. The recovery of the huge economic value embedded in waste electric and electronic equipment (WEEE) has been widely discussed through the last years. The activity is relevant also under an environmental point of view, to limit the consumption of primary critical raw materials and rare metals. Nonetheless, few concerns related to the environmental impacts coming from processes, moreover of recycling, of WEEE have been raised. Thus, in this paper is investigated how, under a twin transition perspective, digital technologies can support the offsetting of these environmental impacts. Furthermore, it is proposed a novel roadmapping tool, called twin transition WEEE roadmapping tool is proposed. It aims at supporting practitioners of the reverse WEEE value chain in identifying the most useful twin transition implementation pathway, moreover in terms of technology selection, for the contingency of their company.

Keywords: WEEE; Twin Transition · Circular Economy · Roadmapping · Environmental Sustainability

1 Introduction

In recent times, it has been widely discussed how waste electric and electronic equipment (WEEE) are increasing in quantity year after year [1]. The recovery of metallic and plastics materials contained in WEEE through adequate processes enables retrivial of the great economic value embedded in WEEE [2]. A large share of this economic value comes from the Critical Raw Materials (CRMs) embedded in WEEE [3]. CRMs can be defined in various ways. Typically, they are materials, usually metallic, that on the one hand bear a great economic and social value, and on the other bear many risks through their usage (in terms of availability, social impacts, environmental impacts, etc.) [4–6]. Policymakers and academicians are paying more and more attention to this topic, moreover for what concerns the recovery of CRMs embedded in WEEE (e.g., [4, 7]). An emblematic example is the CRMs Act issued by the European Commission, where the objective of sourcing 15% of annual European demand for CRMs from recycling [8].

Published by Springer Nature Switzerland AG 2025
P. Sureephong et al. (Eds.): PLM 2024, IFIP AICT 740, pp. 257–270, 2025.
https://doi.org/10.1007/978-3-031-93319-6_20

Nonetheless, it is often argued how recovering secondary metallic from WEEE can bear significant environmental impacts, that should be measured and offset [9–11]. Digital technologies can play a crucial role in enabling WEEE recovery and optimizing its performances, including environmental sustainability (ES) [12–15], under the paradigm of the so called twin transition (TT) [16, 17]. Nonetheless, industrial actors usually meet issues and hurdles when trying to implement (TT), often due to a lack of internal competencies and skills in correctly identifying and implementing the correct digital technologies in accordance to their needs and contingency [18, 19]. A significant help can be provided by roadmapping tools, supporting practitioners during their decision-making processes related to the correct selection of digital technologies to be implemented [20, 21]. In this way, companies can enjoy a further support in respecting requests coming from norms and regulations.

Thus, the research objective of this work is: understanding how a roadmapping tool may support practitioners in pursuing (TT) in the sector of WEEE. Thus, the following two research question is addressed: how can a roadmapping tool support practitioners in pursuing (TT) in the sector of the WEEE collection, treatment, and recovery? According to the authors knowledge, the literature lack of research where the topic of TT in the WEEE RVC is widely analysed. Furthermore, this is the first work where a roadmap for TT specific for WEEE circularity processes is proposed.

The remaining of the paper is structured as follows: in Sect. 2, the adopted methodology is described. In Sect. 3, an overview of the current regulations related to WEEE in Europe is conducted, aiming at showing the relevance of the topic and the direction towards a higher sustainability level that is required. In Sect. 4, a review of the documented applications of digital technologies to the WEEE recovery sector, with the specific aim of improving the ES performance, is conducted, aiming at answering to RQ1. In Sect. 5, a roadmapping tool supporting WEEE sector practitioners in pursuing TT is proposed. The tool is developed by means of Design Science Research (DSR) methodology. In Sect. 6 the tool is validated through a lab-scale use case. Finally, conclusions are drawn in Sect. 7.

2 Methodology

In this paragraph, the methodology adopted to support companies acting in WEEE sector in implementing TT is described.

In [21] by (Perossa et al., 2023), a roadmapping tool, called Twin Transition Cosmetics Roadmapping Tool (TTCRT), was developed by means of an Action research-based methodology. The aim of the TTCRT is supporting practitioners in developing a TT strategy, moreover in the steps of identification of the right digital technologies to be implemented along the product life cycle, to improve its ES performance. The TTCRT is focused on the cosmetics industry. Nonetheless, in the conclusions of [21], it is stated how a possible clue for further research may be the adaptation of the TTCRT to other industrial sectors. The TTCRT is therefore a good starting point for developing an analogous roadmapping tool, focused on WEEE treatment sector.

In order to adapt the TTCRT to WEEE value chain, a twofold review has been conducted: on the one hand, an investigation of the existing European WEEE regulatory framework was performed to understand the current legislative requirements towards companies operating in this sector (paragraph 3). On the other hand, a review of the state of the art of application of digital technologies in the WEEE sector, aiming at improving the environmental sustainability of conducted processes and operations has been performed (paragraph 4). In this way, the relevant peculiarities of the WEEE sector can be analysed and embedded in the new proposed tool.

A further element of analysis has been the identification of the main phases of the WEEE reverse value chain. Indeed, as the TTCRT considers the typical life cycle of a cosmetic good, the proposed tool will consider the stages of WEEE treatment. They have been identified by methodology of literature review and previous knowledge of the authors.

The novel tool is called Twin Transition WEEE Roadmapping Tool (TT-WEEERT). It is developed by means of DSR, following the steps proposed by [22]:

- Problem identification and motivation: procedures of recovery of materials and components from WEEE bears non-negligible environmental impacts. TT can offset these impacts, but expertise in this domain is often missing among practitioners.
- Objectives: in this work is proposed the development of a roadmapping tool, called TT-WEEE-RT, to support practitioners of the WEEE sector in pursuing TT for the WEEE circular processes.
- Design and development: the TT-WEEE-RT will be developed by adapting the existing TTCRT to the WEEE sector.
- Demonstration: TT-WEEE-RT will be applied to a lab-scale automated WEEE disassembly process.
- Evaluation: following the demonstration, some evaluations will be performed, supported by the insight provided by the persons in charge of the mentioned process.

3 Review of European WEEE Regulatory Framework

Several actions and legislations have been put in place by the European Commission in the last decade to improving rate of return of WEEE and supporting efficient reuse of secondary resources, improving the environmental performance of electric and electronic equipment (EEE) sector (e.g.: [23–25]). They are relevant as they highlight the main trends and key points of WEEE and their treatment. The main regulative reference for WEEE in EU is [25]. It embeds the Extended Producer Responsibility (EPR) principle presented in the waste framework directive [26]. [25] applies to six categories of WEEE: (i) temperature exchange equipment, (ii) Screens, monitors, and equipment containing screens having a surface greater than 100 cm2, (iii) Lamps, (iv) Large equipment (any external dimension more than 50 cm), (v) Small equipment (no external dimension more than 50 cm), (vi) Small IT and telecommunication equipment (no external dimension more than 50 cm). Among the WEEE not included in the list there are means of transport for persons or goods. [25] aims at minimizing WEEE disposal into unsorted municipal waste, setting targets for e-waste collection rate for member states (65% of total EEE placed in the market in the previous three years or 85% of total WEEE generated in the year). It also encourages EEE producers to meet minimum WEEE treatment targets, and encourages cooperation between producers and recyclcers to design EEE in line with

EU ErP directive [27]. Another crucial document is the RoHS Directive [28]. It targets the usage of materials that can be harmful for humans and environment in EEE once they become WEEE. Ten substances among heavy metals, flame retardants, and plasticizers are identified as hazardous and their usage regulated. The ErPDirective [27] targets over 40 product categories, including several EEEs. [27] addresses eco-design requirements covering the entire product life cycle. All these Directives witness the willingness of stimulating circularity of the WEEE by the EU.

Further standards have been developed by the Comité européen de normalisation en électronique et en électrotechnique (CENEC), with EN 4555X series. In particular, relevant for CE of WEEE, there are: (i) EN 4552:2020 'General method for the assessment of durability of energy related products'; (ii) EN 4553:2020 'General method for the assessment of the ability to remanufacture energy-related products; (iii) EN 4554:2020 'General methods for the assessment of the ability to repair, reuse and upgrade energy-related products'; (iv) EN 4555:2019 'General methods for assessing the recyclability and recoverability of energy-related products'; (v) EN 4556:2019 'General method for assessing the proportion of reused components in energy related products; (vi) EN 4557:2020 'General method for assessing the proportion of recycled material content in energy-related products'. These standards are symptomatic of a strong need by stakeholders of evaluating circularity of EEEs and WEEEs and, thus, of a strong demand of more circular and more sustainable WEEE. Further relevant standards by the CENEC are EN 50625–1:2014 and EN 50614:2020. They cover collection, transport, re-use, and treatment of WEEE. They concern depollution and its monitoring, monitoring of downstream treatment, and aim at ensuring uniformity in auditing.

4 Review of the Potential Applications of Digital Technologies for WEEE Treatment

Under the perspective of the TT, several digital technologies are addressed in the literature for being used in optimizing ES of circular practices related to WEEE.

Tracing technologies may perform a crucial role in uncovering information of state and conditions of components that are found in WEEE, and potentially also of the materials embedded in them. In this way, it is possible to select those components that can be reused instead of being recycled, saving the environmental impacts related to the recycling process [12]. Obviously, in this kind of information sharing, cloud computing technology has a vital enabling role [29]. At the same time, a crucial tool for gathering this kind of data in a circular value chain context is the Internet of Things [30].

Usage of automated robots to perform disassembly operations allows performing pre-treatment avoiding the risk for human operators to get in contact with hazardous materials potentially embedded in WEEE [15]. Artificial Intelligence algorithms are key enablers for managing and optimising automated disassembly processes [31].

An other relevant role is played by simulation techniques to optimize the ES performance of reverse supply chains and treatment of WEEE. In [14] is provided an example using artificial intelligence-based genetic algorithms for simulation aiming at economic and environmental optimization of WEEE reverse supply chain in Sao Paulo. Another example can be found in [32], where digital twin-based simulation techniques are used to optimize the energy efficiency of a lab-scale smartphone disassembly line. Digital twin is stated in [33] to have contributed to optimising, under an energy efficiency point of view, the refining process of recovered copper. In [32] it is also explored how virtual reality can act in synergy with digitaltwin technology, enhancing the human-simulation interface. In [34] is discussed how process simulation can be applied to quantify resource efficiency and overall environmental performance of e-waste recycling.

Augmented reality is considered in [35] as a tool providing easy-to-read real-time feedback related to recycling processes, to help improvement in resource optimisation. A further technology finding extensive applications in WEEE treatment is additive manufacturing. It can be used, indeed, to produce printed electronics starting from recycled materials [36]. Additive manufacturing is considered as a technology that significantly improves the materials efficiency in manufacturing starting from recycled materials, if compared to traditional machining processes [37].

5 The Twin Transition WEEE Roadmapping Tool

The support that digital technologies provide to improving the ES performance of circularity practices referred to WEEE is well documented, as described in paragraph 3.

Thus, given what already mentioned in the previous paragraphs, an adaptation of the TTCRT to the WEEE sector is proposed here.

5.1 Structure of the TT-WEEE-RT

Like the TTCRT, the TT-WEEE-RT will be conceived to be applied to one product per time. In the context of WEEE, as a product, is meant a given typology of waste, or components that undergoes a series of treatments. The main structure of the TT-WEEERT is the same of the TTCRT, divided into three main sections:

- Know-why: in this section, the main targets and drivers of the company towards sustainability are identified through a questionnaire submitted to the persons working in the enterprise. Identified objectives and drivers will work as a compass in the 5 mplementi of the technologies to be implemented.
- Know-what: in this section, it is evaluated, along the steps of the considered process, what is the current level of ES performance for the considered product, according to some identified performance dimensions. The purpose of this section is identifying the main points of strength and weakness of the considered circularity procedures in terms of ES.
- Know-how: in this section, it is assessed the current level of digital technologies in place, considering only those that have a positive direct or indirect impact on ES. The goal here is defining in which areas TT implementation is more developed and in which areas it is less.

As in the TTCRT, the fillingof the listed threesections constitutes the assessment phase, where the current situation of the considered processes is depicted. Basing on the results of this assessment, is performed the roadmapping phase, where it is defined what are the next technological implementations useful to be performed.

While the TTCRT, in the know-what and know-how phases, considers the life cycle of a cosmetic product, the TT-WEEE-RT needs to consider the possible phases of a WEEE RVC. The proposed phsaes to be considered are the following:

- Collection phase: when the WEEE are collected and sorted and, through reverse logistics channels, are sent to disassembly and recycling plants [38].
- Pre-treatment phase: whenthe variouscomponents are separated, so thatthey can be reused, remanufactured, or their materials recycled in the subsequent phases, or where mechanical processes like shredding or crushing are used for WEEE size reduction [39, 40].
- Remanufacturing or refurbishing: in this phase, the components that can be entirely recovered are remanufactured or refurbished for further usage [41].
- Materials recovery: for those components that can be recovered, thet are sent to recycling plants where the materials embedded in them can be recovered [42].
- Materials refining: to improve the quality of the recovered materials, it is often necessary to refine them [43].
- Production of new EEEs from secondary materials [36].

This list of steps was validated by an academic expert in the field. In some cases, some of the phases can be not covered in the analysis. This depends on the defined boundaries of the analysis. Furthermore, some routes are alternative to others for specific components or materials (e.g., if a component is sent to materials recovery phase, it shall not be sent to remanufacturing phase).

The dimensions of assessment of the ES performance in the know-what and know-how phases are also changed from the TTCRT, to be adapted to the WEEE context. For this reason, the dimensions of ES to be considered have been inspired to the results of the review conducted in paragraph 3 about the possible applications of digital technologies to WEEE industry. The selected dimensions are the following:

- Improved evaluation of components to identify the most appropriate route minimising environmental impacts [12, 29, 30] (reusage is less impacting than remanufacturing and refurbishing, which are less impacting than recycling [48]).
- Minimization of risk on human health due to contact with hazardous materials embedded in WEEE [15].
- Maximization of the materials efficiency [34, 37].
- Maximization of energy efficiency [32, 33].

In the know-how section, are listed the digital technologies to be considered for TT in the WEEE context, as emerged from the review described in paragraph 3:

- Tracing technologies [12].
- Internet of Things [30].
- Automated and collaborative robots [15].
- Data analytics and artificial intelligence [14, 31].
- Simulation technologies [14, 32–34].
- Augmented reality and Virtual reality [32, 35].

- Additive manufacturing [36].

Some of the listed technologies can be implemented also as a support to the other technologies. A scheme representing the TT-WEEE-RT is provided in Fig. 1.

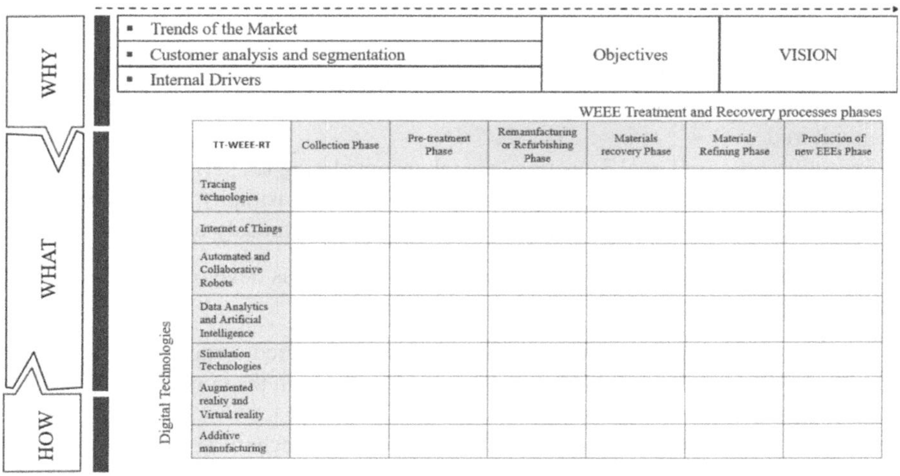

Fig. 1. Scheme of the TT-WEEE-RT adapted from [21]

On the other hand, the know-how section is structured as shown in Fig. 2:

WEEE Treatment and Recovery processes phases

	TT-WEEE-RT	Collection Phase	Pre-treatment Phase	Remanufacturing or Refurbishing Phase	Materials recovery Phase	Materials Refining Phase	Production of new EEEs Phase
Digital Technologies	Tracing technologies						
	Internet of Things						
	Automated and Collaborative Robots						
	Data Analytics and Artificial Intelligence						
	Simulation Technologies						
	Augmented reality and Virtual reality						
	Additive manufacturing						

Fig. 2. Scheme of the know-how section of the TT-WEEE-RT

The know-what section can be compiled according to the following scale, which is similar to the one proposed in [21], that was adapted from [44–46]:

- Not applicable (level X): it is outside study boundaries, or it is not possible to apply the considered technology to the considered context.
- No strategy (Level0): actions arenot takenor are exclusivelydrivenby theneed to be compliant with laws and regulations. No targets are defined.
- Defensive strategy (Level 1): actions with positive impacts on ES are in place but they focus on risk and cost reduction. Mainly the economic dimension is considered. Environmental targets are not defined.
- Accomodative strategy (Level 2): actions are taken considering ES performance and are mainly market driven. Targets are clear but conservative and processes are partially monitored.
- Proactivestrategy (Level 3): actions are takenconsidering ES performance. The company continuously seeks to integrate the sustainability vision by anticipating market requests. Targets are ambitious and processes are continuously monitored.

For filling the know-how table, the following scale was developed. It is similar to the one proposed in [21], which was adapted from [47]:

- Level X: the technology is not applicable in the considered context.
- Level 0: there is no implementation of that technology in the considered RVC phase or its implementation does not consider ES.
- Level 1: pilot actions are in place to implement that technology in the considered phase of the RVC or, alternatively, the technology is well implemented but with no ES objectives. Nonetheless, there are collateral ES benefits resulting from the implementation.
- Level 2: implementation of the technology in that stage of RVC is initiated and ES benefits are observable. Alternatively, the technology is well implemented for other primary purposes, but conscious secondary aspects related ES improvements are enjoyed and measured.
- Level 3: Advanced and mature implementation of the technology in that RVC stage mainly for ES purposes are in place, resulting in clear and measured benefits.

6 Validation and Evaluation

The TT-WEEE-RT has been validated by being applied to a lab-scale disassembly process of printed circuit boards (PCBs) at the Industry 4.0 Lab "Marco Garetti" of Politecnico di Milano. The process is performed with a Cobot and integrates artificial intelligence for image recognition. Through image recognition, it is possible to automatically identify components over the PCBs that are valuable and reusable. Then, the cobot can detach them from the PCB itself. The process is one stage of a three-stages lab-scale macro-process for recovery of materials from WEEE, developed in the context of the Horizon Europe 2020 Research and Innovation Action called TREASURE, co-funded by the European Commission.

The first step of the application of the TT-WEEE-RT led to submit a questionnaire to the researchers working on the process. The main outcomes of the questionnaire led to defining the following objectives and drivers as crucial:

- As external drivers: (i) increase in the demand for scarce resources; (ii) tackling the issue of climate change and environmental pollution; (iii) anticipating ever more stringent rules at national and international levels.
- As internal drivers: (i) willingness to introduce technologies able to optimize ES performances; (ii) availability to hire qualified human resources to optimize the ES performance.

The filling of the know-what and know-why sections was limited to the disassembly process, which takes place in the Lab. Results are illustrated in Fig. 3 and Fig. 4.

	TT-WEEE-RT	Disassembly Phase
Environmental sustainability performances	Improved components evaluation	3
	Minimization of risks for human health	3
	Maximization of materials efficiency	1
	Maximization of energy efficiency	0

Fig. 3. Application of the know-what phase

The Lab-scale proess is able to provide state-of-the-art components evaluation, with the explicit aim of sending to the more sustainable routes (reusage or remanufacturing) the components for whom this is feasible. Explicit targets are present under this perspective. Furthermore, the automated nature of the process leads to totally avoiding any type of direct contact between operators and eventual hazardous materials. Considering the materials efficiency, the process considers it, and actions are performed to improve it, but targets are focusing exclusively on economic returns coming from this efficiency. Furthermore, the process is not able yet to recover all the typologies of potentially valuable components. Finally, actions for maximasation of energy efficiency are totally neglected, even though the energy consumption levels of a cobot are typically not high according to the operators of the Lab.

The only technologies in place are currently collaborative robots and artificial intelligence. They are present at a state-of-the-art level of implementation, and both of them are actively focused on environmental sustainability targets. There is a complete lack of focus on the other considerable technologies.

Concerning the roadmapping phase, the following main points have been addressed:

- Initialization of tracing technologies implementation, in collaboration with partners sending the PCBs to the disassembly process and receiving the components from it, may further support components evaluation.
- Implementation of Internet of Things technologies for smart measurement of the process performances can be a first step for dealing with materials and energy efficiency. Nonetheless, this is deemed by the operators in the lab as being more useful when the process will have a higher technology readiness level.

	TT-WEEE-RT	Disassembly Phase
Digital Technologies	Tracing technologies	0
	Internet of Things	0
	Automated and Collaborative Robots	3
	Data Analytics and Artificial Intelligence	3
	Simulation Technologies	0
	Augmented reality and Virtual reality	0
	Additive manufacturing	X

Fig. 4. Application of the know-how phase

- Implementation of simulation technologies to better evaluate and optimize environmental sustainability parameters of the disassembly process has been deemed by the Lab operators as a useful and feasible possibility. Nonetheless, according to the operators in the Lab, the Technology Readiness Level of the process is still too low for having useful results from Implementing simulation, according to Lab operators.
- Finally, Augmented reality and Virtual reality have been deemed as unuseful for a so low technology readiness level.

After the validation, the TT-WEEE-RT has been evaluated through an interview to the Lab operators. They stated that the objective of the tool of defining the useful possible implementations under a TT perspective is clearly satisfied. Nonetheless, they complained how the tool does not provide an automated mechanism for the selection of the technologies to be implemented, and in the roadmapping phase a lot of work is left to the human effort and interpretation.

7 Conclusions and Future Research

This work tackled the problem of the environmental impacts coming from the circular practices in the WEEE sector. In order to offset these emissions, the solutions provided by the TT pathway are explored. In a first review, the current trends in terms of regulations and legislations of WEEE are analysed, to understand the direction the sector is taking. In a second review, the potentials of the digital technologies in terms of TT applications for WEEE are analysed. It emereged that environmental sustainability performance can be supported by digital technologies moreover in terms of improvement of the reverse routes selection (recycling, remanufacturing, reuse) of the components and materials according to their condition. This enables the selection of environmentally more sustainable routes, according to [48]. A further important element is the minimisation of risks for human health via automation, preventing operators from getting in touch with hazardous materials. Furthermore, it emerged that materials and energy efficiency can both be optimised through digital technologies correct implementation. Subsequently, to support companies in facing the hurdles they typically face when pursuing TT, a roadmapping tool tailored for WEEE sector is proposed. The tool is named TT-WEEE-RT. Finally, a validation of the tool in a lab-scale environment is conducted.

The work presents few limitations. First of all, the validation, by taking place in a lab-scale environment, limits the possibility of application of the tool. Indeed, the only phase of the WEEE reverse value chain (RVC) that can be considered is the disassembly. Furthermore, the possible evaluations that can emerge from the roadmapping phase are limited by the research-oriented environment where the validation takes place. Cues for future research may include two main possibilities. First, as emerged during the evaluation phase, it would be interesting to modify the roadmapping phase of the tool to make the evaluations more automatic, relying less on human opinions and subjectivity. Furthermore, given the adaptability that the TTCRT, conceived for the cosmetics industry, has proved to have in this work in being adapted to different industries and contexts, it would be interesting trying to adapt this typology of tool to further sectors, supporting TT in other industries too.

Acknowledgments. Lorenzo Gandini, PhD Student at Politecnico di Milano, Department of Management, Economics, and Industrial Engineering, and at Università di Pavia, provided a necessary support in the phase of validation of the proposed roadmap. Ricardo Magdalena Zerzuela, post-doctoral reasearcher at Universidad de Zaragoza, ENERGAIA, performed the validation of the phases of the WEEE Reverse value chain. This work has received funding from the European Union's Horizon 2020 research and innovation programme under **grant agreement No 101003587**. In any case, the present work cannot be considered as an official position of the supporting organization, but it reports just the point of view of the authors.

Co-funded by
the European Union

References

1. Shittu, O.S., Williams, I.D., Shaw, P.J.: Global E-waste management: can WEEE make a difference? A review of e-waste trends, legislation, contemporary issues and future challenges. Waste Manag. **120**, 549–563 (2021)
2. Rosa, P., Terzi, S.: Improving end of life vehicle's management practices: an economic assessment through system dynamics. J. Clean. Prod. **184**, 520–536 (2018)
3. Işıldar, A., et al.: Biotechnological strategies for the recovery of valuable and critical raw materials from waste electrical and electronic equipment (WEEE)– a review. J. Hazard. Mater. **362**, 467–481 (2019)
4. Commission, E.: Study on the Critical Raw Materials for the EU 2023 Final Report. European Commission (2023). https://doi.org/10.2873/725585
5. Graedel, T.E., et al.: Methodology of metal criticality determination. Environ. Sci. Technol. **46**, 1063–1070 (2012)

6. Bach, V., et al.: Integrated method to assess resource efficiency – ESSENZ. J. Clean. Prod. **137**, 118–130 (2016)
7. Cimprich, A., et al.: The role of industrial actors in the circular economy for critical raw materials: a framework with case studies across a range of industries. Miner. Econ. **36**, 301–319 (2023)
8. Tâm, T., Và, N. C. Ứ U., Giao, C. Ể N., Ngh, C. & Chu, Ả N B Ụ I. 済無No Title No Title No Title. **01**, 1–23 (2016)
9. Xu, Y., Zhang, L., Yeh, C.H., Liu, Y.: Evaluating WEEE recycling innovation strategies with interacting sustainability-related criteria. J. Clean. Prod. **190**, 618–629 (2018)
10. Ghosh, B., Ghosh, M.K., Parhi, P., Mukherjee, P.S., Mishra, B.K.: Waste printed circuit boards recycling: an extensive assessment of current status. J. Clean. Prod. **94**, 5–19 (2015)
11. Cardamone, G.F., Ardolino, F., Arena, U.: About the environmental sustainability of the European management of WEEE plastics. Waste Manag. **126**, 119–132 (2021)
12. Magrini, C., et al.: Using internet of things and distributed ledger technology for digital circular economy enablement: the case of electronic equipment. Sustain. **13** (2021)
13. Wang, X.V., Wang, L.: Digital twin-based WEEE recycling, recovery and remanufacturingin the background of Industry 4.0. Int. J. Prod. Res. **57**, 3892–3902 (2019)
14. de Oliveira Neto, G.C., et al.: Simulation of electronic waste reverse chains for the sao paulo circular economy: an artificial intelligence-based approach for economic and environmental optimizations. Sensors **23**, 9046 (2023)
15. Lu, Y., Pei, W. Peng, K.: State of the art of automatic disassembly of WEEE and perspective towards intelligent recycling in the era of Industry 4.0. Int. J. Adv. Manuf. Technol. **128**, 2825–2843 (2023) https://doi.org/10.1007/s00170-023-12043-3
16. Montresor, S. Vezzani, A.: Digital technologies and eco-innovation. Evidence of the twin transition from Italian firms. In: DRUID Conference 2022, pp. 1–29 (2022)
17. Ortega-Gras, J.J., Bueno-Delgado, M.V., Cañavate-Cruzado, G. Garrido- Lova, J.: Twin transition through the implementation of industry 4.0 technologies: desk-research analysis and practical use cases in Europe. Sustainability**13** (2021)
18. Stentoft, J., Jensen, K.W., Philipsen, K. Haug, A.: Drivers and barriers for industry 4.0 readiness and practice: A SME perspective with empirical evidence. Proc. Annu. Hawaii Int. Conf. Syst. Sci, pp. 5155–5164 (2019)
19. Horváth, D., Szabó, R.Z.: Driving forces and barriers of Industry 4.0: do multinational and small and medium-sized companies have equal opportunities? Technol. Forecast. Soc. Change **146**, 119–132 (2019)
20. Schumacher, A., Nemeth, T., Sihn, W.: Roadmapping towards industrial digitalization based on an Industry 4.0 maturity model for manufacturing enterprises. Procedia CIRP **79**, 409–414 (2019)
21. Perossa, D., Acerbi, F., Rocca, R., Fumagalli, L., Taisch, M.: Twin Transition cosmetic roadmapping tool for supporting cosmetics manufacturing. Clean. Environ. Syst. **11** (2023)
22. Peffers, K., Tuunanen, T., Rothenberger, M.A., Chatterjee, S.: A design science research methodology for information systems research. J. Manag. Inf. Syst. **24**, 45–77 (2007)
23. European Commission. (Text with EEA relevance) 21.6.2017. **2016**, 48–119 (2018)
24. Journal, O. 2023/2585. **2585**, 7–13 (2023)
25. Parliament, E.: Directive 2012/19/EU of the EUROPEAN parliament and of the council of 4 July 2012 on waste electrical and electronic equipment (WEEE). Off. J. Eur. Union **55**, 38–71 (2012)
26. Commission, E.: Directive 2008/98/EC of the European parliament and of the council of 19 november 2008 on waste and repealing certain directives. Off. J. Eur. Union **L312**, 1–59 (2008)
27. Council, O. F. T. H. E. L 285/10. 10–35 (2009)

28. Union, T., Journal, O. & Union, E. (recast) (Text with EEA relevance). Ec2074/2005 10, 1–21 (2017)
29. Wang, X.V., Wang, L.: A cloud-based production systemfor informationand service integration: an internet of things case study on waste electronics. Enterp. Inf. Syst. **11**, 952–968 (2017)
30. Rocca, R., Rosa, P., Sassanelli, C., Fumagalli, L. Terzi, S.: Industry 4.0 solutions supporting Circular Economy. In: Proceedings of 2020 IEEE International Conference on Engineering. Technology InnovationICE/ITMC2020 (2020).https://doi.org/10.1109/ICE/ITMC49519.2020.9198517
31. Lee, Y. C. J. J. A. G. Science and Technologies for Smart Cities. (2022)
32. Rocca, R., Rosa, P., Sassanelli, C., Fumagalli, L. Terzi, S.: Integratingvirtual reality and digital twin in circular economy practices: a laboratory application case. Sustainability **12** (2020)
33. Nguyen, H.H., Bazhin, V.Y.: Optimization of the control system for electrolytic copper refining with digital twin during dendritic precipitation. Metallurgist **67**, 41–50 (2023)
34. Reuter, M.A., van Schaik, A., Gediga, J.: Simulation-based design for resource efficiency of metal production and recycling systems: cases - copper production and recycling, e-waste (LED lamps) and nickel pig iron. Int. J. Life Cycle Assess. **20**, 671–693 (2015)
35. Sureshkumar, S., Rani, P.K., Agash, C.P., Kumar, B.A. Kaviyaraj, R.: Augmented reality and waste reduction: enhancing the recycling process for mobile e-waste in automotive manufacturing. In: 2023 9th International Confernce Advance Computer Communication System ICACCS 2023, pp. 549–553 (2023). https://doi.org/10.1109/ICACCS57279.2023.10112913
36. Efstratiadis, V.S. Michailidis, N.: Sustainable recovery, recycle of critical metals and rare earth elements from waste electric and electronic equipment (circuits, solar, wind) and their reusability in additive manufacturing applications: a review. Metals (Basel). **12** (2022)
37. Laskurain-Iturbe, I., Arana-Landín, G., Landeta-Manzano, B., Uriarte- Gallastegi, N.: Exploring the influence of industry 4.0 technologies on the circular economy. J. Clean. Prod. **321** (2021)
38. Toffolet, R. WEEE Management. WEEE Recycl. Res. Dev. Policies 1–30 (2016). https://doi.org/10.1016/B978-0-12-803363-0.00001-8
39. Jin, G., Li, W., Wang, S. Gao, S.: A systematic selective disassembly approach for Waste Electrical and Electronic Equipment with case study on liquid crystal display televisions. Proc. Inst. Mech. Eng. Part B J. Eng. Manuf. 231, 2261–2278 (2017)
40. Batinic, B., et al.: Applied WEEE pre-treatment methods: opportunities to maximizing the recovery of critical metals. Glob. Nest J. **20**, 706–711 (2018)
41. Zlamparet, G.I., et al.: Remanufacturing strategies: a solution for WEEE problem. J. Clean. Prod. **149**, 126–136 (2017)
42. Wang, R., Xu, Z.: Recycling of non-metallic fractions from waste electrical and electronic equipment (WEEE): a review. Waste Manag. **34**, 1455–1469 (2014)
43. Guo, X., Zhang, Y. Xu, K.: Metallurgical recovery of metals from waste electrical and electronic equipment (WEEE) in PRC. Met. Sustain. Glob. Challenges, Consequences, Prospect. 151–168 (2016)https://doi.org/10.1002/9781119009115.ch7
44. Baumgartner, R.J., Ebner, D.: Corporate sustainability strategies: sustainability profiles and maturity levels. Sustain. Dev. **18**, 76–89 (2010)
45. Machado, C.G., Pinheiro de Lima, E., Gouvea da Costa, S.E., Angelis, J.J. Mattioda, R.A.: Framingmaturity basedon sustainable operations management principles. Int. J. Prod. Econ. **190**, 3–21 (2017)
46. Schaltegger, S., Lüdeke-Freund, F., Hansen, E.G.: Business cases for sustainability: the role of business model innovation for corporate sustainability. Int. J. Innov. Sustain. Dev. **6**, 95–119 (2012)

47. Santos, R.C. Martinho, J.L. An Industry 4.0 maturity model proposal. J. Manuf. Technol. Manag. **31**, 1023–1043 (2020)
48. EllenMacArthurFoundation (2012).https://www.ellenmacarthurfoundation.org/circular-economy-diagram. Accessed 5 Feb 2024

Virtual Voyages: Enhancing Cultural Heritage Tourism Through VR360 Experiences of Wat Sri Suphan

Suepphong Chernbumroong[1], Perasuk Worragin[1], Supicha Niemsup[2], and Kitti Puritat[2](✉)

[1] College of Arts, Media and Technology, Chiang Mai University, Chiang Mai, Thailand
[2] Department of Library and Information Science, Faculty of Humanities, Chiang Mai University, Chiang Mai, Thailand
kitti.p@cmu.ac.th

Abstract. This paper explores the development and impact of a Virtual Reality (VR) 360 experience centered around Wat Sri Suphan, also known as the Silver Temple, located in Chiang Mai, Thailand. Recognized for its cultural and spiritual heritage, the temple serves as an exemplary case study for leveraging VR technology to transcend geographical barriers, thus making cultural heritage sites more accessible to a global audience. Through a comprehensive methodology that includes design, implementation, user engagement, and educational impact analysis, the paper presents a framework for creating immersive and educational virtual experiences. Results from user engagement and educational impact questionnaires reveal moderate levels of immersion and interactivity, alongside a significant educational impact in enhancing participants' understanding of Wat Sri Suphan's cultural and historical significance. Despite limitations such as a small participant sample size and varied familiarity with VR technology, the study underscores the potential of VR in cultural heritage tourism. It highlights opportunities for future research, particularly in enriching content interactivity and broadening participant diversity. This VR360 project not only contributes to the field of digital heritage preservation but also opens new pathways for cultural exchange, education, and tourism, inviting a reimagining of how we interact with and preserve cultural landmarks worldwide.

Keywords: VR360 · Culture Heritage · Tourism · Virtual Voyages · Wat Sri Suphan

1 Introduction

In the heart of Chiang Mai, Thailand, lies Wat Sri Suphan, a monument that stands as a testament to the region's rich cultural and spiritual heritage. Known colloquially as the Silver Temple, its intricate carvings and use of silver in construction not only showcase exceptional craftsmanship but also symbolize the community's devotion and

Published by Springer Nature Switzerland AG 2025
P. Sureephong et al. (Eds.): PLM 2024, IFIP AICT 740, pp. 271–285, 2025.
https://doi.org/10.1007/978-3-031-93319-6_21

artistic heritage. With the global interest in immersive and authentic cultural experiences escalating, especially in the wake of COVID-19's unprecedented impact on global travel and tourism, the potential of Virtual Reality (VR) technology to bridge the gap between distant audiences and heritage sites has become strikingly evident. This paper explores the design and implementation of a VR360 experience aimed at bringing the splendor of Wat Sri Suphan to a global audience, thereby enhancing cultural preservation, education, and tourism in a time when traditional travel has faced significant challenges.

The advent of VR360 technology offers an unprecedented opportunity to explore cultural heritage sites in a deeply immersive and interactive manner. By allowing users to navigate and experience spaces as if they were physically present, VR360 opens new avenues for education, preservation, and appreciation of historical and cultural nuances that might otherwise remain inaccessible, particularly in an era where the COVID-19 pandemic [10, 11] has limited physical access to such sites. This project seeks to harness the power of VR to transcend geographical boundaries and time zones, enabling people around the world to explore Wat Sri Suphan in all its intricacy and beauty, from the safety and comfort of their own homes.

The creation of a VR360 experience for Wat Sri Suphan presents unique challenges and opportunities, including faithfully representing the temple's architectural and artistic details, ensuring cultural sensitivity and accuracy, and creating an engaging narrative that conveys the site's historical and cultural significance. Through a careful balance of technological innovation and respect for tradition, this paper outlines a framework for developing a VR experience that not only captivates and educates but also fosters a deeper appreciation for Thailand's cultural heritage amid a global pandemic.

In detailing the process from conceptualization through to execution, this paper aims to contribute to the growing field of digital heritage preservation. It underscores the vital role of emerging technologies in making cultural heritage more accessible and engaging to a global audience, thereby opening new pathways for cultural exchange, education, and tourism in the context of the new normal. Through the lens of the Wat Sri Suphan VR360 project, we delve into the possibilities and implications of virtual reality in bridging the past with the future, inviting a reimagining of how we interact with and preserve the cultural landmarks that shape our world, especially during and after the challenges presented by COVID-19.

2 Literature Review

The integration of virtual reality technologies with tourism for cultural heritage has brought about new opportunities for tourists to engage in immersive and engaging experiences. This review of the literature studies the integration of virtual reality technologies, tourism, and cultural heritage, with an emphasis on the use of VR360 experiences to increase tourist participation and advertise historical and cultural attractions.

2.1 Virtual Reality

Virtual reality is a 3D simulation on a computer that creates an appearance of realism. Virtual reality applications in tourism have been proposed in a variety of situations.

Suggested applications include assessing visitor demands and requirements employing entertainment in theme parks, and academic objectives in museums and sites of history [12]. Research supports the theory those who experience favorable sensations and enhanced awareness towards their destination. Virtual reality tourism is described as a "virtual depiction of an existing attraction, location, or experience for tourists intended to serve an introduction to vacation or to enhance people' prior adventures [13]." It plays a significant role in cultural and heritage tourism, greatly impacting tourist contentment and loyalty after their journey [14]. Several cultural and historical destinations have expanded their interactive profiles by including virtual reality applications to improve the experience of visitors before, thereafter, and throughout their trip [15]. Tussyadiah et al. (2018) [16] found that engaging in a virtual environment offers a favorable impression for prospective visitors, increasing their liking for the location and the probability of traveling to that location. Research has focused on the aspects that influence virtual reality visitors' decision to make trips to museums or cultural institutions given that they are more inclined to succeed in doing so [17]. Kim, Bonn, et al. (2017) [18] revealed that visitors reported higher pleasure and favorable intentions for behavior when their encounters were mediated by technology and considered as real. Yung and Khoo-Lattimore (2019) [19] also discovered that authenticity improves pleasure in virtual reality tourist experiences. Consequently, it was apparent that if cultural heritage for tourism was to remain successful, they would need to make investment in technology innovation—specifically, virtual tours.

2.2 Virtual Tour

A virtual tour is a recreation of a real-world place using video images [20]. A series of motion pictures or images, which are augmented by written narratives, voice guidance, or sound effects is the stimulation of virtual tour. The simulated place is designed to accurately recreate the real-site scenario [21]. According to Mah et al. (2019) [22], attraction managers prefer virtual tours for the reason that they call for minimal investment in money, provide resources for studying, and convey both tangible and intangible qualities to tourists. A study by Carvajal, Morita, and Bilmes (2020) [23], museums aim to engage visitors and offer access to knowledge, and virtual tours are a promising means to achieve this objective. Several world-class museums have utilized virtual tour to provide worldwide access to their collections. In 2019, the Louvre Museum collaborated with HTC Vive Arts [24], a digital heritage preservation firm [25], in commemorating the death of Leonardo DaVinci in the 500th anniversary through the use of virtual reality. The Singapore Tourism Board (2020) [26] launched the "Experience Singapore now. Visit later" campaign, enabling travelers worldwide to digitally tour the city-state's prominent sites and neighborhoods. Virtual tours can prevent tourist activities from being suspended throughout times of crisis [27]. They create novel approaches to business and possibilities for diverse organizations in the tourism industry. Various opportunities arise from these developments, such as allowing travelers to explore and educate themselves about diverse destinations even amidst travel restrictions and prohibitions, all from the safety and comfort of their homes [28]. Additionally, they enable museums and various tourist sites to maintain active connections with their audiences [29], while also offering avenues for employment, such as virtual tour guiding roles, for workers like tour guides [30].

2.3 360° Immersive Virtual Reality

The rise in popularity of immersive experiences through three hundred sixty–degree (360°) videos coincided with the availability of affordable VR headsets designed for consumer use, which facilitate a broader field of view and stereoscopic display capabilities [31]. 360° immersive video empowers viewers to actively select their own perspective when experiencing a scene, rather than merely adhering to the director's fixed frame. This capability fosters a personalized and lifelike experience for the viewer [32]. 360° video VR storytelling experiences enable producers to generate a more enduring impact on the viewers through semi-interactive experiences that bridge the gap between game design and film production. It is accomplished by immersing users in the action and narrative, fostering a sense of active participation [33]. Ivkovic et al. (2018) [34] developed an interactive cultural heritage application utilizing 360° videos centered on the Bridges of Sarajevo. Users were encouraged to delve deeper into the narrative of the seven bridges by having the freedom to select the sequence in which they experienced the cultural information. The application incorporated a map interface menu alongside 3D models of the Bridges of Sarajevo, serving as a navigational aid within the environment. According to this study's findings, participants expressed a sense of actually traversing the bridges as they engaged with the video narratives [34]. Based on their work, Selmanovic et al. (2018) [35] present an immersive 360° video application to conserve the bridge diving tradition associated with the Old Bridge in Mostar, Bosnia and Herzegovina. Users engage with 360° videos depicting the bridge's history and the diving tradition, followed by quizzes based on the presented content. While this approach proves effective in crafting immersive cultural heritage solutions experienced through VR headsets, it falls short in establishing a systematic design methodology and evaluating additional motivational and immersion factors that call for consideration.

2.4 Design and Implement of the VR360 Experience for Wat Sri Suphan

Creating a VR360 experience that accurately reflects the cultural and spiritual essence of Wat Sri Suphan necessitated meticulous design and thoughtful implementation. This section outlines the critical phases of the process, ranging from initial design considerations to the final implementation of the VR experience. We utilized methodologies [1, 2] to develop a virtual reality framework tailored for cultural heritage. This framework is comprised of four phases: Design Considerations, Content Creation, Technical Implementation, and Launch and Evaluation. An overview of the Design and Implementation methodology is presented in Fig. 1, with each phase described in detail below:

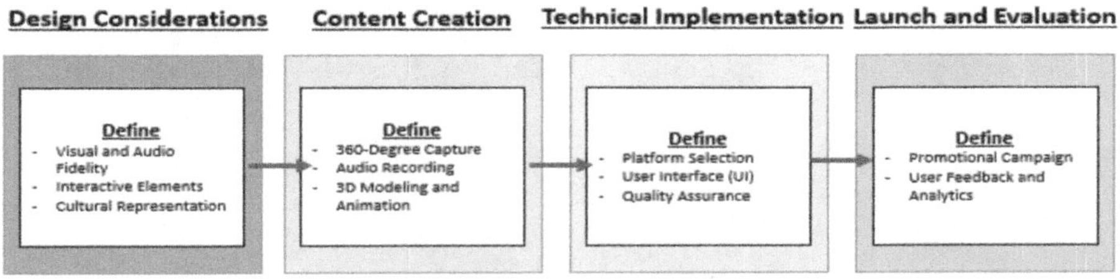

Fig. 1. Overview of Design and Implement methodology.

2.5 Design Considerations Phase

In the design phase, the focus was on striking a delicate balance between technological innovation, user engagement, and cultural authenticity. To achieve the highest level of visual and audio fidelity, state-of-the-art equipment was employed to capture 360-degree visuals and spatial audio within the temple's premises, ensuring an immersive and authentic experience. The integration of interactive elements was pivotal; information hotspots, clickable 3D models of artifacts, and narrated tours were meticulously designed to enrich the learning experience of users. Moreover, in a concerted effort to faithfully represent the temple's cultural essence, we worked closely with cultural consultants to carefully select visuals, narratives, and music that would respect and accurately convey the temple's spiritual significance and historical context. This approach was underpinned by the addition of audio, video, and text elements to the 3D virtual environment, which was considered crucial for designing the concept and enhancing the overall immersive experience.

2.6 Content Creation

The development of content for the VR360 experience at Wat Sri Suphan was characterized by a collaborative and iterative approach, aimed at fostering an engaging and respectful portrayal of this cultural site. Professional photographers and videographers conducted comprehensive sessions within and surrounding the temple premises. They utilized their specialized skills to capture the site in its entirety through 360° visuals, as depicted in Fig. 2. Particular emphasis was placed on lighting conditions and managing visitor flow to minimize disruptions and ensure the highest visual quality.

In conjunction with visual documentation, ambient sounds, chants, and narratives were recorded on-site. This auditory layer contributes to a rich backdrop, significantly enhancing the sense of presence within the virtual environment. Additionally, intricate 3D models of artifacts and architectural elements requiring closer examination or those less accessible to the public were elaborately designed. These models were smoothly incorporated into the VR landscape, enabling users to freely explore these elements devoid of physical constraints. This meticulous integration of auditory and visual elements serves to deepen the immersive quality of the VR experience, allowing for a comprehensive exploration of Wat Sri Suphan's cultural heritage.

Fig. 2. Content Resource for the VR360 Experience

2.7 Technical Implementation

The technical implementation of the VR360 experience for Wat Sri Suphan involved a careful selection of platforms and tools to deliver the meticulously designed virtual environment to users, with a focus on compatibility, accessibility, and ease of use. In this step, we utilized Pano2VR to create the VR360 content. To ensure widespread accessibility, the experience was developed for various platforms, including both standalone VR headsets and web-based VR. This multifaceted approach required specific optimizations for each platform to maintain performance standards and visual integrity.

A pivotal aspect of the project was the creation of a user-friendly interface, meticulously designed to facilitate seamless navigation through the VR experience. This interface was intentionally intuitive, allowing users to freely explore the temple's virtual representation in ways that suited their personal interests. Supporting these efforts was a comprehensive quality assurance and testing regimen implemented across all platforms. This ensured the technical stability of the experience and user comfort and satisfaction.

Fig. 3. (Left) The VR360 experience at Wat Sri Suphan; (Right) Information provided on cultural heritage.

Feedback collected during the initial user testing phases was instrumental in refining and improving the final product, thereby guaranteeing a high-quality experience for all users. An example of the VR360 experience at Wat Sri Suphan is illustrated in Fig. 3.

2.8 Launch and Evaluation

The unveiling of the VR360 experience dedicated to Wat Sri Suphan transcended merely presenting the experience to the public; it marked the initiation of an extensive evaluation strategy aimed at assessing user engagement, educational impact, and overall satisfaction. A targeted promotional campaign was launched to acquaint potential users with the VR experience, specifically reaching out to educational institutions, cultural aficionados, and the broader public interested in virtual tourism. Following the launch, a methodical approach was employed to gather user feedback, serving as a pivotal metric for evaluating the impact of the experience. Furthermore, analytical tools were utilized to gain deeper insights into user behavior within the VR environment. This analysis was instrumental in identifying the experience's most engaging aspects and pinpointing opportunities for further refinements. By combining direct feedback from users with analytical data, a comprehensive understanding of the experience's efficacy and areas necessitating improvement was achieved. The entirety of the VR360 experience is depicted in Fig. 4.

Fig. 4. The VR360 Experience: User-Friendly Interaction with Wat Sri Suphan

3 Methodology Design

The development of the VR360 experience for Wat Sri Suphan leverages the Research and Development (R&D) Methodology [3, 4], a systematic framework that combines rigorous investigation with the application of innovative technologies and design principles.

This approach enables us to create a virtual experience that is not only technologically advanced but also deeply immersive and culturally authentic.

3.1 Participants

The study engaged 25 participants, all of whom were tourists with varying degrees of familiarity with VR technology and interest in cultural heritage sites. These participants were selected to represent a broad spectrum of the potential audience for the VR360 experience, aiming to gather diverse insights into user engagement, usability, and the educational impact of the VR experience. Criteria for selection included age, technological proficiency, and prior experience with cultural tourism, ensuring a representative sample that could provide valuable feedback on the VR experience from multiple perspectives.

3.2 Instruments and Measurement

VR360 application for Wat Sri Suphan

The VR360 Application for Wat Sri Suphan, an immersive cultural exploration tool, combines high-resolution 360° visuals and spatial audio to transport users directly into the heart of one of Thailand's most revered cultural landmarks. Designed with the dual objectives of engagement and education, the application offers an unparalleled virtual journey through the Silver Temple's stunning architecture and intricate craftsmanship. It enriches users' understanding by providing in-depth historical and cultural context, interactive elements for learning about the temple's artifacts, and narratives that bring to life the stories behind its construction and significance. This innovative application stands as a testament to the power of VR technology in making cultural heritage accessible to a global audience, promoting appreciation and understanding across cultural boundaries.

Engagement and Educational Impact Survey

We adapted the post-experience questionnaires, as designed in [5], to measure the level of engagement and the perceived educational value by the participants. The questions were crafted to focus on the immersive quality of the experience, the rate of information retention, and the overall educational impact of the VR experience. The detailed questionnaire is provided in Appendix A for reference.

3.3 Procedure

The procedure adopted in our methodology was carefully crafted to ensure a comprehensive evaluation of the VR360 experience of Wat Sri Suphan. Upon arrival, participants were provided with a detailed briefing, which included signing a consent form, an explanation of the study's objectives, an overview of the VR experience, and an introduction to the data collection instruments they would encounter. Following this, participants individually engaged with the VR360 experience, utilizing Meta Quest 2 headsets for approximately 20 min. This setup allowed them to navigate the virtual environment of Wat Sri Suphan at their own pace. Immediately after this immersive experience, participants were requested to complete the Usability Scale and the Engagement and

Educational Impact Survey, thereby offering immediate feedback on their experience. In parallel, researchers meticulously documented observational data during the VR sessions, aiming to supplement the survey findings with direct observations of user behavior and engagement levels. The concluding stage entailed a detailed analysis of all collected data, including survey responses and observational notes. This process was undertaken to identify patterns, extract insights, and highlight areas needing further enhancement in the VR360 experience.

4 Result and Data Analysis

The integration of virtual reality technologies with tourism for cultural heritage has brought about new opportunities for tourists to engage in immersive and engaging experiences. This review of the literature studies the integration of virtual reality technologies, tourism, and cultural heritage, with an emphasis on the use of VR360 experiences to increase tourist participation and advertise historical and cultural attractions.

4.1 Result of User Engagement

The evaluation of user engagement within the VR360 experience of Wat Sri Suphan yielded insightful feedback from participants, as detailed in Table 1 and illustrated in Fig. 5. The immersion factor, deemed essential for a successful VR experience, achieved a moderate mean score of 3.04 (SD = 0.81). This indicates that, although participants found the experience to be somewhat immersive, there is significant room for enhancement to further captivate the audience. Interactivity, another critical component of engaging VR experiences, received a slightly lower mean score of 2.88 (SD = 0.88), highlighting a distinct opportunity to incorporate more interactive and dynamic elements to create a more engaging and participatory environment. The sense of presence, a measure of how effectively the VR environment replicates real-world presence, garnered a mean score of 2.96 (SD = 0.84), suggesting a necessity for improvements to make the virtual space feel more encompassing and lifelike. However, participants' motivation to learn more about Wat Sri Suphan after the experience was notably stronger, with a mean score of 3.44 (SD = 0.77). This suggests that, despite certain areas requiring enhancements, the VR experience was successful in igniting interest in the cultural and historical significance of Wat Sri Suphan among the participants.

Table 1. Result of User Engagement Questionnaire.

No	Question	Mean (SD)
1	How immersive did you find the VR360 experience of Wat Sri Suphan?	3.04(0.81)
2	How would you rate the interactivity of the VR experience?	2.88(0.88)
3	Did you feel a sense of presence within the virtual environment?	2.96(0.84)
4	Were you compelled to explore more about Wat Sri Suphan after the VR experience?	3.44(0.77)

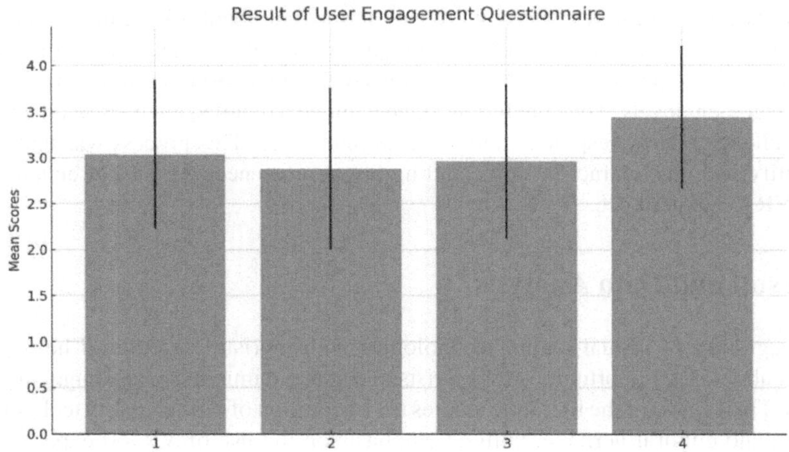

Fig. 5. Diagram of Results from the User Engagement Questionnaire

4.2 Result of Educational Impact

The evaluation of the educational impact of the VR360 experience on participants' understanding of Wat Sri Suphan's history, culture, and architecture offered promising insights, as detailed in Table 2 and depicted in Fig. 6. This assessment concentrated on three principal areas: the content's informativeness, the enhancement of understanding regarding cultural significance, and the clarity in explaining architectural features. Participants awarded the content's informativeness about Wat Sri Suphan's history and culture a mean score of 3.72 (SD = 1.02). This score indicates a relatively high level of satisfaction with the VR experience's conveyance of historical and cultural knowledge, suggesting that the content effectively acquainted participants with the temple's rich heritage.

Regarding whether the VR experience enriched participants' understanding of the temple's cultural significance, the mean score was 3.48 (SD = 1.15). Although still positive, this marginally lower score, in comparison to the informativeness rating, highlights areas for potential enhancement, particularly in deepening participants' comprehension

Table 2. Result of Educational Impact Questionnaire.

No	Question	Mean (SD)
1	How informative did you find the content regarding the history and culture of Wat Sri Suphan?	3.72(1.02)
2	Did the VR experience enhance your understanding of Wat Sri Suphan's cultural significance?	3.48(1.15)
3	How well did the VR experience explain the architectural features of Wat Sri Suphan?	3.56(1.04)

of the temple's cultural relevance. Additionally, the evaluation of how well the VR experience elucidated Wat Sri Suphan's architectural features resulted in a mean score of 3.56 (SD = 1.04). This outcome signifies a favorable response to the architectural insights offered by the VR experience, indicating that participants considered the virtual tour to be educational in gaining an understanding of the temple's architectural nuances.

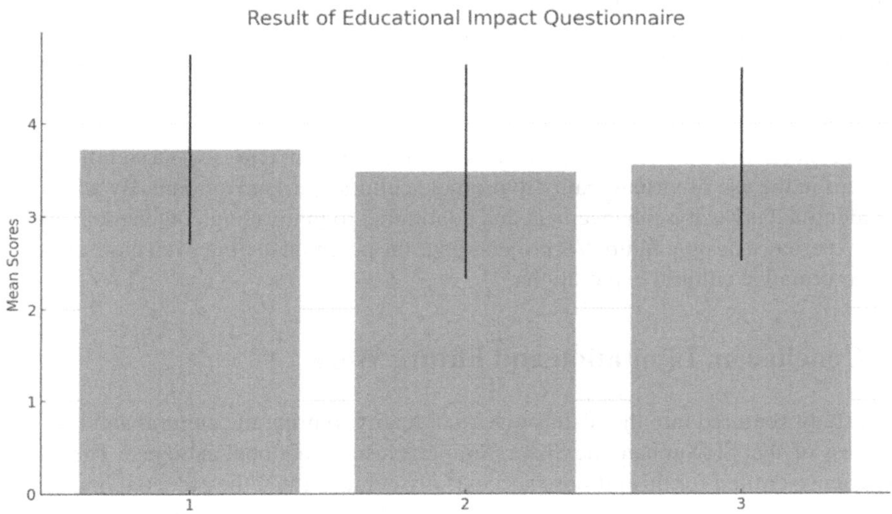

Fig. 6. Diagram of Results from the Educational Impact Questionnaire

5 Discussion

The results of the user engagement and educational impact questionnaires offer valuable insights into the VR360 experience of Wat Sri Suphan, highlighting both its strengths and areas for improvement. The moderate levels of immersion, interactivity, and sense of presence reported by participants suggest that while the VR experience successfully introduced users to the cultural and architectural splendor of Wat Sri Suphan, there are opportunities to enhance the depth and realism of the virtual environment. Notably, the higher motivation to explore more about Wat Sri Suphan post-experience indicates the VR's potential to spark interest in cultural heritage.

The educational impact of the VR experience, as evidenced by the relatively high satisfaction with the content's informativeness and its effectiveness in enhancing understanding of the temple's cultural significance, underscores the potential of VR technology as a powerful tool for cultural education. However, the slight disparity in scores concerning the enhancement of understanding and explanation of architectural features points to the need for more detailed and engaging educational content. These findings align with existing literature that emphasizes the importance of immersive quality and interactivity in maximizing the educational and engagement potential of VR experiences in cultural heritage tourism [6, 7]. Furthermore, they contribute to the growing body of research advocating for VR as a means to transcend geographical barriers and make cultural heritage more accessible to a global audience [8, 9].

Despite the promising results, the study faces limitations, including the relatively small sample size and the participants' varied familiarity with VR technology, which may influence their perceptions of the VR experience. Future work should therefore aim to include a larger, more diverse participant pool and explore the integration of advanced interactive features and narrative techniques to further enhance user engagement and educational outcomes. Additionally, longitudinal studies could provide insights into the long-term impact of VR experiences on users' interest and understanding of cultural heritage.

In conclusion, the VR360 experience of Wat Sri Suphan represents a significant step forward in the use of virtual reality to promote cultural heritage tourism. By addressing the identified areas for enhancement and continuing to innovate in content delivery and user experience design, future VR projects have the potential to offer even more engaging and informative cultural explorations.

6 Conclusion, Limitation and Future Work

This study ventured into the realms of virtual reality to bring the cultural and historical richness of Wat Sri Suphan, the Silver Temple, closer to a global audience. The VR360 experience crafted for this purpose successfully demonstrated the potential of VR technology to enhance cultural heritage tourism by offering an immersive, educational, and engaging exploration of significant cultural sites. Participants' feedback highlighted the effectiveness of the VR experience in fostering a deeper appreciation and understanding of Wat Sri Suphan's cultural heritage, underscoring the invaluable role of VR in making cultural education more accessible and engaging.

Despite the promising outcomes, this study is not without its limitations. The participant sample size was relatively small and may not fully represent the global diversity of potential users of the VR360 experience. Additionally, the varying levels of participants' familiarity with VR technology could have influenced their perceptions and feedback. These factors limit the generalizability of the study's findings and underscore the need for caution in interpreting the results.

Looking ahead, there are several avenues for further research and development to enhance the VR360 experience of cultural heritage sites like Wat Sri Suphan. Future studies should aim to involve a larger and more diverse sample of participants to validate and extend the findings of this study. Exploring advanced interactive features, such as augmented reality (AR) elements and gamification, could offer more engaging and informative experiences. Moreover, developing narrative-driven content that weaves the cultural and historical narratives of heritage sites into compelling storylines may further enhance user engagement and educational impact.

Additionally, longitudinal research could shed light on the long-term effects of VR experiences on users' interest in and understanding of cultural heritage, providing valuable insights for the continuous improvement of VR applications in cultural tourism. Finally, expanding the scope to include multiple cultural heritage sites across different regions could offer comparative insights and foster a more inclusive appreciation of global cultural heritage through virtual exploration.

Appendix

Dimension	Questionnaire
Engagement	How immersive did you find the VR360 experience of Wat Sri Suphan?
	How would you rate the interactivity of the VR experience?
	Did you feel a sense of presence within the virtual environment?
	Were you compelled to explore more about Wat Sri Suphan after the VR experience?
Educational Impact	How informative did you find the content regarding the history and culture of Wat Sri Suphan?
	Did the VR experience enhance your understanding of Wat Sri Suphan's cultural significance?
	How well did the VR experience explain the architectural features of Wat Sri Suphan?

References

1. Montagud, M., Orero, P., Matamala, A.: Culture 4 all: accessibility-enabled cultural experiences through immersive VR360 content. Pers. Ubiquitous Comput. **24**, 887–905 (2020). https://doi.org/10.1007/s00779-019-01357-3
2. Balbi, B., Marasco, A.: Co-designing cultural heritage experiences for all with virtual reality: a scenario-based design approach. Umanistica Digitale. 71–98 (2022). https://doi.org/10.6092/ISSN.2532-8816/13686
3. Barrientos, F., Martin, J., De Luca, C., Tondelli, S., Gómez-García-Bermejo, J., Casanova, E.Z.: Computational methods and rural cultural & natural heritage: a review. J. Cult. Herit. **49**, 250–259 (2021). https://doi.org/10.1016/j.culher.2021.03.009
4. Intawong, K., Scuturici, M., Miguet, S.: A new pixel-based quality measure for segmentation algorithms integrating precision, recall and specificity. In: Computer Analysis of Images and Patterns, pp. 188–195. Springer, Heidelberg (2013)
5. Menkshi, E., Braholli, E., Çobani, S., Shehu, D.: Assessing youth engagement in the preservation and promotion of culture heritage: a case study in Korça city. Albania. Quaest. Geogr. **40**, 109–125 (2021). https://doi.org/10.2478/quageo-2021-0009
6. Han, D.-I.D., Weber, J., Bastiaansen, M., Mitas, O., Lub, X.: Virtual and augmented reality technologies to enhance the visitor experience in cultural tourism. In: Augmented Reality and Virtual Reality, pp. 113–128. Springer, Cham (2019)
7. Chernbumroong, S., Ariya, P., Yolthasart, S., Wongwan, N., Intawong, K., Puritat, K.: Comparing the impact of non-gamified and gamified virtual reality in digital twin virtual museum environments: A case study of Wieng Yong House Museum. Thailand. Heritage. **7**, 1870–1892 (2024). https://doi.org/10.3390/heritage7040089

8. Atzeni, M., Del Chiappa, G., Mei Pung, J.: Enhancing visit intention in heritage tourism: the role of object-based and existential authenticity in non-immersive virtual reality heritage experiences. Int. J. Tourism Res. **24**, 240–255 (2022). https://doi.org/10.1002/jtr.2497
9. Park, H., Kim, J., Bang, S., Woo, W.: The effect of applying film-induced tourism to virtual reality tours of cultural heritage sites. In: 2018 3rd Digital Heritage International Congress (DigitalHERITAGE) held Jointly with 2018 24th International Conference on Virtual Systems & Multimedia (VSMM 2018). IEEE (2018)
10. Intawong, K., Olson, D., Chariyalertsak, S.: Application technology to fight the COVID-19 pandemic: lessons learned in Thailand. Biochem. Biophys. Res. Commun. **534**, 830–836 (2021). https://doi.org/10.1016/j.bbrc.2020.10.097
11. Intawong, K., Ong-artborirak, P., Boonchieng, W.: Seamless electronic medical record for health management system in emergency patients. In: 2021 Joint International Conference on Digital Arts, Media and Technology with ECTI Northern Section Conference on Electrical, Electronics, Computer and Telecommunication Engineering. IEEE (2021)
12. Guttentag, D.A.: Virtual reality: applications and implications for tourism. Tour. Manag. **31**, 637–651 (2010). https://doi.org/10.1016/j.tourman.2009.07.003
13. Kim, M.J., Lee, C.-K., Jung, T.: Exploring consumer behavior in virtual reality tourism using an extended stimulus-organism-response model. J. Travel Res. **59**, 69–89 (2020). https://doi.org/10.1177/0047287518818915
14. Lin, L.-P. (lynn), Huang, S.-C. (lucy), Ho, Y.-C.: Could virtual reality effectively market slow travel in a heritage destination? Tour. Manag. **78**, 104027 (2020). https://doi.org/10.1016/j.tourman.2019.104027
15. Errichiello, L., Micera, R., Atzeni, M., Del Chiappa, G.: Exploring the implications of wearable virtual reality technology for museum visitors' experience: a cluster analysis. Int. J. Tourism Res. **21**, 590–605 (2019). https://doi.org/10.1002/jtr.2283
16. Tussyadiah, I.P., Wang, D., Jung, T.H., tom Dieck, M.C.: Virtual reality, presence, and attitude change: empirical evidence from tourism. Tour. Manag. **66**, 140–154 (2018). https://doi.org/10.1016/j.tourman.2017.12.003
17. Lee, H., Jung, T.H., tom Dieck, M.C., Chung, N.: Experiencing immersive virtual reality in museums. Inf. Manag. **57**, 103229 (2020). https://doi.org/10.1016/j.im.2019.103229
18. Kim, M.J., Bonn, M., Lee, C.-K.: Seniors' dual route of persuasive communications in mobile social media and the moderating role of discretionary time. Asia Pac. J. Tour. Res. **22**, 799–818 (2017). https://doi.org/10.1080/10941665.2017.1331925
19. Yung, R., Khoo-Lattimore, C.: New realities: a systematic literature review on virtual reality and augmented reality in tourism research. Curr. Issues Tourism. **22**, 2056–2081 (2019). https://doi.org/10.1080/13683500.2017.1417359
20. Osman, A., Wahab, N.A., Ismail, M.H.: Development and evaluation of an interactive 360 virtual tour for tourist destinations. J. Inf. Technol. Impact. **9**, 173–182 (2009)
21. Spielmann, N., Mantonakis, A.: In virtuo: How user-driven interactivity in virtual tours leads to attitude change. J. Bus. Res. **88**, 255–264 (2018). https://doi.org/10.1016/j.jbusres.2018.03.037
22. Mah, O.B.P., et al.: Generating a virtual tour for the preservation of the (in)tangible cultural heritage of Tampines Chinese temple in Singapore. J. Cult. Herit. **39**, 202–211 (2019). https://doi.org/10.1016/j.culher.2019.04.004
23. Loaiza Carvajal, D.A., Morita, M.M., Bilmes, G.M.: Virtual museums. Captured reality and 3Dmodeling. J. Cult. Herit. **45**, 234–239 (2020). https://doi.org/10.1016/j.culher.2020.04.013
24. louvre-Auteur: The Musée du Louvre Partners with HTC Vive Arts for its First Virtual Reality Experience - Espace presse du musée du Louvre, https://presse.louvre.fr/stronglemusee-du-louvre-lance-sa-premiere-experience-de-realite-virtuelle-en-partenariat-avec- htc-vive-artsnbsp-strong-1063000191514/

25. Arts, V.: Our Mission. https://www.vivearts.com/
26. CloudPanoBYBehram, September: Virtual Tour 360 software: Explore the world from home,https://www.cloudpano.com/blog/virtual-tour-360-software-explore-the-world- from-home
27. Guha, A.: Virtual tours in the times of COVID-19: stay home, yet see the world - TAN, https://travelandynews.com/virtual-tours-in-the-times-of-covid-19-heres-how-you-can-stay-home-yet-see-the-world/
28. UNESCO: Virtual Museums on Underwater Cultural Heritage Respond to the Covid-19 Crisis, https://unesdoc.unesco.org/ark:/48223/pf0000374227
29. Sallent, M.: Tourism in Africa: Virtual safaris kick in as countries slowly open to tourists. https://www.un.org/africarenewal/web-features/coronavirus/tourism-africa-virtual-safaris-kick-countries-prepare-reopen-tourists
30. COVID-19 and opportunities for VR based tourism economy. https://www.accessibletourism.org/?i=enat.en.news.2176
31. Argyriou, L., Economou, D., Bouki, V.: Design methodology for 360° immersive video applications: the case study of a cultural heritage virtual tour. Pers. Ubiquitous Comput. **24**, 843–859 (2020). https://doi.org/10.1007/s00779-020-01373-8
32. Marougkas, A., Troussas, C., Krouska, A., Sgouropoulou, C.: Virtual reality in education: a review of learning theories, approaches and methodologies for the last decade. Electronics (Basel). 12, 2832 (2023). https://doi.org/10.3390/electronics12132832
33. Page,D.:Howvirtualrealitywillchangethecinematicexperience (2015). https://www.theguardian.com/culture-professionals-network/2015/jan/30/virtual-reality-cinema-experience-vr
34. Fay, P.K.: Philosophy documentation center: building bridges in Sarajevo: the plenary papers from CTEWC 2018. J. Soc. Christ. Ethics. **41**, 185–186 (2021). https://doi.org/10.5840/jsce20214118
35. Selmanovic, E., Rizvic, S., Harvey, C., Boskovic, D., Hulusic, V., Chahin, M., Sljivo, S.: VR video storytelling for intangible cultural heritage preservation (2018). https://doi.org/10.2312/GCH.20181341

A Comparative Study on the Use of Artificial Intelligence and Other Approaches for Biomimetic Design: Application in Packaging Innovation for Fragile Products

Alice Araujo Marques de Sá(✉), Améziane Aoussat, and Nicolas Maranzana

Arts et Métiers Institute of Technology, LCPI, HESAM Université, 75013 Paris, France
alicearaujoms@gmail.com

Abstract. Biomimetics combines biological knowledge with the technical inventive capacity for developing innovations. Specialists underscore that new technologies, such as Artificial Intelligence (AI), will revolutionize this field. Therefore, this study assessed the application of AI and classical approaches for searching and retrieving biological inspiration in biomimetic design, more specifically, for the development of a packaging design for fragile food products. A multi-methodological approach was adopted, consisting of five creative sessions, based on the following resources: scientific literature database; biomimetic database; mind mapping; and two AI chatbots (ChatGPT-3.5, and Microsoft Copilot). The results were analyzed according to the session duration; quantity, relevance, and applicability of the retrieved biological attributes. The recurring attributes found throughout multiple sessions were examined regarding their morphology, structure, and pertinence for an egg packaging design. The advantages and limits of the five approaches were discussed. Comparatively with classical approaches in biomimetic design, AI seems to require an even greater participation of human experts to ensure prompt quality and to conduct a critical evaluation of the generated propositions, assessing their pertinence to the desired context. Thus, it is possible to conjecture that AI will not completely replace other biomimetic design resources, but can be used complementarily. AI represents a valuable asset in the designer's toolkit for discovering and analyzing biological information, facilitating problem-solving, enhancing the proposition of alternatives, and streamlining creative sessions. Additional investigation and experimentation with other classical biomimetic approaches and AI tools are recommended, including studying their combination throughout the creative process.

Keywords: Biomimetics · Bioinspiration · Artificial Intelligence · Product Design · Biomimetic Methods · Packaging Design · Bioinspired Design

1 Introduction

The last decades have been defined by the growth of porosities between scientific and applied disciplines, resulting in new technologies and products. One of the approaches of remarkable interdisciplinarity consists of drawing inspiration from nature to guide

Published by Springer Nature Switzerland AG 2025
P. Sureephong et al. (Eds.): PLM 2024, IFIP AICT 740, pp. 286–303, 2025.
https://doi.org/10.1007/978-3-031-93319-6_22

the development of innovations such as bioinspired products, biologically-based mathematical models, intelligent systems, and algorithms that emulate natural patterns [1].

In this perspective, a field of notable convergence of expertise and know-how is biomimetics, which combines biological knowledge and technical inventive capacity in domains such as design, engineering, and architecture [2–7]. In biomimetics – instead of directly using or domesticating nature – living forms, processes, and systems are analyzed, transposing their properties to human activities. Multidisciplinary teams engage in an intricate process of observation and research to establish analogies that are the foundations of new designs [8–10]. Indeed, nature, having undergone 3.8 million years of evolution, represents an immense repository to be further investigated. Given this profusion of potentially inspiring elements, it may be challenging to successfully conceive innovative designs. Thus, it is essential to explore different approaches in biomimetics, including the opportunities offered by new technologies, such as Artificial Intelligence (AI), especially considering the increasing digitization and complexification of products, services, and systems [9, 11, 12].

In this article, AI will be considered as a technology system based on deep learning and large sets of networks and language models that emulate human intelligence, including its capacity for perception, reasoning, learning, and problem-solving [13–15]. This technology is capable of operating generatively, producing a range of outputs for specific tasks based on an initial objective or framing question, which is configured as a prompt. Thus, content, models, images, texts, forecasts, and recommendations are generated for a given context. Interaction with these systems is facilitated since it is based on natural language [14, 16].

The present study was driven by the following research questions: How can relevant biological attributes be selected for biomimetic design? Are new technologies - such as AI - useful tools in the biomimetic design process?

In summary, the objective of this study was to assess the application of AI and other classical creative approaches for searching and retrieving biological inspiration for biomimetic design, more specifically, for the development of a packaging design for fragile food products.

2 Biomimetics and Artificial Intelligence

Investigations of the international scenario on biomimetics indicate a rising interest in this area with a substantial set of research focusing on product design, architecture, engineering, material sciences, robotics, programming, navigation, and healthcare [17–19].

However, as indicated by several specialists [20–24], in a problem-based biomimetic design approach (Fig. 1), there are numerous challenges correlated to steps of high abstraction levels, such as the transfer of technical problems to a biologically-based formulation; the identification and selection of the most fitting and inspiring biological attributes from the myriad of existing species, and the following abstraction of their specificities into the creative domain. Therefore, engaging in biomimetic design requires time, a suitable methodological framework, and the participation of expert practitioners [20–24].

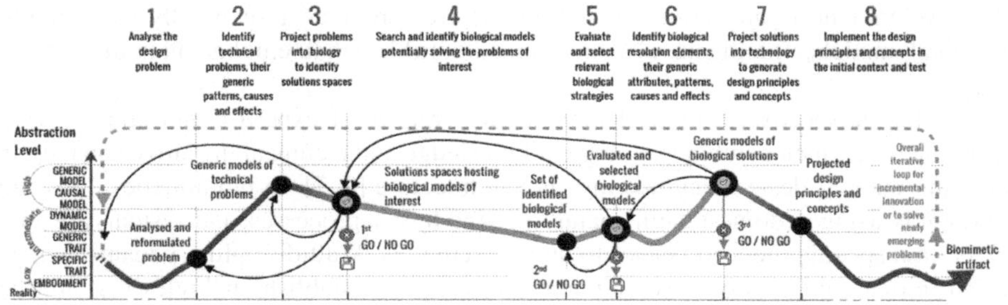

Fig. 1. The Technology-Pull Interdisciplinary Biomimetic Process illustrates that many challenges are associated with transitioning from steps 2 to 7, especially regarding the rising levels in abstraction [17].

Multiple tools and processes structure the biomimetic design approach, assisting in the assimilation of biological information and facilitating communication between multi-professional teams. Some examples are: BioTRIZ, Ask Nature, E2B Thesaurus, Biomimetree, and Linkage [6, 8, 20, 23, 25]. Comprehensive approaches encompassing databases, natural systems, subsystems, and the integration of textual information with visual representations can generate valuable insights. Considering the implications associated with biomimetic innovations, it is pertinent to investigate new technologies that have been shifting paradigms of ideation, concept generation, prototyping, manufacturing, distribution, and consumption, that is, Industry 4.0, and notably, Artificial Intelligence (AI) [26].

As defined by IBM [13, 26], AI integrates computer science with robust datasets, also incorporating algorithms, machine learning, and deep learning. Thus, AI is capable of undergoing training, adaptation, and evolution, enabling problem-solving and pattern recognition [27, 28]. These technologies are becoming ubiquitous in everyday life, allowing for customization of content and products to the target audience with a greater degree of interaction [9, 26]. In design, engineering, and architecture, AI has the potential for quickly generating a large set of text and images and also for analyzing data [27, 29, 30].

Though specialists assert that biomimetic design processes, tools, and databases are not yet fully integrated with AI systems, it is presumed that such techniques will soon revolutionize the field [1, 8]. Nonetheless, many works are being undertaken in medicine, engineering, robotics, aerospace, and architecture [8, 31, 32]. More developments are also expected in assessing the efficiency and efficacy of sustainable materials [33]. Additionally, it is worth highlighting some studies and contributions that combine the field of biomimetics with AI and that show the plurality of works that have been carried out, such as: a) the use of machine learning algorithms and systems to organize the literature on biomimetics [34]; b) the creation of models to help identify and select inspiring biological attributes, such as the BERT-based BioDesign model [8] and the PeTaL system [32]; c) algorithms for integrated automation of AskNature data [35]; d) processes for image morphing based on StyleGAN [36]; e) the BiOMiG AI-based platform created for biological and functional research, including a material database and a methodological framework for ecodesign [37]; and f) the comprehensive framework to assist industrial designers in using ChatGPT for biomimetic projects [38].

3 Method, Design Brief and Creative Sessions

Exploratory and descriptive research was conducted with a multi-methodological approach composed of five steps. The context and requirements of the biomimetic design case were defined: develop an egg packaging box. Such a design brief was chosen to better control the level of complexity of the project, in view that this study focused on a preliminary evaluation of multiple methodological paths, that are composed of many different phases, used for searching and retrieving biological attributes for biomimetic design. It is worth mentioning that, in classical product design processes, the scale, multifunctionality, and the number of components can also represent a high degree of complexity, which, when correlated with the biomimetic approach, can be even more significant [39, 40] due to the impressive body of biological knowledge that can be investigated by the design team [12]. Also, it is frequent that, in product design training courses, experimentation is conducted focusing on a concise proposal, based on studying the shape and functions of a simple object used in everyday life [40–42]. The egg box, although simple, raises interesting challenges surrounding the necessity of protecting delicate food items against mechanical impacts, while also taking into account the ease of use of the packaging; the optimal modular internal distribution; and the possibility of stacking for transportation and storage. Based on the steps proposed by Fayemi [6], Graeff [7], and Graeff et al. [6, 7, 21], the design problem was stipulated and abstracted. The design-based brief was then reformulated in the following biological terms: *"How does nature protect, store, support, involve, and/or stack ?"*. These keywords, connected to the product's intended functionality, allowed organizing the subsequent creative sessions. Each session was based on the adoption of a different resource to search for potentially inspiring biological attributes, namely: Scopus database; Ask Nature; mind mapping; ChatGPT-3.5, and Microsoft Copilot. The selection of such tools was driven by their accessibility and availability (e.g., the human, material, and bibliographic resources of the laboratory where the sessions were held).

3.1 Criteria for Evaluating Biological Attributes

In this study, biological attributes are defined as the distinctive characteristics (physical features and properties) of organisms, natural principles, and elements that reveal potential for application in biomimetic design. For conducting the creative sessions, it was necessary to establish parameters for selecting what was considered pertinent and applicable biological attributes. The following categories were adopted: a) Level 1 - low applicability. The biological attributes require a high level of research and development to be properly transposed into a technical context. Also, they are based exclusively on physicochemical aspects of materials, and/or they focus on bioutilization (biodesign) and not biomimetics [43]; b) Level 2 - moderate level of applicability. An average need for abstraction, research, and development is required. In the medium term, such biological attributes may be implemented; and c) Level 3 - high applicability. Easily transposable forms and/or materials that present high performance against impacts and/or efficient modular organization. In addition to the level-based analysis classification, a categorization was conducted according to the following factors: form, material, process, and a combination of those aspects. To comparatively analyze the creative sessions, the following parameters were established: research duration; quantity, and relevance of biological attributes. Moreover, a set of ten of the most recurring potentially inspiring biological attributes found throughout the five sessions was examined regarding their morphological and structural characteristics and their level of applicability for an egg packaging design.

3.2 Session 1: Scientific Literature – Scopus

The first research session consisted of investigating a scientific database, as it can be one of the main sources for retrieving specialized biological knowledge. Based on the Technology-Pull Interdisciplinary Biomimetic Process [6, 7, 21], a set of keywords was chosen, which allowed the creation of the following search prompt: *"[(nature OR biology) AND (biomim* OR bioinsp* OR bio-insp* OR biodesign) AND (protect OR involve OR store OR structure)]"*. This initial search generated a total of 7,837 documents. A filter was applied to select only scientific articles over 10 years (2014–2024). Such period represents a panorama of research also including recent years. Given that the set of documents was still very significant (4.064 documents) a new filter by subject areas was applied, corresponding to: a) Arts and Humanities; b) Social Sciences; and c) Agricultural and Biological Sciences, resulting in 106 documents. The choice for such correlated fields was a necessary procedure, in view that the Scopus platform does not provide a specific filter by subject for the Design domain [17]. The metadata was extracted in ".CSV" format, and then, the VOSviewer [44] software was used to create a visualization of co-occurring keywords that could indicate possible paths for new biomimetic egg packaging design.

3.3 Session 2: Biomimetic Database – Ask Nature

Among the repositories oriented toward biomimetic design practices is the Ask Nature database. This open-access platform organizes transferrable biological knowledge for technological projects, allowing to search for a variety of biological attributes and also providing many examples of pre-existing biomimetic cases and applications [35, 45, 46]. The platform undergoes continuous updates and features an extensive compilation of over 2,000 "biological strategies". These strategies are systematically categorized using a taxonomic model, that distributes them into groups, subgroups, and functions [1, 8]. The following keywords were adopted for this session: protect, involve, support, packaging, store, and structure. Thus, six searches were conducted in Ask Nature, using each word individually. It was necessary to constantly verify the overlap of results on the platform, recording them, one by one, in a spreadsheet before obtaining the correct total number of the set of biological attributes. This resulting set was then analyzed according to the criteria from Sect. 3.1, and all the biological attributes that were categorized as Level 3 were considered pertinent. Such elements were illustrated in a word cloud using the WorditOut Platform [47].

3.4 Session 3: Designer – Mind Mapping

The third session was carried out with a designer who had previous training in biomimetic design processes. Among the classic methods for generating early-stage product design, this session focused on the use of mind mapping, which was conducted using the Miro platform [48]. Such an approach contributes to generating a set of ideas and allows for visually organizing them and establishing connections between elements in a set [9, 49]. The session was carried out based on the same biomimetic guiding question that was previously described. The pertinence of the biological attributes that were found was examined according to the applicability criteria from Sect. 3.1, and all elements associated with Level 3 were showcased in a word cloud.

3.5 Sessions 4 and 5: Artificial Intelligence – ChatGPT-3.5 and Microsoft Copilot

The last sessions were conducted with the chatbots ChatGPT-3.5 [50] and Microsoft Copilot [51]. They allow for collecting and interpreting conversational prompts of users to provide feedback, supporting the creation and selection of alternatives, assisting in ideation and decision-making by designers [9, 38]. Each interaction with the AIs began by asking for a description of biomimetic design. Then, the main prompt consisted of: "*Describe what a biomimetic design packaging for a set of eggs would be like. That is, create a packaging for similar-shaped, round, fragile objects. The packaging must protect from impacts, optimize the internal distribution of elements, and be easy to use, facilitating storage and stacking. Detail and describe the natural inspiration of your biomimetic approach, suggesting at least 5 different possible natural inspirations for such project*". Following this prompt, a series of iterations was performed, always asking for an additional five suggestions after each interaction with the AI. Once again, the selection of the biological attributes was guided by criteria already mentioned, and the results classified as Level 3 were compiled in a word cloud.

4 Results and Discussion

The main results regarding the duration of each session are illustrated in Table 1. The scientific literature session was the most time-consuming and resulted in the lowest number of relevant biological attributes. On the other hand, both AI sessions were the most efficient at directly retrieving inspiring elements, but they provided a limited set of relevant attributes when compared to Ask Nature and Mind Mapping sessions.

4.1 Session 1: Scientific Literature – Scopus

This session was considered the most complex approach when adopted at the very beginning of the process for searching and retrieving biological attributes. This is due to its multiple required steps and also to the large quantity of highly specialized documents provided, posing a challenge in extracting pertinent information. The adoption of keywords that represented the desired functions for the packaging design (e.g., protect, store, involve, structure) did not consistently lead to documents highlighting specific biological

Table 1. Comparative results of the five sessions for searching and retrieving biological inspiration.

Categories	Scopus	Ask Nature	Mind Mapping	ChatGPT-3.5	M. Copilot
Session duration	5 h30	4 h30	2 h	1 h30	1 h30
Total of natural elements	25	268	38	58	29
Pertinent natural elements	6	54	28	13	9

attributes considered pertinent to the design brief. This could be due to the formatting of the search prompt, which could be reformulated. Nonetheless, one of the possible approaches to facilitate this process, when using scientific literature at the beginning of the search for biological attributes (initial steps indicated in Fig. 1, Phase 4) is to use information visualization software, to create keyword maps. Following the description in Sect. 3.2, after downloading metadata from Scopus, 646 keywords were obtained, which were used to create co-occurrence maps. However, when maps were automatically generated in VOSViewer, they did not indicate relevant biological attributes (Fig. 2). So, the full list of keywords had to be manually filtered, selecting one by one, on the software keyword selection tab. Only then it was possible to separate generic words (e.g., sustainability, 4D printing, optimization, creative design process) from the set of specific biological attributes considered pertinent to the design brief of the present study (Fig. 3).

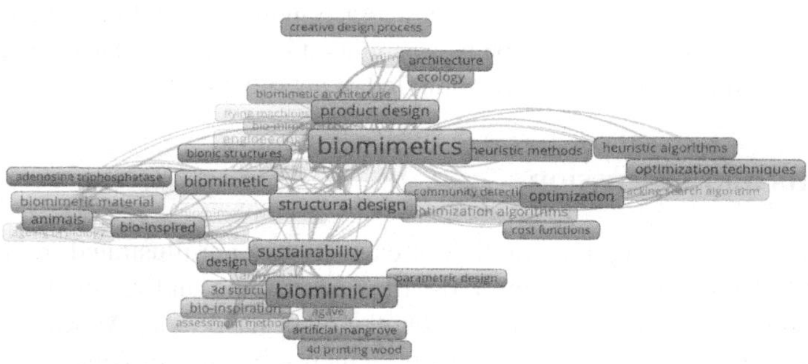

Fig. 2. Word cloud generated automatically. Clusters are organized by subjects and colors (based on link strength and word frequency). The diagram shows few biological attributes for egg packaging design.

Figure 3 exhibits some biological attributes that are not conducive to egg packaging design, either because they have an overly broad meaning, or they cannot be transposable to the required functions of protection and storage (e.g., ecology, evolution, ecosystem, membranes, swarm intelligence). However, six specific biological attributes were categorized as having a high applicability level for a packaging design for their particular

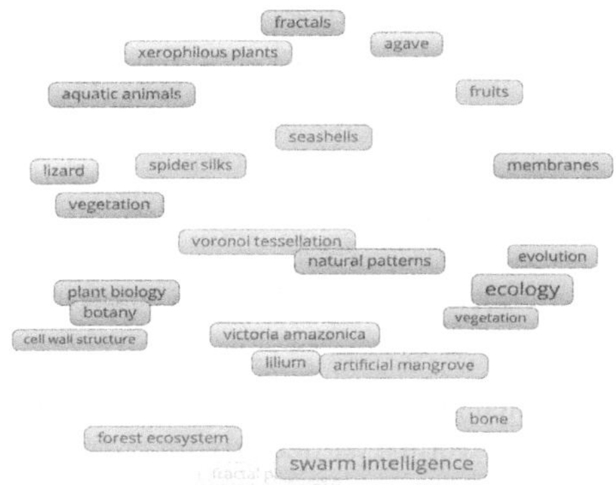

Fig. 3. Illustration of manually filtering the keywords. More specific biological elements can be observed.

structural and mechanical resistance as well as their forms, which integrate modularity and efficiency in material utilization, namely: seashells [52], agave [53], Voronoi tessellation [54], cell wall [55], *Victoria amazonica* [56], and bones [57].

It was concluded that scientific literature should be adopted preferably in the intermediate and final steps of Phase 4 of the biomimetic process indicated in Fig. 1 [21]. As suggested by Graeff [58], this approach should be conducted collaboratively with biologists and other specialized professionals so they can assist the creative team in properly understanding and transposing the biological information. Also, as stated by Zhao et al. [34], it can be pertinent to investigate the use of AI and machine learning algorithms in gathering and organizing literature on biomimetics.

4.2 Session 2: Biomimetic Database – Ask Nature

The Ask Nature database was the second longest session. Nonetheless, it showcased the most significant primary set of biological attributes, with a total of 268 elements. It resulted in the largest andmost varied group of attributes classified as having a high pertinence for applicability in the design case, totaling 54 biological attributes which are illustrated in Fig. 4. A common denominator of these elements is their ability to disperse impacts, their modular structure, their efficient and economical use in materials, and the frequent hierarchical organization of porous materials and shapes ranging from the microscopic to the macroscopic scale.

During the session, it was remarked that the documentation of biological knowledge lacks consistency, with variations in the entries for describing different biological attributes, also regarding the use of images and schematic illustrations. This is likely due to how experts from different fields fill the platform, without adhering to a specific protocol. The Ask Nature search bar occasionally retrieves keywords that do not directly correspond to a specific biological attribute, but are instead used to structure explanatory text. The platform features an interface that was considered more intuitive and appropriate for the biomimetic process when compared to the other approaches assessed in the

Fig. 4. Group of 54 pertinent biological attributes for an egg-packaging design [46].

present study. This was expected given that it is a dedicated database for bio-inspired initiatives. Such ease of use is noteworthy, particularly for creative team members engaged in biomimetic design initiatives that only have initial or intermediate levels of biological knowledge and/or experience in the biomimetic design processes. But, it is worth highlighting that even if Ask Nature undergoes continuous updates, it still represents an infinitesimal portion of all the cataloged biological species known to date [6, 35]. Thus, this approach should be complemented by investigating scientific literature databases and collaborating with specialized professionals to access more biological information.

4.3 Session 3: Designer – Mind Mapping

The third session, conducted with the designer, resulted in 38 biological attributes for the creative brief. A group of 28 of these attributes was classified as having a remarkable level of applicability for egg packaging design (Fig. 5). It is possible to observe that a variety of marine animals; botanical organisms and elements; and avian species were retrieved. They were correlated to functions of protection against external stimuli and mechanical impacts.

Fig. 5. Set of 28 biological attributes obtained from the mind mapping session.

The entire process was facilitated by the previous experience of the designer, and it can be assumed that, in the absence of these skills, the outcomes could be quite different [8]. In this scenario, it would be essential to invest much more time in professional training to obtain a set of relevant biological attributes. It is possible to suppose that the active participation of the designer in the biomimetic process can allow for the integration of a variety of creative resources, inherent to this domain, that can be explored complementarily to the approaches included in the present study. Some examples include brainstorming techniques, mood boards, sketches, digital modeling, and the construction of three-dimensional prototypes [24, 59, 60].

4.4 Sessions 4 and 5: Artificial Intelligence – ChatGPT-3.5 and Microsoft Copilot

The sessions with the AI chatbots were efficient concerning the time spent on the retrieval of biological attributes. The chosen initial search prompt was considered pertinent given the multiple alternatives extracted. Nevertheless, it would be worth to experiment with other prompt configurations, such as requesting more biological attributes by specifying geographical regions, biomes, taxonomic levels, and scientific names. Throughout the process, both AI tools needed to be constantly reoriented. Requesting solution sets at each iteration was a suitable approach. However, when this is done successively, AI tends to provide non-pertinent suggestions, such as abiotic factors and non-biomimetic elements (e.g., bouncy spheres, pebbles, origami, and amphoras). To overcome this issue, it was necessary to request the exclusion of these groups. Regarding the retrieved biological attributes, ChatGPT-3.5 provided a primary set of 58, and Copilot a group of 29 general possible solutions for the design brief. After evaluating their applicability, a set of 13 biological attributes for ChatGPT and a total of 9 for Microsoft Copilot was obtained (Figs. 6 and 7). Chat-GPT often suggested classic biomimetic examples such as the pinecone, honeycomb, turtle shell, and lotus plant. Copilot also tended to focus on well-known biological attributes such as pomelo, pangolin, and exoskeleton.

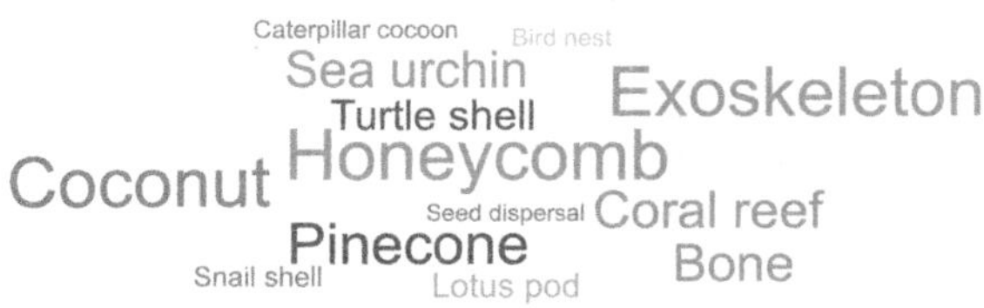

Fig. 6. Set of 13 inspiring elements for a biomimetic design for egg packaging extracted with Chat-GPT.

Fig. 7. Set of 9 elements collected in the interaction with Microsoft Copilot.

Chat-GPT offered more complete descriptions of the biological attributes, comparatively to the other chatbot, providing a report including the general biological characteristics, the recommended materials, and the possible packaging shape. Microsoft Copilot offered more concise descriptions. This chatbot frequently provided low-usefulness biological attributes for the context of packaging design, such as: "*Seashells: they have a variety of shapes, colors, and textures that reflect their origin and function, and are also biodegradable and rich in calcium*" [51]. ChatGPT-3.5 provided a similar, but more detailed description of the same element, that could allow for transposition into biomimetic design: "*Shell Structure: The spiral-shaped shell provides natural compartments that can cradle each egg securely, preventing movement and minimizing the risk of breakage during transport or storage [...] also allowing for efficient stacking and nesting, optimizing storage space. [...]. The shell consists of multiple layers, each with a specific function. Mimicking this layered construction can provide additional strength and durability, enhancing protection for the eggs [...].*" [50].

Also, while both AIs suggested relevant biological attributes, their description was often insufficient or even incorrect for the design brief. For example, the information about sea urchins found on Ask Nature [43], states that this biological attribute was retrieved and considered pertinent to the egg packaging design brief because it is associated with mechanical resistance due to a thin shell composed of interlocking plates that disperse impact preventing it from cracking. However, although both AIs suggested the same biological example, they focused exclusively on the sea urchins' long external spines which, despite protecting the animal against predators, are not pertinent to the egg packaging context [50]. The AIs suggested many biological attributes that were focused only on aesthetic or visual aspects (e.g., "*bioluminescent organisms for integrating gentle, organic lighting to create an aesthetically pleasing and visible protective layer around the eggs*" [43]). These findings are corroborated by Zhu et al. [1], which underscored the lack of technicality in the knowledge of retrieval of biological information. Both tools also tended to focus on the description of the physicochemical properties of materials on a microscopic scale, while the design briefing corresponded to the search for biological forms at a macroscopic level. The AIs recurrently recommended bio-assistance and bio-fabrication without specifying the processes and materials or only loosely describing them (e.g., "*spider silk: drawing inspiration from the strength and flexibility of spider silk, the packaging could incorporate lightweight but robust web materials that offer both protection and eco-friendliness*" [50]).

The results found in this session converge with the study of Hegazy, Saleh [30], and Fillipi [9], in which they stated that AI-generated conceptual design can present limited controls and customization; lack of consideration for structural design feasibility (in regards to materials, structural elements, and proportions); and some inconsistency of outcomes. Given what was previously presented, using AIs for searching and retrieving biological attributes should be either carried out through collaborating with experts in the field or by verifying all provided information regarding biological knowledge in other scientific databases or Ask Nature.

4.5 Recurring Biological Attributes

During the multiple sessions, it was verified that some biological attributes were retrieved multiple times. Table 2 illustrates a set of ten of the most frequently found attributes, their description, and applicability, evaluating them according to the criteria presented in Sect. 3.1. Such recurrence may provide valuable insights for the proposed design brief.

When analyzing the set of ten recurrent biological elements and the information gathered throughout the five sessions, some shared patterns can be perceived, such as: a) the recommendation of spongelike materials and biomaterials with porous properties filled with air or empty spaces allowing for good impact dissipation; b) the use of simple geometrical shapes in a complex tessellation to provide resistance; c) the organization in a modular pattern for optimized distribution of internal contents; d) the pertinence

Table 2. Set of ten recurrent biological attributes found in multiple sessions [68–77].

Sessions	Biological Element	Image	Description	Applicability
1,2,3, 4,5	Bone Microstructure	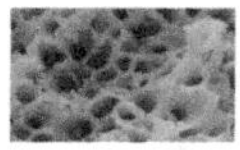	Resist strain by reorganizing the trabeculae structure that scales allometrically within physiological limits creating a mass-efficient pattern [46, 50]	Level 2, material-focused
2,3,4,5	Honeycomb		Organized in a lightweight, structurally resistant, and space-efficient hexagonal tessellation [46, 50]	Level 3, form-focused
1,2,4,5	Spider Silk		Webs are flexible, lightweight, and remarkably strong facing nonlinear stress. Their geometry integrates a minimalistic use of materials [46, 50]	Level 1, material-focused
1,4,5	Cacti and Xerophytes		The overall geometry is subdivided into ribs to provide self-shading. Possess water-retention capacity, and protective spines and tend to grow in a modular geometry [46, 50]	Level 1, form-focused
1,2,3	Giant Water Lily		The underside of the leaves is made of a radial web of air-filled ribs that sustain and distribute weight through the structure of the plant [46, 50]	Level 3, form-focused
2,4,5	Exoskeleton		Made of chitin fibers embedded in a highly crossed matrix. By changing thickness, stiffness, and fiber orientation, it can adapt to strain and load, protecting the insect [46, 50]	Level 3, form and material-focused
2,4,5	Turtle shell	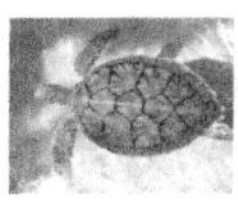	A two-layer keratin structure, a collagen layer, and a bony core disperse forces when under impact. The geometric arrangement creates a pattern that allows the curvature of the shell [46, 50]	Level 3, form and material-focused
2,3,5	Pomelo		The multiple-layered thick skin absorbs impact by rupturing air and fluid-filled intercellular spaces, similar to a compressible foam. Inside, modular slices are distributed in a radial pattern [46, 50]	Level 3, material and form-focused
2,4,5	Sea Urchin		Long spines and an oblate thin shell of calcium carbonate protect the soft insides of the animal. The shell is composed of small interlocking plates distributing the impact of a predator's bite [46, 50]	Level 3, material and form-focused
2,3,4	Seed dispersal		Many species of plants produce pods with unique shapes and seed dispersal mechanisms (by gravity, wind, bursting, water, and animals). Pods are modular, carrying various seeds at a time [46, 50]	Level 3, form and process-focused

of multifunctionality; and e) and the preference for the use of two types of materials – whether highly biodegradable for composting or, on the other hand, very durable materials to allow reuse. So, it is believed that using one or multiple of these biological attributes as inspiration for egg packaging design proposals is of interest.

Table 2 also reveals that, when analyzing the group of recurring elements, 9 of the set of 10 were suggested by at least one of the AI chatbots. This indicates that AIs can provide interesting paths for the design brief when corroborated with information from other classical biomimetic design approaches. Supplementary exploration of the prompt configuration and extending the interaction period with the AIs could provide an even greater diversity of biological information.

4.6 Overall Discussion

Regarding the five creative sessions, it was noted that some of the approaches are more comprehensive in terms of their versatility during the stages of the biomimetic process than others. Also, some processes and tools are preferable for specific stages of the biomimetic creative approach, such as scientific literature, which is very complex and time-consuming to carry out and analyze. Exploring this resource in the later stages of the research for inspiring biological elements is recommended.

Many results of the sessions were not directly conducive to an egg packaging design or would require an extensive redesign to be viable, especially when it came to the alternatives found in the literature research sessions and the experimentation with the two AIs. It was not surprising that the AskNature database showed the most promising results, given its direct association with the biomimetic field, even when considering its occasional lack of consistency and the somewhat limited number of biological datasheets. The creative session with the designer also showed good applicability and quantity of alternatives in the desired context. It would be necessary to verify whether the result is the same with another group of professionals, given that, in the present case, the designer already had training in biomimetics. It was expected that the AIs would represent an advantage in terms of the time required to obtain results. However, the chatbots had to be reoriented several times to retrieve the required information. The description of the natural elements was also limited and inconsistent, especially regarding its structural specifications, converging with what Hegazy, Saleh [30], and Fillipi [9] underscored in their studies. This research was carried out with ChatGPT-3.5, as it was available free of charge. Nonetheless, it is interesting to speculate that if the same creative session were to be carried out with the latest model, ChatGPT-4, the results obtained might be different. Especially given that this new version enables multimodal interaction, including images and graphics as input data, possibly facilitating both problem-driven and solution-based biomimetic design processes [6, 12, 61, 62]. Also, ChatGPT-4 can process a greater complexity of data than its predecessor. It provides a more immersive interaction and longer detailed descriptions, which may facilitate understanding of the inspiring biological attributes [61, 63]. However, both models occasionally provide inconsistent and incorrect answers and can reproduce underlying biases in the dataset on which they were trained [61, 63, 64]. Thus, critical analysis by biomimetic experts and designers is essential.

It is recommended that further studies be carried out comparing the use of ChatGPT-4 with GPT-3.5 in the context of packaging design. Concerning the recurring biological attributes found, it was noted that using alveolar structures with a gradient of different sizes and scales, integrating layers with different mechanical properties, and adopting modular patterns are the main characteristics that could guide an innovative and efficient biomimetic egg packaging.

As observed during all the creative sessions, the participation of specialists is paramount for obtaining significant results in each examined pathway. Encouraging greater involvement of biologists, material scientists, and other specialists in teams dedicated to biomimetic projects, which are frequently composed of designers, architects, and engineers, is essential [7, 60]. Fostering dialogue between these professionals and analyzing their contributions to each of the biomimetic creative paths can improve the degree of applicability, relevance, and innovation of the biomimetic solutions.

It is speculated that soon, AI tools will be more integrated into the biomimetic design process, with multiple research and projects in current development, as mentioned in Section 2. Further contributions and opportunities lie in the development of AIs and other tools integrating different principles and approaches of biomimetic design, for example, focusing on more complex biomimetic levels (e.g., natural systems) and other frameworks, such as the solution-based biomimetic design process [12, 65].

5 Conclusion and Perspectives

As determined by the research questions, this study demonstrated the use of different creative resources and tools that can be adopted to identify and extract inspiring biological attributes for design projects. It was observed that new technologies, especially AIs, are relevant to the biomimetic design process, however, such resources must be adopted with the assistance of specialists so that their use is meaningful. It is believed that AI technologies will soon be even better integrated with biomimetic resources, enabling more advanced searches about organisms and their inspirational functions. As stipulated in the general objective of the research, an analysis was undertaken of the application of two AIs and a set of three other classic creative approaches to the development of a biomimetic packaging design for fragile food products. It was evident that the "traditional" approaches to the biomimetic design process (investigation of scientific literature, Ask Nature database, and mind mapping) resulted in a broader and more diverse set of inspiring biological attributes. At present, artificial intelligence resources from chatbots such as Microsoft Copilot and ChatGPT-3.5 still have limitations when it comes to offering a set of alternatives directly adapted to the biomimetic context of packaging design for fragile products, in terms of the number of biological attributes suggested, the adequate description of their functions and their relevance and applicability to the context of the selected project.

Comparatively, with classical methods and tools in biomimetic design, AI is a recent technology, so it seems to require even greater participation of human experts to ensure prompt quality and to conduct a critical evaluation of the generated propositions, assessing their pertinence to the desired context. Thus, it is possible to conjecture that AI will not replace other biomimetic design resources, but can be used complementarily with them. AI represents significant potential as a valuable asset in the designer's toolkit, aiding in discovering and analyzing biological information, facilitating problem-solving, and also enhancing the efficiency of creative sessions. It is worth mentioning that human-AI collaboration in design concept generation can surpass the boundaries of what the algorithm directly offers. Designers can use knowledge and inspiration provided by the AI to reframe and better develop their endeavors, adapting their framing questions, in an iterative process [1, 8, 9, 28, 30, 66]. However, when using such technologies, it is essential to consider the full scope of issues related to obtaining the information and using it in further industrial production to avoid nefarious impacts on creativity and design. This is particularly significant since such technology is not yet fully regulated, and its ethical and cultural implications are still in debate [30, 67].

As a perspective for continuing the present work, additional investigation and experimentation with complementary biomimetic approaches and other AI systems (e.g., text-to-image, AI-generated multimedia) are recommended. It is also necessary to research the combination of multiple tools and resources, such as the ones concerned in the present work, to better understand the optimal integration of these elements in the facilitation of the biomimetic process. It would also be worth studying how AIs can help in integrating the properties of multiple biological attributes to create efficient, multifunctional solutions. Finally, studying how different professional profiles (e.g., designers, biologists, engineers, architects, social scientists) partake in these creative processes and how they interact differently with each tool is also relevant.

References

1. Zhu, Q., Zhang, X., Luo, J.: Biologically inspired design concept generation using generative pre-trained transformers. J. Mech. Des. **145**, 1–12 (2023)
2. Ceebios, Muséum d'Histoire Naturelle, EuraMaterials: BiOMIg: Bio-inspired materials open innovation generator (2021). https://www.ceebios.com. Accessed 13 Mar 2024
3. ISO: ISO 18458:2015 (2015). https://www.iso.org/. Accessed 13 Mar 2024
4. Pawlyn, M.: Biomimicry in architecture. Riba publishing, Newcastle upon Tyne (2016)
5. Helfman Cohen, Y., Reich, Y.: Biomimetic Design Method for Innovation and Sustainability. Springer International Publishing, Cham (2016)
6. Fayemi, P.-E.: Innovation par la conception bio-inspirée: proposition d'un modèle structurant les méthodes biomimétiques et formalisation d'un outil de transfert de connaissances. Dissertation, École Nationale Supérieure des Arts et Métiers (2016)
7. Graeff, E.: Innovation bio-inspirée: modélisation d'un processus interdisciplinaire de conception biomimétique outillé et intégration d'un nouvel acteur, le Biomiméticien. Dissertation, École Nationale Supérieure des Arts et Métiers (2020)
8. Sun, F., Xu, H., Meng, Y., Lu, Z.: A BERT-based model for coupled biological strategies in biomimetic design. Neural Comput. Appl. **35**, 2827–2843 (2023)
9. Filippi, S.: Measuring the Impact of ChatGPT on fostering concept generation in innovative product design. Electronics **12**, 3535 (2023)
10. Amilton, J.V.A.: Methods and Processes in Bionics and Biomimetics: the Technological Revolution for Nature. Editora Blucher, São Paulo (2018)
11. Benyus, J.M.: Biomimétisme: Quand la nature inspire des innovations durables. Rue de l'Échiquier, Paris (2011)
12. Baumeister, D., Tocke, R., Dwyer, J., Ritter, S., Benyus, J.: Biomimicry Resource Handbook: A Seed Bank of Best Practices. CreateSpace Independent Publishing Platform, Missoula, Montana (2014)
13. IBM What is Artificial Intelligence (AI)? https://www.ibm.com/topics/artificial-intelligence. Accessed 13 Mar 2024
14. International Organization for Standardization (2022) ISO/IEC 22989:2022. https://www.iso.org/standard/74296.html. Accessed 3 May 2024
15. McKinsey. What is AI (artificial intelligence)? https://www.mckinsey.com/featured-insights/mckinsey-explainers/what-is-ai. Accessed 3 May 2024

16. Britannica. Artificial intelligence (AI). Definition, Examples, Types, Applications, Companies, & Facts. https://www.britannica.com/ Accessed 6 May 2024
17. Sá, A.A.M., Viana, D.M.: Design and biomimicry: a review of interconnections and creative potentials. Biomimetics **8**, 1–16 (2023)
18. Lepora, N., Verschure, P., Prescott, T.: The state of the art in biomimetics. Bioinspir. Biomim. **8**, 013001 (2013)
19. Ceebios, Ministére de la Transition Écologique et Solidaire: Biomimétisme en France - Etat des lieux (2018). https://ceebios.com. Accessed 13 Mar 2024
20. Wanieck, K., Fayemi, P.-E., Maranzana, N., Zollfrank, C., Jacobs, S.: Biomimetics and its Tools. Bioinspired Biomim Nanobiomaterials **6**, 1–52 (2017)
21. Graeff, E., Anneline, L., Raskin, K., Maranzana, N., Aoussat, A.: Biomimetics from practical feedback to an interdisciplinary process. Res. Eng. Des. **32**, 1–27 (2021)
22. Fayemi, P.-E., Wanieck, K., Zollfrank, C., Maranzana, N., Aoussat, A.: Biomimetics: Process, tools and practice. Bioinspir. Biomim. **12**, 011002 (2017)
23. Vincent, J.F.V., Bogatyreva, O.A., Bogatyrev, N.R., Bowyer, A., Pahl, A.-K.: Biomimetics: its practice and theory. J. R. Soc. Interface. **3**, 471–482 (2006)
24. Anneline, L., Graeff, E., Maranzana, N., Raskin, K., Aoussat, A.: How do designers impact the biomimetic concepts typology? Paper presented at the International Conference on Engineering and Product Design Education. Via University (2020)
25. Graeff, E., Maranzana, N., Aoussat, A.: Linkage, an online tool to support interdisciplinary biomimetic design teams. J. Mech. Des. **143**, 1–11 (2021)
26. Aphirakmethawong, J., Yang, E., Mehnen, J.: An overview of artificial intelligence in product design for smart manufacturing. Paper presented at the 27th International Conference on Automation and Computing (ICAC) (2022)
27. Lee, K.: A systematic review on social sustainability of artificial intelligence in product design. Sustainability **13**, 2668 (2021)
28. Wang, L., Zhang, H.-C., Wang, Q.: On the concepts of artificial intelligence and innovative design in product design. IOP Conf. Ser. Mater Sci. Eng. **573**, 012095 (2019)
29. Hanafy, N.O.: Artificial intelligence's effects on design process creativity: A study on used A.I. Text-to-Image in architecture. J. Build. Eng. **80**, 107999 (2023)
30. Hegazy, M., Saleh, A.: Evolution of AI role in architectural design: between parametric exploration and machine hallucination. MSA Eng J **2**, 262–288 (2023)
31. Mendez, E., Marcum, E.: Deep Learning with Biomimicry. Jour of Des and Scien (2019)
32. Shyam, V., Eggermont, M., Hepp, A.F.: Biomimicry for Aerospace : Technologies and Applications. Elsevier Science, Amsterdam (2022)
33. Rane, N., Choudhary, S., Rane, J.: Enhancing Sustainable Construction Materials Through the Integration of Generative Artificial Intelligence. SSRN Eletr Journ 1–25 (2024)
34. Zhao, Y., et al.: Data Driven Techniques for Organizing Scientific Articles Relevant to Biomimicry. In: Proceedings of the 2018 AAAI/ACM Conference on AI, Ethics, and Society, pp. 347–353. ACM, LA USA (2018)
35. Vandevenne, D., Verhaegen, P.-A., Dewulf, S., Duflou, J.R.: A scalable approach for ideation in biologically inspired design. AI EDAM **29**, 19–31 (2015)
36. Deng, Z., Lv, J., Liu, X., Hou, Y.: Bionic design model for co-creative product innovation based on deep generative and BID. Int. J. Comput. Intell. Syst. **16**, 8 (2023)
37. BiOMIg. In: EuraMaterials. https://euramaterials.eu/. Accessed 5 Mar 2024
38. Lascari, G.: An Approach to Incorporate Generative Artificial Intelligence as a Means to Include Biomimicry within Industrial Design. Master's Thesis. University of Auburn (2023)
39. Baxter, M.: Product Design. CRC Press, Florida (2018)
40. Cardoso, R.: Design for a Complex World. Ubu Editora, São Paulo (2022)
41. Norman, D.A.: The Design of Everyday Things. Basic Books, New York (2013)

42. Bonsiepe, G.: Design, Culture and Society, 1st edition. Editora Blucher, São Paulo (2011)
43. Aish, A., Sun, J.S.: Bioinspire-Museum: Document de cadrage. MNHN, Paris (2020)
44. VOSviewer. VOSviewer. https://vosviewer.com/. Accessed 9 Mar 2024
45. Deldin, J.-M., Schuknecht, M.: The AskNature Database: Enabling Solutions in Biomimetic Design. In: Goel, A.K., McAdams, D.A., Stone, R.B. (eds.). Springer London, London (2014)
46. AskNature - Innovation Inspired by Nature. www.asknature.org/. Accessed 13 Mar 2024
47. WordItOut. https://worditout.com/. Accessed 14 Mar 2024
48. Miro Visual Workspace for Innovation. https://miro.com/. Accessed 14 Mar 2024
49. Cross, N.: Design Thinking: Understanding How Designers Think and Work. Berg Publishing, Oxford (2011)
50. ChatGPT. https://chat.openai.com. Accessed 13 Mar 2024
51. Bing AI. Bing. https://www.bing.com. Accessed 13 Mar 2024
52. Selcuk, S.A., Sorguç, A.: Exploring complex forms in nature through mathematical modeling: a case on turritella terebra. Paper presented at the 27th Education and Research in Computer Aided Architectural Design in Europe Conference. Istanbul Technical University (2009)
53. Barbosa de Araújo, R., et al.: Biomimetic Application Potential of Agave sisalana Mechanical Properties, Lightness, Resistance Strategies, and Life Cycle for Digital Fabrication. Paper Presented at the 19th Senses & Sensibility Conference. IADE – Universidade Europeia, 27–29 November 2019 (2022)
54. Mele, E., Fraldi, M., Montuori, G., Perrella, G.: Non-conventional structural patterns for tall buildings: from diagrid to hexagrid and beyond. Paper presented at the DCEE 5th International Workshop on Design in Civil and Environmental Engineering. Sapienza University of Rome (2016)
55. AskNature. Structural Composition Provides Strength in Changing Conditions. Biological Strategy. https://asknature.org. Accessed 14 Mar 2024
56. AskNature. Ribbed Structure Provides Support. www.asknature.org/ Accessed 9 Mar 2024
57. Fine Strands Stabilize Bones. AskNature. www.asknature.org. Accessed 14 Mar 2024
58. Graeff, E., Maranzana, N., Aoussat: Biomimetics, where are the biologists? J. Eng. Design **30**(8–9), 289–310 (2019)
59. Zhu, Q., Luo, J.: Generative transformers for design concept generation. J. Comp. Inf. Sci. Eng. **23**, 041003 (2023)
60. Letard, A.: Contribution au développement du cadre méthodologique de la conception biomimétique: intégration de l'expertise des profils formés au Design pour favoriser le déploiement de l'approche dans les pratiques de conception et d'innovation. Dissertation, École Nationale Supérieure des Arts et Métiers (2021)
61. Prakash, C.: Chat GPT 3.5 Vs Chat GPT 4.0. www.medium.com/. Accessed 6 May 2024
62. ChatGPT 3.5 et ChatGPT 4. www.anthemcreation.com. Accessed 6 May 2024
63. Takagi, S., Watari, T., Erabi, A., Sakaguchi, K.: Performance of GPT-3.5 and GPT-4 on the Japanese Medical Licensing Examination: Comparison Study. JMIR Med Educ **9**, 48002 (2023)
64. Taloni, A., et al.: Comparative performance of humans versus GPT-4.0 and GPT-3.5 in the self-assessment program of American Academy of Ophthalmology. Sci. Rep. **13**, 18562 (2023)
65. Saint-Sardos, A., et al.: Bioinspire-explore: taxonomy-driven exploration of biodiversity data for bioinspired innovation. Biomimetics **9**, 63 (2024)
66. Velázquez Fernández, H.: A natural artificial intelligence? Some notes on the computational biomimicry of human intelligence. Metafís y Persona **15**, 31–38 (2023)
67. Palmieri, A.: Midjourney experimentation: representing Nature on a macro scale. Sci. Res. Inf. Technol. **13**, 181–188 (2023)
68. AskNature Team. Fine Strands Stabilize Bones. AskNature. https://www.asknature.org/. Accessed 14 Mar 2024

69. Honeycomb. Unsplash. https://unsplash.com/. Accessed 14 Mar 2024
70. Shaull, L.: Spider Web in Aitkin County, Minnesota. https://commons.wikimedia.org/. Accessed 14 Mar 2024
71. Karwath, A.: Golden Barrel Echinocactus grusonii. Wikimedia Commons. https://commons.wikimedia.org/. Accessed 14 Mar 2024
72. Wiersma, D.: Victoria amazonica Leaf Underside. Science Photo Library. https://www.sciencephoto.com/. Accessed 14 Mar 2024
73. Conrad, J.: Discarded exoskeleton (exuviae) of dragonfly nymph. Wikimedia Commons. https://commons.wikimedia.org/. Accessed 14 Mar 2024
74. Diaz, A.A.C.: Sea Turtle. Unsplash. https://unsplash.com/. Accessed 14 Mar 2024
75. Pomelo Grapefruit. Freepik. https://br.freepik.com/. Accessed 14 Mar 2024
76. Green Mexican Sea Urchin Strongylocentrotus Drobachiensis. Sea Shell Supply. https://www.seashellsupply.com/. Accessed 14 Mar 2024
77. AskNature Team. Seeds Dispersed Across the Sea: Sea-bean. AskNature. https://www.asknature.org/. Accessed 14 Mar 2024

Designing Augmented Reality Games for the Future of Cultural Heritage Tourism: The Case of Wua-Lai Communities

Suratchanee Yolthasart[1], Suepphong Chernbumroong[2], Songpon Khanchai[3], and Kitti Puritat[4(✉)]

[1] School of Tourism, Chiang Rai Rajabhat University, Chiang Rai, Thailand
[2] College of Arts, Media and Technology, Chiang Mai University, Chiang Mai, Thailand
[3] ChiangMai University, Chiang Mai, Thailand
[4] Department of Library and Information Science, Faculty of Humanities, Chiang Mai University, Chiang Mai, Thailand
kitti.p@cmu.ac.th

Abstract. This study examines the impact of "Silver Craft," an Augmented Reality (AR) game, on enhancing the cultural heritage tourism experience in the Wua-Lai communities, known for their intricate silver craftsmanship. Amidst the evolving post-COVID-19 tourism landscape, we explore the potential of AR to engage tourists by providing immersive experiences that highlight the rich cultural narratives of Wua-Lai. Utilizing the Mechanics-Dynamics-Aesthetics (MDA) framework, the game's design focuses on user experience, cultural engagement, and educational content. The research adopted a comprehensive Research and Development (R&D) approach, recruiting 40 first-time visitors to Wua-Lai who used the Android version of the game. Participants' feedback, collected via questionnaires, indicates that while the game was successful in enhancing the overall tourism experience and cultural understanding, there were areas needing improvement, particularly in technical performance. The findings suggest a promising avenue for leveraging AR in cultural tourism, albeit with considerations for broader technological inclusivity and continual user-driven refinements. Future work entails expanding the technological scope, increasing the participant pool, and exploring AR applications in other cultural contexts, aiming to contribute to the sustainable development of heritage tourism.

Keywords: Augmented Reality · Cultural Heritage · Digital Tourism · Wua-Lai Communities

1 Introduction

As the world gradually emerges from the shadow of the COVID-19 pandemic, the tourism industry is poised to redefine its future, with a keen focus on innovation, safety, and sustainability. This pivotal moment presents an unparalleled opportunity to integrate advanced technologies such as Augmented Reality (AR) into tourism experiences,

P. Sureephong et al. (Eds.): PLM 2024, IFIP AICT 740, pp. 304–316, 2025.
https://doi.org/10.1007/978-3-031-93319-6_23

particularly in culturally rich and historically significant destinations. Among these, the Wua-Lai communities in Thailand stand out as a beacon of cultural heritage, known for their exquisite silver craftsmanship, traditional Lanna architecture, and vibrant Saturday walking street markets. However, like many cultural heritage sites worldwide, Wua-Lai has faced the dual challenges of preserving its unique cultural identity while adapting to the changing dynamics of post-pandemic tourism.

This paper explores the transformative potential of AR and game design in revitalizing the tourism sector, with a specific focus on enhancing the cultural heritage experiences of destinations like the Wua-Lai communities. It delves into how AR games can offer immersive and interactive explorations of cultural heritage, enabling visitors to engage with the history, traditions, and stories of the Wua-Lai communities in new and exciting ways. This innovative approach not only promises to enrich the tourist experience but also plays a critical role in the economic recovery and sustainability of such destinations in the post-COVID-19 era [3, 4].

The incorporation of AR and game design in cultural heritage tourism represents a dynamic solution to engage digital-native tourists and foster a deeper connection with the heritage sites. By crafting AR experiences centered around the rich narratives, artifacts, and landscapes of places like Wua-Lai, developers can create engaging, participatory narratives that bring the community's heritage to life, offering a contrast to traditional tourism models. Moreover, in the context of post-pandemic recovery, AR games provide a safe, contactless mode of exploration that aligns with health guidelines and visitor well-being, showcasing how technology can be leveraged to address current challenges while enhancing cultural appreciation and understanding.

This paper aims to contribute to the discourse on the future of tourism by examining the role of AR game design in cultural heritage contexts, with a particular emphasis on the Wua-Lai communities. Through an exploration of design principles, case studies, and the broader implications for the tourism industry, this study seeks to illuminate the pathways through which technology and tradition can converge to create meaningful, engaging, and safe tourist experiences in the post-COVID-19 landscape.

2 Literature Review

The innovative potential of augmented reality (AR) has garnered attention as a novel means to enhance visitor experiences and insights at cultural heritage locations. This review of the literature focuses on the integration of augmented reality, cultural heritage tourism, and the design of games, with a particular emphasis on augmented reality games' capacity to influence the evolution of cultural tourism.

2.1 Augmented Reality

Augmented reality refers to technology that enriches the physical environment by overlaying virtual, computer-generated elements within the real world. With the advent of advanced technology, the world has stepped into embracing this novel and inspiring mode of interaction [7]. It possesses the capability to either augment or diminish the visibility of physical objects, substituting them with alternative content [8]. Consequently, augmented reality holds the capacity to enhance users' perception of reality by enriching their visual experience of the real world or by presenting entirely artificial environments that diverge from reality [9]. Through the seamless integration of computer simulations into actual environments, AR constructs an enriched perspective by complementing users' surroundings with digital content. This facilitates the combination of physical and virtual realms, promoting cohesion between the two worlds [10]. It surpasses other media forms, namely video and audio, due to its capacity to merge "a live view in actual time with virtual computer-generated pictures, generating real-time augmented reality experiences [11]. With these attributes in mind, AR has garnered widespread acclaim for its capacity to generate more elaborate, immersive content, enriching interaction with and understanding of the surrounding world [12]. The experience of augmented reality can be encountered via various display methods, including head-mounted devices, handheld displays, and projectors [13]. The primary benefit lies in the user's freedom from being on a computer or cellphone screen continuously, allowing for on-the-go usage [7]. Although AR can be encountered across various devices and advanced wearable technologies seamlessly integrate digital content into reality, smartphones persist as the most prevalent AR device owing to their widespread availability and accessibility [14]. Within the realm of tourism, smartphones have emerged as the favored AR platform, attributed to their straightforward integration, cost-effectiveness, reliability, and capability to support effective AR applications [15].

2.2 Augmented Reality in Tourism

The tourism sector has evolved over time, often in tandem with technological advancements, focusing its efforts on captivating tourists' interest, arousing curiosity, and fulfilling their expectations [16]. The advent of technological innovations, including AR, has exerted a profound influence on all tourism enterprises. The heightened awareness and adoption of these technologies have transformed travel behaviors by fundamentally altering how tourists seek information, make decisions [17], buy a product of tourism, and study reviews [18]. With the proliferation of technology in tourism, the boundary between tourist experiences and everyday life has progressively faded. This integration has been termed spill-overs [14], prompting extensive research into the effect of AR spill-overs on travel encounters [19]. Examining the travel behaviors of American tourists, Wang et al. (2016) noted a significant integration of smartphone usage into their travel experiences, forming an integral aspect of their daily routines and interactions. The study suggested that the widespread availability of technology has fostered a culture in which tourists increasingly depend on technology to sustain their social connections and fulfill work and educational obligations, thereby impacting various aspects of their daily lives [14]. According to He et al. (2018), incorporating AR within the tourism sphere can enrich consumer experiences, enhance their attitudes, and foster a greater inclination towards positive behavioral intentions [20]. Tarantino et al. [8] developed an AR guide system known as GENnArí, which creates customized multiday itineraries aimed at maximizing the attractiveness and points of interest (PoIs) while minimizing the distance traveled by users [21]. The PazinAR project was created as an interactive tour guide for the city of Pazin, with the aim of enhancing the city's tourism offerings. The application was developed using the Unity platform, with augmented reality functionality integrated through the addition of the Vuforia SDK. Installed on the ASUS ZenBook 14 smartphone, the application incorporates all its features, which were thoroughly tested. Marker recognition was examined, showing a positive outcome [22].

2.3 Augmented Reality in Cultural Heritage Tourism

Augmented reality has found applications across diverse domains, spanning from medicine and education to the automotive industry, healthcare, and tourism [23]. Specifically, its efficacy has been demonstrated in the management and conservation of cultural heritage. AR serves several primary purposes within cultural heritage contexts, which are enhancing traveler experiences, facilitating reconstruction and travel [24], aiding in conservation and preservation efforts [25], and revitalizing past events [26]. Digitization is the predominant technology utilized across these applications, serving as a supplementary approach for safeguarding cultural heritage assets and augmenting conventional conservation practices [27]. In their work, Kim et al. [2] introduced a context-aware mobile AR system designed for a cultural heritage site in South Korea. This system was created using semantic web technology, which incorporates interoperability among five web databases [28]. ARCHEOGUIDE, set in Olympia, Greece, sought to explore the intersection of AR and Greek heritage. The project centered on the archaeological site of Olympia, where practical and demonstrative studies were conducted. Utilizing a system tailored for guided tours, ARCHEOGUIDE [29] facilitated exploration of the site through portable devices including laptops, tablets, and smart phones. The Vítica project was developed in two stages: the initial phase concentrates on creating an application to revitalize cultural heritage, while the subsequent phase centers on assessing the educational impact of employing mobile apps for this purpose. In this Colombian initiative, the digital content and AR project intersect with heritage and culture, with a particular emphasis on the socio-economic sector known as "Guayaquil" in nineteenth-century Medellín's downtown district. The aim of AR is to provide users with an immersive experience where the boundaries between reality and virtuality dissolve, fostering a seamless environment where mobile devices like phones, tablets, and screens showcase the experience, eliminating the need for previously complex equipment [30].

2.4 Designing and Implement Augmented Reality Games

Utilizing the MDA Framework. We employed the Mechanics-Dynamics-Aesthetics (MDA) framework [1, 2], which serves as a foundational approach in game design. This framework delineates the interplay between the rules (mechanics), system behavior (dynamics), and the player's emotional experience (aesthetics). When applied to the design of AR games for cultural heritage, the MDA framework ensures that educational and engagement goals are achieved through carefully crafted gameplay.

*Mechanics.*The mechanics in an AR location-based game designed for cultural heritage tourism, such as in the Wua-Lai communities, involve the integration of GPS technology, AR visuals, and mobile computing. Players interact with the game through their mobile devices, which serve as a window to a layer of digital content superimposed on the real world. Key mechanics include location tracking, AR object recognition, and interactive challenges that unlock cultural narratives or facts.

Dynamics. The dynamics emerge from player interaction with the game mechanics and the environment. For a game inspired by Pokémon GO, dynamics could include exploring real-world locations to discover virtual artifacts, solve puzzles related to the community's history, or participate in virtual reenactments of cultural events. This layer encourages physical movement and exploration, driving the player to visit multiple points of interest within the Wua-Lai communities.

Aesthetics. The aesthetic experience in AR games for cultural heritage tourism aims to evoke a sense of wonder, curiosity, and connection with the culture being explored. By leveraging AR and location-based gameplay, players can experience a seamless blend of the virtual and real worlds, where historical figures come to life and traditional crafts are demonstrated in augmented reality, all while navigating the actual streets of Wua-Lai.

Implementing AR Location-Based Games. To successfully implement an AR location-based game, drawing inspiration from the game design principles of Pokémon GO, tailored for the cultural and historical richness of the Wua-Lai communities, a comprehensive, step-by-step approach is required. This involves Content Development, Technical Development, User Interface Design, and Testing and Iteration, each with its own set of critical tasks and considerations.

In the development of content: we collaborated with local historians, artisans, and cultural experts to curate material that is both informative and engaging. This collaboration aimed to ensure that the content would serve as a solid foundation for the challenges and narratives featured within the game. Specifically, our focus was on promoting the heritage of silver craftsmanship, a cultural hallmark of the Wua-Lai communities renowned for its significance and craftsmanship, as shown in Figs. 1 and 2. This thematic focus not only aims to highlight an essential aspect of Wua-Lai's cultural identity but also seeks to immerse players in the rich traditions and history behind their renowned silver craft.

Fig. 1. The Heritage of Silver Craftsmanship in Wua-Lai Communities.

Fig. 2. The Monster Design Based on Silver Craftsmanship.

Technical Development. we utilized the Unity game engine in conjunction with AR development platforms like ARKit and ARCore, to craft immersive augmented reality experiences. To enhance the localization and mapping capabilities critical for our game, we integrated Mapbox. This combination allowed us to precisely use GPS technology for location-based triggers, enabling the game to unlock content as players explore specific sites within the Wua-Lai communities. Unity's versatile environment, paired with Mapbox's advanced mapping solutions, facilitated the creation of a seamless and engaging user experience. This approach ensured that as players navigate the real-world streets of Wua-Lai, they encounter enriching cultural narratives and challenges, anchored accurately to the community's landmarks and significant locations through GPS coordinates.

User Interface Design. we adopted a user-centric approach to ensure that navigation is intuitive, enabling players to focus on the cultural dimensions of the experience. The interface features a culturally-inspired map pin that is prominently displayed over significant locales, incorporating local iconography to foster a deep connection with cultural points of interest. Set against a live backdrop of Wua-Lai's landmarks, rendered in a style that pays homage to the community's traditional architecture, the game's environment is effectively integrated with the real world. Players navigate the experience through an in-game avatar, adding a personalized touch to their exploratory journey. Navigation is further streamlined by vivid lines that guide players from one objective to the next, ensuring a focused and directional gameplay experience. The interface also includes a discrete tracker for cultural artifacts, encouraging players to discover and engage with the historical narratives behind each item. Designed to be non-obtrusive, the menu buttons provide easy access to settings and information without compromising the immersive experience. Altogether, these interface elements harmonize to not only facilitate a seamless and engaging gaming experience but also to enhance players' understanding and appreciation of the Wua-Lai communities' cultural heritage. The user interface is illustrated in Fig. 3.

Fig. 3. The User Interface in the Silver Craft Mobile Game.

Testing and Iteration. Our testing and iteration process involves rigorous trials with a diverse array of user groups, an essential step in refining gameplay and ironing out any technical glitches. The recent testing phase at Wat Sri Suphan, known as the Silver Temple of the Wua-Lai community, provided valuable insights. As users engaged with our AR game in this culturally significant setting, their interactions with the application allowed us to observe firsthand the harmony between the virtual content and the physical environment. This phase was crucial in ensuring that our representation of the temple's intricate silver craftsmanship and the narratives weaved into our game resonated accurately and respectfully with the cultural significance of the location. By continuously iterating based on direct user feedback and expert consultation, we aim to finetune the player experience to be both educational and respectful of the rich cultural heritage encapsulated within the historic walls of Wat Sri Suphan (Fig. 4).

Fig. 4. Testing the Game in Wat Sri Suphan.

3 Research Methodology

This study employs the Research and Development (R&D) methodology to systematically create and evaluate an AR location-based game tailored for the Wua-Lai communities' cultural heritage tourism. The R&D approach is iterative, beginning with exploratory research, followed by the development and testing of the AR game, and culminating in an analysis of its effectiveness and potential for scalability.

3.1 Participants

For this research, we sought to enroll a group of 40 participants, each a traveler within the Wua-Lai communities and a first-time visitor to the area. The selection was limited to those using Android mobile phones to ensure compatibility with the AR application, specifically designed for the Android operating system. Targeting tourists engaged in Thai tourism, we aimed to evaluate the AR game's potential to enrich their cultural exploration. Recruitment efforts were concentrated at local tourism information centers, through travel agencies that facilitate tours of Wua-Lai, and via online platforms targeting prospective visitors to the region. Clear criteria were communicated to ensure participants understood the commitment, including engagement with the AR application during their visit and subsequent feedback provision. Ethical protocols were rigorously followed, with participants providing informed consent, retaining the right to withdraw at any stage, and being assured of the confidentiality of their identities and responses, in line with the ethical research standards concerning human subjects.

3.2 Instruments and Measurement: Questionnaires

Mobile Application: Silver Craft. The "Silver Craft" mobile application emerges as a pioneering tool in the realm of Augmented Reality (AR), meticulously designed to submerge tourists in the profound heritage of silver craftsmanship intrinsic to the Wua-Lai communities. This application adeptly integrates digital interactivity with the tangible world, orchestrating an educational odyssey through the art of silver crafting. It is equipped with interactive tutorials that adeptly navigate users through the traditional techniques and historical lore enveloping Wua-Lai's renowned silver crafts. To infuse an element of engagement, the application embraces gamification strategies, including challenges and quizzes, aimed at fostering a deeper exploration and enriching the learning experience. For the purpose of this study, the application has been developed exclusively for the Android mobile platform.

Satisfaction Questionnaire. In this study, we have adapted the Satisfaction Questionnaire, as delineated in [5, 6], to critically evaluate the levels of satisfaction and the cultural enrichment engendered by the tourism experience facilitated through the AR application. The questions are intricately crafted, centering on assessing the quality of the user experience, the degree of engagement and enjoyment, and the overarching cultural enhancement imparted by the AR game. For comprehensive insights, the detailed questionnaire is made available in Appendix A for reference.

3.3 Research Procedure

The research on the effectiveness of the "Silver Craft" AR game in enriching the cultural tourism experience in Wua-Lai was meticulously planned and executed through several distinct phases. Initially, the study was anchored by clearly defined objectives, with the primary goal of evaluating the AR game's impact on tourists' engagement with local heritage. The participant pool was carefully selected based on specific criteria, focusing on first-time visitors to Wua-Lai who are Android smartphone users. Recruitment was conducted via local tourism boards, social media, and travel agencies to ensure a diverse sample of 40 participants. Before their actual engagement with the AR game, participants underwent an orientation session to familiarize themselves with its features and navigation. They were then encouraged to explore Wua-Lai using the "Silver Craft" app, which is designed to overlay digital interactivity onto their physical journey, thereby deepening their cultural immersion. Following their experience, participants provided feedback through meticulously developed questionnaires, adapted from established models, to gauge satisfaction and cultural enrichment. The data collected were rigorously analyzed, employing statistical methods for quantitative responses and thematic analysis for qualitative feedback, to distill significant patterns and responses indicative of the AR game's impact. The culmination of this research process involved synthesizing the findings into actionable insights and recommendations, aimed at enhancing the "Silver Craft" app and its utility in promoting cultural heritage through tourism.

4 Result and Data Analysis

The Satisfaction Questionnaire distributed to 40 participants after they interacted with the "Silver Craft" AR game provided significant insights into three crucial dimensions: User Experience, Engagement and Enjoyment, and Cultural Enhancement. In terms of User Experience, the game was rated moderately well for ease of launch and operation (mean = 3.17, SD = 0.95) and intuitive navigation (mean = 3.25, SD = 1.05), though a slightly lower score for technical performance (mean = 2.97, SD = 0.91) indicates room for improvement in minimizing technical issues. The Engagement and Enjoyment dimension received notably higher scores, suggesting that the game successfully kept participants engaged (mean = 3.75, SD = 1.17), enhanced the enjoyment of their visit to Wua-Lai (mean = 3.82, SD = 1.10), and made the experience more interactive and fun (mean = 3.97, SD = 1.09). In terms of Cultural Enhancement, participants felt that the AR game significantly augmented their understanding of Wua-Lai's cultural heritage (mean = 3.72, SD = 1.08), provided informative and relevant content (mean = 3.90, SD = 1.12), and helped them feel more connected to the community (mean = 3.57, SD = 1.17). These results underscore the "Silver Craft" AR game's potential to enrich cultural tourism experiences by boosting engagement, enjoyment, and cultural understanding, despite some technical challenges that need addressing (Fig. 5 and Table 1).

Table 1. Result of Satisfaction Questionnaire (n = 40).

Dimension	No.	Questionnaire item	Mean (SD)
User Experience	1	The AR game was easy to launch and operate	3.17 (0.95)
	2	Navigating through the AR game was intuitive and straightforward	3.25 (1.05)
	3	I experienced no technical issues while using the AR game	2.97 (0.91)
Engagement and Enjoyment	4	The AR game kept me engaged throughout the experience	3.75 (1.17)
	5	I enjoyed using the AR game during my visit to Wua-Lai	3.82 (1.10)
	6	The AR game made my visit more fun and interactive	3.97 (1.09)
Cultural Enhancement	7	The AR game enhanced my understanding of Wua-Lai's cultural heritage	3.72 (1.08)
	8	The content provided by the AR game was informative and relevant	3.90 (1.12)
	9	I felt more connected to the Wua-Lai com- munity after using the AR game	3.57 (1.17)

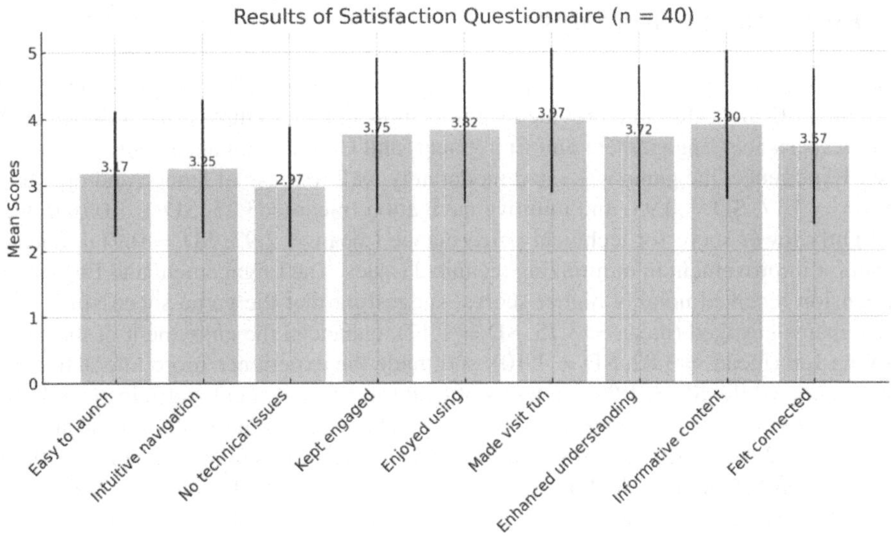

Fig. 5. Results of Satisfaction Questionnaire.

5 Discussion

The "Silver Craft" AR game has demonstrated significant potential in enhancing the cultural tourism experience in Wua-Lai through immersive engagement with the local heritage of silver craftsmanship. The research findings indicate that the game effectively increases tourists' engagement and enjoyment, as evidenced by high scores in the Engagement and Enjoyment dimension. Moreover, the AR game plays a pivotal role in Cultural Enhancement, facilitating a deeper understanding and connection to Wua-Lai's cultural heritage among first-time visitors. Despite its successes, the User Experience scores point towards areas needing improvement, particularly in addressing technical issues encountered during gameplay. This feedback is invaluable for refining the game, ensuring it not only entertains but also educates in a seamless manner.

6 Conclusion, Limitation, and Future Work

In conclusion, the study highlights the significant potential of AR technology, particularly through the "Silver Craft" AR game, in enhancing the cultural tourism experience within the Wua-Lai communities. By offering an immersive and educational platform, the game effectively deepens tourists' engagement with and appreciation for the local heritage of silver craftsmanship. However, the research faces limitations, notably its exclusive focus on Android users and a relatively small participant sample, which may not capture the full spectrum of user experiences. Future research should consider a broader technological scope and a larger, more diverse participant pool to provide a more comprehensive understanding of AR's impact on cultural tourism. Additionally, exploring the development of AR experiences that encompass other facets of Wua-Lai's rich cultural heritage, coupled with ongoing refinements based on user feedback, could further enrich the tourism experience. Addressing the technical challenges identified in this study will be essential for maintaining the relevance and effectiveness of AR applications like "Silver Craft" in the evolving landscape of cultural heritage tourism.

References

1. Husniah, L., Fannani, F., Kholimi, A.S., Kristanto, A.E.: Game development to introduce Indonesian traditional weapons using MDA Framework. Kinet. Game Technol. Inf. Syst. Comput. Netw. Comput. Electron. Control. **4**, 27–36 (2018)
2. Chernbumroong, S., Nadee, W., Jansukpum, K., Puritat, K., Julrode, P.: The effects of gamified exhibition in a physical and online digital interactive exhibition promoting digital heritage and tourism. TEM J. 1520–1530 (2022)
3. Intawong, K., Olson, D., Chariyalertsak, S.: Application technology to fight the COVID-19 pandemic: Lessons learned in Thailand. Biochem. Biophys. Res. Commun. **538**, 231–237 (2021)
4. Intawong, K., Ong-artborirak, P., Boonchieng, W.: Seamless electronic medical record for health management system in emergency patients. In: 2021 Joint International Conference on Digital Arts, Media and Technology with ECTI Northern Section Conference on Electrical, Electronics, Computer and Telecommunication Engineering. IEEE (2021)
5. Lehto, A., Luostarinen, N., Kostia, P.: Augmented reality gaming as a tool for subjectivizing visitor experience at cultural heritage locations—case Lights On! J. Comput. Cult. Herit. **13**, 1–16 (2020)
6. Capecchi, I., Bernetti, I., Borghini, T., Caporali, A., Saragosa, C.: Augmented reality and serious game to engage the alpha generation in urban cultural heritage. J. Cult. Herit. **66**, 523–535 (2024)
7. Cibilić, I., Poslončec-Petrić, V., Tominić, K.: Implementing Augmented reality in tourism. Proc. ICA. **4**, 1–5 (2021)
8. Azuma, R.T.: A survey of augmented reality. Presence: Teleoperators and Virtual Environments **6**, 355–385 (1997)
9. Kipper, G., Rampolla, J.: Augmented reality: An emerging technologies guide to AR. Syngress Media, Rockland, MA (2012)
10. Flavián, C., Ibáñez-Sánchez, S., Orús, C.: The impact of virtual, augmented and mixed reality technologies on the customer experience. J. Bus. Res. **100**, 547–560 (2019)
11. Scribd: https://www.scribd.com/document/53412786/Success-Factors-for-Augmented-Reality-Business-Models. Last accessed 24 April 2024
12. Cranmer, E.E., tom Dieck, M.C., Fountoulaki, P.: Exploring the value of augmented reality for tourism. Tour. Manag. Perspect. **35**, 100672 (2020)
13. Van Krevelen, D.W.F., Poelman, R.: A survey of augmented reality technologies, applications and limitations. Int. J. Virtual Real. **9**, 1–20 (2010)
14. Wang, D., Xiang, Z., Fesenmaier, D.R.: Smartphone use in everyday life and travel. J. Travel Res. **55**, 52–63 (2016)
15. Chung, N., Han, H., Joun, Y.: Tourists' intention to visit a destination: the role of augmented reality (AR) application for a heritage site. Comput. Human Behav. **50**, 588–599 (2015)
16. Özkul, E., Kumlu, S.T.: Augmented reality applications in tourism. Int. J. Contemp. Tour. Res. 107–122 (2019)

17. Wang, X., Love, P.E.D., Kim, M.J., Wang, W.: Mutual awareness in collaborative design: an augmented reality integrated telepresence system. Comput. Ind. **65**, 314–324 (2014)
18. Ukpabi, D.C., Karjaluoto, H.: Consumers' acceptance of information and communications technology in tourism: a review. Telemat. Inform. **34**, 618–644 (2017)
19. Han, D.-I.D., Tom Dieck, M.C., Jung, T.: Augmented Reality Smart Glasses (ARSG) visitor adoption in cultural tourism. Leis. Stud. **38**, 618–633 (2019)
20. He, Z., Wu, L., Li, X.: When art meets tech: the role of augmented reality in enhancing museum experiences and purchase intentions. Tour. Manag. **68**, 127–139 (2018)
21. Tarantino, E., De Falco, I., Scafuri, U.: A mobile personalized tourist guide and its user evaluation. Inf. Technol. Tour. **21**, 413–455 (2019)
22. Tominić, K.: Development of the application "Pazin AP" (2021)
23. Arena, F., Collotta, M., Pau, G., Termine, F.: An overview of augmented reality. Computers **11**, 28 (2022)
24. Okanovic, V., et al.: Interaction in eXtended Reality applications for cultural heritage. Appl. Sci. (Basel) **12**, 1241 (2022)
25. Merchán, M.J., Merchán, P., Pérez, E.: Good practices in the use of augmented reality for the dissemination of architectural heritage of rural areas. Appl. Sci. (Basel) **11**, 2055 (2021)
26. Boboc, R., Duguleană, M., Voinea, G.-D., Postelnicu, C.-C., Popovici, D.-M., Carrozzino, M.: Mobile augmented reality for cultural heritage: following the footsteps of Ovid among different locations in Europe. Sustainability. **11**, 1167 (2019)
27. Salleh, S.Z., Bushroa, A.R.: Bibliometric and content analysis on publications in digitization technology implementation in cultural heritage for recent five years (2016–2021). Digit. Appl. Archaeol. Cult. Herit. **25**, e00225 (2022)
28. Kim, H., Matuszka, T., Kim, J.-I., Kim, J., Woo, W.: Ontology-based mobile augmented reality in cultural heritage sites: information modeling and user study. Multimed. Tools Appl. **76**, 26001–26029 (2017)
29. Vlahakis, V., et al.: Archeoguide: an augmented reality guide for archaeological sites. IEEE Comput. Graph. Appl. **22**, 52–60 (2002)
30. Hincapie, M., Diaz, C., Zapata, M., Mesias, C.: Methodological framework for the design and development of applications for reactivation of cultural heritage: case study cisneros marketplace at Medellin. Colombia. J. Comput. Cult. Herit. **9**, 1–24 (2016)

Advancing Smart Process Planning: Automated Generation of the Bill of Process for Rotational Parts Using Graph Neural Networks

Selin Kesler(✉) and Oliver Lohse

Siemens AG, Munich, Germany
{selin.kesler,oliver.lohse}@siemens.com

Abstract. Mass customized production significantly complicates Computer-Aided Process Planning (CAPP), particularly when it comes to managing diverse manufacturing processes. In the current CAPP systems, process planning relies on the experience of the process planner. The latest achievements in using deep learning technologies for smart process planning systems have been limited by the data representation and the availability of labeled datasets. The contribution of the proposed paper is twofold: it provides a new method for the automated generation of the Bill of Process (BOP) based on graphical representation of the 3D-CAD-model and a new labeled dataset. The proposed method for automated generation of the BOP uses Graph Neural Network (GNN) and its inherent structure to extract the geometrical and informational attributes of the 3D-CAD-model. The components are represented as graphs, which encapsulates the relation between the machining features with their Product Manufacturing Information (PMI) such as surface finishes, shape and positional tolerances. To address the lack of accessible and adequate data, a new automatically generated dataset is proposed, which is built of manufacturable rotational parts with PMIs and machining features following the DIN-norms. The dataset is constructed of ISO-10303-242-STEP files and their corresponding BOPs. The presented GNN approach is trained and validated on the proposed dataset and the outcomes are examined. The resulting BOP details the operations for each machining feature, which are then grouped together into operation blocks. The outputted BOP is undisputed and manufacturable since each feature in the given part is assigned a specific operation.

Keywords: Process planning · Applied AI in manufacturing · Artificial Intelligence Bill of Process · Operation selection

1 Introduction

In today's rapidly evolving industries, automation of Computer-Aided Process Planning (CAPP) systems has become non-negligible in production processes [26]. The integration of digitalization has emerged as a key factor for higher productivity, cost reduction, and improved quality [24]. The making of a new component starts in Computer-Aided

Published by Springer Nature Switzerland AG 2025
P. Sureephong et al. (Eds.): PLM 2024, IFIP AICT 740, pp. 317–329, 2025.
https://doi.org/10.1007/978-3-031-93319-6_24

Design (CAD) system, which is followed by process planning and then ends in Computer-Aided Manufacturing (CAM) system [22]. The steps for manufacturing of a part is determined in process planning stage, in which the geometrical and design properties of the part is taken into consideration with the available machining and technological capabilities in the factory. The selection and sequencing of machining operations are the main objectives of the process planning, which returns a plan with one or multiple machining operations in a sensible sequence. The dimensions, tolerances, surface finish requirements and geometrical dependencies of the features on the part to be manufactured play an important role in the decision making process in BOP generation [22]. Therefore, there are multiple combinatorial factors that need to be investigated. Today, many solutions rely on the experience of the process planners, their knowledge of the plant specifications like available machines and the machine capabilities [26]. This knowledge about the factory and manufacturing specifications are also present in the historical BOPs of this plant, as all these specifications are already considered for the previous BOPs.

This paper presents a new approach to automate the BOP generation, that uses the advantages of the Graph Neural Network (GNN) structure to represent the geometrical and informational attributes of the 3D-CAD-model, while using the available historical BOPs to extract the experience and knowledge of a CAM-programmer to a certain degree. To narrow the wide manufacturing range within a defined scope, the proposed method is constrained to rotational parts. To overcome the limitation of the publicly available data, a dataset built of rotational 3D-CAD-model and BOP pairs is proposed. The goal of this paper is not to generate a complete functioning BOP to be used directly in the production, but to examine the possibility of deep learning capabilities in a near real production scenario for the selection and sequencing of the operations directly based on the 3D-CAD-models.

The paper is organized as follows. The Sect. 2 reviews the existing methods for BOP generation automation in detail and explains their shortcomings. It is followed by a brief introduction to the GNNs and the previous work in 3D-CAD domain based on GNNS. In 3, the guidelines for generation of a synthetic dataset is defined and the proposed automated BOP generation algorithm is explained in detail. The Sect. 4 presents the results of the proposed method and is followed by the Sect. 5 for summary and outlook.

2 State of Art

To automate the BOP generation process, there have been various approaches over the years, from which the most relevant ones are explained in the Section 2.1. The general idea behind the graph neural networks and their current usages with the 3D-CAD-model based applications are described in the Sect. 2.2.

2.1 BOP Generation

The traditional way of BOP creation relies on the experience of the process planners, where the knowledge is stored in the existing BOPs. There are researches to automate the BOP generation process and reducing the dependency on the expertise knowledge.

Due to the lack of publicly available data, the existing methods from literature are fundamentally incomparable in a common metric.

A knowledge-based method is proposed, which uses feature-based 3D-CAD-models to create corresponding BOPs. Rule-based decisions are made to create a BOP based on the existing design features, which is based on on a knowledge library that contains information such as available tools, manufacturing process parameters and materials. The work plan contains various information such as manufacturing processes, tools and feed rates. The analyzed method deals exclusively with rotational parts and is validated by incomplex example parts. The limitation of this method lies in the manual generation and attendance of the knowledge library, which needs a manual effort each time a new parameter wants to be added or when there needs be a slight change in the rules [21].

To reduce the manual effort and build a more dynamic solution, a multi-layer perceptron (MLP) based method is developed for the selection of possible manufacturing operations for rotational parts. The method doesn't contain a direct relation to the 3D-CAD-model and therefore a manually created list containing the feature type, dimension, tolerance and surface finish for a design feature is passed as an input to the MLP. The output is one or sequential combination of multiple process stages for the given feature. This results in exponential growth in the possible output classes if a new process stage is added. Heuristic rules and rules of thumb from process planning are used to pre-structure the neural network. The method is validated by using a case study and the generalization capabilities are not examined in a larger scale dataset [3].

Another approach examines the prismatic 3D-CAD-models with PMIs to extract their associated BOPs. The 3D-CAD-model is converted into a 3D voxel model, in which the individual voxels contain information about geometry, tolerances and surface finishes. This information serves as input for a 3D Convolution Neural Network (3DCNN). The 3DCNN is trained using a 3D-CAD-model dataset consisting of 385 prismatic components with associated manufacturing operations. In order to increase the generalization capability, 60 copies of each prismatic component were created and randomly rotated in 3D CAD space, resulting in 21,000 3D-CAD-models, which are not publicly available for further examination. The 3DCNN returns a set of manufacturing operations for every feature with the associated tool diameter. One shortcoming of the developed approach is scalability, as the 3D voxel models lose detailed information about the original 3D-CAD-models due to the low voxel resolution, while a higher voxel resolution results in higher computational costs [27].

2.2 Graph Neural Network

Graph Neural Networks (GNNs) are designed to process data represented as networks, where entities are nodes (N) and their relationships are edges (E), forming a graph representation as $G = (N, E)$. In a GNN architecture, each layer operates by passing messages between nodes and aggregating information from neighboring nodes. This message passing step allows each node to update its own representation based on the features of its neighbors. This process is guided by learnable functions, enabling nodes to refine their features using both local and global information from the graph. The updated node representations are then combined through pooling or aggregation functions, which extract collective insights from the entire neighborhood. This enables GNNs to effectively learn representations that encode the graph's structure at different scales, making them suitable at tasks such as node classification and link prediction. By iteratively exchanging and aggregating messages, GNNs are capable of capturing intricate patterns and dependencies within graph-structured data [23].

The advantage of the GNNs for capturing the 3D-CAD-model information is proven in feature detection methods that are built as node classification tasks. These methods are based on the boundary representation (B-Rep) of the 3DCAD-model and results in each face in the 3D-CAD-model to be classified with a pre-defined feature class [1, 2].

3 Method

The proposed method is divided into two subsections. In the Sect. 3.1, the systematic synthetic data generation for training and testing purposed is described, while the Sect. 3.2 explores the proposed graph neural network method for the BOP generation. One of the main challenges, that limits the researches is the lack of available data, which contains the 3D-CAD-model and the corresponding BOP. This problem arises due to the confidentiality of these kind of data in manufacturing companies. The problem with the manufacturer specified 3D-CAD-model and BOP pairs is that they are not only manufacturer specific but also a variance in the BOPs is seen between the BOPs from different process planners in the same company.

3.1 Dataset Generation

As the objective of this paper is to propose a method to automatically generate a BOP for a given 3D-CAD-model with PMIs, a dataset is needed to train and test the method on. Due to the lack of publicly available data in the literature, the need of generation of a dataset arises. A labeled synthetic dataset is created, in which the manufacturability of the generated 3D-CAD-model and therefore the direct relation to the real industrial application is set a criteria to be fulfilled. To ensure the manufacturability and the logicalness of the generated parts, DIN-norms and machining handbooks are followed during the automated generation of the 3D-CAD-models and their BOPs.

The feature classes that are used in the generated models and their grouped labels for feature classification are listed in the Table 1. There are 22 machining feature classes and 11 grouped label classes for the features. N.N stands for no norm specified, which means that these features are not limited by the given norms and have more flexibility and is only bounded by the surrounding geometrical attributes.

The feature generation on the rotational part is an iterative process and is based on multiple cylinders that are stacked after each other. Each generated feature on the cylinders is dynamic and depend on the size of the cylinder that it is being applied to. Depending on the applied feature, one or multiple PMIs are added based on the DIN-norm. The process starts by determining how many cylinder segments the shaft should consist of in total, which are then assigned with cylinder segment types from the classes torque, bearing seat and unspecified cylinder. The random decisions on the selection of the cylinder segments must take the listed rules into account. All cylinder segments that are not specified directly as torque or bearing seat, become unspecified cylinder segments and have more degree of freedom in the geometry and design feature assignment.

Table 1. Mapping of Geometry and Design Features to Labels

Feature	Label	Norm for Geometry	PMIs
Keyway Slot	A	DIN 6885–1 [13]	Dimensions [12]
Hole (Radial)	B	N.N	N.N
Hole (Axial)	B	N.N	N.N
Hole (Axial)	B	N.N	N.N
Oil Hole	B	N.N	N.N
Circlip Groove	E	DIN 471 [10]	N.N
O-Ring Groove	E	DIN 3601 [14]	N.N
Undercut	E	ISO 18388, Form H [18]	Surface finish [16]
Groove	E	N.N	N.N
Flat	FL	N.N	N.N
Chamfer	K	N.N	N.N
Fillet (Radius)	R	N.N	N.N
Slot for SHC	S	DIN 6885–1/-2 [13]	Dimensions [12]
Deep Hole	TB	N.N	N.N
SHC Involute Flanks	V	DIN 5480 [9]	Runout [9] Description [19]
SHC Serration Flanks	V	DIN 5481 [11]	Runout [11]
SHC Splined Shaft	V	DIN 14 [7]	Symmetry [7] Description [19]
Gear	V	DIN 13–28 [5]	Concentricity [5] Runout [5]
Cylinder (Planar)	F	DIN 5418 [8]	Runout
Cylinder (Shell)	Z	N.N	N.N
Centering Bore	B + K	DIN 332, Form B [6]	Dimensions [6]
Tapered Shaft End	K	DIN 1448 [4]	Surface finish [15] Linearity [12]

1. Shafts must have either no torque features or at least 2 torque features.
2. External components must be mountable on the shafts.
3. Geometry and design features on the shaft must be manufacturable.
4. Shafts with torque features must have at least one bearing seat or there may be a maximum of 2 bearing seats in total.
5. Bearing seats with a shaft length of 300 mm or more must not be more than 30 percent of the total length.

In the end of the segment definition, a list of cylinder segments is defined, which has the following structure. BER stands for bearing seats, TORQ for torque cylinders and CYL for non-specified cylinders. The length of the model and the assignments of the segments are made randomly within the limitations of defined guidelines to ensure variety in the dataset.

BER	TORQ	BER	CYL	TORQ	CYL

The next step is to select the geometry and design features for each individual cylinder segment. A diameter range between 20 and 110 and length range between 2 and 200 is defined in which each of the geometry and design features should be located. First, based on the previously run segment definition, it is analyzed which cylinder segment type is selected for the currently processed cylinder. For this purpose, all suitable geometry and design features are selected from the associated feature database. Suitable means that the constraint properties applicable to the cylinder segment are fulfilled. The dimensional ranges for diameter and length defined at the beginning, are the only available constraint properties. A random geometry or design feature is then selected from the suitable geometry and design features, which is toleranced using manufacturing information from standards, manufacturer specifications or normed guidelines from the literature as shown at the Table 1. This manufacturing information is integrated into the 3D-CAD-model in the form of PMIs during the creation step of the 3D-CAD-model. If a normal cylinder segment is provided in the segment definition, a cylinder with randomly chosen dimensions is selected and assigned with other machining features such as holes and slots.

Each created feature is labeled and segmented by labeling of the corresponding faces. To access the face information of the 3D-CAD-model, the Boundary Representation (B-Rep) of the model needs to be used, which contains topological information such as faces and edges. Therefore, the models are used in ISO-10303-242-STEP [20] format to have access to these information.

For indicating the feature type, the abbreviation in the label column from the Table 1 is used. For the segmentation, each feature type gets a counter, starting with 1 for the feature at the minimum x-coordinate and ascending to the maximum x coordinate. An example of the labeled and segmented features is shown in the Fig. 1. The labeling of the faces for classification and segmentation plays a crucial role for the matching of the features with the corresponding PMIs, which are also directly assigned to the faces.

After the 3D-CAD-model is generated, labeled and segmented by the features, the corresponding BOPs for each part are created automatically. The Table 2 shows the decision process for the selection of the manufacturing process for the features. Depending on the feature type and their associated PMIs, the manufacturing process is determined, which has the abbreviations (Abr.) as listed at the table. L stands for the length and D for the depth of the feature, which are used for the ratio calculation for deep hole drilling.

Each feature must be assigned to at least one process, but it may be assigned to multiple processes. Different than the combinational approach in the previous works [3], each manufacturing process is managed separately, therefore it allows more freedom in adding and removing new operations. As all the features and their attributes such as PMIs are determined in the 3D-CAD-model generation process, the features can be matched with their corresponding manufacturing processes as shown in the Table 2. Each 3D-CAD-model gets a complete BOP assigned in the following.csv format. To ensure an

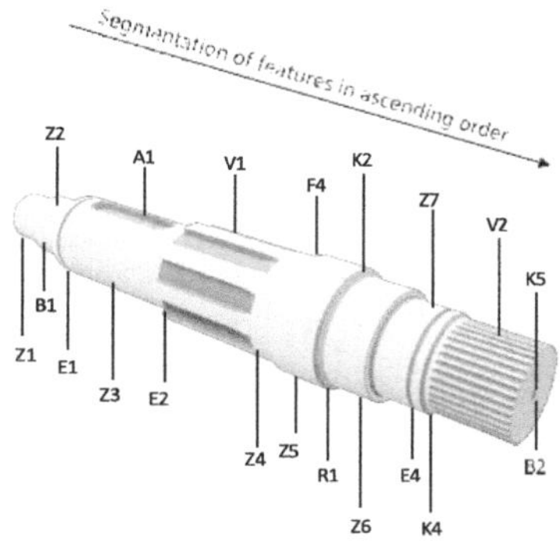

Fig. 1. Labeled and Segmented Feature-Based 3D-CAD-Model

Table 2. Manufacturing Processes and Features.

Manufacturing Process	Abr.	Manufacturing Features
Turning	TR	Groove, Chamfer, Fillet, Cylindrical Flat Surface, Cylindrical Shell Surface
Milling	ML	Pocket, Flat, Groove, Gear
Drilling	DR	Hole (Axial), Hole (Radial), Deep Hole ($3 \leq L/D \leq 30$)
Deep Drilling	DD	Deep Hole ($L/D \geq 30$)
Gear Milling	GM	Gear
Grinding (Hard Turning)	HT	When primary manufacturing process does not meet quality requirements
Gear Grinding	GG	When high quality requirements are needed for gear
Hardening	HD	Gear
Turning	TR	Groove, Chamfer, Fillet, Cylindrical Flat Surface, Cylindrical Shell Surface
Milling	ML	Pocket, Flat, Groove, Gear

easy mapping between the models and the BOPs, they get the same name assigned as file name.

3.2 Graph Neural Network Structure

To incorporate the geometrical and design information of the 3D-CAD-model, a graph structure is used to represent the parts. Each node corresponds to a segmented feature and there exists an edge between nodes if the faces of the segmented features share a common edge in the 3D-CAD-model. An example constellation of a graphical structure is presented in the Fig. 2. To ensure simplicity in the visual representation, only a section of the 3D-CAD-model is represented in the graph, in which each node is highlighted with the color of the corresponding feature in the model.

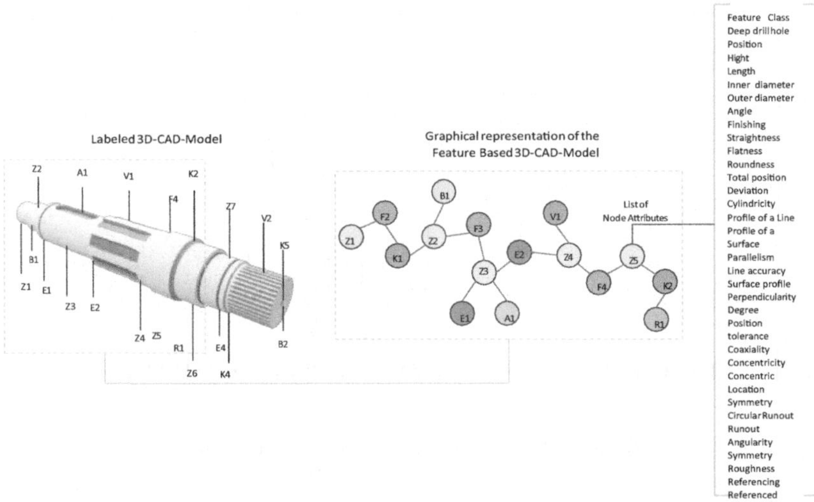

Fig. 2. Graph Representation of the 3D-CAD-Model with Node Attributes

Only one edge type exists in the generated graph structure, which represents the neighborhood characteristics between features. Each node is assigned with a node attribute vector of a fixed size, which has the listed properties from the Fig. 2. The node attributes can be grouped into the following five main categories.

1. Feature class: represents the labeled feature class based on the label list introduced in the Table 1
2. Geometrical attributes of the feature: contains the pre-defined geometrical attributes of the faces belonging to the feature
3. Form and positional tolerances: is a collection of all the available PMI from the 3D-CAD-model
4. Surface finish: contains Ra, Rz and Rmax values, if exists
5. Reference surface configuration: gives information about if the feature is a reference surface or uses a reference surface

To retrieve the graph representation and the node attributes, the ISO-10303-242-STEP [20] format is used, as it is proven to be suitable for B-rep and PMI extraction [2]. After generation of the graphical structure, the automatically generated BOPs need to be encoded as labels output for the multi-class classification task. Therefore, the generated BOPs are one-hot-encoded as shown in the Fig. 3, which is also the output format of the proposed GNN method as the multi-class classification.

If a feature is assigned to an operation, the operation is encoded with 1 and all the other non-assigned operations get 0. As each feature needs to be assigned to at least one operation, each node, representing a machining feature needs to be classified as at least one operation. For each node, an embedding vector is returned, which goes through a softmax function to return a likelihood score for each operation class. A threshold of 0.5 is used to decide on the selection of the operation class based on the returned likelihood score.

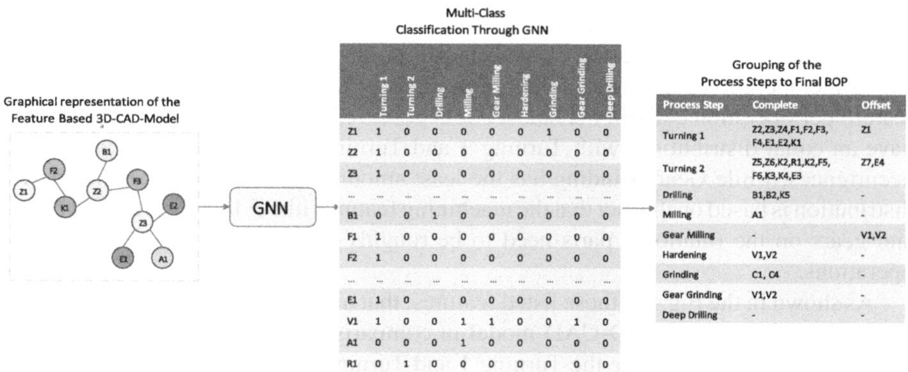

Fig. 3. General Structure of the Proposed Model

The focus of the GNN models selected for the proposed method are the GraphSAGE (SAmple and aggreGatE) [17] And GIN (Graph Isomorphism Network) [25]. The GraphSAGE is selected as it takes into account the degrees of the nodes in the neighborhood by an aggregation function, allowing it to capture the local neighborhood information in a more comprehensive manner. GIN is also proffered to be examined, as the isomorphic approach shows promising results in capturing the general graph structure rather than only focusing on the nodes [25]. This allows it to incorporate the graph representation in node classification tasks. By choosing these two GNN architectures, it is aimed to examine the if focus on local neighborhood or focus on the general graph structure brings more advantage for the use-case of BOP generation.

To test the outcomes of the sleeted GNN methods, the segmented features are grouped together into the operation classes to build the final BOP. Each feature is added to the process step, for which the classification resulted in 1 after the threshold layer. The BOP is constructed with three columns as shown in the Fig. 3, in which the first column is for the process step, second column for the features, that are finished at that process step and don't need any additional machining. Third column is for the features, which are processed at the given process step but not completely finished. They get an offset assigned, meaning that they need to be machined finely in the coming steps like grinding. All the features in the offset column must be present in at least one coming process step.

4 Validation

For the validation of the proposed method, 4192 3D-CAD-models are generated and saved as labeled ISO-10303-242-STEP [20] models. Their corresponding BOPs are created accordingly. The dataset is published in the following link[1]. For the validation purposes, available data is spitted into training and test set with a 70–30 percent distribution. The test set is completely separated by training set and therefore the generalization performance of the model is examined in the validation runs.

[1] https://github.com/selinkesler/BOPGNN.

The created dataset is analyzed in the Fig. 4 in two main aspects, machining feature class distribution and operation class distribution throughout the dataset. The Fig. 4a shows the operation class distribution, representing the number of assigned machining feature per operation class in the dataset. It is observed that the operation classes don't have an equal distribution, with Turning 1 and Turning 2 having the most number of occurrences, while Gear Grinding has the least amount of representation. This uneven distribution is based on the fact that the machining features like cylinder surfaces, grooves and gears on the rotational parts need to be roughly processed in one of the turning operations.

As shown in the Fig. 4b, these listed features, that need to be turned, constitute larger proportion of the overall 3D-CAD-model in comparison to the other features such as slots and holes and therefore the Turning 1 and Turning 2 classes overweight the other operation classes. There is a correlation between the feature class and operation class distribution.

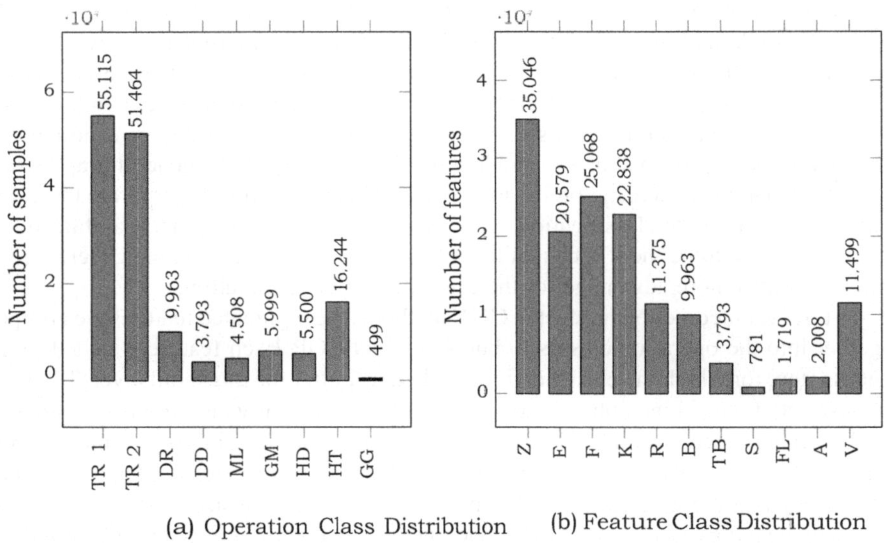

Fig. 4. Dataset Analysis

The Table 3 shows the achieved accuracy overall and single operation class accuracy for the selected GNN models, GraphSAGE [17] and GIN [25], on the test dataset, for which the number of correct classifications are divided to the total number of samples belonging to that class. The first training and test setup is conducted with an equal weight distribution of each operation class and each model is trained for 200 epochs. It is observed that although the SAGE model is built only 27.86% trainable parameters in comparison to the GIN model, it outperforms the GIN approach in every single class accuracy and in overall accuracy by 23.1%. Therefore the desired comparison, mentioned in the Sect. 3.2, between lying focus on local neighboring attributes via SAGE [17] or on general graph structure via GIN [25] for node classification results in the advantage of local attributes over general graph representation. For the next validation runs, only the SAGE model is used, since better results are achieved. Due to the uneven distribution of the nodes per operation class as shown in the Fig. 4a, it is observed that some classes deteriorate under the disadvantage of limited representation in BOPs in the generated dataset. To minimize this disadvantage of the less represented operation classes, a weight assignment is introduced for each operation class to help in the training phase. To find the optimum weight distribution, an ablation study is conducted and the following weights at the Table 3 for each operation class is chosen. With the assignment of the new weights in the training process, it is observed that the single operation class accuracies and therefore the overall accuracy for the test dataset improves in comparison to the equally distributed weights.

Table 3. Single and Overall Class Accuracies.

Class	GIN [25]	SAGE [17]	Weights	SAGE weighted
Turing 1	80.19	**98.37**	0.97	98.23
Turning 2	78.97	97.31	0.97	**98.33**
Milling	83.41	**99.40**	0.95	99.12
Gear Milling	55.93	99.76	0.95	**99.82**
Drilling	94.04	99.33	0.95	**99.37**
Hard Turning	88.91	98.34	0.95	**98.56**
Hardening	55.93	**99.76**	0.95	99.59
Deep Drilling	94.57	96.52	1.0	**97.88**
Gear Grinding	86.00	95.51	1.0	**97.13**
Overall Score	79.45	97.88	-	**98.21**
# Parameters	1335305	372105	-	372105

The best performing SAGE model is built of two convolutional layers with batch normalization layer after each convolution. The output is given to the dropout layer with a drop out probability of 40%. The final layer is a fully connected layer for the mapping of the operation classes. A learning rate of 1e-3 with a 1e-4 weight decay is used with Adam optimizer to achieve the best test results.

5 Conclusion

A GNN-based method is presented, which takes rotational 3D-CAD-model in ISO-10303-242-STEP [20] format and returns the corresponding BOP. As there is no publicly available datasets to train and test the BOP generation capabilities, a synthetically generated dataset is introduced containing 3D-CAD-models with normed PMIs together with their corresponding BOPs. The proposed GNN method is tested on the presented dataset and an overall accuracy of 98.21% is achieved with weighted classes.

For future work, it is important to extend the dataset of 3D-CAD-models and BOPs with new and more complex features and with prismatic parts with a focus in milling operations specifically. It is critical to train and test the proposed method with customer data and extend the BOP creation with variables such as workstation numbers, setup and machining time, quality check procedures and alternatives.

References

1. Cao, W., et al., (eds.): Graph Representation of 3D CAD Models for Machining Feature Recognition With Deep Learning, International Design Engineering Technical Conferences and Computers and Information in Engineering Conference, vol. Volume 11A: 46th Design Automation Conference (DAC) (2020). https://doi.org/10.1115/DETC2020-22355
2. Colligan, A.R., Robinson, T.T., Nolan, D.C., Hua, Y., Cao, W.: Hierarchical cad net: Learning from b-reps for machining feature recognition. Computer-Aided Design **147**, 103226 (2022). https://doi.org/10.1016/j.cad.2022.103226
3. Deb, S., Ghosh, K., Paul, S.: A neural network based methodology for machining operations selection in computer-aided process planning for rotationally symmetrical parts. J. Intell. Manuf. **17**(5), 557–569 (2006). https://doi.org/10.1007/s10845-006-0026-0
4. DIN: 1448-1:1970-01: Conical shaft ends with external thread; dimensions (1970- 01). https://doi.org/10.31030/1083585
5. DIN: 13-28:1975-09: Iso metric screw threads; coarse and fine screw threads from 1 to 250 mm screw thread diameter; root cross-sections, tensile stress cross-sections and pitch angles (1975-09). https://doi.org/10.31030/1068159
6. DIN: 332-1:1986-04: 60° centre holes; types r, a, b, and c (1986-04). https://doi.org/10.31030/2006582
7. DIN: 14-1:1987-02: Iso-metric screw thread; screw threads below 1 mm diameter; basic profile (1987-02). https://doi.org/10.31030/2046559
8. DIN: 5418:1993-02: Rolling bearings; dimensions for mounting (1993-02). https://doi.org/10.31030/2515193
9. DIN: 5480-1:2006-03: Involute splines based on reference diameters - part 1: Generalities (2006-03). https://doi.org/10.31030/9640250
10. DIN: 471:2011-04: Retaining rings for shafts normal type and heavy type (2011-04). https://doi.org/10.31030/1744825
11. DIN: 5481:2019-04: Serration splines (2019-04). https://doi.org/10.31030/3030745
12. DIN: 6885-2:2021-06: Drive type fastenings without taper action, parallel keys, keyways deep pattern for machine tools part 2: Dimensions, tolerances, mass (2021-06). https://doi.org/10.31030/3236422
13. DIN: 6885-1:2021-11: Drive type fastenings without taper action, parallel keys, keyways deep pattern part 1: Dimensions, tolerances, mass (2021-11). https://doi.org/10.31030/3302865
14. DIN ISO: 3601-1:2013-11: Fluid power systems- o-rings - part 1: Inside diameters, cross-sections, tolerances and designation codes (iso 3601-1:2012 + cor. 1:2012) (2013-11). https://doi.org/10.31030/2055597
15. Fritz, A., Hoischen, H.: Technisches Zeichnen: Grundlagen, Normen, Beispiele, darstellende Geometrie ; Lehr-, U¨ bungsund Nachschlagewerk f u¨ r Schule, Fortbildung, Studium und Praxis. Cornelsen-Verlag
16. Grollius, H.W.: Technisches Zeichnen im Maschinenbau. Carl Hanser Verlag, Mu¨ nchen, 5., aktualisierte auflage edn. (2023)
17. Hamilton, W.L., Ying, R., Leskovec, J.: Inductive representation learning on large graphs, http://arxiv.org/pdf/1706.02216v4

18. ISO: 18388:2016-02: Technical product documentation (tpd) - relief grooves types and dimensioning (2016-02)
19. ISO: 6413:2018-09: Technical product documentation - representation of splines and serrations (2018-09)
20. ISO: 10303-242:2022-12: Industrial automation systems and integration - product data representation and exchange - part 242: Application protocol: Managed model-based 3d engineering (2022-12)
21. Krishna, V., NVS, S., Surendra Babu, B.: Feature based modeling and automated process plan for turned components. Adv. Prod. Eng. Manage. **63**, 153–162 (2011)
22. Soori, M., Asmael, M.: Classification of research and applications of the computer aided process planning in manufacturing systems. Independ. J. Manage. Prod. **12**(5), 1250–1281 (2021). https://doi.org/10.14807/ijmp.v12i5.1397
23. Kipf, T.N., Welling, M.: Semi-supervised classification with graph convolutional networks (2017)
24. Trstenjak, M., Opetuk, T., Cajner, H., Tosanovic, N.: Process planning in industry 4.0—current state, potential and management of transformation. Sustainability **12**(15), 5878 (2020). https://doi.org/10.3390/su12155878
25. Xu, K., Hu, W., Leskovec, J., Jegelka, S.: How powerful are graph neural networks?, http://arxiv.org/pdf/1810.00826v3
26. Xu, X., Wang, L., Newman, S.T.: Computer-aided process planning – a critical review of recent developments and future trends. Int. J. Comp. Integr. Manufact. **24**(1), 1–31 (2011). https://doi.org/10.1080/0951192X.2010.518632
27. Zhang, Y., Zhang, S., Huang, R., Huang, B., Yang, L., Liang, J.: A deep learning-based approach for machining process route generation. The Int. J. Adv. Manufact. Technol. **115**(11–12), 3493–3511 (2021). https://doi.org/10.1007/s00170-021-07412-9

Key Design Conflict Identification for Complex Product Based on Causal Discovery

Chen Zheng[1,2](✉), Yang Yu[1], Qiongsheng Zheng[2], Shiqi Gao[1], and Benoît Eynard[3]

[1] School of Mechanical Engineering, Northwestern Polytechnical University, 127 West Youyi Road, Xi'an Shaanxi 710072, People's Republic of China
chen.zheng@nwpu.edu.cn

[2] Guangdong Provincial Key Laboratory of Advanced Welding Technology for Ships, CSSC Huangpu Wenchong Shipbuilding Company Limited, Guangzhou, People's Republic of China

[3] Department of Mechanical Systems Engineering, UMR7337 Roberval, Université de Technologie de Compiègne, CNRS, 60319, 60203 Compiègne Cedex, CS, France

Abstract. Modern product design has become a collaborative activity of crowd intelligence, which inevitably brings an unprecedented increase in design conflicts. The accurate identification of key design conflicts can significantly reduce costs and speed up the design process; on the contrary, it will not only cause a lot of iterations but also become a major hidden problem for product quality. In order to accurately identify the key design conflicts, designers must identify the causes and consequences of the conflicts. With the support of causal discovery, the paper aims to identify the causes and consequences of design conflicts through the perception of the design conflict formation process and achieve the accurate identification of the key design conflicts.

First, a deep survey relating to current studies on product development and conflict identification is conducted. Second, the proposed causal discovery-based design conflict detection approach is presented. Finally, the authors draw a conclusion and point out the direction for future research.

Keywords: Product Design · Design Conflict Identification · Causal Discovery

1 Introduction

Modern product design has become a collaborative process of crowd intelligence, which inevitably leads to an unprecedented increase in design conflicts. According to a survey conducted by Ford Motor Company, more than 40% of collaborative design tasks are faced with various design conflicts [1]. The Boeing Company reported that the design of the Boeing 777 was conducted by 238 integrated design teams, which were composed of more than 8,000 designers, including design, manufacturing, quality, maintenance, etc., a total of 12,000 design conflicts were discovered [2]. As a result, it is crucial to accurately identify the key design conflicts during the collaborative design process. Patterson et al. pointed out that the accurate identification of the key design conflicts enables a 50% reduction of the product development lead time and cost, and product

Published by Springer Nature Switzerland AG 2025
P. Sureephong et al. (Eds.): PLM 2024, IFIP AICT 740, pp. 330–340, 2025.
https://doi.org/10.1007/978-3-031-93319-6_25

quality can be improved by more than 30%. On the contrary, undetected design conflicts will not only cause a lot of iterations but also become a major hidden problem for product quality [3].

Ge et al. pointed out that the accurate identification of key design conflicts requires a thorough consideration of various context factors of the conflicts, such as the object, content, type, degree, and location of the conflict, as well as an in-depth study on the causes and consequences of the conflict [4]. Hirshorn stated that design conflicts can be identified and eliminated accurately if all designers understand the causes and consequences of design conflicts correctly [5]. However, most of the existing studies identified the design conflict based on explicit conflict data and ignored the underlying causes of conflicts behind such conflict data. Therefore, how to find the root causes of design conflicts and their consequences from complex design contextual information is still a challenge for accurate identification of the key design conflicts.

Causal discovery is a process in which an effect has already occurred, by analyzing the causal relationship between variables in a system or dataset, to explore the underlying causes behind the effect, and then uncover the cause-and-effect structure. It has been widely used in biomedicine, management, and social sciences. With the support of causal discovery, the paper aims to identify the causes and consequences of design conflicts through the perception of design contextual information and achieve the accurate identification of the key design conflicts. First, by adopting the context-awareness technology, the design contextual information, such as the domain background, internal and external design context, design intentions, and choices of different designers when the design conflict occurs can be perceived. Then, the root causes and their possible consequences of the design conflict can be accurately determined based on the perceived information by using the causal discovery technology. The key design conflict during the design process can finally be identified.

The rest of the paper is organized as follows. In Sect. 2, a deep survey on current product development is conducted. Section 3 presents the proposed causal discovery-based design conflict detection approach. Finally, the authors draw the conclusion and point out the future research direction in Sect. 4.

2 Literature Review

2.1 Collaborative Design

The current approach to product design is viewed as a collaborative endeavor involving designers from diverse disciplines. Products in industries such as aviation, aerospace, and shipbuilding are evolving into highly complex systems, typically characterized by significant scale and a multitude of subsystems or components. Consequently, the integration of various disciplines including mechanical, electrical, electronics, optics, thermal, and others has become more pronounced. According to a survey conducted by the authors' research team in an airplane manufacturing company, the design of the fuselage involves design teams such as mechanical structure, aerodynamics and flight mechanics, composite materials, flight control, etc. The parameters, such as electrical and mechanical interface, gravity center, inertia moment, etc., which are closely related to the fuselage, should satisfy the design requirements of each other. Moreover, stakeholders involved in various stages of design, including manufacturing, assembly, reliability and quality, etc., are required to engage in collaborate efforts throughout the design process. According to reports conducted by Dörfler and Baumann, a large amount of rework has been found in the design of the Airbus A380 due to neglect of the wire assembly in the fuselage and the pressure-bearing of the wings [6]. As a result, it took nearly ten years for the first Airbus A380 to enter service. In order to avoid a similar situation, together with the aerospace center in Germany and the Netherlands, Boeing began to focus on the upstream and downstream of the design phase, such as manufacturing, quality, maintenance, etc., when developing new aviation products [2, 7].

Various studies have shown that crowd intelligence from different disciplines brings many advantages when dealing with design tasks for complex products. In terms of environmental impact, Liu et al. pointed out that in the product design process, the sustainability of materials and energy should be considered to reduce dependence on limited resources and provide more sustainable solutions [8]. ElMaraghy et al. emphasized the necessity of integrating social and economic factors into product design, and they advocated for using the product remanufacturing process as a vital indicator in evaluating design solutions [9]. Zheng et al. also highlighted the significance of collaborative design and pointed out design process should consider the requirements from various stages of products. Meanwhile, they proposed methods to express design-oriented requirements and transfer manufacturing knowledge to designers [10]. Woolley et al. pointed out that the integration of multi-discipline, multilevel, and multi-objective should be achieved for the collaborative design of complex products. During this process, the global design goal is replaced by the interests of the different designers, and differences in designers' backgrounds inevitably lead to an unprecedented increase in design conflicts [11].

2.2 Design Conflict

A large amount of research on concurrent engineering, collaborative design and design conflict has been conducted since the late 1990s. However, according to El-damnhoury et al. [12], the theories and methods that support the identification and resolution of key design conflicts still lag far behind the growing demand for collaborative design. The delivery of the "HAL Tejas" fighter of the Indian Air Force was delayed by four years because a large number of design conflicts have been found and more than 400 design modifications have been conducted [13].

Researchers have explored the causes and consequences of design conflicts from different perspectives, including psychology, management, and organizational behavior. Onselen et al. collected 32 conflict cases from design professionals and analyzed them using a grounded theory approach. Then, they found that there are broad value differences among designers, such as economic worth versus design ethics and work ethics versus intrinsic motivational drivers, which may generate value-based conflicts [14]. Milliken proposed a group composition model based on dispersion theory, focusing on the distribution of individual characteristics within a group. According to their perspective, group members will autonomously disperse and form different or-ganizations based on differences in social characteristics such as race, gender, nationality, age, etc., leading to conflicts between organizations [15]. However, this model does not take into account the combined effects of multiple social characteristic attributes in the process of conflict formation. Building on this view, Poon et al. have put forward the group conformity theory from a social psychology perspective, suggesting that conflicts arise from the common differences in one or more social attributes between organizations, and each group member will assemble in multiple organizations based on their own different social characteristics and attributes [16]. In addition, Um et al. proposed the implementation of projects between different organizations involves significant differences among participants in terms of decision criteria, strategic importance, roles and responsibilities, and problem-solving approaches, which may cause various types of conflicts, such as cognitive and emotional conflicts [17]. One example related to the Design for Manufacturing (DFM) approach will illustrate the point. While engineers hold the keys to feasibility and how the technology works, designers hold the keys to the customer/market need and principles toward a delightful user experience. If the two sides (design and engineering) don't come together in perfect harmony, it could result in a functional feature that customers don't enjoy and experiences that's inefficient and expensive. In summary, conflicts among designers may arise from various factors, which can lead to disagreements, communication difficulties, collaboration barriers, and ultimately affect the progress and outcomes of projects.

Therefore, due to the complexity of conflict content and degree, and the diversity of conflict types and locations, in order to identify the key design conflicts, it is necessary to fully consider the design contextual information when the conflicts occur, and find out the causes of conflicts and their consequences.

Identification of Key Design Conflicts. Most current conflict identification methods are based on the compatibility of design parameters. There are two main methods to achieve conflict detection: rule-based and constraint-based. The rule-based conflict identification approaches utilize manually defined rule sets and logical expressions to match and evaluate design parameters, aiming to identify potential conflicts. The constraint-based approaches rely on the design structure matrix (DSM) and use mathematical modeling and solver techniques to detect incompatible situations in the design process. Both methods require predefined rules and constraints to help designers to identify potential conflicts. The rule-based approaches based on axiomatic design concepts focus on the logical relationships between design parameters, while the constraint-based approaches based on DSM emphasize the constraint relationships between parameters. The two complement each other and help improve the accuracy and efficiency of conflict recognition. However, they cannot further explore the root causes and consequences of conflicts, making it difficult to judge the seriousness extent and impact on product quality. Hirshorn indicated that one of the most important reasons that it is difficult for designers to identify the key design conflict is the difficulty in finding out the causes and consequences of design conflicts [5]. Therefore, it is necessary to consider the internal and external contextual factors when conflicts occur, and to deeply explore the causes and consequences of conflicts.

Resolution of Key Design Conflicts. Resolution of design conflicts requires designers to extract and analyze the conflict's content, type, phenomenon, degree and location, by considering their indicators, experience, rules, knowledge, preferences, goals and interests, and finally make suitable decisions to solve the conflict. Currently, design conflict resolution approaches focus on the elimination of design conflicts caused by data structure or semantic differences. In order to avoid the conflict caused by semantic differences in different disciplines, Topp et al. developed an ontology that provides a unified semantic expression for the data in different domains such as robots, manufacturing, and product development [18]. Qiao et al. proposed a semantic sharing approach for designers and multi-disciplinary engineers [19]. Zheng et al. developed a domain ontology for robot manufacturing system design, avoiding design conflicts caused by data semantic differences among the stakeholders in different domains [20].

Reasoning-based, rule-based and multi-objective decision-making approaches are also used for conflict resolution. Conflict resolution approaches based on the reasoning mechanism adopted the design conflict solutions that have often been adopted and proven effective by the experts in various domains in long-term practice to construct the expert systems, design knowledge base, or case base. The reasoning-based conflict resolution approach generally is divided into case-based reasoning (CBR), knowledge-based reasoning (KBR), and model-based reasoning (MBR). First, as a typical rule-based conflict resolution method, TRIZ theory formulates 76 standard conflict resolution rules for addressing design conflicts [21]. Second, since designers should evaluate the conflicted design solutions based on the design goals and technical indicators of their respective domains when resolving conflicts, conflict resolution can also be considered a multi-objective decision-making process. Therefore, the existing multi-objective decision-making methods, such as multi-objective optimization algorithm, fuzzy comprehensive evaluation, data envelopment analysis and game theory are also applied in conflict resolution in product design.

The resolution of design conflicts requires stakeholders of the conflict to compromise with each other through negotiation or arbitration. However, according to the study conducted by Maier et al., all stakeholders of the conflict will not intend to communicate or negotiate instinctively when the conflict occurs [22]. Although current conflict resolution methods based on the explicit conflict data can reduce or avoid design conflicts to a certain extent, however, due to their different domain backgrounds, internal and external design context, they can hardly understand each other's design motivations, plans and choices behind the data. Moreover, the rejection of communication further intensifies the estrangement between different stakeholders, making it more difficult to reach an agreement on conflict resolution during conflict negotiation. If the stakeholders resolve the design conflict by compromising with each other based on the explicit conflicted data, but do not consider the cause and consequences of this conflict, certain parameters that are closely related to product quality might be modified, which might lead to the failure of product development.

2.3 Summary

Although a lot of research has been carried out on design conflicts in large and complex products, the following problems still exist when identifying and resolving the key design conflicts:

(1) If all stakeholders fully understand the causes and possible consequences of design conflicts, the conflict can be correctly understood and eliminated. Although many technologies and methods have been proposed for the identification and resolution

of design conflicts, however, there are still few studies on how to find out the cause and consequence of the design conflicts.

(2) Most of the current research uses explicit conflict data as the starting point when identifying and resolving design conflicts. However, due to the different domain backgrounds, internal and external design context, the stakeholders are unable to understand each other's design motivations and choices hidden behind the conflict data. If the stakeholders do not consider the cause and consequences of this conflict, certain important parameters that are closely related to product quality might be modified, which might lead to the failure of product development.

3 Key Design Conflict Identification for Complex Product

In order to accurately identify the key design conflicts, this paper proposed a new key design conflict identification approach for the development of complex products by using causal discovery technology. Based on the perceived contextual information, such as the domain backgrounds, internal and external design context, the cause of the conflict and its possible consequences can be accurately identified. The framework of the proposed approach with relevant technologies will be presented hereafter.

3.1 Context-Aware Technique

Awareness of various contextual information when conflicts occur is a prerequisite to help designers to discover the causes of design conflicts and their possible consequences. Context-aware technology (or context-aware computing), proposed by Schilit, aims to achieve efficient and accurate identification and prediction of the behavior, intention and motivation of the object through the acquisition and calculation of the current contextual information of the individual or group [23]. The data and information about the context of the sensing object can be collected either through the active provision of the object, or the perception of the sensor device. Shimomura et al. obtained the behavioral preferences from the questionnaires provided actively submitted by users [24]. Pirayesh et al. used information communication modules to collect contextual information such as users, suppliers, and external environments to help the designers accurately obtain real-time changing market information [25]. Wang et al. adopted a variety of contextual information, such as product operation logs, industrial sensor networks, and product data models to discover users' intentions to assist the service design [26].

Therefore, by adopting the context-aware technique, the design contextual information, such as the domain background, internal and external design context, design intentions and choices of different designers when the design conflict occurs can be perceived and analyzed, thus finding the cause of the conflict, and its possible consequence. The process of contextual information gathering and representation is illustrated in Fig. 1.

First, considering that the contextual information relating to the conflict will not only change with the different perspectives of designers, showing a high degree of dependence on the time and space they are in; at the same time, they also show multimodality and may exist in a structured form. Different perception methods should be used to gather the aforementioned various types of contextual information, which might be represented in

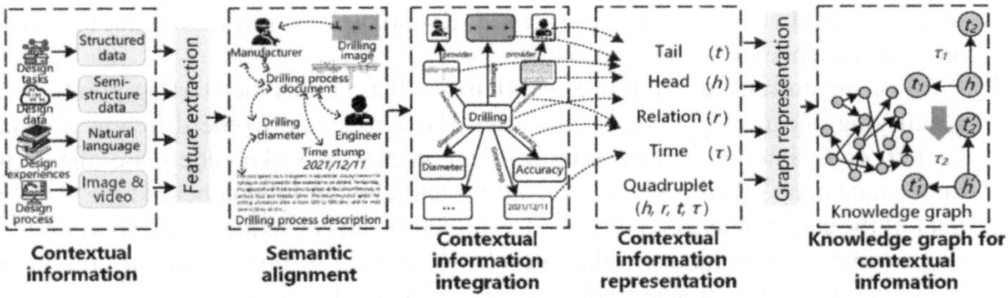

Fig. 1. Contextual information gathering and representation process.

structured, semi-structured data or unstructured text, pictures, and even natural language. For the contextual information expressed in semantically unified structured data (e.g., OWL, RDF, etc.), its semantic features can be directly extracted because this type of information contains rich semantic information. For the semi-structured contextual data acquired through sensors (e.g., XML, JSON, etc.), because it cannot express semantics explicitly, an ontology that enables the representation of the semi-structured information architecture and its instances should be constructed by extracting the semantic information hidden in data tags and structures. The RDF document obtains its semantic features. Moreover, the design experience, quality manual and other information expressed in natural language, a multi-lingual natural language processing tool, such as Polyglot, is used to perform entity recognition on the natural language expression. For the contextual information expressed in image modalities such as pictures or videos, the existing domain picture data sets can be integrated with the Mask R-CNN algorithm and long and short-term memory neural network to identify, segment and extract entities and their relationships in pictures or videos. Through the above process, the semantic information contained in the situation elements can be perceived. Second, the obtained multi-modal contextual information should be aligned and integrated. Distance-based semantic similarity calculation is used to complete the semantic alignment process of entities in each semantic network. Third, the quadruplet (h, r, t, τ), where the head node (h) represents the contextual element entity, the relationship node (r) represents the association between nodes, the tail node (t) represents the spatial or modal information, and the time node (τ) represents time information, is used to represent the aligned contextual information. This quadruplet can be finally transformed into the representation of knowledge graph for further processing to find out the causal relationship between different nodes by adopting the causal discovery techniques, the details of which will be presented in the next subsection.

3.2 Causal Discovery Technique

A large number of studies on causal discovery have found that common data analysis methods such as clustering, correlation, and regression only rely on statistical correlation of data and lack consideration of causal relationships between data. Moreover, existing artificial intelligence methods such as machine learning regard data processing as a black box, which lack explainability, and the overreliance on data fitting may also result in learning the wrong causality across data. Causal discovery is an emerging research field developed on the basis of statistics and data science and was first proposed by Judea Pearl [27]. The principle of causal discovery is to mine the causal relationship between variables from complicated data when a phenomenon has already occurred and to discover the underlying reasons behind the phenomenon. It has been applied in fields such as medicine, ecology, and social sciences. For example, Prosperi et al. developed a clinical intervention model based on causal discovery to predict the therapeutic effects of different drugs [28]. Sugihara et al. explored the causal relationship between the number of different species in the ecosystem by causal discovery [29]. Ratledge et al. used a causal discovery method to evaluate the impact on the regional economy brought by power grid access [30]. Liu et al. constructed a causal discovery framework based on incomplete observation data to mine the causal relationship between the alleged facts and judgment results in legal files to provide support to make more accurate judgments [31].

In summary, the context-aware technique can achieve efficient and accurate acquisition of contextual information, such as the current background, status, and environment of cognitive objects, while causal discovery can explore the causal relationships between variables from massive and complex information. Therefore, by integrating the context-aware technique and causal discovery, the root cause of the conflict and its possible consequences can be found from the complex contextual information when the conflict occurs. Accurate identification and reasonable resolution of key design conflicts can finally be achieved.

Figure 2 represents the key conflict identification process based on the causal discovery technique. First, after the knowledge graph which represents the contextual information is completed, the Greedy Fast Causal Inference (GFCI) algorithm combined with timing constraints to construct the causal association between the nodes in the graph. Since there are hidden variables in the design contextual information that are not perceived or cannot be measured, which may lead to false causal relationships, the GFCI algorithm can be used to describe the cause and effect between variables by considering the hidden variables. Considering the time node (τ) as the timing constraint, that is, the "effect" cannot occur before "cause", a Partial Ancestor Graph (PAG, Partial Ancestor Graph) considering hidden variables is constructed to represent the causal relationships between different nodes. Then, since the quadruplet (h, r, t, τ) has been used to provide a unified semantic expression for the contextual information that generates conflicts, it can be used Semantic Web Rule Language (SWRL) to conduct the reasoning process and find out the root causes of the conflict and its possible consequences. By judging the extent and severity of the conflicts, the key design conflict can be finally identified and thus a more powerful support for the reasonable resolution of conflict can be provided.

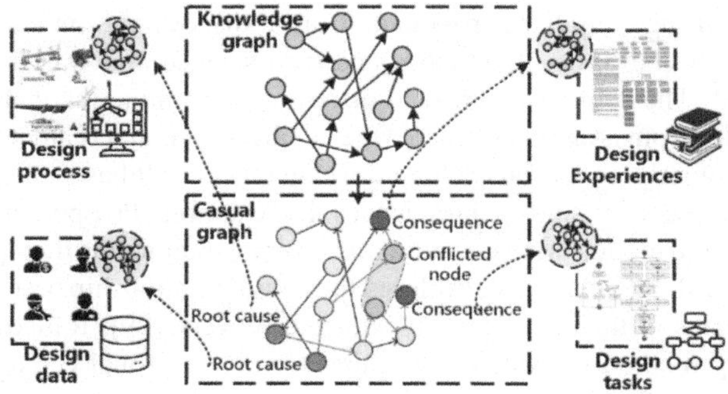

Fig. 2. Key conflict identification process based on the causal discovery technique.

4 Conclusion

The paper aims at the accurate identification of the key design conflicts during the collaborative design process by using the causal discovery technique. First, a deep survey on current product development is conducted. Then, the framework and relevant techniques of the proposed causal discovery-based design conflict detection approach by considering the limitations of current studies are discussed. Future research work can be generally divided into two parts. On the one hand, the conflict resolution approach should be further studied once the key design conflict has been identified. On the other hand, in order to demonstrate the effectiveness of the key design conflict identification approach, the proposed approach should be implemented into an industrial case study based on a complex product design and compared with existing collaborative design methods.

Acknowledgement. This work was supported by the Natural Science Basic Research Program of Shaanxi (Grant No. 2023-JC-YB-299), the National Natural Science Foundation of China (Grant No. 72471193) and the Aeronautical Science Foundation of China (Grant No. 2024Z072053001).

References

1. Ford Motor Company, Measuring the impact of knowledge management. (2015)
2. Xu, J., Su, H., Zhang, H.: Research on design collaboration of aircraft digital mock up for suppliers. Journal of Physics: Conference Series **2006**(1), 012066 (2021). https://doi.org/10.1088/1742-6596/2006/1/012066
3. Patterson, K., Grenny, J., McMillan, R., Switzler, A.: Crucial confrontations: tools for resolving broken promises, violated expectations, and bad behavior. McGraw-Hill Book Company, New York, USA (2012)
4. Ge, X., Leifer, L., Shui, L.: Situated emotion and its constructive role in collaborative design: a mixed-method study of experienced designers. Des. Stud. **75**, 1–41 (2021). https://doi.org/10.1016/j.destud.2021.101020
5. Hirshorn, S.R.: NASA systems engineering handbook. DIANE Publishing (2020)
6. Dörfler, I., Baumann, O.: Learning from a drastic failure: the case of the Airbus A380 program. Ind. Innov. **21**(3), 197–214 (2014)

7. Ciampa, P.D., et al.: A collaborative architecture supporting AGILE design of complex aeronautics products. In: 8th AIAA/ISSMO Multidiscip. Anal. Optim. Conf., pp. 4138–4156. Denver, USA (2017). https://doi.org/10.2514/6.2017-4138
8. Liu, Y., Yang, M., Guo, Z.: Reinforcement learning based optimal decision making towards product lifecycle sustainability. Int. J. Comput. Integr. Manuf. **35**(10–11), 1269–1296 (2022). https://doi.org/10.1080/0951192X.2022.2025623
9. ElMaraghy, H., ElMaraghy, W.: Adaptive cognitive manufacturing system (ACMS)–a new paradigm. Int. J. Prod. Res. **60**(24), 7436–7449 (2022). https://doi.org/10.1080/00207543.2022.2078248
10. Zheng, C., et al.: A requirement-driven architecture definition approach for conceptual design of mechatronic systems. Integr. Comput. Aided. Eng. **26**(4), 361–382 (2019)
11. Woolley, A.W., Chabris, C.F., Pentland, A., Hashmi, N., Malone, T.W.: Evidence for a collective intelligence factor in the performance of human groups. Science **330**(6004), 686–688 (2010). https://doi.org/10.1126/science.1193147
12. Eldamnhoury, E., Healy, L., Fruchter, R.: End-to-end framework in support of virtual design-engineering-manufacturing-construction space exploration. In: Work. Intell. Comput. Eng., Berlin, Germany (2021)
13. Abbott, G.: The tejas story: the light combat aircraft project. Air Power Hist. **59**(2), 49 (2012)
14. van Onselen, L., Valkenburg, R., Snelders, D.: Value-based conflicts experienced by junior design professionals in collaborative practice. CoDesign. **18**(4), 466–482 (2002). https://doi.org/10.1080/15710882.2020.1854314
15. Milliken, F.J.: Diversity and corporate social responsibility: exploring the potential connections between top management team/board diversity, CSR, and workforce diversity. In: Oxford Handb. Corp. Soc. Responsib. Psychol. Organ. Perspect., pp. 96–123. Oxford University Press, Oxford, UK, (2019). https://doi.org/10.1093/oxfordhb/9780198802280.013.5
16. Poon, G.M.Y., Murnighan, J.K.: The interaction effects of faultline on group perception and group viability. Acad. Manag. Proc. **1**, 16156 (2019)
17. Um, K.H., Oh, J.Y.: The mediating effects of cognitive conflict and affective conflict on the relationship between new product development task uncertainty and performance. Int. J. Project Manage. **39**(1), 85–95 (2021)
18. Topp, E.A., et al.: Ontology-based knowledge representation for increased skill reusability in industrial robots. In: IEEE/RSJ Int. Conf. Intell. Robot. Syst., pp. 5672–5678. IEEE (2018). https://doi.org/10.1109/IROS.2018.8593566
19. Qiao, L., Qie, Y., Zhu, Z., Zhu, Y., Zaman, U.K., Anwer, N.: An ontology-based modelling and reasoning framework for assembly sequence planning. Int. J. Adv. Manuf. Technol. **94**, 4187–4197 (2018). https://doi.org/10.1007/s00170-017-1077-4
20. Zheng, C., et al.: Knowledge-based engineering approach for defining robotic manufacturing system architectures. Int. J. Prod. Res. **61**(5), 1436–1454 (2022). https://doi.org/10.1080/00207543.2022.2037025
21. Ilevbare, I.M., Probert, D., Phaal, R.: A review of TRIZ, and its benefits and challenges in practice. Technovation **33**(2–3), 30–37 (2013)
22. Maier, A.M., Eckert, C.M., Clarkson, P.J.: Factors influencing communication in collaborative design. J. Eng. Des. **32**(12), 671–702 (2021). https://doi.org/10.1080/09544828.2021.1954146
23. Schilit, B., Adams, N., Want, R.: Context-aware computing applications. In: 1st Work. Mob. Comput. Syst. Appl., pp. 85–90. IEEE (1994). https://doi.org/10.1109/WMCSA.1994.16
24. Shimomura, Y., Nemoto, Y., Ishii, T., Nakamura, T.: A method for identifying customer orientations and requirements for product–service systems design. Int. J. Prod. Res. **56**(7), 2585–2595 (2018). https://doi.org/10.1080/00207543.2017.1384581

25. Pirayesh, A., et al.: Model- driven requirements elicitation for manufacturing system development. In: Enterp. Interoperability Smart Serv. Bus. Impact Enterp. Interoperability, pp. 343–368. Wiley Online Library, Hoboken, USA (2018). https://doi.org/10.1002/9781119564034.ch42
26. Wang, Z., Chen, C.-H., Zheng, P., Li, X., Khoo, L.P.: A novel data-driven graph-based requirement elicitation framework in the smart product-service system context. Adv. Eng. Informatics **42**, 983–993 (2019). https://doi.org/10.1016/j.aei.2019.100983
27. Pearl, J.: The seven tools of causal inference, with reflections on machine learning. Commun. ACM **62**(3), 54–60 (2019). https://doi.org/10.1145/3241036
28. Prosperi, M., Guo, Y., Sperrin, M., Koopman, J.S., Min, J.S., He, X.: Causal inference and counterfactual prediction in machine learning for actionable healthcare. Nat. Mach. Intell. **2**(7), 369–375 (2020). https://doi.org/10.1038/s42256-020-0197-y
29. Sugihara, G., et al.: Detecting causality in complex ecosystems. Science **338**(6106), 496–500 (2012). https://doi.org/10.1126/science.1227079
30. Ratledge, N., Cadamuro, G., Cuesta, D.L.B., Stigler, M., Burke, M.: Using machine learning to assess the livelihood impact of electricity access. Nature **61**(7936), 491–495 (2022). https://doi.org/10.1038/s41586-022-05322-8
31. Liu, X., Yin, D., Feng, Y., Wu, Y., Zhao, D.: Everything has a cause: leveraging causal inference in legal text analysis. In: Annu. Conf. North Am. Chapter, pp. 1928–1941. Assoc. Comput. Linguist. (2021). https://doi.org/10.18653/v1/2021.naacl-main.155

A Proposition to Leverage Productization, Modularity, and Sharing in Design to Improve the Lifecycle Performance of Rural Healthcare Delivery System

Renu Tewari and Vishal Singh(✉)

Department of Design and Manufacturing, Indian Institute of Science Bangalore, Bengaluru, Karnataka 560012, India
singhv@iisc.ac.in

Abstract. The traditional Public Healthcare Centers (PHCs) in most developing countries, including India, suffer from inadequate infrastructure, operations, and maintenance. Consequently, rural healthcare delivery suffers, which remains a major concern. To address the shortage of quality healthcare facilities, several countries have resorted to using Mobile Health Units (MHUs) to deliver healthcare services to other remote and rural areas. However, the standard MHUs suffer from their limitations due to inherent inefficiencies resulting from frequent relocations, high operational costs, and makeshift arrangements. As a result, the Rural Healthcare Delivery System (RHDS) suffers inefficiencies over the lifecycle. To address these challenges, and to improve the lifecycle performance and efficiency of the RHDS, this paper presents a novel approach building on the productization and modularity of the PHCs and the MHUs. The proposed solution adopts a systems approach, seeking enhanced resource efficiency at the macro level. The hypothesis posits that minor tweaks in design and operations of these PHCs and MHUs through the concept and modularity and sharing in design can provide greater resource efficiency in the RHDS. An exploratory case study conducted in Southern Bangaluru and literature review. These studies laid the foundation for the proposed RHDS. The study concludes that modularity and sharing in design concepts can be leveraged for a synergistic integration between PHCs and MHUs to create a novel RHDS with enhanced resource efficiency.

Keywords: Built environment · Modular construction · Mobile Health Units

1 Introduction

Globally, delivering essential public services such as quality healthcare remains a significant challenge, particularly in underdeveloped and developing countries. The challenge is even more prominent in populated and geographically diverse countries such as India, where a substantial rural-urban disparity exists in healthcare delivery, compounded by demographic, epidemiological, and health transitions [1]. The government

Published by Springer Nature Switzerland AG 2025
P. Sureephong et al. (Eds.): PLM 2024, IFIP AICT 740, pp. 341–355, 2025.
https://doi.org/10.1007/978-3-031-93319-6_26

has made significant efforts to bridge these gaps by introducing new interventions, policies, and outreach programs, such as telemedicine [2], health insurance [3], Mobile Health Units [4, 5], and deployment of community health workers [6]. Nevertheless, the efforts to bridge the rural-urban healthcare divide are primarily driven by technological and service-oriented initiatives, with less progress on another critical aspect of healthcare, i.e., the physical infrastructure or the built healthcare facilities (HCFs). Despite all the developments in telemedicine, health insurance and the deployment of community health workers, it is essential to recognize that adequate physical infrastructure is central to facilitating effective rural healthcare delivery.

The physical infrastructure in rural areas comprises traditional Public Healthcare Centers (PHCs) and the standard MHUs. While the PHCs often suffer from inadequate infrastructure, operations, and maintenance, the standard MHUs suffer from their own limitations due to inherent inefficiencies resulting from frequent relocations, high operational costs, and makeshift arrangements. As a result, the Rural Healthcare Delivery System (RHDS) suffers inefficiencies over the lifecycle. To address these challenges, and to improve the lifecycle performance and efficiency of the RHDS, this paper presents a novel approach building on the productization and modularity of the PHCs and the MHUs.

The proposed solution adopts a systems approach, seeking enhanced resource efficiency at the macro level. The current research postulates that a synergistic integration of static healthcare facilities and Mobile Health Units at the design and operational level can lead to enhanced resource efficiency in RHDS. This research acknowledges that modular spatial solutions offer opportunities for enhancing RHDS. These modular spaces encompass both fixed spaces like PHCs and mobile spaces like MHUs. Consequently, there is a unique opportunity to merge and leverage the advantages of stationary and mobile modular spaces, which require systematic integration into the RHDS. The proposed approach is founded on the principle that RHDS ought to expand beyond fundamental public health amenities and services to encompass specialized healthcare services. Consistent with the Astana Declaration, it is argued that rural and remote communities have the same needs and rights for access to the state-of-the-art healthcare services that are available to their urban counterparts in their vicinity.

2 Background Study

To improve RHDS, the revitalization of built PHCs is indispensable. In the current practice, this revitalization involves construction, renovation, and reconstruction, often using traditional in-situ construction methods. Nonetheless, conventional onsite construction and renovation of established HCFs face numerous challenges, including inefficiencies and issues related to quality, safety, and sustainability [7]. Further, employing traditional construction approaches for renovating, and reconstructing HCFs can have adverse impact, rendering the facility non-operational for extended periods during remodeling process [8]. Additionally, renovating obsolete HCFs using in-situ construction is financially burdensome and non-feasible considering the waste concerns [9]. Hence, there is a need to explore alternative solutions for designing, constructing, operating, and renovating built HCFs. Over time, modular construction technology has emerged as a promising alternative in enhancing the design, construction, and operations of built HCFs [10].

Modular construction, especially prefabricated prefinished volumetric construction (PPVC), has emerged as an alternative solution in the architectural, engineering, and construction (AEC) industry [11]. Modular construction entails the prefabrication of building components offsite, their subsequent transportation to the construction site, and the assembly into a well-furnished infrastructure. Modular construction technology has gained global attention due to its numerous advantages over traditional construction, including reduced construction time, enhanced building quality, workmanship, and productivity, and most importantly, reduced waste generation [12]. Despite these advantages, the adoption of modular construction remains limited due to reported reasons such as the need for more design guidelines, lack of investors and trainers, poor public perception, lack of structural reliability, transportation issues, and intensive capital investment [12]. As a result, PPVC is still in its infancy, and much of the knowledge regarding PPVC still needs to be explored.

While PPVC is considered state-of-art in the AEC industry, modularity and modular volumetric units have been explored for habitable spaces for quite some time, ranging from modular bathrooms, kitchens, dormitories, and official cabins to other use cases such as mobile homes, caravans, and disaster relief shelters. More recently, in the healthcare context, there has been growing emphasis on "space on wheels" or MHUs to deliver essential public services in rural and remote regions. In rural healthcare, the MHUs currently operated by the government and Non-Governmental Organizations (NGOs) can be seen as a volumetric modular solution. These MHUs are custom vehicles equipped with medical equipment, medicines, and healthcare professionals. These independent MHUs successfully provide healthcare services to the most vulnerable sections of society. However, the weaknesses associated with the logistical challenges and high operational cost leave several aspects of last-mile delivery unresolved [4, 13].

3 Research Questions and Methodology

This paper revisits modular spatial solutions in healthcare from a systems perspective. Globally, MHUs-based solutions have been conceived, implemented, and investigated as standalone solutions that lack physical connection with built PHCs, which are static and often outdated in rural and remote regions. The point of departure in this proposal is the systematic integration of the static and mobile modular spaces in the RHDS. The primary premise is that systematically integrating modular HCFs and MHUs will lead to a resource-efficient RHDS with greater agility, flexibility, and accessibility to variable healthcare requirements in rural and remote communities. The main research questions are:

R.Q. 1: What is the present condition of rural healthcare facilities (HCFs) in India, and what key factors influence the determination of resource efficiency within these facilities?

R.Q. 2: What alternative solutions in AEC industry have been previously investigated to mitigate the last-mile service gap in the Rural Healthcare Delivery System?

R.Q. 3: Can productization and modularity lead to a systematic integration of static HCFs and MHUs for improved resource efficiency in the Rural Healthcare Delivery System?

3.1 Research Methodology

The research data were gathered through a literature review and desktop analysis to evaluate the current development trends and capabilities of constructed HCFs and MHUs. The literature review encompasses both published and grey literature sources. Published literature includes articles from electronic databases such as Google Scholar, PubMed, Science Direct, and BioMed Central. Grey literature includes Government reports, webpages, reports from NGOs, editorials, commentaries, and investigative news articles. A breakup of the reviewed literature is shown in Table 1. It is important to note that the authors consulted multiple documents throughout the study, but due to space constraints, Table 1 includes only a few significant literature.

Table 1. The breakup of reviewed literature

Theme	Type	Notable sources
Status of RHDS in India	Published literature Grey literature	[14–17]
Status of PHCs in India	Published literature	[18–20]
	Grey literature	[21]
Operation of MHUs	Published literature	[4, 22–25]
	Grey literature	[26, 27]
Modularity in AEC and HCFs	Published literature	[8, 11, 28–30]
	Grey literature	[31]
Design and resource efficiency	Published literature	[32, 33]
	Grey literature	N/A

Descriptive abduction has been used to analyze the collected data. Descriptive abduction imparts empirical data with theoretical underpinnings to describe a new phenomenon, which, in this case, is the system-level architecture of the RHDS. Abduction reasoning is a widely used method to generate the best possible explanation of the patterns or trends of any system under investigation.

4 Results and Discussion

4.1 Status of the Current Rural Healthcare Facilities

In urban India, modern, sleek healthcare facilities (HCFs) are prevalent, while in rural India, healthcare facilities such as PHCs struggle to establish their presence as shown in Fig. 1. Rural PHCs in India typically contend with challenges such as inadequate investment, poor maintenance and upgrades, technology obsolescence, and a lack of flexibility and adaptability to changing spatial and service requirements. Additionally, PHCs in rural areas face limited opportunities for structural transformation and are illequipped to keep pace with emerging healthcare technologies and demographic transitions.

Like the Indian context, the situation is dire even in developed countries such as the UK, which grapples with unprecedented backlog in care and physical infrastructure maintenance [8, 34]. Many National Health Service (NHS) buildings in the UK are decades old, some dating to the Victorian period as mentioned in a previous study [35]. The lack of adaptability and ill-maintenance of rural HCFs became increasingly evident and pronounced during the COVID pandemic. Obsolete buildings and poorly maintained HCFs can impede the quality of healthcare services. Moreover, the neglect to modernize traditional HCFs and PHCs, coupled with their inability to accommodate modern healthcare equipment, can deprive local communities of the potential benefits of contemporary medical technologies. Also, in response to demographic and epidemiological transitions, our constructed healthcare facilities should exhibit greater agility, responsiveness, flexibility, and adaptability to meet diverse service delivery requirements. Hence, it is essential to recognize that efforts should not only focus on improving medical practices, introducing modern technologies, and increasing staffing levels or introducing more service-oriented solutions but also enhancing the capabilities of the rural PHCs to make them more robust, adaptable, resilient, and future-ready. However, PHCs are static infrastructure equipped with bare minimum services and are fulfilling the last mile service delivery gap in rural and remote regions. This static infrastructure has the potentiality of providing not only basic primary care but also the secondary care to the residents.

Fig. 1. Primary Healthcare Centres (Image source: Google images)

4.2 Review of the Current Trends/Solutions

This section briefly reviews the emerging trends in modular construction and healthcare-built environment and thereby discusses future possibilities.

Modular Construction and Emerging Trends. Over the years, modular construction technology has made significant strides in the field of Architecture, Engineering, and Construction technology [11]. Modular solutions built using this technology have found applications in various contexts. In the healthcare sector, modular solutions have shown huge significance in increasing capacity during high demand or to continue providing services while the current healthcare facility is being renovated or refurbished. These modular solutions encompass a spectrum varying from standalone modular units to complete hospitals constructed using volumetric modules.

The importance of modular construction as a technology and urgent need for these modular-built facilities was also universally recognized during the Covid-19 pandemic when several healthcare facilities such as isolation centres, testing centres, accommodation for healthcare workers, and other quarantine facilities were constructed worldwide to address the massive shortage of infrastructure, healthcare facilities, and quarantine centres [35]. Similarly, in response to Wuhan Outbreak, Houshenshan and Leishanshan hospitals in China were swiftly constructed using steel-based prefabricated modular units within 10 days [36]. Additionally, during the pandemic, Italian architects deployed shipping containers called Connected Units for Respiratory Ailments (CURA). These CURA pods could be affixed to existing built healthcare facilities, serving either as standalone units or extensions, ensuring uninterrupted medical care [37]. Similarly, the Lei Yue Mun Quarantine facility in Hong Kong, accommodating incoming patients from overseas with 352 rooms, dramatically reduced construction waste to a mere 5%, compared to the 20% typically generated by traditional methods [38]. Thus, modular systems played a pivotal role in global preparedness during crises like the COVID-19 pandemic, demanding swift emergency responses and resilient healthcare infrastructure.

It is noteworthy that modular facilities have been utilized in healthcare long before the pandemic. For instance, Bridges et al. documented the successful implementation of modular laboratories as cost-effective assets in public health campaigns against malaria in Mali and Zambia [39]. Moreover, the Royal Infirmary of Edinburgh in the UK established tailored modular facilities within repurposed mobile units [34]. Similarly, mobile health units (MHUs) represent another vital modular solution, enhancing access to public healthcare, particularly in rural, challenging, underserved, and unreached areas, both nationally and internationally. MHUs bring healthcare services directly to populations, notably in rural, vulnerable, and underserved regions. Table 2 represents the various use cases of these MHUs in India and overseas as detailed in a previous study [30]. These MHUs have proved to be effective in fulfilling last mile service delivery gap by providing diverse services such as diagnostic care, curative care etc.

However, the efforts are still in the infancy stage and are limited by human, structural, logistic, operational, and financial constraints. For instance, when the government or NGOs deploy MHUs, the ancillary facilities around them are often makeshift, consisting of temporary tents for purposes such as registration rooms, pharmacy rooms, and restrooms. This situation leaves the end-user experience unresolved. This emphasizes the need for a thorough socio-economic analysis to ensure the successful operation of MHUs. Also, these endeavors are driven by the need to address immediate and urgent healthcare challenges rather than focusing on long-term improvements in the rural healthcare physical service infrastructure. However, the authors contend that our efforts should be directed at more than just constructing new infrastructure or enhancing capacity through temporary solutions. Instead, the emphasis should be on sustainable infrastructure utilization, thereby achieving resource efficiency. This argument forms the basis of the proposed solution approach.

4.3 Need for Modularity and Sharing in Current RHDS Architecture

The need for rural healthcare facilities to be more agile, flexible, and up to date with emerging technologies has already been emphasized. The concept of modularity can

Table 2. Literature on the design and implementation of MHUs as service delivery system

Year	Country	Project/Authors	Short Description
2009	India	Smile on Wheels-Smile foundation	Doorstep OPD services are provided. Examination and consulting are provided under a retractable temporary roof installed beside the MHU
2015	India	Mobile eye surgical unit in India	Cataract surgeries are performed in remote areas of Tamil Nadu using a pair of buses–(1) a preparatory MHU, and (2) a surgical MHU; connected by a vestibule
2015	India	Siemens Sanjeevani Mobile clinic	MHU retrofitted on a minibus to commute and serve municipal region in Mumbai targeting migrant labourers, daily wage earners, and others
1982	India	Relief on the move-HelpAge's mobile health service	Provides healthcare to destitute elders through MHUs consisting of a doctor, pharmacist, and social worker
2014–2015	India	Varisthajana Swasthya Sewa Abhiyan	MHUs provide vital medical services to the elderly at their homes, such as medical consultations, medicine distribution, basic diagnostic tests, special health camps, and palliative care
2016	India	Kumar M	MHUs deployed in three districts of Jharkhand for diagnostics and laboratory tests
2006	India	Deen Dayal Mobile Health Clinic	MHUs integrated with Global Positioning System (GPS) for tracking accessibility in remote areas. Each MHU van has a doctor, staff, necessary appliances, and provides medicines as part of the programme

(*continued*)

Table 2. (*continued*)

Year	Country	Project/Authors	Short Description
2003	Greece	Labiris G. et al	A four-wheel-drive vehicle converted into an MHU to provide conventional medical and telemedical services to remote glaucoma patients
201–16	India	Dhanwantari	Provision of awareness, diagnostic and curative services through MHUs in the rural community
Key is sues and gaps	MHU operations are fraught with operational challenges in most of the above cases. For instance, for several days a month, the MHUs remain idle and unused when not camped at a service site. There is limited research and understanding of their operational efficiency in RHDS		

play a crucial role in bridging this gap by modernizing healthcare infrastructure and making it more responsive to emergency as well as occasional demands. Modularity as a concept arises from the concept design phase where in the product architecture is formed. Modular product architecture allows the decomposition of the product into independent subassemblies and components. The independent units are modules which self-contained units, encapsulating a specific set of functions within the product [40]. This approach allows modules to be easily interchanged, added, or removed without affecting other parts of the system. Hence, modularity as a concept garnered huge attention in manufacturing industries due to its benefits such as reusability, maintainability, and most important, variability and scalability through standardization in product development [41]. Standardization offers flexibility for using these common elements across various projects, addressing uncertainties in market demands for product variety. Hence, modularity positively impacts the entire product life cycle, providing several means to adapt to technological changes and benefits like flexibility, easy maintenance, and recyclability, resulting in high life cycle efficiency [42].

Apart from modularity, sharing in design (SiD) is an important concept in design science research, aiming to create resource-effective products. Sharing in design involves identifying elements/structures between different sub-systems to achieve more functions with fewer elements, thus enhancing resource effectiveness [32]. Multifunctional products are a prominent example demonstrating the concept of sharing in design. Over the years, sharing in design as a concept has been applied by many researchers unconsciously or consciously in designing artifacts to make them more resource-effective. Chakrabarty discussed the categories of sharing in design, including structure sharing (SS), function sharing (FS), multi-mode integration (MMI), and structure redundancy, based on mapping between functions and structures [32]. The study concluded that structure-sharing and multimodal integration categories of sharing in design contribute significantly to making products more resource-effective. Chakrabarty et al., in their study, gave a comprehensive explanation of function and structure sharing and outlined a detailed methodology to assess the degree of structure sharing and resource effectiveness [33]. The authors presented a formula to assess the resource effectiveness as:

$$\text{Resource effectiveness} = \frac{\text{No. of main functions in theproduct}}{\text{No. of structures at the lowest level of abstraction}} \quad (1)$$

We argue that "resource efficiency" is more appropriate to justify the formula than "resource effectiveness." We argue that the formula calculates "resource efficiency" instead of "resource effectiveness." This argument is based on inspecting the distinction between the term's 'efficiency' and 'effectiveness.' Efficiency measures how well something is done, whereas effectiveness measures how useful something is [43]. Since the formula measures how well the structures are used to get more functions per unit structure, the formula assesses resource efficiency. Therefore, "resource efficiency" will be measured as:

$$\text{Resource efficiency} = \frac{\text{No. of main functions in the product}}{\text{No. of structures at the lowest level of abstraction}} \quad (2)$$

Structure sharing is often seen as an efficient design strategy wherein physical elements within a design serve multiple functionalities simultaneously. In design context, structure sharing contributes to the creation of integrated products by enabling one structure to perform multiple functions effectively. Hence, integral product architecture is claimed to have high resource efficiency as compared to modular product architecture. However, if we look from the product portfolio level, modularity may offer the potential to introduce flexible functions that can be added or removed from the temporary integration of structure based on demand thus offering changeability at the local or product level and enhanced resource efficiency at the portfolio level (collection of products). This argument proposes a new perspective on resource efficiency not at just the product level but at the portfolio level (collection of products. By introducing flexible functions via interaction among different modules, overall resource efficiency may improve at the portfolio level.

In the context of RHDS architecture, viewing complementary developments such as built healthcare facilities as independent modular products may open innovative opportunities for applying modularity concepts. This unlocks new avenues for design flexibility, construction efficiency, adaptable utilization of the existing physical infrastructural facilities, and resource efficiency. The concept of modularity can build adaptable PHCs in two ways. Firstly, by breaking down PHCs into independent and interchangeable modules through a process known as modularization [40]. Each module consists of interchangeable components, thus serving one or more specific functions. Secondly, integrating external modules to a parent structure such as PHCs (public healthcare centers) can create a hybrid infrastructure with flexible functions that can be added or removed from the parent structure as needed. The primary advantage of modular product architecture is its ability to facilitate mass customization through various combinations of these modules thus incorporating multiple flexible functions [42]. In contemporary modular construction approaches in healthcare, like PPVC modules, the original modularity is frequently forfeited upon assembly, rendering these modules irreplaceable and fixed in place without disassembling the entire structure. Nonetheless, the suggested approach permits module replacement without impacting the entirety of the product architecture, thereby achieving lifecycle modularity. It also allows the sharing of modules among different structures thereby achieving more flexible functions at the portfolio level with fewer structures, thus enhancing resource efficiency at the system or portfolio level as per the formula mentioned above. Moreover, the entire infrastructure can be upgraded by adding new flexible functions through the addition of new modules, enhancing the capabilities of the built healthcare facility. Therefore, the primary objective of modularity is to reduce complexity while maintaining customizability and enabling sharing.

4.4 Need to Synergistically Integrate Static PHCs and Mobile MHUs via ModuLarity and Sharing

Traditionally, built facilities can incorporate the concept of life-cycle modularity by utilizing existing modules sourced externally. An illustrative example is the design of a personal computer, which is an assembly of existing modules with standard interfaces, including components like the motherboard, a graphic card, and a USB port. Therefore, the proposed RHDS architecture aims to leverage the benefits of already established MHUs in healthcare to bring life cycle modularity to PHCs by seamlessly integrating them with detachable external modules like MHUs as needed. As an analogy, the PHCs are like the computers, while the MHUs are the external devices and hard drives that can connect via a standard interface to expand functional capability of the main structure. This approach offers the opportunity to enhance the functionality of existing PHCs by adding or replacing various external modular units over the facility's lifecycle, thereby maintaining life cycle modularity. This external modularity facilitates sharing these modules amongst multiple PHCs, thereby enhancing system-level resource efficiency. Modules can also be shared among MHUs and PHCs based on requirement. Additionally, this approach could address current infrastructural and logistical challenges associated with the MHUs.

Variants of the idea of integrating static HCF with MHUs have been explored in the past. Globally, there have been instances and examples of solutions that enable the coupling of temporary spatial solutions with existing static structures. However, the motivations for doing so may have differed from the proposal outlined in this article. In the current RHDS, PHCs and MHUs are developed and operated independently. However, by bridging this divide through design innovation, new opportunities emerge for reimagining the built HCFs and their role in fostering smart and inclusive communities.

4.5 Proposed System-Level Approach to Synergistically Integrate Built Facilities and Modular Health Units via the Concept of Modularity

The approach in this study proposes a minor tweak in the design of MHUs. Instead of a standard fully integrated MHU, akin to an ambulance, the proposed MHUs consist of modular trailer units with specialized services, which can be moved and pulled by a detachable prime mover, e.g., a semi-truck.

At the RHDS level, the proposed concept envisions a decentralized network comprising PHCs and a fleet of specialized MHUs that are mobile and can be relocated and connected to various PHCs. Each PHC is equipped with multiple docking stations, resembling rural hubs or base stations, to accommodate independent incoming modules that can be easily attached in a plug-and-play fashion. The relationship between the MHUs and PHCs can be likened to how cargo trailers (mobile units) attach to warehouses (buildings). This plug-and-play flexibility empowers each PHC to enhance its capabilities and healthcare services by accommodating MHUs with various specialized functions at different times. The stationed MHUs deliver specialized services to rural areas, complemented by well-equipped host PHCs that offer carefully designed waiting areas, preparation spaces, restrooms, utilities, and other essential auxiliary functions. The specialized MHUs can be removed and replaced throughout the building's lifecycle, thereby ensuring the life cycle modularity of the local HCF. From a sharing-indesign perspective [22], the proposed approach enables multimodal integration (MMI), which emphasizes the ability of a structure or system to perform multiple functions, switching between different modes based on specific needs or circumstances. In this scenario, MHUs can transition between modes, serving either as an extension of an existing facility or operating autonomously as needed. That is, the MHU as a modular structure, embodies multimode integration within itself. It seamlessly integrates different modes of operation within a single module, offering versatility and flexibility. The modules can be tracked and monitored using an Internet of Things (IoT) infrastructure. Using IoT, it is possible to track and monitor a fleet of MHUs with specialist functions that can serve a network of PHCs in various locations, creating shared resources. For example, one MHU could be docked to a specific PHC for a few days to several months. This arrangement may also facilitate maintenance and operations. This proposed concept aims to meet the unfulfilled demand for basic and specialized healthcare services in rural areas. In contrast to the current model where PHCs and MHUs function as separate entities, the proposed approach advocates for the utilization of modular spaces shared between PHCs and MHUs. This fosters interdependencies within the system and enhances the resource efficiency of the RHDS through the Modular Multimodal Integration category of sharing in Design [22, 23]. Several factors drive the proposition to integrate these two previously independent units:

Resource Efficiency: By sharing modular spaces and resources, the proposed system can enhance resource utilization, reducing duplication and promoting efficiency in healthcare services delivery.

Flexibility: The plug-and-play capability enables PHCs to adapt to changing healthcare needs by easily incorporating MHUs with specialized functions, ensuring the healthcare facility remains flexible and responsive.

Scalability and Customization: This approach can bridge the gap in RHDS by allowing PHCs to host specialized MHUs, ensuring broader range of healthcare services to the local population.

Sustainability: Life cycle modularity ensures the long-term sustainability of the healthcare facility, as it can evolve and expand its services without major structural changes.

Systemic Dependencies: The integration of these units promotes systemic dependencies, fostering collaboration and a more holistic approach to healthcare delivery.

The concept proposed presents multiple benefits, however, it is also accompanied by practical challenges. The most critical point of success is the Proof-of-Concept implementation, which requires resources, community engagement, and external partner support. Technological barriers include the need for robust infrastructure, seamless integration of disparate systems, and the requirement of hardware and software systems. Logistic barriers may arise from complex coordination of various stakeholders in the planning and deployment of logistics in diverse geographical areas. This requires carefulness in supply chain management, transportation, and other on-ground operations. Financial constraints are another challenge as securing funding for PoC implementation and sustaining it throughout the project lifecycle is critical. Budgetary limits may extend beyond the initial estimated amount to cover running operational costs, maintenance expenses, and other unforeseen complexities which require careful financial planning and resource allocation. Another risk is the reluctance of service providers in participation which may be due to apprehensions about the viability of the project, competing priorities, or disagreement over terms and conditions. Community engagement is another risk, as the success of PoC heavily relies on active participation from target communities. Achieving successful community engagement requires overcoming language barriers, cultural differences, and accessibility issues which require strategies in community outreach and capacity-building initiatives. Lastly, the fluid nature of the socio-political environment at pilot sites poses significant risks with the possibility of regulatory changes and policy shifts impacting the project halfway through.

5 Conclusions and Future Research

Since the rural healthcare ecosystem is constantly changing, the resources must evolve with time to satiate the emerging needs of the growing population. However, traditional HCFs are neither adaptable nor easy to maintain or upgrade. Hence, designing flexible and adaptable HCFs has been encouraged over the years. Modular, prefabricated construction has been adopted widely owing to its significant advantages compared to traditional construction. A range of modular solutions has emerged in healthcare, from standalone units to full-fledged hospitals for emergency reasons. Synergistic integration of static HCFs with mobile MHUs offer scalable, resource-efficient, flexible, and adaptable solutions, thus offering new possibilities for growth considering the life cycle modularity perspective. The article presents a novel standpoint on transforming the existing physical structure in the Rural Healthcare Delivery System (RHDS) through the synergistic integration of built facilities with MHUs. However, future research needs to investigate several challenges, especially the social desirability of the proposed intervention. A proof of concept is currently under development with a pilot site identified for deployment in collaboration with healthcare delivery partners. Apart from that, stakeholder analysis, Delphi studies, and field studies in the pilot site are planned to assess the social desirability of the proposed intervention.

References

1. Mohan, P., Sethi, H., Reddy, K., Bhan, M.: Designing primary healthcare systems for future in India. J. Fam. Med. Prim. Care **8**(6), 1817–1820 (2019). https://doi.org/10.4103/jfmpc.422.19
2. Chandwani, R.K., Dwivedi, Y.K.: Telemedicine in India: current state, challenges and opportunities. Transform. Gov. People, Process Policy **9**(4), 393–400 (2015). https://doi.org/10.1108/TG.07.2015.0029
3. Madhukumar, S., Gaikwad, V.: Awareness and perception regarding health insurance in Bangalore rural population. Int. J. Med. Public Heal. **2**(2), 18–22 (2012). https://doi.org/10.5530/ijmedph.2.2.5
4. Chanda, S., Dogra, V., Randhawa, S.: The design, operations, and feasibility of primary healthcare service delivery through mobile medical units: a way for equitable health access and achieving universal health coverage in low resource settings. Int. J. Heal. Syst. Implement. Res. **3**(2), 40–47 (2019)
5. Vashishtha, V.: Reach the Unreached – A Systematic Review on Mobile Dental Units. J. Clin. Diagnostic Res. 5–8 (2014). https://doi.org/10.7860/jcdr/2014/8688.4717
6. Bhatia, K.: Community health worker programs in India: A rights-based review. Perspect. Public Health **134**(5), 276–282 (2014). https://doi.org/10.1177/1757913914543446
7. Karthik, S., Sharareh K., Behzad, R.: Modular Construction vs. Traditional Construction: Advantages and Limitations: A Comparative Study, InCreative Construction e-Conference, pp. 11–19 (2020). https://doi.org/10.3311/ccc2020-012
8. Clough, I.: The case for more modular facilities in the NHS. Br. J. Heal. Care Manag. **28**, 3–11 (2022). https://doi.org/10.12968/bjhc.2021.0104
9. Jain, S., Singhal, S., Jain, N.K.: Construction and demolition waste generation in cities in India: an integrated approach. Int. J. Sustain. Eng. **12**(5), 333–340 (2019). https://doi.org/10.1080/19397038.2019.1612967
10. Silander, M.K., Sarkilahti, K., Torkki, A., Peltokorpi, P., Tarkkanen, A.: Perspectives of mass customisation and modularisation in health service delivery : a scoping review. Int. J. Healthc. Technol. Manag. **17**(4), 255–227 (2019)
11. Di Pasquale, J., Innella, F., Bai, Y.: Structural Concept and Solution for Hybrid Modular Buildings with Removable Modules. J. Archit. Eng. **26** (2020)
12. Hwang, B.G., Shan, M., Looi, K.Y.: Key constraints and mitigation strategies for prefabricated prefinished volumetric construction. J. Clean. Prod. **183**, 183–193 (2018). https://doi.org/10.1016/j.jclepro.2018.02.136
13. Hill, C.F., Powers, B.W., Jain, S.H., Bennet, J., Vavasis, A., Oriol, N.E.: Mobile health clinics in the era of reform. Am. J. Manag. Care **20**(3), 261–264 (2014)
14. Chillimuntha, A.K., Thakor, R.: Disadvantaged rural health – issues and challenges: a review. Natl. J. Med. Res. **3**(1), 80–82 (2013). https://doi.org/10.1108/JHLSCM.07.2013.0023
15. Vij, V.: Health Care Infrastructure in India : Need for Reallocation and Regulation. Res. Rev. Int. J. Multidiscip. **4**(03) (2019)
16. Taqi, M., Bidhuri, S., Sarkar, S., Ahmad, WS., Wangchok, P.: Rural Healthcare Infrastructural Disparities in India: a Critical Analysis of Availability and Accessibility. J. Multidiscip. Res. Healthc. **3**(2), 125–149 (2017). https://doi.org/10.15415/jmrh
17. Faizi, S.K.N.: Universal health coverage - There is more to it than meets the eye. Journal of Family Medicine and Primary Care **6**(1), 169–170 (2017). https://doi.org/10.4103/jfmpc.jfmpc
18. Dey, S., Chattopadhyay, S.: Assessment of quality of primary healthcare facilities in west Bengal. Int. J. Res. Geogr. **4**(2), 22–33 (2018). https://doi.org/10.20431/2454-8685.0402003

19. Sahoo, P.M.: Analysis of public health-care facilities in rural India. Facil. **41**(13/14), 910–926 (2023). https://doi.org/10.1108/F-07-2022-0098
20. Sriram, S.: Availability of infrastructure and manpower for primary health centers in a district in Andhra Pradesh, India. J. Fam. Med. Prim. Care **7**, 1256–1262 (2018). https://doi.org/10.4103/jfmpc.jfmpc
21. Mohanty, U.: Issues and Challenges of Primary Health Care Facility (2016)
22. Kumar, M., Kiran, A., Kujur, M.: Assessment of mobile medical units functioning in Jharkhand, India. Int. J. Community Med. Public Heal. **4**, 878–880 (2016). https://doi.org/10.18203/2394-6040.ijcmph20160921
23. Khanna, A.B., Narula, S.A.: Mobile medical units—can they improve the quality of health services in developing countries? J. Health Manag. **19**(3), 508–521 (2017). https://doi.org/10.1177/0972063417717900
24. Nair, A., Raikwar, A.A., Wassey, M.A., Dogra, V.: Beneficiary satisfaction with mobile medical units in the state of Andhra Pradesh. Int. J. Community Med. Public Heal. **8**(1), 366–371 (2021)
25. Abbasi, S., Mohajer, H., Samouei, R.: Investigation of mobile clinics and their challenges. Int. J. Heal. Syst. Disaster Manag. **4**, 1–5 (2016). https://doi.org/10.4103/2347-9019.175669
26. Kothandaraman, P., Mookerjee, S.: Healthcare for All: Narayana Hrudayalaya, Bangalore (2007)
27. Sathya, S., Central, S.: Sri Sathya Sai Central Trust Sri Sathya Sai Mobile Hospital Emphasizing the importance of Spirituality in holistic health (2023). https://sssmt.org.in/pages/mobile.hospital
28. Smolova, M., Smolova, D.: Emergency architecture. Modular construction of healthcare facilities as a response to pandemic outbreak. E3S Web Conf. **274** (2021). https://doi.org/10.1051/e3sconf/2021.27401013
29. Gatheeshgar, P.: Development of affordable steel-framed modular buildings for emergency situations (Covid-19). Structures **31**, 862–875 (2021). https://doi.org/10.1016/j.istruc.2021.02.004
30. Akhtar, M.H., Ramkumar, J.: Primary Health Center: Can it be made mobile for efficient healthcare services for hard to reach population? A state-of-the-art review. Discov. Heal. Syst. **2**(1) (2023). https://doi.org/10.1007/s44250.023.00017
31. https://www.bdcnetwork.com/hga-and-boldt-company-devise-prefabricatedtemporary-hospital-manage-surge-capacity-during-viral
32. Chakrabarti, A.: Sharing in design-categories, importance, and issues". Int. Conf. Eng. Des. **1**, 21–23 (2001)
33. Chakrabarti, A., Singh, V.: A method for structure sharing to enhance resource effectiveness. J. Eng. Des. **18**(1), 73–91 (2007)
34. Clough, I.: Building back smarter - Part two: Practical considerations for the commissioning and delivery of modular facilities. Br. J. Heal. Care Manag. **27**(11), 272–281 (2021). https://doi.org/10.12968/bjhc.2021.0138
35. Clough, I.: Placing modular facilities in a resource-limited healthcare system. Br. J. Healthc. Manag. **28**, 22–31 (2022). https://doi.org/10.12968/bjhc.2021.0149
36. Luo, H., Liu, J., Li, C., Chen, K., Zhang, M.: Ultra-rapid delivery of specialty field hospitals to combat COVID-19: Lessons learned from the Leishenshan Hospital project in Wuhan. Autom. Constr. **119** (2020)
37. Harrouk, C.: Carlo Ratti Converts Shipping Containers into Intensive-Care Pods for the COVID-19 Pandemic
38. Zhang, Z., Pan, W., Zheng, Z.: Fighting Covid-19 through fast delivery of a modular quarantine camp with smart construction. Proc. Inst. Civ. Eng. Civ. Eng. **174**(2), 89–96 (2021). https://doi.org/10.1680/jcien.20.00025

39. Bridges, D.J.: Modular laboratories - cost-effective and sustainable infrastructure for resource-limited settings. Am. J. Trop. Med. Hyg. **91**(6), 1074–1078 (2014). https://doi.org/10.4269/ajtmh.14.0054
40. Gershenson, J.K., Prasad, G.J., Zhang, Y.: Product modularity: Definitions and benefits. J. Eng. Des. **14**(3), 295–313 (2003). https://doi.org/10.1080/0954482031000091068
41. Ulrich, K.T., Eppinger, S.D.: Product Design and Development (2018). https://doi.org/10.4337/9781784718152.00017
42. Ishii, K.K.: Modularity: A Key Concept in Product Life-cycle Engineering (1998)
43. Gharfalkar, M., Ali, Z., Hillier, G.: Measuring resource efficiency and resource effectiveness in manufacturing. Int. J. Product. Perform. Manag. **67**(9), 1854–1881 (2018). https://doi.org/10.1108/IJPPM.11.2017.0282

Overview of MBSE and PLM Integration from a Traceability Perspective: A Mechatronic Case Study

Yana Brovar[1](✉), Jöran Frey[2], Sophia Salas Cordero[3], Dominik Knoll[1], Clément Fortin[1], Matthieu Bricogne[4], Benoit Eynard[4], and Felix Nyffenegger[2]

[1] Skolkovo Institute of Science and Technology, Moscow, Russia
yana.brovar@skoltech.ru
[2] OST –Ostschweizer Fachhochschule, Rapperswil, Switzerland
[3] ISAE-SUPAERO, Université de Toulouse, Toulouse, France
[4] Laboratoire Roberval, Université de Technologie de Compiègne, Compiègne, France

Abstract. Mechatronic systems, combining mechanical, electrical and software technologies, require the collaboration of different engineering disciplines. In this complex interdisciplinary space, the importance of Model-Based Systems Engineering (MBSE) and Product Lifecycle Management (PLM) comes to the fore. MBSE and PLM represent different approaches that are inherently interrelated, within the broad field of product design and development. The integration of these two approaches is crucial for ensuring seamless information exchange and collaboration throughout the entire product lifecycle; it involves connecting different tools, methodologies and data formats. Several challenges hinder the seamless integration of these approaches. These challenges include mismatched data representations, semantic inconsistencies, varying granularity of information and different levels of abstraction. Overcoming these challenges requires a systematic approach where traceability plays a key role in establishing and maintaining connections between data, ensuring consistency of information throughout the product lifecycle. By highlighting the importance of traceability on a vending machine use-case, this paper contributes to the evolving discourse on effective methodologies for harmonizing different engineering practices in the development of mechatronic systems.

Keywords: Systems Engineering · Traceability · Model-Based Systems Engineering · Product Lifecycle Management · Multiple-Domain Matrix

1 Introduction

In the realm of mechatronic systems, the design process stands as a critical juncture where the integration of Model-Based Systems Engineering (MBSE) and Product Lifecycle Management (PLM) holds immense potential for enhancing the development and

Published by Springer Nature Switzerland AG 2025
P. Sureephong et al. (Eds.): PLM 2024, IFIP AICT 740, pp. 356–370, 2025.
https://doi.org/10.1007/978-3-031-93319-6_27

management of complex products [1]. MBSE and PLM represent two indispensable approaches within the engineering and product development industry. The successful integration of these domains is imperative to facilitate seamless information exchange and foster collaboration throughout the entire product lifecycle. Traceability serves as the thread that links disparate elements within the product development process, ensuring a consistent flow of information, and maintaining the integrity of the system model throughout its lifecycle. A comprehensive understanding of the gaps and challenges in traceability provides a solid foundation for devising effective solutions to enhance integration.

Traceability in MBSE refers to the ability to establish and maintain the relationships between different artifacts and elements throughout the system development process [2]. It involves capturing and documenting the dependencies, associations, and connections between requirements, models, components, and other system elements. Traceability provides a means to understand the rationale behind design decisions, track changes, and ensures that all system requirements are adequately addressed. Tools that support system modeling languages, enable the creation and visualization of system models, enabling the establishment of traceability links between different model elements. For instance, some tools can provide traceability matrices, which are tabular representations that show the relationships between different artifacts, such as requirements, design elements, and test cases. Each row and column in the matrix correspond to a specific artifact, and the cells contain information indicating the relationships or dependencies between them. The theoretical framework of traceability matrices encompasses a comprehensive view of traceability, facilitating the mapping of requirements to design elements, design elements to implementation artifacts, and implementation artifacts to test cases. Such a visual overview of the traceability links facilitates stakeholders to assess the coverage and completeness of traceability. However, while traceability matrices excel at capturing high-level system requirements, architecture, and behavior, they may lack the geometric and spatial information necessary for comprehensive integration with CAD models.

During the detailed design stage, high-level system requirements may be translated into detailed specifications and designs. Thus, traceability between MBSE and PLM plays a crucial role during this stage by establishing and maintaining the connections between various design artifacts, ensuring consistency, and enabling effective design validation and verification. By establishing trace links between design elements and the corresponding requirements, it becomes easier to verify that the design meets the requirements. Over the past decade, PLM vendors have progressively incorporated MBSE capabilities into their products. This evolution initially focused on connecting the system design phase with detailed design. However, significant enhancements are still needed to ensure this integration to operate efficiently. The main reason behind this appears to be the capabilities of PLM systems and their associated tools. So far, they serve as robust platforms to model and analyze complex products, with a particular focus on spatial representation and product structure. However, PLM solutions lack the representation of product functions and behaviors. Simultaneously, they aim to offer functionalities for establishing and overseeing traceability across the entire product lifecycle, that leads to the increasing complexity of the PLM solutions.

The structure of this paper is the following: Sect. 2 provides an overview of current practices to handle MBSE-PLM traceability; while Sect. 3 provides methodology utilized in the use case presented in Sect. 4; Sect. 5 discusses the presented approach and provides conclusions.

2 Literature Review

2.1 Current Practices

A strategy for handling traceability between MBSE and PLM involves embracing a unified master-model which is meant to contain any product related information within a PLM system for all phases of a product's life. This section discusses current practices in handling traceability between MBSE and PLM, and delves into specific methodologies and standards utilized in this context, such as: Core Product Model (CPM), Requirement-Functional-Logical-Physical (RFLP) approach, and Open Services for Lifecycle Collaboration (OSLC). These practices are aimed at addressing the complexities of modern product development, enhancing traceability, interoperability, and collaboration across engineering disciplines.

Core Product Model. Is a conceptual model with broad semantics, providing means to articulate essential characteristics of PLM information [3]. The impetus for developing CPM stemmed from the demand for next-generation product development systems to effectively handle large and diverse data flows. At its core, the primary entity within the model is the 'artifact,' representing a distinct element within a product. Amongst others, this artifact entity includes three entities defining its function, form and behavior. The specific function entity elaborates the intended purpose of the artifact based on engineering requirements and stakeholder needs. In the case of software components, this entity describes the specific tasks, processes, or functionalities that the software is designed to perform. The form entity embodies the design solution to implement the function, detailing aspects such as geometry, material, and feature. For non-physical system components like software, the form entity may encompass aspects such as code structure, architecture, and configuration. The behavior entity elucidates how the form of an artifact is derived from its function [3]. For software components, this entity describes how the software behaves under different operating conditions, including its responses to various inputs, events and user interactions.

To comprehensively capture assembly and system-level tolerance information and establish kinematic constraints, the CPM framework was expanded with the Open Assembly Model (OAM) [4]. This extension emphasized information requirements for part features and assembly relationships, employing the data model from the Standard for the Exchange of Product data (STEP), based on ISO 10303 [5]. This standard provides a vendor-independent format for the computer-interpretable representation and exchange of product data, enhancing the basic CPM data structure with entities such as 'kinematic pair,' 'assembly,' 'assembly feature,' and 'tolerance'. The 'assembly' entity, comprising 'parts' and 'subassemblies,' encapsulates information regarding assembly relationships and component composition. 'Geometric tolerancing' is crucial for mechatronic systems' design, enabling assembly analysis and controlling the variability of all dimensions. This extension is essential to properly define products and systems.

CPM, along with its extensions like OAM, contribute to the NIST Information Modeling Framework. This framework specifies product design information and knowledge, facilitating semantic interoperability between CAD/CAE/CAM systems and capturing the evolution of product families. Building upon this framework, various attempts have been made to enhance CPM and create more applicable Enterprise Systems (ES). Examples include ontological approaches like OntoSTEP [6] and OntoSTEP-NC [7]. OntoSTEP serves as a model for translating STEP schema and instances into the ontology web language (OWL), seamlessly integrating geometry representation via STEP with a semantic model and non-geometrical concepts of function, behavior, and requirements based on CPM/OAM [8]. However, one challenge with OntoSTEP is that it not only represents semantic information related to product features but also includes structural data without semantics, reserved for future implementation. OntoSTEP-NC, on the other hand, aims to facilitate system interoperability and a data closed-loop between design and manufacturing processes. Its formalization relies on the integration and reuse of knowledge embedded in existing standards for product data (ISO 10303 [5]) and manufacturing data as defined in STEP-NC standard (ISO 14649 [9]).

Requirement-Functional-Logical-Physical Approach. Traceability is a fundamental objective in systems engineering, and the RFLP [9] approach is a great example aiming to achieve it. By connecting requirements, functional elements, logical representations (representations of system elements and their relationships), and physical components (actual hardware and software implementations) in a detailed manner, it enhances the ability to trace and validate the entire system development lifecycle. This approach has been developed by MIL-STD 499B "Military Standard – Systems Engineering" (first version 1974) that was then replaced by IEEE/1220 [10]. The RFLP approach emerged as a response to the dynamic landscape, offering a comprehensive methodology for system development and can be found within the 3DEXPERIENCE platform [11].

A key strength of the RFLP approach lies in its commitment to portray holistically system design. By integrating MBSE principles within product life-cycle management tools, it seeks to create a digital twin, providing a thorough understanding of the system of interest. The RFLP approach ensures linkages between requirements, functional elements, logical representations, and physical components. This interconnectedness is crucial for successful system development and validation. The RFLP approach supports Requirement-in-Loop (RiP) during simulation processes. It enables the evaluation of parameters like masses or dimensions, ensuring that virtual prototypes meet specified requirements. The established links within the RFLP approach become instrumental in validating and tracing system development processes.

Open Services for Lifecycle Collaboration. To establish a collaborative engineering environment and facilitate seamless data exchange between different engineering tools and domains, particularly software, the Open Services for Lifecycle Collaboration (OSLC) was originally proposed by IBM and is now a standard managed by the OASIS group [12]. The OSLC approach is characterized by vendor independence, enabling tools to employ diverse technologies for data management. This flexibility allows users across multiple disciplines to utilize tools implementing OSLC to establish connections between digital artifacts efficiently. This process, while straightforward, proves powerful, fostering the continuous flow of information in the digital stream by providing traceability and notifications of changes. The OSLC approach is rooted in a model representation based on the Resource Description Framework (RDF) standard, used for representing interconnected data. Communication within OSLC occurs through a Representation State Transfer (RESTful) protocol [13], a technology used for distributed information systems. Unlike relying on automated or analytical approaches to link data across tools, OSLC supports proficient users by providing an accessible interface methodology for manually creating linkages between artifacts originating from various ontologies. The specifications within OSLC are designed for different integration scenarios, simplifying vendor-independent heterogeneous tool integration. The standard, built on web communication standards and ontology definitions, comprises an OSLC Core specification and a series of Domain specifications. The core specification establishes fundamental concepts and rules for integration methodologies, ensuring coherence among different domain specifications. The domain specifications focus on specific lifecycle topics, such as requirements management, change management, configuration management, and architecture management.

2.2 Integration of MBSE and PLM in the Context of Mechatronic Design

Collaboration is another cornerstone for successful development of mechatronic systems [14, 15]. MBSE and PLM integration fosters collaborative workflows by connecting different engineering disciplines, ensuring that mechanical, electrical, and software engineering teams can work together seamlessly, sharing and updating information to accelerate decision-making and mitigate the risks of miscommunication/errors.

The importance of traceability derives from the fact that design of mechatronic systems often involves iterative changes and updates [16]. The integration of MBSE and PLM addresses the objectives of managing changes and versions effectively, facilitating version control and change management, ensuring that changes are synchronized among different system models and that design decisions are linked to specific requirements. By facilitating robust version control, the integration ensures that each design iteration is systematically recorded, enabling engineers to trace back and comprehend the evolution of the design over time. Engineering changes within mechatronic systems can emanate from various facets - being mechanical adjustments, electrical enhancements, or software updates. The seamless integration of MBSE and PLM should orchestrate the change management, allowing teams to comprehensively record the engineering changes, ensuring that alterations align with the broader design objectives and comply with specific requirements and regulations. While the benefits are substantial, there are many challenges in achieving an efficient integration of MBSE and PLM, such as data representation mismatches and semantic inconsistencies, to overcome these challenges till this date there is a strong need of a systematic and carefully planned integration process.

2.3 SysML and MagicGrid®: A Comprehensive Framework for MBSE

Creating robust system models requires a comprehensive modeling environment comprising of a modeling language, a modeling tool, and a modeling methodology. While the modeling language and tool define the modeling capabilities, the methodology dictates how the model is structured. SysML, the Systems Modeling Language [17], is a general-purpose graphical modeling language with the capabilities to define, analyze, develop, and validate complex systems. Below, the subset of SysML diagram types essential to the case study is briefly introduced (Sect. 4).

The requirements diagram organizes and defines system requirements. It offers a multitude of relation types between elements to support traceability from stakeholder needs through subsystem requirements to specific components or functions. A use case diagram outlines system services and participating actors, providing a solution neutral black box view on the use cases of a system and elaborating its context. An activity diagram visualizes the dynamic behavior of a system by illustrating activity sequence and decision-making with the use of decision nodes. The block definition diagram (bdd) depicts systems or components as so-called blocks. Connectors defined with interface blocks can then be used to model flows e.g., information, energy or matter flow. An internal block diagram (ibd) provides a detailed insight into the internal structure of a block from a bdd, illustrating how different subsystems or components within a system communicate with each other. It can also be used to link requirements with components.

While the modeling language addresses syntax and semantics, an additional methodology like the MagicGrid® framework [18] is useful to structure the model effectively. It incorporates a range of modeling activities as shown in Fig. 1. MagicGrid® utilizes SysML as the modeling language and embraces a top-down approach, sharing similarities to the Object-Oriented Systems Engineering methodology (OOSEM) which is a best practices approach [19].

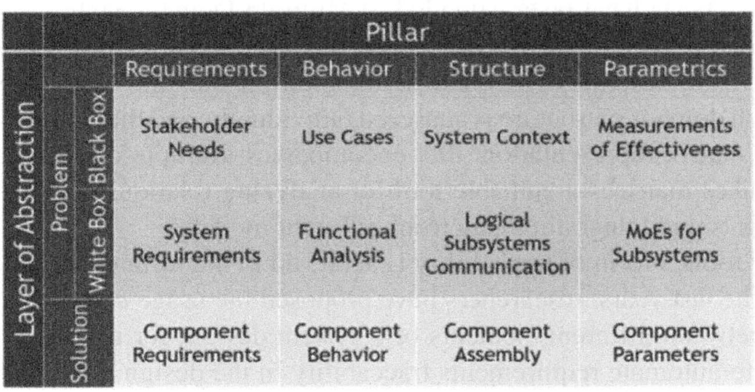

Fig. 1. The MBSE MagicGrid®: Framework [18].

The framework defines development steps for the model. The Blackbox layer in the problem domain focuses on identifying stakeholder needs, describing use cases, and estimating the flow of materials and information within the system. This stage is solution-neutral, emphasizing the understanding of what needs to be achieved rather than how. Subsequently, in the White Box layer, more specific activity diagrams defining functions are generated. From these functions, subsystems are then derived, marking a transition from conceptual activities to more concrete, physical subsystems. In the

Solution Domain, these subsystems are then further structured into specific components, defining their interactions within the system.

Integration of the verification and validation process into the systems model is achieved through the so-called Parametric Pillar, by employing Measures of Effectiveness which are essentially parameters (e.g. energy consumption) to measure a system's effectiveness [20]. They are continuously monitored via parametric diagrams which use mathematical functions on parameters of other diagrams. In addition, test cases are built into requirement diagrams, these test cases are used to validate individual requirements.

2.4 Design Structure Matrix and Multiple-Domain Matrix

A Design Structure Matrix (DSM) is an NxN matrix that maps the interactions among the set of N elements [21]. It is a network modeling tool used to represent the elements comprising a system and their interactions, thereby highlighting the system's architecture or designed structure. These elements can take the form of physical components, design teams, systems, design parameters or any other items where an interaction or interface occurs. Different interactions may be considered, such as energy, spatial, material, or information. Eppinger and Browning [21] have demonstrated the utility of DSM-based methods across various domains, highlighting its versatility and effectiveness in managing complex systems. One of the key features of DSMs is their ability to show the direction of flow. Interactions below the diagonal represent feed-forward interactions, indicating flows from one element to another, while interactions above the diagonal signify feedback loops. This directional information is crucial, especially for understanding time or decision-based sequences within a system.

Moreover, DSMs have been extended to Multiple-Domain Matrices (MDM) [22], expanding their applicability across different industries and domains. MDM is a method used to analyze the structure of a system across multiple domains simultaneously. In this approach, each domain's structure is analyzed individually, and the results are condensed into a single matrix representation, that encompasses multiple domains at once. This versatility makes matrices a valuable tool for analyzing relationships between various system elements and integrating data from different models.

In MBSE tools a form of traceability is achieved in the form of so-called derivation matrices. In the last years, researchers have proposed the usage of DSMs to display the traceability between different elements of a system design; for instance [23] proposes an approach to automate requirements traceability in the design of CubeSats, and [24] presents an open-source tool for the creation of DSMs from MBSE models.

3 Methodology

We propose the use of Multiple-Domain Matrix as a tool to maintain traceability, creating structured links between systems requirements and physical system elements by extracting data from Model-Based Systems Engineering diagrams and PLM systems. Firstly, the system components contributing to fulfilling each requirement are identified from the MBSE diagrams (see blue arrow in Fig. 2). These components span hardware and software subsystems, interfaces, or other relevant entities, for example environmental elements. MDM encompases both requirements and subcomponents, ensuring a comprehensive view of the system.

To identify system interfaces at the subcomponents domain (see red arrow in Fig. 2), we use the common taxonomy of interactive systems modeling proposed by Pimmler and Eppinger [25] where they consider four important types of interaction. The first type of interaction defines the need for adjacency or orientation between two components (structural connection). The other three types are associated with an exchange (flow) of material (or matter), energy, and information. In turn, Crawley et al. [26] extended the four-dimensional taxonomy with an introduction of a 'spatial' type of interface relationship, that "captures absolute or relative location or orientation". In a DSM a cell may be marked with I, M or E when there is an Informational, Material or Energy flow, respectively. All these types of interactions are inherently unidirectional, in that they have a specific flow direction and must be read from column to row in the DSM. The off-diagonal element of the matrix is marked with S when two given components have a structural relationship between each other. The matrix element is marked with Sp when the relationship between two components is spatial. Both structural and spatial relationships are bi-directional. However, it's worth noting that spatial and structural relationships are often more comprehensively represented in PLM systems rather than MBSE, however, their representation tends to be implicit.

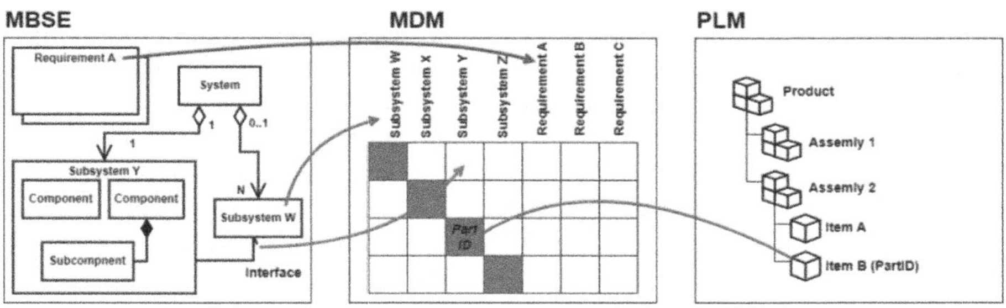

Fig. 2. MBSE and PLM integration through MDM representation

To add requirement domain, project requirements are extracted from MBSE diagrams (see pink arrow in Fig. 2). These requirements encompass the functional specifications outlining what the system is expected to achieve or perform. Then, a systematic approach is used to map each requirement to the specific components that fulfill it. This ensures comprehensive coverage by analyzing functional dependencies between requirements and components. Such linkage is represented by the 'unit' positioned at the intersection of the row corresponding to a requirement title and column that is associated with system component that shall satisfy this requirement. Additionally, if a requirement pertains to interfaces between two subsystems, it is mapped to both respecting subsystems to ensure comprehensive integration.

To establish a link MBSE and PLM systems through MDM, the Part ID (corresponding to EBOM) is positioned in the diagonal of the matrix (see green arrow in Fig. 2). This integration facilitates seamless navigation between functional requirements and their realization in the system design. By connecting the MBSE system and the Bill of Material, which integrates the structure of part masters related to CAD models, a comprehensive traceability tool is formed, offering a holistic perspective of the system development process from conception to realization.

By leveraging the MDM's capability to map requirements to specific components, engineers and designers gain invaluable insights into how each design decision contributes to fulfilling the overall system objectives. This interconnectedness between function and form empowers stakeholders to make informed decisions throughout the design lifecycle, ensuring alignment with stakeholder needs and design specifications. Section 4 represent change scenarios and how the proposed approach can be used.

4 Case Study

The following case study was conducted to explore the integration of the MBSE with the PLM approach. A model that was developed for training and research purposes is used as the basis for this case study. This model represents a vending machine, this system was selected for its potential for functional and requirements changes, all while maintaining a level of simplicity that allows for effective modeling and analysis. Developed using the MagicGrid® framework, this vending machine model encapsulates both mechanical, electronic and software aspects, focusing specifically on the product dispensing subsystem. The identified Product dispensing system components are Gripper Motor, Gripping mechanism, Force sensor, Portal, Motors, Motor controller, Programmable Logic Controller (PLC), and Product compartment. The Gripping mechanism, along with other components such as the Gripper Motor and Force sensor, contributes to the seamless operation of the vending machine. Through the utilization of the Gripping Mechanism, the vending machine can accurately detect, grasp, and deliver products to users, enhancing the overall user experience.

Within the component's domain, the MDM represents decomposition of its components in correspondence to the block definition diagram and definition of interfaces. The identically labelled rows and columns of the MDM correspond to the identified Product dispensing system components. Furthermore, this decomposition has been complimented by adding environmental components, such as Gripped object and overall environment.

The information about functions and requirements was gathered to create and model the subsystems leading to a holistic representation of the system. Traceability in between layers is achieved with the usage of derivation matrices that define relationships from artifacts and elements using the derived attribute. The modeling depth was determined based on the guideline that the lowest model entity shall only encompass one domain, either mechanics, software, or electronics. The modeling focused exclusively on the product dispensing subsystem of the vending machine, which includes mechanical and electronic parts.

In addition to tracing requirements, Fig. 3 represents interfaces between system components at the intersection of columns and rows. This complements the SysML representation since it does not explicitly define ports and connectors for the purpose of structural or spatial interfaces; as ports and connectors in SysML are primarily associated with representing flows of material, data, signals, or control between system components. In the MDM a cell may be marked with I, M or E when there is an Informational, Material or Energy flow, respectively. All these types of interactions are inherently unidirectional, in that they have a specific flow direction and must be read from column to row in the MDM. The off-diagonal element of the matrix is marked with S when two given components have a structural relationship between each other. The matrix element is marked with Sp when the relationship between two components is spatial. Both structural and spatial relationships are bi-directional.

											SSR-1.2	SSR-1.10	SSR-1.10
	Gripper Motor	Gripping Machanism	Force sensor	Portal	Motors	Motor controller	PLC	Product compartmen	Gripped object	Environment	Universal Product Handling Capability	Product weight	Data rate
Gripper Motor	00373396.1	S					I S			E S			
Gripping Machanism	E S	00373397.1						M Sp	Sp	Sp		1	
Force sensor		I S	00373395.1										
Portal				00373398.1	E S								
Motors				S	00373641.1	I S				E S		1	
Motor controller					I S	00373640.1	I S			E S			1
PLC	I S	S	I S			I S				I Sp			1
Product compartment		Sp											
Gripped Object		Sp		Sp						Sp	1	1	
Environment	S	M Sp			S	S	Sp		Sp				
Universal Product Handling Capability													
Product weight											1		
Data rate													

Fig. 3. MDM of the Product dispensing system

4.1 First Scenario: Engineering Change Traceability

The lowest modeled instance in the case of the Gripper is the Motor controller, the Motor, and the Gripper mechanism, which corresponds with the defined guideline of having just one engineering domain in the lowest modeling instances.

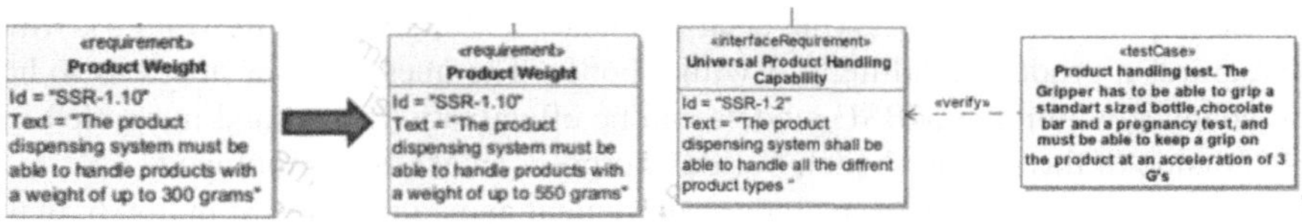

Fig. 4. Changed requirement and testCase

The following example portrays the traceability involved in a scenario where there is a requirement change from the MBSE models to the PLM system. Let's say Company "XY" specializes in vending machines and currently focuses on models designed predominantly for the distribution of food products. However, a recent market analysis revealed a substantial market potential for vending machines capable of dispensing both food and beverages. Considering this new market insight, the development team needs to analyze how this new requirement impacts the current design of the vending machine. An MBSE model, which was created during initial development is utilized. First, the new requirement is introduced as a stakeholder need in the black box domain. Then, the impact on the system structure needs to be systematically analyzed. For this purpose, traceability matrices are a valuable tool as they help to determine the specific components within the model influenced by the change. In our case Different requirements are affected by the change. For example, the maximum product weight, which was previously set to 300 g, has now been increased to 550 g (Fig. 4). In addition, a new test case for the product dispensing system will be introduced to test the dispensing of beverages.

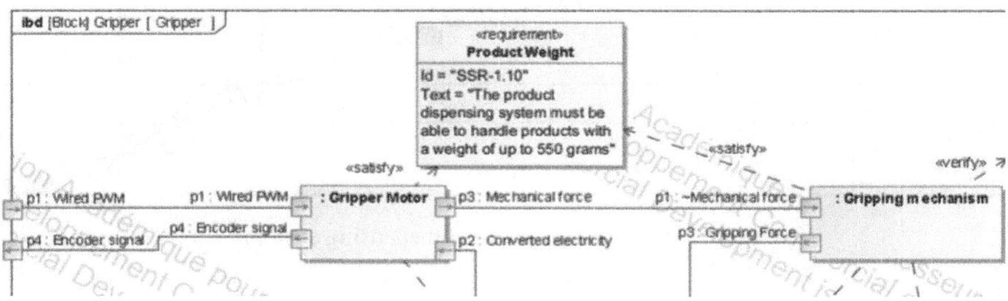

Fig. 5. Ibd of the Gripper with the modified requirement

The change in this subsystem requirement now triggers an association that was established at the subsystem level for the Gripper. Two components are associated with this requirement: the motor for the Gripper, which was originally designed for a product weight of 300 g and, as a result, no longer fulfills the requirement SSR-1.10. In addition, this requirement change led to a modification in a test case designed for the interface requirement SSR-1.2 Universal Product Handling Capability (Fig. 5), which now requires a product handling test with a bottle. The question now arises as to how these changes within the MBSE models can be effectively tracked and integrated into the PLM side of the system model. For this purpose, an MDM is employed.

Locate the row corresponding to the requirement name in the MDM in Fig. 6. In this case, find the requirement related to the maximum product weight, which was previously limited to 300 g but has now been increased to 550 g. Navigate to the intersection unit within the MDM where the row representing the requirement and the columns representing the affected components intersect. Identify the columns in the MDM that represent the components or parts affected by the requirement change. In this example, these columns are the gripping mechanical structure and the Motors. The redesign of the Motor has revealed that the currently used Motor is too weak. The test case universal product handling capability has revealed that bottles tip over in the existing mechanism,

	Gripper Motor	Gripping Machanism	Force sensor	Portal	Motors	Motor controller	PLC	Product compartmer	Gripped object	Environment	SSR-1.2 Universal Product Handling Capability	SSR-1.10 Product weight	SSR-1.10 Data rate
Gripper Motor	00373396.1	S					I S			E S			
Gripping Machanism	E S	00373397.1						M Sp	Sp	Sp		1	
Force sensor		I S	00373395.1										
Portal				00373398.1	E S								
Motors				S	00373641.1	I S				E S		1	
Motor controller					I S	00373640.1	I S			E S			1
PLC	I S	S	I S			I S				I Sp			1
Product compartment		Sp											
Gripped Object		Sp		Sp						Sp	1	1	
Environment	S	M Sp			S	S	Sp		Sp				
Universal Product Handling Capability													
Product weight											1		
Data rate													

Fig. 6. Traceability of Engineering Changes through the MDM-based approach

as a result a modification was made to the gripping mechanism which better supports the bottle with an additional notch as shown in Fig. 7.

Fig. 7. Engineering change of the Gripper in accordance with the stakeholder need modification.

4.2 Second Scenario: Requirement Change Traceability and Impact on Interfaces

	Gripper Motor	Gripping Machanism	Force sensor	Portal	Motors	Motor controller	PLC	Product compartmer	Gripped object	Environment	SSR-1.2 Universal Product Handling Capability	SSR-1.10 Product weight	SSR-1.10 Data rate
Gripper Motor	00373396.1	S					I S			E S			
Gripping Machanism	E S	00373397.1						M Sp	Sp	Sp		1	
Force sensor		I S	00373395.1										
Portal				00373398.1	E S								
Motors				S	00373641.1	I S				E S		1	
Motor controller					I S	00373640.1	I S			E S			1
PLC	I S	S	I S			I S				I Sp			1
Product compartment		Sp											
Gripped Object		Sp		Sp						Sp	1	1	
Environment	S	M Sp			S	S	Sp		Sp				
Universal Product Handling Capability													
Product weight											1		
Data rate													

Fig. 8. Requirement change traceability through the DSM-based approach

A second scenario considering a requirement change is presented in this section. The change that is addressed relates to the data transfer rate between the Motor controller and the PLC. Initially set at 1 Mbps, the requirement has been updated to support real-time data exchange at a minimum rate of 2 Mbps. This requirement pertains to the informational interface between the two components. This interface is represented at the intersection of the corresponding rows and columns of Motor controller and PLC. In the MDM present in Fig. 8, the requirement change is documented by updating the intersection cell corresponding to the rows representing the Motor controller and PLC columns. In that case, we can trace the requirement to the matrix cell at the intersection of the corresponding rows and columns, where the requirement value could be placed and possibly linked with the value at the domain-specific tool. In this context, it is important to emphasize the importance of a domain-specific tool that creates the link to the logical domain, especially since the interfaces in a M-CAD model often do not have an explicit and coherent representation. By using this method, we provide clarity and consistency in managing requirements changes during the development process.

5 Discussion and Conclusion

The scenarios presented in the case study demonstrate a systematic approach for analyzing the impact of engineering and requirement changes on the system design. The Multi-domain matrix approach allows for the representation and storage of different elements (such as requirements, components, interfaces) and their corresponding trace links. These artifacts are represented in matrix form, facilitating their storage and retrieval. The MDM approach provides a mechanism for tracing requirements to specific system elements.

Multi-domain matrices can facilitate the integration of MBSE and PLM by providing a structured approach to bridge the gap between different levels of abstraction of a system. MBSE primarily focuses on capturing system operations and requirements, while PLM emphasizes on the physical form or spatial aspects of the system. Multi-domain matrices offer a mean to correlate requirements with physical components, as well as structural connections, that are related to form as well, and interactions, that implicitly correspond to system functional behaviors. By mapping functional behaviors and requirements to corresponding physical components, these matrices provide a clear understanding of how system functions translate into physical forms and structures. This allows for traceability and alignment between the requirements defined in MBSE and the physical configurations managed in PLM.

The integration of MBSE with PLM, supported by MDM, offers a holistic view of the system development process. From conceptualization to realization, stakeholders can track the evolution of requirements, understand their implications on system components, and ensure alignment with stakeholder needs and design specifications. MDM encourages collaboration across different domains and disciplines by providing a common framework for representing different levels of systems abstraction. This facilitates communication and coordination among people with diverse expertise, leading to better-integrated solutions. Overall, the MDM approach offers a practical and efficient mean to manage changes, particularly in complex systems engineering projects. Continuously updating and maintaining the requirement-component mapping is essential as project requirements evolve or changes occur in the system architecture. This ongoing refinement ensures alignment between functional specifications and system elements. MDM systems, recognized for their scalability and flexibility, support this process effectively. Its focus on simplicity, traceability, integration, and collaboration makes it a valuable complement to traditional frameworks like RFLP, because it provides clear dependency visualization. While the RFLP approach provides a structured framework for capturing requirements, functions, and physical components, the explicit representation and tracing of interface requirements may vary depending on how interfaces are represented within the logical and physical domains.

It may require additional considerations or extensions to ensure comprehensive traceability of interface requirements throughout the system development process. For instance, in the logical domain, interfaces may be represented indirectly through functions or interactions between functions. While interface requirements themselves may not be explicitly represented, the interactions between functions can imply certain interface requirements. However, if interface requirements are not explicitly captured or if the representation of functions does not adequately reflect interface dependencies, tracing interface requirements may be challenging. In the physical domain, interfaces are typically represented more explicitly when discussing physical components and their connections. Interface requirements may be inferred from the connections between physical components or from specific properties associated with those connections. However, if interface requirements are not directly linked to physical components or if the representation of physical components lacks granularity, tracing interface requirements may still be limited.

In contrast, the MDM approach presented in this paper allows for interface requirements to be explicitly captured within the cells of the matrix, indicating the dependencies or interactions between different components or interfaces. As requirements evolve or new interfaces are introduced, the matrix shall be updated accordingly to reflect these changes. Integrating tools or methodologies specifically designed for interface management and traceability may also enhance the effectiveness of the RFLP approach in this regard. Thus, integrating elements from both approaches can yield a robust development framework that adeptly addresses both requirement management and system design considerations.

References

1. Gerhard, D., et al.: MBSE-PLM Integration: Initiatives and Future Outlook. In: IFIP International Conference on Product Lifecycle Management, pp. 165–175. Springer, Cham (2022)
2. Gotel, O., et al.: Traceability fundamentals. Software and systems traceability, 3–22 (2012)
3. Fenves, S. J.: A core product model for representing design information. US Department of Commerce, Technology Administration. National Institute of Standards and Technology (2001)
4. Sudarsan, R., Fenves, S.J., Sriram, R.D., Wang, F.: A product information modeling framework for product lifecycle management. Comput. Aided Des. **37**(13), 1399–1411 (2005)
5. Pratt, M.J.: ISO 10303, the STEP standard for product data exchange, and its PLM capabilities. Int. J. Product Lifecycle Manage. **1**(1), 86–94 (2005)
6. Barbau, R., et al.: OntoSTEP: Enriching product model data using ontologies. Computer-Aided Design **44**(6), 575–590 (2012)

7. Danjou, C., et al.: Manufacturing knowledge management based on STEP-NC standard: a closed-loop manufacturing approach. Int. J. Comput. Integr. Manuf. **30**(9), 995–1009 (2007)
8. Fiorentini, X., et al.: An Ontology for Assembly Representation. NIST Interagency/Internal Report (NISTIR). National Institute of Standards and Technology, Gaithersburg (2007)
9. ISO 14649-1:2003: Industrial automation systems and integration - Physical device control - Data model for computerized numerical controllers - Part 1: Overview and fundamental principles. https://www.iso.org/standard/34743.html. Accessed 23 Feb 2024
10. Kleiner, S., Kramer, C.: Model based design with systems engineering based on RFLP using V6. In: Proceedings of the 23rd CIRP Design Conference, pp. 93–102. Springer, Berlin (2013)
11. Doran, T.: IEEE 1220: for practical systems engineering. Computer **39**(5), 92–94 (2006)
12. Dassault Systèmes: 3DEXPERIENCE platform. Available at: https://www.3ds.com/3dexperience. Last accessed 23 February 2024
13. OSLC community Website. Available at: http://open-services.net/. Last accessed 23 February 2024
14. Richards, R.: Representational state transfer (REST). Pro PHP XML and Web Services, 633–672 (2006)
15. Bricogne, M., et al.: Concurrent versioning principles for collaboration: towards PLM for hardware and software data management. Int. J. Product Lifecycle Manage. **7**(1), 17–37 (2014)
16. Guérineau, J., et al.: Organizing the fragmented landscape of multidisciplinary product development: a mapping of approaches, processes, methods and tools from the scientific literature. Res. Eng. Design **33**(3), 307–349 (2022)
17. Zheng, C., et al.: Interface model-based configuration design of mechatronic systems for industrial manufacturing applications. Robo. Comp. Integr. Manufact. **59**, 373–384 (2019)
18. Friedenthal, S., et al.: A practical guide to SysML: the systems modeling language. Morgan Kaufmann (2014)
19. Morkevicius, A., et al.: Towards a common systems engineering methodology to cover a complete system development process. In INCOSE International Symposium **30**(1), 138–152 (2020)
20. Karban, R., et al.: The OpenSE Cookbook: a practical, recipe based collection of patterns, procedures, and best practices for executable systems engineering for the Thirty Meter Telescope. In: Modeling, Systems Engineering, and Project Management for Astronomy VIII vol. 10705, pp. 319–341. SPIE (2018)
21. Morkevicius, A., et al.: MBSE grid: a simplified SysML-based approach for modeling complex systems. In INCOSE International Symposium **27**(1), 136–150 (2017)
22. Eppinger, S.D., Browning, T.R.: Design structure matrix methods and applications. MIT press, Cambridge (2012)
23. Maurer, M., Lindemann, U.: The application of the Multiple-Domain Matrix: Considering multiple domains and dependency types in complex product design. In: 2008 IEEE International Conference on Systems, Man and Cybernetics, pp. 2487–2493. IEEE (2008)
24. Strobbe, C., et al.: Open-source CubeSat MBSE Approach to Address Traceability: Power Subsystem. In: 2023 IEEE International Systems Conference (SysCon), pp. 1–8. IEEE (2023)
25. Pons, W., et al.: Design structure matrix generation from open-source MBSE tools. In: 2021 IEEE International Symposium on Systems Engineering (ISSE), pp. 1–8. IEEE (2021)
26. Pimmler, T.U., Eppinger, S.D.: Integration analysis of product decompositions. In: International Design Engineering Technical Conferences and Computers and Information in Engineering Conference, vol. 12822, pp. 343–351. ASME (1994)
27. Crawley, E., et al.: System architecture: strategy and product development for complex systems. Prentice Hall Press (2015)

A Proposed Product Data Model for Vaccine Lifecycle Management

Allam Sanae[1](✉), El Bouzekri El Idrissi Adiba[1], Sekhari Seklouli Aicha[2], and Ouzourt Yacine[2]

[1] National School of Applied Sciences, Science Engineer Lab Energy, Chouaïb Doukkali University, El Jadida, Morocco
allam.s@ucd.ac.ma

[2] DISP Laboratory, University of Lumière Lyon 2, Lyon, France

Abstract. Vaccines play a pivotal role in protecting against diseases, especially in pandemics like the COVID-19 pandemic. However, the success of rapid vaccine development in such crises has underscored the necessity of implementing a systematic approach that encompasses the entire vaccine lifecycle and involves all stakeholders. Yet, managing their lifecycle poses challenges. This paper addresses the need for an improved approach to vaccine data management through the proposal of a Product Lifecycle Management (PLM) model tailored to vaccine lifecycle management.

Based on a review of literature and existing models, we have developed an integrated framework to address vaccine development, production, distribution, and administration. This model aims to fill existing gaps and is built upon criteria that are defined based on the needs of the vaccine lifecycle for data modeling and the PLM approach.

In conclusion, through the proposed integrated model, this paper aims to address existing gaps in vaccine lifecycle management, emphasizing the necessity of a systematic approach to effectively navigate the challenges. Additionally, this study introduces various alternative product data models, outlines specific criteria for evaluating them within this context, and offers insights into potential directions for future research.

Keywords: Vaccine lifecycle management · Product data model · Integration · Data management

1 Introduction

Vaccines stand as one of the most powerful tools in modern medicine, playing a pivotal role in preventing infectious diseases and safeguarding public health worldwide [1]. From the eradication of smallpox to the mitigation of Coronavirus disease, vaccines have demonstrated their unparalleled ability to save lives and reduce the burden of disease [2]. However, behind the scenes of vaccine development, production, and distribution lies a complex matrix of processes, regulations, and stakeholders.

Published by Springer Nature Switzerland AG 2025
P. Sureephong et al. (Eds.): PLM 2024, IFIP AICT 740, pp. 371–384, 2025.
https://doi.org/10.1007/978-3-031-93319-6_28

Effective management of vaccines across their lifecycle is crucial not only for ensuring their safety and efficacy but also for addressing global health challenges such as pandemics and for emerging infectious diseases. Despite the undeniable importance of vaccines, the management of vaccine data often faces significant challenges. Current practices are characterized by fragmentation, with data scattered across disparate systems and stakeholders. This fragmentation not only hampers efficiency but also complicates decision-making processes and impedes timely responses to emerging threats [3].

Moreover, the lack of integration and traceability in vaccine data management poses risks to regulatory compliance and public health outcomes. In light of these challenges, there is an urgent need for a comprehensive approach to vaccine data management, one that addresses the unique requirements of vaccine lifecycle management and fosters collaboration among stakeholders [4, 5].

Therefore, this paper proposes a novel product data model specifically tailored to vaccine lifecycle management. This model seeks to integrate all product formats and Bill of Materials (BOM), providing a unified view of vaccine components and streamlining data exchange across the entire lifecycle.

Furthermore, the model emphasizes collaboration and coordination among multidisciplinary teams and external partners, recognizing the importance of shared knowledge and expertise in vaccine development and distribution. Additionally, traceability and integrity are paramount, with the model designed to ensure robust tracking and monitoring of vaccine components to meet regulatory requirements and mitigate risks. By providing a unified framework for vaccine data management, this proposed model aims to address the challenges inherent in current practices and pave the way for more efficient, transparent, and effective vaccine management.

2 Related Works

The studies combining the vaccine lifecycle management and data modeling are scarce. Firstly, in this study [6] the authors did based on the Model Drive architecture (MDA) approach addressed vaccine quality data and interoperability challenges in the vaccine supply chain. Despite its significance, contribution especially for addressing this gap and considering the vaccine development and its complexities but the study did only consider ERP systems as the single source of information and the scope view did explicitly include all the vaccine lifecycle stages.

Additionally, the study [7] introduced an integrated approach aimed at harmonizing stakeholders across vaccine development, production, distribution, and vaccination into a cohesive system. This approach emphasizes the establishment of an innovation ecosystem and a virtual platform to facilitate knowledge circulation, technology sharing, and innovation within the vaccine development domain. However, despite these contributions, the study lacks the proposal of a vaccine data model. Before implementing a complex system or virtual platform, it is crucial to gain a thorough understanding of the data requirements, including data sources, data flows, and data structures. A robust vaccine data model would provide the necessary foundation for a system and will be better to be developed first.

Therefore, our study will aim to fill this gap by proposing a comprehensive vaccine data model that can effectively manage data throughout the entire vaccine lifecycle, addressing the needs of stakeholders at every stage. By developing such a model, we can ensure seamless integration, efficient data exchange, and enhanced decision making processes within the vaccine development ecosystem. This approach will not only strengthen the foundation for the proposed integrated system but also contribute significantly to advancing the field of vaccine development and management.

3 Vaccine Lifecycle Management: Overview and Characteristics

The vaccine industry's emphasis greatly shapes the specific portrayal of its main product, the vaccine. The vaccine lifecycle encompasses several crucial stages essential for successful development, distribution, and administration. However, literature tends to focus less on the comprehensive view of the entire vaccine lifecycle, often concentrating on individual steps or subsets without fully recognizing their role within the larger system.

By analyzing published works, we can broadly categorize the vaccine lifecycle into the following phases, as illustrated in Fig. 1 (see Fig. 1):

- Research and development (R&D) stage: This phase involves initial research, identification of vaccine candidates, preclinical testing, clinical trials (phases II, III, and I), regulatory approval, and licensure.
- Manufacturing stage: After regulatory approval, the vaccine moves into production. This includes scaling up manufacturing processes, ensuring quality control, and meeting regulatory standards for production.
- Distribution stage: Once manufactured, the vaccine needs to be distributed to various locations. This involves logistical planning, storage considerations (such as cold chain requirements), and transportation networks.
- Administration and utilization stage: In this final phase, the vaccine is administered to the target population. This involves vaccination campaigns, healthcare infrastructure utilization, monitoring for adverse events, and ensuring equitable access to vaccination.

The lifecycle of vaccines is marked by a series of intricate characteristics (see Fig. 2) that underscore their critical role in global health. Foremost, vaccines are held to an exceptionally high standard, subjected to stringent regulatory scrutiny owing to their profound implications for public health. Ensuring their safety and efficacy is paramount, and this necessitates rigorous quality control measures that span from the sourcing of raw materials to the delivery of the final product. Alarmingly, this quality control phase consumes a staggering 70% of the entire manufacturing cycle time, highlighting its pivotal importance.

Furthermore, the journey from vaccine inception to distribution is neither swift nor straightforward. Vaccine development is an arduous, multifaceted process that can extend over years, if not decades, encompassing numerous intricate stages. This complexity is compounded by the economic dynamics of the vaccine industry, which operates within a fluctuating demand landscape, often resulting in production limitations. These limitations are further exacerbated by the unique storage and distribution requirements of vaccines, which often necessitate specialized infrastructures and logistics.

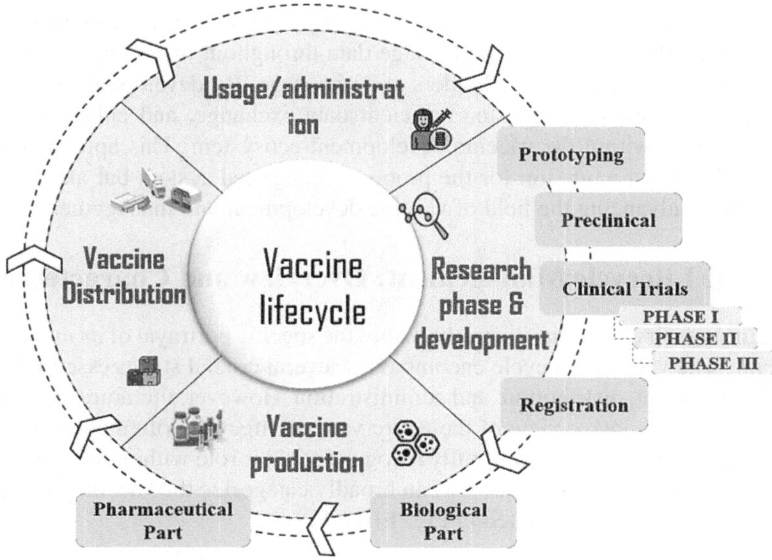

Fig. 1. The vaccine lifecycle

Yet, beyond the scientific and logistical challenges, the vaccine landscape is fraught with profound ethical considerations. The design and execution of clinical trials must adhere to stringent ethical standards, ensuring informed consent and transparency at every juncture. Addressing vaccine hesitancy and advocating for equitable access are paramount, especially in regions grappling with limited infrastructure. Indeed, the distribution and administration of vaccines present formidable logistical hurdles, with challenges ranging from maintaining the cold chain integrity to navigating transportation and storage obstacles, particularly in reaching remote and underserved populations.

4 Product Data Model: State of Art

4.1 Definition

A product data model, as defined by various sources, serves as a structured representation encompassing essential information and attributes associated with a specific product or a range of products within an enterprise. Originating in the early 1990s pioneers like Dupinet, Mony, and Krause laid the foundation for these models, later advanced by Tichkiewitch and Bernard to manage diverse product information in multi-actor, multi-view contexts. The models presented by these experts aim to manipulate information for distinct purposes, such as structuring data (e.g., FBS PPRE and IPPOP), facilitating reuse (e.g., patterns), or enhancing exchange (e.g., AP 214 of STEP or PLCS) [8].

In alignment with this, this work [9] describes a product data model as a structured representation of all essential information and attributes related to a product within an organization. This study [10] emphasizes the support for Product Lifecycle Management

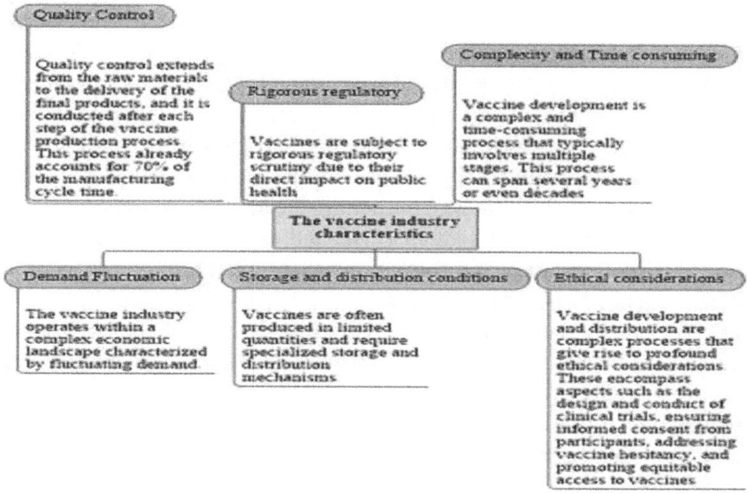

Fig. 2. Vaccine industry characteristics

(PLM) functions throughout the product life, highlighting that a well- developed product data model can be instrumental in mechatronic system design.

4.2 Examining Some Literature Models: Contributions and Limitations

In this section, several studies have been conducted in the context of product lifecycle management (PLM), addressing various gaps, from the lack of holistic approaches to challenges in data integrity and quality.

Firstly, the focus of several studies was to establish a framework with a holistic view. [11] emphasized the importance of product lifecycle modeling, exploring the interconnections between different stages and proposing a Web-based integration framework to streamline data distribution. Their research highlighted essential elements necessary for seamless integration of information across the entire product lifecycle, thereby promoting collaboration. However, the model they proposed has limitations, such as the lack of support for dynamic data and potential interoperability issues.

To address these limitations, this study [12] proposed a multidimensional data model aiming to rectify deficiencies in traditional Bill of Materials (BOMs). They introduced a specialized dimension for dynamic batch information and emphasized the importance of considering fundamental aspects in product structure and production processes. It is worth noting that the study primarily focuses on the production stage, potentially overlooking other stages of the lifecycle. Then, some studies focus only on enhancing quality management throughout the product lifecycle, exemplified by [13] with their multi-layered quality data model. Other studies, like [14] and [15] concentrated on data management and traceability systems, respectively, to streamline processes within supply chains. Meanwhile, returning to the concept of a holistic approach that covers the entire lifecycle of a product, [16] contributed by introducing dynamic data domain-based approaches and encompassing various domains including concept design, structural design, detailed design, process planning, manufacturing, and sales and service. This was achieved through the process of mapping. However, the limitations faced by this work include complex mapping relations in multi-phase data integration leading to redundant data, causing issues with consistency and integrality throughout the product lifecycle. On the other hand, [17] propose unified meta-models to support seamless data integration across lifecycle stages, mitigating the adverse effects of isolated BOMs. However, semantic and syntax issues in UML and questions are raised regarding the meta-model's ability to express all information on a product throughout its lifecycle and the potential challenges in implementation.

Recently, to remedy these limitations, [18] presented comprehensive data management models based on product morphology. Their approach integrates conceptual, digital, and physical product forms to provide a holistic PLM approach. By encompassing various dimensions of product representation, from its abstract conceptualization to its tangible physical manifestation, Liu et al. aimed to create a unified framework that captures the entirety of the product lifecycle. One of the notable achievements of [18] approach is its guarantee of enhanced data integrity throughout the product lifecycle. By integrating conceptual, digital, and physical product forms, the models ensure consistency and coherence across various stages of product development, manufacturing, and service. Moreover, their approach facilitates seamless data exchange and interoperability between different phases of the product lifecycle, thereby reducing the risk of errors and inconsistencies commonly encountered in traditional PLM systems. However, despite the advancements in data integrity and integration, Liu et al.'s approach has its limitations.

The comprehensive nature of the data management models may introduce complexity, making it challenging to implement and maintain in real-world industrial settings. Additionally, the focus on product morphology and representation may overlook crucial aspects of collaboration and communication among stakeholders involved in the product lifecycle (Table 1).

5 Needs for Vaccine Lifecycle Data Management and Data Modeling

Currently, vaccine data management operates within fragmented silos, lacking cohesive integration across the lifecycle. This disjointed approach hampers efficiency and transparency, hindering seamless collaboration and decision-making among stakeholders [3]. Regulatory compliance is often a challenge, with disparate systems struggling to meet stringent standards, potentially compromising the safety and efficacy of vaccine management.

Data integrity issues further exacerbate these challenges, with concerns surrounding the accuracy, completeness, and security of information throughout the lifecycle. Coordination and collaboration efforts are often hindered by disjointed communication channels and fragmented workflows, impeding optimal resource allocation and response to emerging challenges [7]. Therefore, the current landscape underscores the pressing need for a transformative shift towards integrated approaches, such as Product Lifecycle

Table 1. Some proposed models their contribution and limits.

Article title	Auteur	Type	Year	The Gaps / problems addressed	Contribution	Limits
A Conceptual Framework For Product Lifecycle Modeling	Qilin Shu, Chengen Wang	Conferenc e paper	2006	- Lack of a holistic approach in existing PLM models - Limited Collaboration Across Enterprises - Inefficient Information Integration: The need for a unified approach to model, manage, and utilize product information is emphasized - The need for Web to enhance product lifecycle information infrastructure	The research underscores key elements of product lifecycle modeling, explores relationships between stages, and suggests a Web-based integration framework for distributed data sources. Identified key elements of a product life cycle modelling to integrate all the information and support the networked manufacturing mode.	- Real-Time Data - Challenges - Interoperability Issues - Data Volume and Complexity - (evolution of technology)
Research on PLM multidimensional data model	WANG Zhen, WANG Lifengt, RUI Yuxiangt, LI Congxin	Research article	2006	- Limitations of Traditional BOM Data Model - the inadequacy of the traditional Bill of Materials (BOM) data model in meeting the requirements of Product Lifecycle Management (PLM), particularly in facilitating efficient data query and capturing - Inability to Manage Inspection Data by ERP and PDM - The complexities and limitations of integrating various subsystems such as PDM, ERP, MES, CRM, SCM, and OA, emphasizing the need for a more comprehensive approach to PLM implementation	- The work contributes by presenting a holistic multidimensional data model that addresses current challenges in PLM - The study introduces a specialized dimension for dynamic batch information, and emphasizes the importance of considering fundamental aspects in product structure and production processes.	The need for a holistic perspective
Data model for quality in product lifecycle	Xiaoqing Tang ,HuYun	Research article	2008	- Need for Integrated Quality Data Model	The authors contribute by proposing a multi-layered quality data model framework, offering integration techniques for modeling quality data, and presenting a systematic approach for managing quality data in the product lifecycle.	The need for a holistic perspective

(*continued*)

Table 1. (*continued*)

Product data model for PLM system	Yumei Li & Li Wan & Tifan Xiong	Research article	2011	- Existing models may be static and fail to incorporate dynamic data - Existing models focus solely on product data without considering related resource information crucial for product evolution in PLM environments - Existing models often struggle with managing data integrity, security, and traceability, despite attempts to ensure a single data source for product information.	Li's contribution involves the utilization of data domains, encompassing a concept design domain, a structural design domain, a detailed design domain, a process planning domain, a manufacturing domain, and a sales and service domain, achieved through the process of mapping.	Complex mapping relations in multi-phase data integration lead to redundant data, causing issues with consistency and integrality throughout the product lifecycle
A Product Lifecycle Data Management Framework Based on Resource Meta-model	Shaojun Qin, Hongming Cai, Lihong Jiang	Conference paper	2012	aim to find solutions to support the integration and unified management of distributed and heterogeneous product lifecycle data complexity, lack of flexibility, and configurability in the existing models	The authors introduce a product lifecycle data management framework based on a resource meta-model. This framework organizes product lifecycle information as resources, enabling flexible integration and configuration. Additionally, they propose an information accessing control mechanism to ensure secure and authorized access to data. Through a case study, they demonstrate the effectiveness of their approach in seamlessly integrating and managing heterogeneous data throughout the product lifecycle	This limitation implies challenges in handling various types of product data efficiently. A potential gap in addressing the dynamic changes and advancements that products undergo throughout their lifecycle.
Data Model in PLM System to Support Product Traceability	Dharmendra K. Mishra, Aicha Sekhari, Sebastien Henry, Yacine Ouzrout	Research article	2017	- Manual information management within the oxygen gas supply chain - Lack of Tracking Mechanism	The authors address the gap of manual information management within the oxygen gas supply chain by proposing a traceability system that automates data collection and tracking processes	The absence of a holistic approach to the product lifecycle

(*continued*)

Table 1. (*continued*)

Research on Product Lifecycle Model-Based on UML	Zihang Xia, Jiping Yang Sheng Yin and Tao Hu	Rearch article	2021	- The absence of a unified meta-model that spans the entire life cycle of a product - The adverse effects of isolated bills of materials (Boms) in different application systems - Maintain the consistency of the model and data integration	Xia proposes a unified meta-model that provides support for data meta-models across product design, process manufacturing, sales and operations, and maintenance services stages.	Semantic and Syntax Issues in UML Questions are raised regarding the meta-model's ability to express all information on a product throughout its lifecycle and the potential challenges in implementation
A Data Management Approach Based on Product Morphology in Product Lifecycle Management	Gang Liu Rongjun Man and Yanyan Wang	Research article	2021	- The study identifies a gap in addressing the critical factor of "product form," emphasizing its significance for product existence and data evolution - Existing studies mainly integrate conceptual and digital or digital and physical products, neglecting a comprehensive description of the evolution process - Consistency and integrality of product data throughout the full lifecycle are deemed challenging to control, leading to potential issues in maintaining data accuracy and completeness - Integration of multi-phase data introduces complexity and redundancy due to complex mapping relations - A holistic approach is needed; single data sources or multi-phase models are insufficient for a comprehensive product data model	Gang suggests three product data management forms and introduces a comprehensive data management model that encompasses these three product forms: conceptual products, digital products, and physical products. This model offers a novel approach to handling product lifecycle data.	- The absence of explicit mention regarding collaboration and coordination between stakeholders in the presented models and methodologies : Lack of Stakeholder Interaction Model - Simplified Treatment of Physical Product Data - The study introduces global object mapping to achieve the morphological evolution path, but the complexity and potential challenges of implementing this mapping method are not thoroughly addressed

Management (PLM) and robust data models, to streamline processes, enhance regulatory compliance, and foster collaboration across the vaccine lifecycle. Transitioning from fragmented silos to integrated approaches in vaccine data management is pivotal, and the strategic adoption of Product Lifecycle Management (PLM) and a robust data model offers numerous benefits. Unlike the current disjointed approach, this transformative shift promises a cohesive framework where data seamlessly traverses every stage of the vaccine lifecycle. This holistic integration is crucial for revolutionizing the efficiency, transparency, and agility of vaccine management processes.

A smooth and uncomplicated implementation process is key. It is vital that the integration of PLM and the data model occurs with minimal disruption to ongoing operations, ensuring a seamless transition for all stakeholders involved [19]. This requires careful planning, robust change management protocols, and comprehensive training initiatives to empower users with the necessary skills and knowledge to utilize the new system effectively. Regulatory compliance is nonnegotible. The system must be meticulously designed and rigorously tested to meet stringent regulatory standards, ensuring full alignment with prevailing guidelines and mandates governing vaccine development, manufacturing, distribution, and administration. Adherence to these regulations not only safeguards public health but also instills confidence in the integrity and reliability of vaccine management processes.

Data integrity is another critical consideration. The system must prioritize the accuracy, completeness, and security of data throughout its lifecycle, safeguarding against data corruption, unauthorized access, and other potential threats. This involves implementing robust data governance frameworks, employing advanced encryption protocols, and instituting stringent access controls to mitigate risks and uphold the sanctity of data. Effective coordination and collaboration are essential [7]. Given the inherently complex and multifaceted nature of vaccine development, distribution, and surveillance, seamless collaboration across multidisciplinary teams and global partners is necessary. This requires establishing clear communication channels, standardized workflows, and collaborative platforms to facilitate real-time information exchange, foster cross-functional collaboration, and optimize resource allocation.

Overall, the transition towards integrated vaccine lifecycle management systems demands a holistic approach that encompasses technical, regulatory, and organizational dimensions. By addressing these imperatives diligently and with foresight, organizations can unlock unprecedented levels of efficiency, transparency, and efficacy in vaccine management, ultimately contributing to the advancement of global health initiatives and the wellbeing of populations worldwide.

In our pursuit of effective vaccine lifecycle management, we have formulated a structured framework (Fig. 3) comprising five integral layers. Beginning with the Actors Layer, we identify the pivotal stakeholders, from researchers to patients, who shape the vaccine journey. The Data Layer follows, emphasizing meticulous data collection and management throughout development, production, and distribution phases. We then focus on the Systems and Applications Layer, leveraging tailored software for research, production, and supply chain management. Standardizing data representation across systems forms the Data Model Layer, ensuring consistency and integrity.

Finally, at the core lies the PLM System Layer, a centralized platform facilitating compliance, collaboration, and traceability. This comprehensive framework enables streamlined processes, mitigates risks, and upholds vaccine quality and safety, bolstering our commitment to public health advancement.

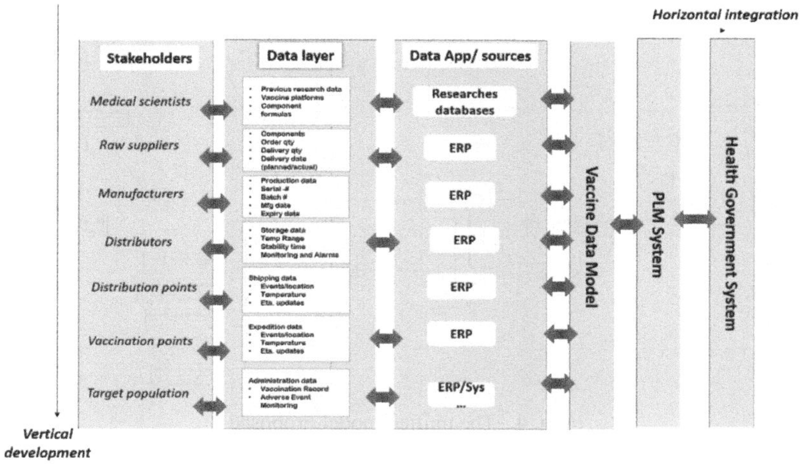

Fig. 3. Framework and data architecture

6 Proposed Data Model

Our data model revolves around the vaccine product in both its digital and physical forms, integrating its Bills of Materials (BOMs) as well. The aim is to ensure comprehensive and coherent management of all vaccine-related data, from conception to distribution.

This involves linking the various stakeholders involved in the process, including researchers, manufacturers, regulators, and other key actors, to ensure effective collaboration throughout the vaccine lifecycle.

In the context of vaccine lifecycle management, our proposition aims to bridge critical gaps to ensure seamless and efficient vaccine management operations. Firstly, we focus on integrating all product formats and Bill of Materials (BOM) associated with vaccine development, encompassing both physical and digital formats.

Additionally, our proposition prioritizes traceability and integrity of the BOM, ensuring that every component is meticulously tracked and verified throughout the entire lifecycle, thus enhancing transparency, regulatory compliance, and ultimately, the safety and efficacy of vaccine (Fig. 4).

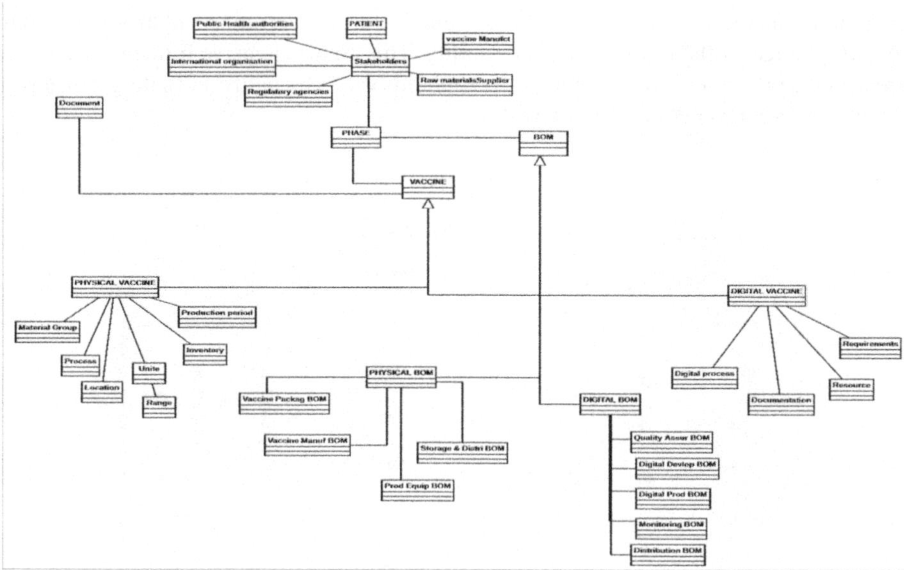

Fig. 4. The unified model proposed

7 Conclusion

In conclusion, the proposed product data model for vaccine lifecycle management represents a significant step towards addressing the complex challenges inherent in vaccine development, production, and distribution.

By integrating all product formats and Bill of Materials (BOM), prioritizing collaboration and coordination, and emphasizing traceability and integrity throughout the lifecycle, this model offers a comprehensive solution to enhance transparency, efficiency, and regulatory compliance. Implementing this model has the potential to streamline processes, accelerate decision-making, and ultimately, improve the safety and efficacy of vaccines.

Moving forward, further research and validation efforts will be essential to refine and optimize the proposed model, ensuring its effectiveness in meeting the evolving needs of vaccine management in an increasingly complex and dynamic global landscape.

Disclosure of Interests. The authors have no competing interests to declare that are relevant to the content of this article.

References

1. Cook-Moreau, J., Mehring, M., Buxeraud, J., Juvin, S.: L'essentiel sur les vaccins. Actual. Pharm. **55**(559), 16–22 (2016). https://doi.org/10.1016/j.actpha.2016.07.005
2. Sarlangue, J.: Histoire des vaccinations, de la variole à la Covid-19. Perfectionnement Pédiatrie **5**(1), 72–83 (2022). https://doi.org/10.1016/j.perped.2022.01.012

3. Kuczenski, B.: Life cycle fragments: Computational methods for analyzing and sharing life cycle assessment models. In: Ames, D.P., Quinn, N.W.T., Rizzoli, A.E. (eds.) 7th International Congress on Environmental Modelling and Software, San Diego, California, USA. International Environmental Modelling and Software Society (iEMSs), pp. 1–15 (2014). https://doi.org/10.13140/2.1.4373.2486
4. Quinn, N.W.T. (eds.): International Environmental Modelling and Software Society, pp. 107–113. iEMSs (2014). https://www.scopus.com/inward/record.uri?eid=2-s2.0
5. Moalla, N., Bouras, A., Ouzrout, Y.: Production data integration in the vaccine industry. In: 5th IEEE International Conference on Industrial Informatics (INDIN 2007), pp. 603– 608. IEEE, Vienna (2007). https://doi.org/10.1109/INDIN.2007.4384826
6. Stiefel, J., Zimmer, J., Schloßhauer, J.L., Vosen, A., Kilz, S., Balakin, S.: Just keep rolling?—an encompassing review towards accelerated vaccine product life cycles. Vaccines **11**(8), 1287 (2023). https://doi.org/10.3390/vaccines11081287
7. Moalla, N., Bouras, A., Ouzrout, Y., Neubert, G.: A MDA approach for product data quality throughout vaccine product lifecycle. In: Proceedings of the ASME 2008 International Design Engineering Technical Conferences and Computers and Information in Engineering Conference, vol. 3, pp. 1293–1301. ASME, Brooklyn (2008). https://doi.org/10.1115/DETC2008-49932
8. Ilin, I., Levina, A., Frolov, K.: Innovative ecosystem model of vaccine lifecycle management. J. Open Innov.: Technol. Mark. Complexity **8**(1), 5 (2022). https://doi.org/10.3390/joitmc8010005
9. Le Duigou, J.: A Framework for modeling PLM systems in extended enterprises. application to mechanical SMEs. Ph.D. Thesis, École Centrale de Nantes (2010). https://theses.hal.science/tel-00487196v1
10. Sanae, E.B., El Idrissi, A., Sekhari Seklouli, A., Noredine, E.K.: Evaluating product data modeling for vaccine lifecycle management. In: 15th International Conference on Software, Knowledge, Information Management and Applications (SKIMA 2023), pp. 261–266. IEEE, Kuala Lumpur (2023). https://doi.org/10.1109/SKIMA59232.2023.10387363
11. Zheng, C., Bricogne, M., Le Duigou, J., Eynard, B.: Mechatronic design process: a survey of product data model. Procedia CIRP **21**, 282–287 (2014). https://doi.org/10.1016/j.procir.2014.03.176
12. Zhang, W., Fan, Y.: A conceptual framework for product lifecycle modeling. In: 1st International Conference on Innovative Computing, Information and Control (ICICIC 2006), vol. 1, pp. 486–489. IEEE, Beijing (2006). https://doi.org/10.1109/ICICIC.2006.190
13. Wang, Z., Wang, L., Rui, Y., Li, C.: Research on PLM multidimensional data model. In: 2006 International Technology and Innovation Conference (ITIC 2006), pp. 1499–1504. IEEE, Hangzhou (2006). https://doi.org/10.1049/cp:20061004
14. Tang, X., Yun, H.: Data model for quality in product lifecycle. Comput. Ind. **59**(2–3), 167–179 (2008). https://doi.org/10.1016/j.compind.2007.06.011
15. Qin, S., Cai, H., Jiang, L.: A product lifecycle data management framework based on resource meta-model. In: IEEE Asia-Pacific Services Computing Conference (APSCC 2012), pp. 305–308. IEEE, Guilin (2012). https://doi.org/10.1109/APSCC.2012.53
16. Mishra, D.K., Sekhari, A., Henry, S., Ouzrout, Y.: Data model in PLM system to support product traceability. In: Ríos, J., Bernard, A., Bouras, A., Foufou, S. (eds.) PLM 2017. IFIP Advances in Information and Communication Technology, vol. 517, pp. 607–616. Springer, Cham (2017). https://doi.org/10.1007/978-3-319-72905-3_54
17. Li, Y., Wan, L., Xiong, T.: Product data model for PLM system. Int. J. Adv. Manuf. Technol. **55**, 1149–1158 (2011). https://doi.org/10.1007/s00170-010-3130-4
18. Xia, Z., Yang, J., Yin, S., Hu, T.: Research on product lifecycle model based on UML. In: 2020 Asia Conference on Geological Research and Environmental Technology. IOP Conference

Series: Earth and Environmental Science, vol. 632, 052078. IOP Publishing (2021). https://doi.org/10.1088/1755-1315/632/5/052078
19. Liu, G., Man, R., Wang, Y.: A data management approach based on product morphology in product lifecycle management. Processes **9**(7), 1235 (2021). https://doi.org/10.3390/pr9071235
20. Mostefai, S., Bouras, A., Batouche, M.C.: Data integration in a PLM perspective for mechanical products. Int. Arab J. Inf. Technol. **2**, 141–147 (2005). https://api.semanticscholar.org/CorpusID:14315702

Towards a Personalized Knowledge Graph Thanks to Gamification

Pierre Miroite[1(✉)], Mickaël Gardoni[1,2], and Geneviève-Anaïs Proulx[3]

[1] École de Technologie Supérieure, 1100 Rue Notre Dame Ouest, Montréal, QC H3C 1K3, Canada
pierre.miroite.1@ens.etsmtl.ca, mickael.gardoni@etsmtl.ca
[2] INSA Strasbourg, 24 Bd. De La Victoire, 67000 Strasbourg, France
[3] Moment Factory, 6250 Avenue du Parc, Montréal, QC H2V 4H8, Canada
genevieveanaisp@momentfactory.com

Abstract. Knowledge management aims to capitalize on knowledge and makes it accessible within an organization or group. However, it can be difficult to engage all members of the organization in its realization, particularly in terms of access to existing knowledge, and the sharing of new knowledge. Gamification is being considered as a way of making an impact on this commitment. By following the DSR method, a visual representation of an organization's knowledge including gamification mechanics has been prototyped. It aims to be personalized for each member of an organization, using the progression mechanics, so they could easily know which knowledge has been acquired, and which knowledge has to be. The model is also driven towards collaboration within the organization. The prototype presently faces software limitations but could easily be used to verify future user interest. Other gamification mechanics and elements still need to be explored more in depth to be added to the prototype, such as players' profiles.

Keywords: Knowledge Management · Gamification · Knowledge Graph

1 Introduction

Knowledge management has become a critical issue for organizations seeking to increase their performance and competitiveness in an ever-changing environment. Indeed, the ability to identify, capture, share, and use internal knowledge and skills is a key factor for success in the knowledge economy [1]. Several methods and tools exist to implement knowledge management within an organization. These notably include the MEREX method, which allows for the formalization and structuring of explicit knowledge, as well as communities of practice, which promote the sharing and creation of tacit knowledge within a group of individuals sharing a common interest [2]. However, one of the major challenges of knowledge management lies in user engagement. Knowledge sharing is not always natural and can be perceived as a waste of time or energy. Gamification, which involves using game mechanics in non-game contexts, can be a promising solution for

Published by Springer Nature Switzerland AG 2025
P. Sureephong et al. (Eds.): PLM 2024, IFIP AICT 740, pp. 385–397, 2025.
https://doi.org/10.1007/978-3-031-93319-6_29

improving the engagement and motivation of participants [3]. This article therefore proposes a conceptual model that combines the MEREX method, communities of practice, and gamification to improve the capitalization and formalization of knowledge within organizations, and to increase engagement in the purchase and access of said knowledge.

2 Knowledge Management

2.1 Fundamentals of Knowledge Management

Knowledge management is a crucial process that enables individuals, groups, and organizations to effectively identify, share, use, and store their knowledge and skills. It differs from simple data management by its anchoring in a specific context and its interpretive character, and from know-how, which is recognized by peers in a given field. Two types of knowledge coexist and interact within an organization:

- Explicit knowledge: This is theoretical and formalizable knowledge that is easily translated into procedures, manuals, or guides. It forms the basis for training and reference documents.
- Tacit knowledge: This is abstract and complex knowledge that is difficult to formalize. It is derived from experience, expertise, and individual know-how. It is often transmitted informally through mentoring, coaching, or direct observation [4].

Knowledge management is based on structured and iteratives processes that are: the acquisition of knowledge inside or outside an organization, the storage of this knowledge followed closely by its updating, if necessary, the sharing of knowledge within the organization, and finally its use in problem solving or decision.

Two main strategies guide the implementation of knowledge management:

- Codification: This aims to document and formalize explicit knowledge to make it easily accessible and transferable [4]. This involves creating training materials, databases, or practical guides.
- Personalization: This encourages informal interactions and exchanges between employees to promote the sharing of tacit knowledge, which is more difficult to formalize [5]. Approaches such as coaching, mentoring, and communities of practice can be implemented to support this strategy.

By combining these two approaches and adapting knowledge management to its specific context, an organization can unlock the full potential of its intellectual capital and sustainably increase its performance and competitiveness.

2.2 Communities of Practice

In addition to knowledge codification and personalization strategies, communities of practice and communities of jobs, also known as knowledge communities, are valuable levers for knowledge management within an organization. Unlike project groups, which have a circumscribed objective of completing a specific task, communities of practice and communities of jobs focus on the sharing, production, and maintenance of knowledge in a specific domain.

These communities are distinguished by their informal nature. They are created spontaneously or in a planned manner, and their operation is based on the volunteerism and commitment of their members. The main motivation of participants is their common interest in a specific topic and the desire to progress collectively in this area [6]. Whether they are enthusiasts or people seeking to solve a common problem, community members come together to develop and share their knowledge.

Unlike communities of jobs, which bring together individuals sharing the same profession or position, communities of practice are organized around a common interest and aim to share their knowledge. The profession is therefore not necessarily a criterion for entry, thus promoting diversity of perspectives and expertise within the community.

2.3 MEREX Method for Knowledge Capitalisation

To support communities in capitalizing on their knowledge, the MEREX method can be used. Its main objective is to collect, organize, and disseminate the knowledge and know-how of an organization's members in order to make them accessible and reusable by the entire organization [7]. The originality of the MEREX method lies in its foundation on several key principles. First, it values the experience of the organization's actors as the main source of knowledge, thus recognizing the crucial role of human capital in collective success [8]. This approach then emphasizes the formalization of knowledge in the form of synthetic and structured sheets called "MEREX sheets." These sheets allow the presentation and organization of knowledge in a clear and concise manner, thus facilitating their consultation and subsequent use. They provide a concrete and rapid response to a problem in a given context, and it then becomes easier to categorize them and update them as needed [9].

In terms of knowledge organization, MEREX sheets are organized according to a thematic and tree-like logic, allowing for the creation of a knowledge map of the organization. This organization facilitates navigation and information search, thus promoting collaboration and knowledge sharing between different departments. Finally, the dissemination of capitalized knowledge is an essential element of the MEREX method. To guarantee the accessibility and reusability of knowledge, the method recommends the use of different media and communication channels, such as intranets, documentary databases, collaborative platforms, and internal training.

By relying on these fundamental principles, the MEREX method offers many advantages to the organizations that adopt it. First, it improves collective performance by facilitating the sharing and reuse of knowledge, which can lead to better operational efficiency and cost reduction [10]. In addition, knowledge capitalization helps to reduce the risk of errors by capitalizing on lessons learned from the past and disseminating them to all employees.

3 Gamification

3.1 Basic Gamification Principles

There are several definitions of gamification, one of the most comprehensive being the implementation of game design elements into non-game contexts [11]. The goal is to increase the motivation and engagement of the person to whom it is proposed. In this context, a game is understood as a codified structure with rules to follow and a clearly defined goal [12]. It also includes notions such as rewarding users and providing direct feedback on actions, allowing them to know the consequences of their actions almost instantly. More general definitions are also given, for example by Werbach [13], who describes gamification as the process of making activities more like a game, or by Marczewski [14], who defines it as the use of game elements and design metaphors to solve problems. The goal is not necessarily to create a game for the purpose of entertainment, but rather to create engagement with users in order to encourage them to adopt a desired behavior.

But gamification isn't effective on its own, adding gamification to a system won't turn it automatically more efficient. Kapp [15] insists on the fact that a gamified system needs to be qualitative by answering the needs it promised to answer; adapted to the users to whom it is proposed to; and appropriate to the context in which it is applied, all this before being gamified. We define the system as a model, process, or even an object that will be gamified and given to users.

3.2 Gamification for Knowledge Management

Gamification can influence knowledge management through the various elements available. If we refer to the model presented in Fig. 6, these different components will influence the engagement and motivation of users towards a communicated goal. The goal here is to carry out any action related to knowledge management. And for the realization of these actions to have the most positive impact, it is better that it is enjoyable for the user. It must therefore influence the user's intrinsic motivation, hence the interest in using gamification [16]. Knowing the factors of motivation is therefore useful when you want to create a gamified system. For example, to stimulate motivation through recognition, you can use badges, or more generally, rewards, which can take different forms. They can be awarded when a task has been successfully completed, or when a challenge has been met [12]. They can be related to extrinsic motivation if they serve as a social marker, or to intrinsic motivation if they are used to trigger ambition, for example. They can therefore be linked to different motivational factors.

Still on the topic of rewards, according to Chou [17], there are six types of contextual rewards:

- Fixed action reward: Awarded after completing a predefined objective. Ex: having contributed 10 new pieces of knowledge to the system
- Random reward: Same principle, except that the nature of the reward is unknown, only what needs to be done to obtain it (same example applies)
- Sudden reward: Obtained suddenly without being warned of its nature or the actions required to obtain it. Ex: having been the fastest to find knowledge searched by someone else
- Lottery reward: Awarded to only a small group on a random draw basis after participating in an action. Ex: having been selected to test the new knowledge management system
- Social treasure: Offered by other users of the system and can only be obtained from others. Ex: being designated the best knowledge creator of the year

- Cumulative reward: Received after accumulating other rewards, or multiple times the same reward. Ex: having contributed 5 pieces of knowledge in the same week

Varying the proposed elements and their nature allows to avoid redundancy and to leave a constant part of novelty in a knowledge management system that will not, or less quickly, decrease over time [18].

3.3 Several Gamification Models

DMC Model. In their literature review, Friedrich et al. [19] categorize gamification elements into three categories: game dynamics, game mechanics, and game components. Components are the most concrete forms of gamification (points, quests, levels), and they trigger game mechanics, which are equivalent to incentives. These mechanics in turn give rise to dynamics, which are the most abstract forms of gamification. This categorization makes it possible to identify what one wants to trigger in the person to whom gamification is proposed, and to choose the relevant components (Figs. 1, 2, 3 and 5)

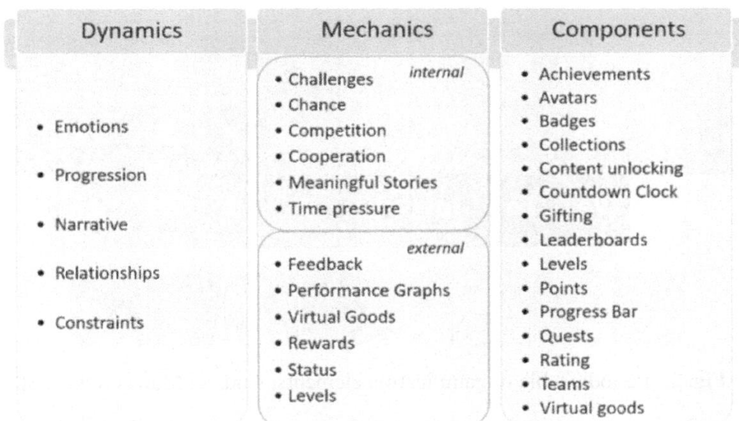

Fig. 1. Dynamics, mechanics et components of gamification [19]

In addition to the models, it is necessary to ensure that gamification is effective when it is implemented. While the goal, rules, motivation, direct feedback on actions, and rewards are essential, three other things must also be ensured. According to Kapp [20], the system must first be reliable and functional before it is gamified. It must meet the expectations of the user, but above all, it must meet the expectations it promises to meet. If the system is not effective, gamifying it will not improve things. Secondly, it must be adapted to the users, who do not all have the same appetite for technology or games. It is therefore necessary to target the right audience, or to do the opposite and work on increasing their appetite for games. Finally, the system must be adapted to the environment in which it will be implemented. For example, if the system is based on competition, while users are used to working collaboratively, it is likely to be unsuccessful.

Periodic Table of Gamification Elements. Andrzej Marczewski focused more particularly on the relationship between different gamification mechanics and elements that can be found from one system to another. However, he proposes a representation centered on the user profiles that he has already identified [21]. The table consists of 52 game mechanics divided into 8 distinct categories. One is dedicated to rewards, one is more general, and 6 others correspond to a predefined player profile. This rigorous organization allows a holistic visualization of the intrinsic relationships between the mechanics and to apprehend their impact on user engagement and satisfaction.

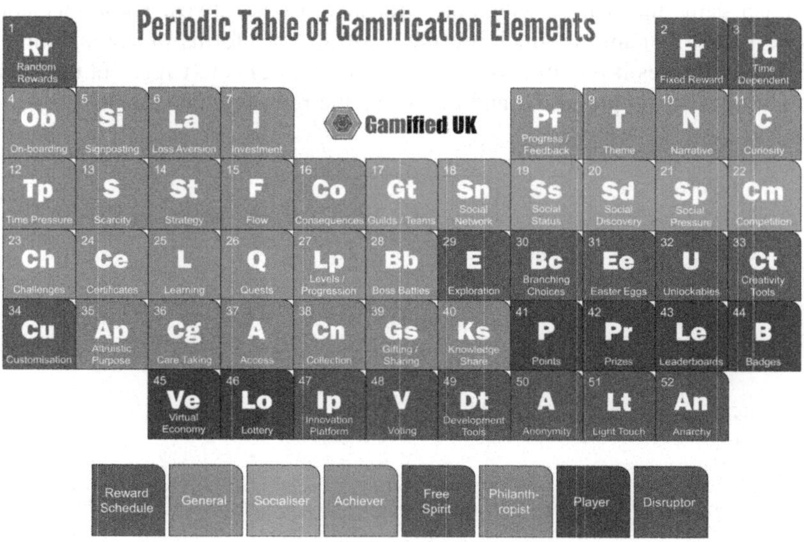

Fig. 2. Periodic table of gamification elements, Andrzej Marczewski [22]

This table is an important source of inspiration for the integration of mechanics into a gamification system and is in line with the DMC model of Friedrich presented previously. The user profiles are based on the Hexad Scale [21] which offers a more diversified categorization spectrum than the categorization proposed by Bartle [23] which separated players into 4 profiles:

- Killers: They seek to dominate other players and only progress to be better than them (they act on people).
- Achievers: They seek to fulfill all objectives to succeed and obtain the best score, they only care about their level (they act in the environment).
- Explorers: They want to learn more about the environment or the system that is proposed to them, they tend to want to look for flaws or understand how things work (they interact with the environment).

- Socializers: They ultimately want to learn more about what others are doing in the environment, or they want to meet other people to move forward with (they interact with people).

Marczewski's [22] differentiation is more precise and allows for a more advanced personalization of a gamified system. It is based on the intrinsic and extrinsic motivations of users, and the intrinsic motivations of relatedness, competence, and autonomy, with the addition of purpose as identified by Edward et al. [24]. Thus, we obtain a motivation associated with each profile:

- Philanthropists who are motivated by intrinsic purpose and that are willing to give without expecting a reward in return.
- Socializers who are motivated by relatedness and mainly want to interact with other users.
- Free Spirits who seek autonomy and freedom, so they could express without any external control.
- Achievers who are motivated by competence by seeking progress. It could be by completing tasks or overcome difficult challenges within a system.
- Players who are motivated by extrinsic rewards which lead them to realize anything within a system, independently of the type of the activity, to earn some.
- Disruptors who just want to trigger change, or even chaos, by disturbing the system directly or through others. It could be negatively or positively, to either benefit their selves or disturb others. [14]

Knowing what motivates each users allows a "system administrator", or any organization in charge of creating a gamified system, to know how to address to the proper users, or to select via the gamification's mechanics which users they aim to impact. This way of thinking is synched with Kapp's sayings [20] concerning the fact a system has to be adapted to the users it is addressed to.

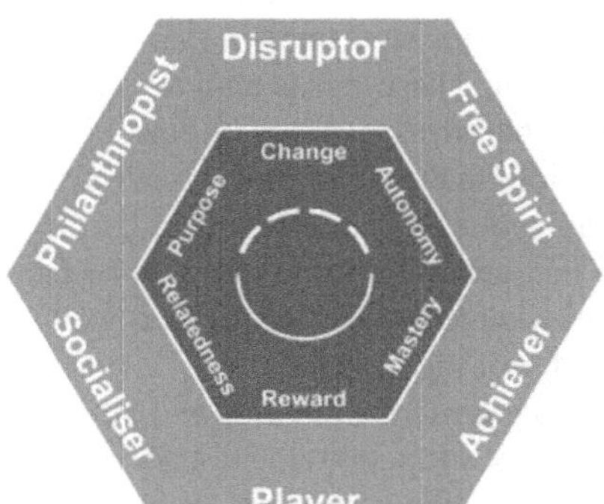

Fig. 3. Gamification user types Hexad [14]

4 Proposition of Gamification Integration into Knowledge Management

Gamification will be used here for the purpose of motivating access to and sharing of knowledge within an organization. Based on the DMC model and the DSR method, the goal is to propose a personalized visual knowledge map for each member of an organization. This map would be based on the knowledge capitalized within the organization's communities of practice. It would meet the needs identified in the research of:

1. Knowledge creation by the people to whom a gamified knowledge management process is proposed.
2. Knowledge acquisition by these same people who are offered this process [25]

The use of gamification combined with the capitalization of knowledge from communities using MEREX sheets is done within the framework of the following short- and medium-term objectives:

- Promote collaboration and the transmission of individual knowledge in order to make it collective.
- Facilitate the formalization of knowledge for sharing.
- Create a personalized visual representation of knowledge acquired and to be acquired to motivate users to access and share knowledge.

4.1 DSR Method

The DSR method will be used to create an artifact that extends the limits of organizational capabilities. Artifacts obtained using the DSR method can be of various types and can therefore take the form of software, a process, a model, or a method. The DSR process also aims to generate knowledge on how things can and should be built, arranged or designed, generally by humans, to achieve a set of desired objectives [26]. A conceptual representation of the process is shown in Fig. 4.

On one side, there is the environment where the problem arises, and which is composed of elements already encountered, namely the organization, its members and the technology in place. It is also in this environment that we find the components of a need to be filled by the organization. These needs also take into account the existing, such as the technologies used or the infrastructures in place. On the other side, there is the knowledge base which provides material for the realization of the DSR. This is where we find the knowledge already acquired or accumulated by the organization, the methods already in place, but also the new knowledge that will be capitalized. We draw from it the so-called applicable knowledge that will be used, as well as the needs, for the Design part that is found at the center of the model.

It is in this design part that artifacts, theories and all other novelties are conceived. Each artifact created undergoes a functional and iterative evaluation to ensure its relevance to the environment. Once this is the case, it becomes applicable and usable by the environment, and is considered as a new resource available in the organization's knowledge base.

Concerning the artifact's creation, it is planned to apply the same method given by Peffers et al. [27] which consists of 6 parts:

1. Identify the problem and the motivations for solving it.
2. Define the objectives to be achieved with the artifact.
3. Create and develop the artifact.
4. Use and apply the artifact.
5. Evaluate the artifact.
6. Communicate about the artifact.

The problem to be solved and the objectives to be achieved are identified in Sect. 4. The artifact creation phase is currently underway and will be developed in the following section.

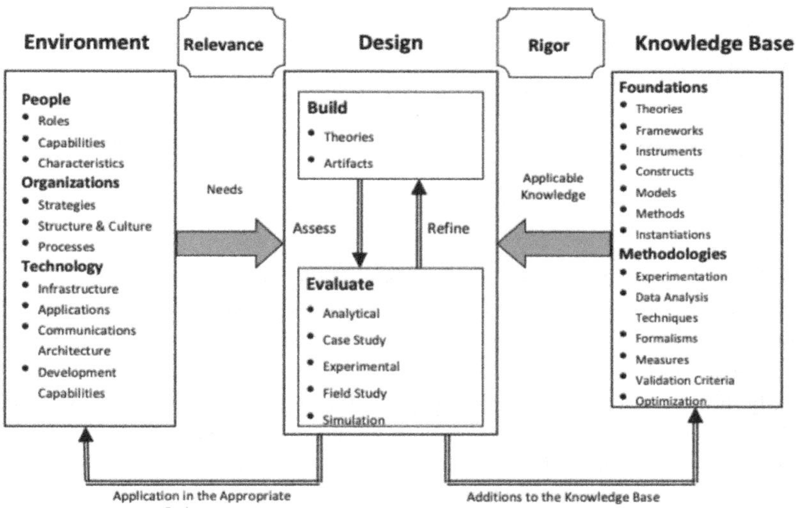

Fig. 4. DSR Model, adapted from Hevner et al. (2004) [26]

4.2 Artifact Proposition

The visual representation of knowledge to be acquired must therefore be based on an existing knowledge base within the organization. This will add to and update this mapping within a system that it will control and will also be responsible for proposing gamification adapted to its users. This is in order to motivate them to complete this mapping.

To find this progression mechanism, it would involve proposing to each user their own instance of the cartography in order to be able to see at any time which knowledge remains to be validated, or which is already validated, thanks to real-time feedback. The user would also be able to add new knowledge to the system or update existing knowledge. A functional diagram of the artifact would therefore be as follows:

The two cartographies would therefore be interconnected so that the knowledge base is always up to date. In addition, it would be possible to share access to knowledge

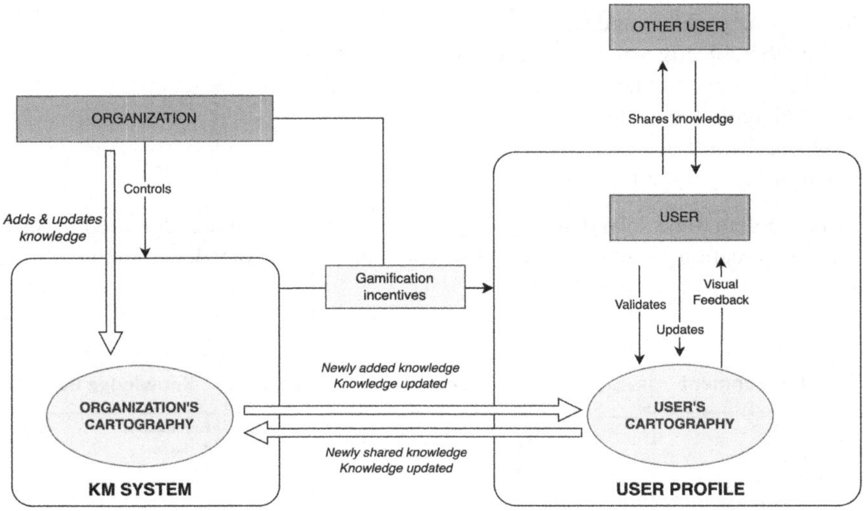

Fig. 5. Functional diagram of the knowledge cartography

between users. Because let us remember they would not all necessarily be accessible from the start in order to remain aligned with the progression dynamic. However, it is also not a question of limiting users in their access to knowledge if it is essential for their work. To promote knowledge sharing and thus meet one of the objectives mentioned previously, it has been decided to make it possible to share access to knowledge. As in a MEREX sheet, the author is mentioned, it would also be visible even if a knowledge sheet is not yet accessible. This would therefore encourage a user to exchange with another user within their organization.

As for the gamification system proposed by the system and the organization, it is imagined offering challenges linked to the different player profiles, and therefore to propose adapted rewards accordingly. Thus, users would earn points according to their predominant player profile and would obtain privileges as rewards. For example, if a person has a predominant Achiever profile, they would be rewarded by having access to even more knowledge that would not normally be accessible.

Conceptual tests have been carried out using the Capacities organizational knowledge management software to represent a knowledge map. Its structure allows the creation of objects, knowledge sheets, tags, and to link them together. These links are then automatically represented visually by the software as shown in Fig. 6. It therefore becomes easier to create links between MEREX sheets according to the theme they deal with. In order to respect the progression dynamic, levels are to be assigned to the sheets. For the moment, they are present for information purposes only, but in the future, it could be possible to determine them with the help of the members of the communities, or by following a method of quantifying the difficulty of a field of expertise [28].

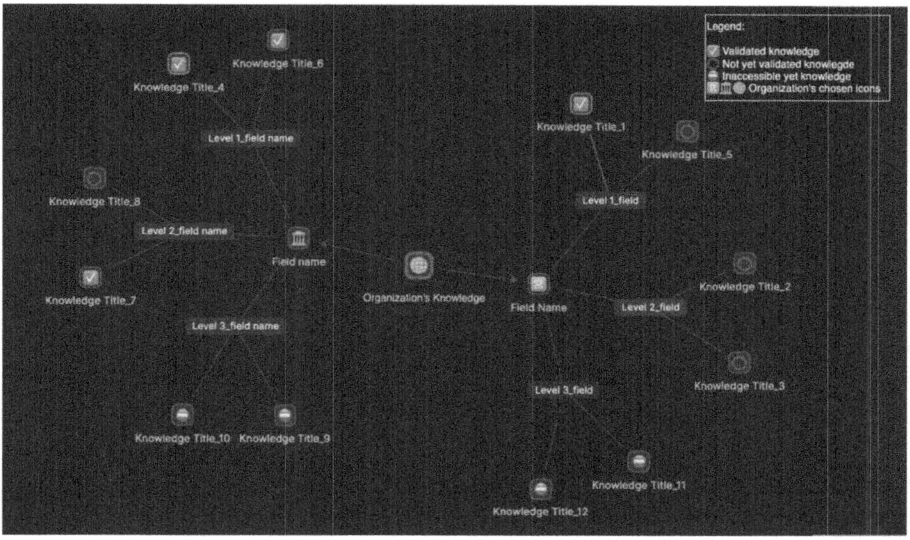

Fig. 6. Visual representation of interactions between knowledge cards on capacities.io (with additional information)

4.3 Software Limits & Future Work

For the moment, the software is still limited and does not allow the creation of instances of the same map, it is only possible to share MEREX sheets with other users, but there again the access management remains limited. If a person does not have access to a sheet, it is impossible for them to know its existence and its links with the other sheets. It is therefore not yet possible to show that "*inaccessible yet*" MEREX cards exist within the system for a user who does not have access to them. However, this representation could subsequently be used to validate or not the interest of sharing knowledge within an organization. As for the notion of validating knowledge, there is currently no precise rule that defines this validation. It is the user who, on his own, can modify the icon of a sheet and thus obtain his own visual feedback. A knowledge card is therefore currently validated if the user decides so.

Concerning gamification, it is currently the progression dynamics and the cooperation mechanics that will be tested. But it is planned to explore more in depth the notions of player profile in order to propose rewards and mechanics adapted to the users. T engage users on a more social side of knowledge sharing, peer-reviewing is also considered, enhancing the quality of shared-knowledge within an organization. It can also be linked to gamification mechanics as the Socializer has been identified as a user- profile [14].

5 Conclusion

In this paper, we explored the potential of gamification for knowledge management within organizations. We began by presenting the basic concepts of knowledge management, before focusing on the potential benefits of communities of practice and how the

MEREX method could be combined with them to formalize the knowledge generated. We then presented gamification models that can be used to increase user motivation in the context of knowledge management. Finally, we presented the DSR model that we used to propose an artifact. The artifact is based on the knowledge capitalized within the organization's communities of practice using the MEREX method. It provides a personalized visual representation of the knowledge acquired and to be acquired and uses gamification mechanics to motivate users to share and access knowledge.

The proposed artifact would meet the following objectives:

- Promote collaboration and the transmission of individual knowledge in order to make it collective.
- Facilitate the formalization of knowledge for sharing.
- Create a personalized visual representation of knowledge acquired and to be acquired to motivate users to access and share knowledge.

The artifact is still under development, but it has the potential to be a valuable tool for organizations that are looking to improve their knowledge management practices and engage their members in participating.

References

1. Alavi, M., Leidner, D.E.: Knowledge management and knowledge management systems: Conceptual foundations and research issues. MIS Q. **25**(1), 107–136 (2001)
2. Wenger, E.: Communities of practice: Learning, meaning, and identity. Cambridge University Press (1998). https://doi.org/10.1017/CBO9780511803932
3. Deterding, S., Dixon, D., Khaled, R., Nacke, L.: From game design elements to gamefulness: defining "gamification". In: Proceedings of the 15th International Academic MindTrek Conference: Envisioning Future Media Environments, pp. 9–15. (2011)
4. Gardoni, M., Navarre, A.: Pratiques de gestion de l'innovation : guide sur les stratégies et les processus. Montréal: École de technologie supérieure (2017)
5. Nonaka, I., Takeuchi, H.: The knowledge-creating company: how Japanese companies create the dynamics of innovation. Oxford University PressNew York, NY (1995). https://doi.org/10.1093/oso/9780195092691.001.0001
6. Tremblay, P.-O.: Un chaos organisé : Les communautés de métier chez Ubisoft. HEC Montréal (2008)
7. Corbel, J.-C.: La méthode MEREX : capitaliser l'expérience pour améliorer la performance. Éditions Hermès, Paris (1995)
8. Dalkir, K.: Knowledge management in theory and practice. Routledge, London (2011)
9. Hamadache, K.: Le knowledge management : fondements et gestion de projet (2006). https://www.memoireonline.com/10/06/219/m_knowledge-management-fondements-et-gestion-de-projet.html
10. Zack, M.H.: Developing a knowledge strategy. John Wiley & Sons, California (2005)
11. Swacha, J.: Gamification in knowledge management motivating for knowledge sharing. **12**, 150-160 (2015)
12. Ďuriník, M.: Gamification in knowledge management systems. Central European J. Manage. (2015). https://doi.org/10.5817/CEJM2014-2-3
13. Werbach, K.: (Re) Defining gamification: a process approach. In: Spagnolli, A., Chittaro, L., Gamberini, L. (eds.) Persuasive Technology, pp. 266–272. Springer International Publishing, Cham (2014). https://doi.org/10.1007/978-3-319-07127-5_23

14. Marczewski, A.: User types. In: Even Ninja Monkeys Like to Play: Gamification, Game Thinking & Motivational Design. CreateSpace Independent Publishing Platform, pp. 69–84 (2015)
15. Kapp, K.M., Blair, L., Mesch, R.: The Gamification of Learning and Instruction Fieldbook: Ideas into Practice (Pfeiffer) (2014)
16. Shpakova, A., Dörfler, V., MacBryde, J.: Changing the game: a case for gamifying knowledge management. World J. Sci. Technol. Sustain. Dev. **14**(2/3), 143–154 (2017). https://doi.org/10.1108/WJSTSD-01-2017-0002
17. Chou, Y.: Actionable gamification: beyond points, badges, and leaderboards. Packt Publishing Ltd (2019)
18. Hamari, J., Koivisto, J., Sarsa, H.: Does gamification work?--a literature review of empirical studies on gamification. In: 2014 47th Hawaii International Conference on System Sciences (pp. 3025–3034). IEEE (2014)
19. Friedrich, J., Becker, M., Kramer, F., Wirth, M., Schneider, M.: Incentive design and gamification for knowledge management. J. Bus. Res. **106**, 341–352 (2020). https://doi.org/10.1016/j.jbusres.2019.02.009
20. Kapp, K.M.: The Gamification of learning and instruction : game-based methods and strategies for training and education (Pfeiffer) (2012)
21. Tondello, G.F., Wehbe, R.R., Diamond, L., Busch, M., Marczewski, A., Nacke, L.E.: The gamification user types hexad scale. In: Proceedings of the 2016 Annual Symposium on Computer-Human Interaction in Play (pp. 229–243) (2016)
22. Marczewski, A.: Gamification mechanics and elements. In: Even Ninja Monkeys Like to Play: Gamification, Game Thinking & Motivational Design. CreateSpace Independent Publishing Platform, pp. 165–177 (2015)
23. Bartle, R.: Hearts, clubs, diamonds, spades: players who suit MUDs. J. MUD Res. (1996)
24. Deci, E.L., Ryan, R.M.: Intrinsic motivation and self-determination in human behavior. In: Springer Science Business Media (2013)
25. Sampaio, M.C., Sousa, M.J., Dionísio, A.: The use of Gamification in Knowledge Management Processes: A Systematic Literature Review. J. Rev. Global Econ. **8**, 1662–1679 (2019). https://doi.org/10.6000/1929-7092.2019.08.150
26. Brocke, J., vom, Hevner, A., Maedche, A.: Introduction to Design Science Research pp. 1-13 (2020). https://doi.org/10.1007/978-3-030-46781-4_1
27. Peffers, K., Tuunanen, T., Rothenberger, M.A., Chatterjee, S.: A design science research methodology for information systems research. J. Manage. Inf. Syst. **24**(3), 45–77 (2007). https://doi.org/10.2753/MIS0742-1222240302
28. Zhang, X., Thomson, V.: A knowledge-based measure of product complexity. Comput. Ind. Eng. **115**, 80–87 (2018)

The Role of Modularity in a Factory in Space

Farouk Abdulhamid(✉), Brendan P. Sullivan, Monica Rossi, and Sergio Terzi

Politecnico Di Milano, Via Raffaele Lambruschini 4/B, 20156 Milano, Italy
farouk.abdulhamid@polimi.it

Abstract. Ever since the advent of the space race in the late 1950s, which culminated with the historic human moon landing of the Apollo 11 lunar mission crew, extraterrestrial exploration continues to gain interest in the scientific community. This is further highlighted by the commercialization of the space economy which has led to the emergence of innovative companies such as Space X, Redwire, Blue Origin, Thales Alenia Space, etc. While the relevance of space explorations continues to increase, so does the danger and complexity related to these activities as demonstrated by the fatal incidents of the Soyuz 11 and Columbia spacecraft. Therefore, the safety concerns -crew and payload- must be addressed. These concerns are inherently tied to launch capabilities, which are constrained by mass, volume, and eventually cost. For this reason, the Factory in Space (FIS) concept was introduced. FIS presents the possibility of manufacturing, assembly, servicing, repurposing, repair & refurbishing, and maintenance outside of the earth's atmosphere. Furthermore, ambitious missions in the future would not be feasible utilizing the current method of launching a fully functional and integrated system on a single vehicle. Gradually, space systems are being built with a certain level of modularity. This would allow basic platforms to be launched and later built/assembled in space. This study aims to evaluate the role of modular space systems in the ecosystem of FIS.

Keywords: Modularity · On-orbit manufacturing · Space debris · Satellites

1 Introduction

The continuous growth of the space economy has led to the search for innovative ways to improve and maintain space systems such as satellites and space stations. Constraints such as launching capability and limited access to space mean utilizing conventional on-ground processes and technologies is quite challenging. Furthermore, the safety of the crew and payload is of paramount importance during space missions, and it is essential to provide the necessary supplies to ensure their survival. Hence the imperative for locally sourced solutions through on-orbit servicing, manufacturing, and assembly, which are the fundamentals of the concept of Factory in Space (FIS) [1]. FIS, being a concept that encompasses several capabilities under its umbrella, has the potential to support a transition to a more resilient and sustainable space exploration. For example, servicing

Published by Springer Nature Switzerland AG 2025
P. Sureephong et al. (Eds.): PLM 2024, IFIP AICT 740, pp. 398–408, 2025.
https://doi.org/10.1007/978-3-031-93319-6_30

activities include life extension, repair, and de-orbit among other functions. Manufacturing activities include the production of new components from raw or recycled material and assembly activities include the construction of components made in-situ or transported from earth. This benefit of FIS has been further emphasized in a White House 2022 National Strategy report [2].

Furthermore, this wide range of FIS activities has the potential to improve and benefit human spaceflight. Servicing operations such as deorbiting and refueling reduce collision chances and provide a more robust communication infrastructure respectively. Moreover, transitioning maintenance activities from astronauts (Hubble space telescope repair mission [3]) to robots (development and demonstrations of robotic arms [4–6]) would reduce the cost of operating space systems in the long run [7].

Space systems manufactured and/or assembled in orbit will have significant technical benefits over those done traditionally on the ground. These systems would be spared from fairing constraints [8]. The volume of the launch vehicle's fairing dictates the size of the launched systems [9], hence FIS serves as a way to circumnavigate these constraints with on-orbit assembly. The modular design and assembly of the ISS by the international space community has demonstrated the benefit of on-orbit assembly, as the ISS -the size of a football field- could not have been commissioned in a single launch [10]. NASA and its partners are currently exploring FIS concepts relating to on- orbit assembly to replace deployable structures with components to be assembled, thus maximizing packaging efficiency [11], as well as the introduction of modularity during the assembly of traditional satellites [12]. In addition to on-orbit assembly, space practitioners are exploring the utilization of the enormous potential of on-orbit manufacturing. Made In Space, Inc. Has been investigating the manufacturing of materials in space terrestrial and extraterrestrial applications [13]. A system manufactured and assembled entirely in space would be considerably different from the one manufactured on Earth. FIS could take advantage of materials that could not be exposed to air, and bypass the geometric, size, and structural limitations imposed by gravitational forces and launch constraints [14]. Most of the robotic platforms planned for manufacturing activities in space are very reminiscent to the technologies utilized for on-orbit assembly. After a component is manufactured, several maneuvers are required to enable its activation on a space system. Several terrestrial maneuvers such as joining, and welding is feasible in orbit. The evaluation and inspection of components manufactured in space would be of paramount importance as the complexity of components increases [15]. However, compared to assembly and servicing activities, manufacturing in orbit is still in the nascent stages [15]. For example, the first 3D printers were installed in the ISS in 2014 with a permanent Additive Manufacturing Facility installed later in 2016[16], comparably the ISS was assembled much earlier and became operational in the year 2000[17]. This paper employs a narrative literature review to investigate the role of modularity in FIS. The paper starts with an introduction and has 3 subsequent sections. The introduction section provides general background on the motivation for FIS. The second section is a literature review that explores relevant literature on the development of FIS In the third section the concept of modularity is introduced. Furthermore, modular spacecraft are discussed and finally, the role of modularity in FIS is explored. The final section provides a general overview of the paper and presents avenues for potential future research.

2 Factory in Space (FIS)

Adventurous forays of humans deep into space make it even more challenging to predict and prepare for all possible breakdowns and accidents that could happen [18]. A favorable solution would be the adoption and integration of manufacturing and maintenance platforms on space missions. FIS introduces the concept of manufacturing, assembly, integration, repair, and maintenance of components and supplies in space [19, 20]. FIS is a concept that has existed since the late 1900s [21], when the altitude control system aboard the Solar Maximum Mission (SMM) failed, and a crew of astronauts was dispatched on a repair mission on the National Aeronautics and Space Administration's (NASA) space shuttle in April 1984[22]. This led to the development of the Hubble Space Telescope (HST), with in-space servicing taking a fundamental role in the mission objectives [23]. The HST was designed such that astronauts could replace defunct components in orbit to keep the telescope functional and extend its mission life. The servicing missions maintained the full mission capabilities and extended their mission life beyond the initial 15-year span [3, 24]. Stoor [25] concluded that the HST validated the cost reduction goal and was a crucial lesson for future on- orbit manufacturing missions. Following the HST, many efforts have been placed into developing technology that can facilitate autonomous in-space servicing and manufacturing. The "Orbital Express" - initiated in 2007 – developed with the Defence Advanced Research Project Agency (DARPA), successfully demonstrated robot satellite services, including autonomous assembly [26]. Furthermore, the development of robotic arms using proprietary technology by both NASA and the European Space Agency (ESA) further demonstrates the keen interest in developing autonomous technology for on-orbit applications [27]. While several studies have reported previous efforts to develop technologies for on-orbit servicing, manufacturing, and assembly [27–30], one key technology that has been identified is AM.

It is possible to draw parallels between these efforts and the transformative technology of AM. The advancement of AM techniques allows for a rapid prototyping technology that provides ready-to-use parts directly from stock material [31]. This has allowed for the development of AM technology for FIS applications. NASA launched the "AM in Space" initiative, which led to the first installation of an AM system in the ISS under the 3D Printing in Zero-G project [16]. Furthermore, in 2016, the Additive Manufacturing Facility (AMF) was installed. NASA's Restore-L program (launched in 2016) was renamed On-orbit Servicing, Assembly, and Manufacturing Mission (OSAM-1&2) in 2020. The OSAM-1 spacecraft infrastructure will refuel satellites, assemble antennas, and manufacture beams [32]. At the same time, OSAM-2, a technology demonstration mission, was successfully concluded in 2023[33]. The European Union (EU), on the other hand, is developing FIS concepts as part of their projects on AM in the aerospace industry [34]. While the European Space Agency (ESA) is working on the continuous development of the ISS by providing spare parts, the EU mainly focuses on technology that leads to sustainability and circularity. In 2013, the EU, in collaboration with ESA and a British company, MTC, launched the AM Zero Waste and Efficient Production of High-Tech Metal Products (AMAZE) project [35]. AMAZE used the AM techniques developed in the US to demonstrate in- situ manufacturing on extraterrestrial environments such as the moon and asteroids. The project culminated with the establishment of four pilot-scale factories across the EU [36]. Makaya et al. [37] summarized ESA's activities in space manufacturing. Another recent effort is the work from the China Academy of Space Technology [38], where they developed an AM system called Space-based Composite Material 3D Printing System, which uses carbon-fiber reinforced composite to autonomously print objects. In 2020, the AM system was successfully tested in LEO aboard China's Long March 5B heavy-lift carrier rocket. Similarly, several researchers [18, 31, 39] have reported on the development of various AM techniques for in-space manufacturing. Figure 1 summarizes some of the possible activities of FIS.

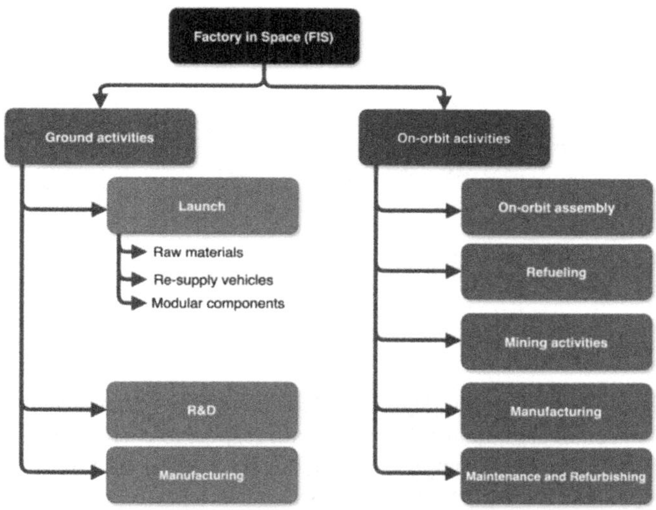

Fig. 1. Breakdown of FIS activities.

Furthermore, private entities play an important in the realization of FIS [34]. This is evident through the works of several private companies such as Redwire's (Made in Space) pivotal role in the development of the AMF aboard the ISS and later that of the OSAM, Northrop Grumman's development of Mission Robotic Vehicle (MRV), and deployment of Mission Extension Vehicle (MEV), and Thales Alenia Space's (TAS) developing a servicer satellite with manufacturing capabilities with a demonstration launch planned for 2026. The first FIS mission of companies is reported in Fig. 2 [19]. Canceled and dormant missions have been categorized differently. Companies with FIS-related research activities without an announced launch date are grouped under "not announced".

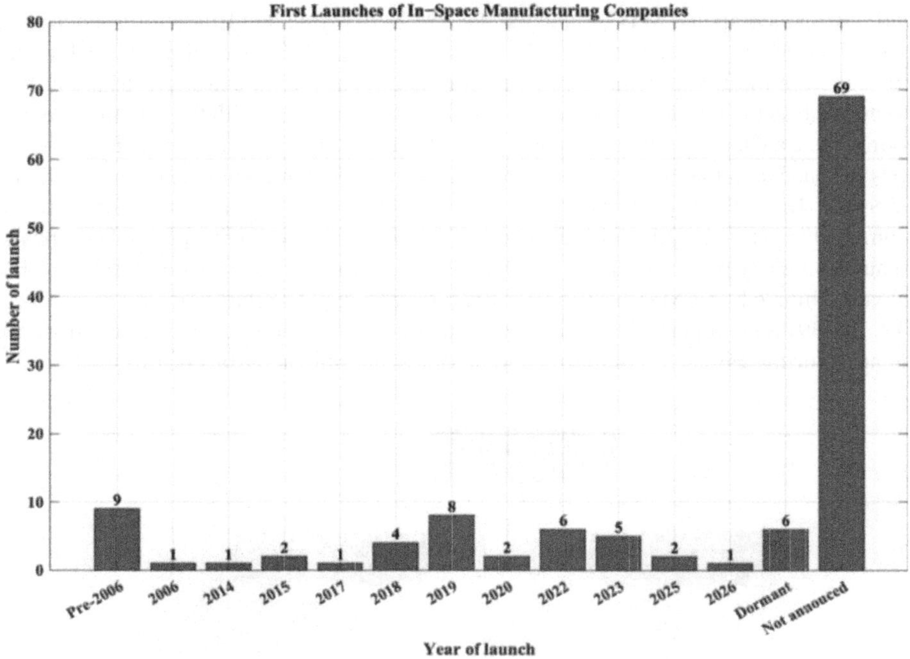

Fig. 2. First launches of in-space manufacturing companies [19].

3 Modularity

Modularity generally refers to the degree to which units can be separated and reassembled into new sets of relatively autonomous units with respect to each other. The modularity of components reduces their complexity by reducing a single unit to different levels of mutually independent parts [40]. Modular systems must adhere to component standardization and plug-and-play principles to facilitate smooth interaction between the modules [41]. Modular spacecraft design dictates that a modular system comprising standardized components with unique functionality be serially manufactured, connecting subsystems with other components through standardized interfaces, and subsequently constructing spacecraft by integrating smaller subsystems.

3.1 Modularity in Space Systems

The search for sustenance and long-term stability in the space sector has seen several stakeholders attempt to reinvent themselves through mergers and acquisitions, small satellites, and mega-constellations among other solutions. However, one radical solution that promises a new approach and sustainable business potential is the concept of modular satellites [42]. Introducing plug-and-play subsystems into satellite design would lead to new ventures of interoperability and significantly increase mission flexibility [43]. Production time and associated costs would be reduced, and at the same time, the resilience and functionality of space platforms would be increased.

Space-based systems are strategic assets that work in harsh conditions, which makes them fragile and vulnerable. Furthermore, their timeline from design to deployment is usually very long and expensive. Assembly time could span years for a satellite that lasts a maximum of 15 years [41]. Platform manufacturers utilize customized systems limiting spacecraft design options, in a process that is long, expensive, and complex. The current manufacturing cycle compromises the resiliency of space systems, as a significant portion of production costs go to the suppliers of the customized systems. This cost for customization is due to the lack of interoperability between the suppliers since there is no single standard for hardware and software communication. Integrating components from a myriad of manufacturers presents a challenge that until now could result in significant increases in production time and costs. Enabling space systems scale production such that the faulty components are replaced in a shorter duration by increasing the manufacturer's ability to reproduce the same components in a fraction of time without compromising on the quality is one way to increase space system resilience. 'Maintaining a robust infrastructure to service or quickly replace disabled satellites will be critical. These are all ingredients of an effective deterrence by denial.' [44]. Plug-and-play designs foster easier cooperation with various manufacturers. The plug-and-play node could enable satellite modularity and flexible architecture, reducing costs and time of satellite manufacturing enabling satellite scale production, effectively achieving a complete 'plug-and-fly' solution. Furthermore, this solution also enables faster technology cycle and the testing of new payloads, by lowering the cost of the satellites and allowing faster and automatic integration of new systems into existing ones [43]. Modular space systems could significantly alter space travel by reducing some of the constraints related to vehicle launch. For instance, only one module is needed to commence a mission and can be built incrementally while already deployed similar to the development of the International Space Station (ISS). This allows stakeholders to make expansion decisions based on market demands and reallocation of assets as needed [42].

In the transition to a modular system, engineers need to rethink the way satellites are designed and make them faster to build and easier to launch. For instance, payloads could be independent from buses governed by a shared set of standards. This could lead to a quick integration ahead of launch with the assurance that they would operate as intended. Such progress could unearth new capabilities such as greater production efficiency and design flexibility. However, significant hurdles remain for a transition to modular space systems. Such as the development and alignment of technologies and standards that would lead to cost savings and more importantly, functional reliability of the modular components [45]. While modular space systems are still largely theoretical [42], the development of other core concepts such as satellite servicing, on- orbit manufacturing and assembly, and spacecraft connectors are accelerating the realization of modular space systems.

3.2 Role of Modularity in FIS

Modularity and FIS share a symbiosis rooted in their core definitions, as modular space systems - designed to be upgraded and refueled – could be launched with fewer modules and fuel than needed for the entire mission duration. The space system, for instance, a satellite, could then be gradually built and refueled at certain intervals depending on the mission needs and requirements. This symbiosis could lead to the reduction of hundreds of kilograms in launch weight and could also positively impact the launch vehicle throw-weight margins. This ability to augment or interchange satellite parts and components on-orbit improves the redundancy of the whole system. Additionally, launch schedule adherence would be improved and systems could be launched without critical components as FIS would later install the required parts.

This enhanced flexibility would birth a paradigm shift in the assurance and sustainability of space missions, reducing the requirement for lengthy rigorous testing through the reliance on the newfound capability to install and remanufacture faulty components on-orbit [7].

A combination of modular spacecraft and FIS would change the design and manufacturing of space vehicles. For instance, it is not possible to build a large spacecraft with a single rocket launch due to structural and dimensional limitations as well as inadequate propulsion. However, it would be possible to build such a spacecraft by employing modular designs and on-orbit manufacturing and assembly (concepts of FIS) [46]. The potential of FIS is significantly increased if the synergies with modular designs are properly harvested. FIS capabilities have the potential to transform the entire space domain. Transforming traditional manufacturing standards to the design for the serviceability and modularity of payloads would also lead to an increase in the potential customer base for servicing missions. Such modularity and standards could help foster and grow a new ecosystem of users and suppliers with components that could be significantly different from the current generation of custom-designed satellites, each with its bus and subsystems. Large pieces of systems that can be modularly connected or the raw material to construct those pieces can be launched into space to augment existing facilities [15].

In FIS, modular payloads would have lower lifecycle costs when operated from persistent platforms with other payloads that can be switched out or upgraded intermittently. Payload switching and upgrading would also provide flexibility to allow space practitioners to react quickly to market demands and/or prevent oversaturation and limit losses if a market does not materialize. Payloads that can be more easily swapped out, or more easily manipulated by robotic components, with interfaces that make operating on them simpler, are essential for repair, replacement, and integration operations. Finally, it can be concluded that the future of the developing technologies of FIS and modular spacecraft are intertwined, the success of the latter could accelerate the development of the latter and vice versa.

4 Discussion and Conclusion

The paper discusses the alignment of modular spacecraft design and the concept of FIS. The biggest inspiration and demonstration of such synergy was the incremental development and assembly of the ISS. The development of the ISS further fostered collaboration between several international bodies and private entities. Therefore, the synergy of modularity and FIS could not only transform the operation of space travel but could also lead to increased collaboration between space practitioners, especially suppliers. This could allow for the development of standardized interfaces for space system components to foster a plug-and-fly solution. Considering the limitation of resources in the space environment, space explorations need to be self-sufficient and regenerative by design, hence the concept of FIS. Significant future value could arise from functions that occur entirely in orbit, such as on-orbit servicing, research and development, and manufacturing. Super-heavy launch vehicles, such as SpaceX's Starship, could provide an avenue for companies to establish a factory in space.

While the concept of FIS and modular spacecraft seem promising, they are still in very nascent stages and the rapid advancement of science and technology means suitable infrastructure could be ready very soon. Some recommendations for future work considering the progress of cutting-edge technologies such as AM, Artificial Intelligence (AI), digital twin, and Internet of Things (IoT) among others are discussed below:

1. **Module Standardization**: Space systems module division could imitate the computer industry by having a unified standard. The lack of unified standards means most modular technologies remain in the testing phase. Furthermore, the modular approaches need to differentiate special modules from the basic ones such that there is a clear distinction between modules used to support basic operations such as power, control, and computation modules, and the special modules which are usually mission specific. Module standardization would also lead to standard design size, shape, and capability which is likely to enhance the practical applications of modular technology [41].
2. **Human-computer interaction**: The continuous development and demonstrations of autonomous technology for FIS such as the Mission Extension Vehicle (MEV) [47] means that the boundary of the relationship between humans and machines would affect the security and robustness of autonomous systems. Autonomy has taken prominence in the research of FIS. Autonomous platforms can operate independently of external controls in dynamic environments, resulting in improved reaction time and performance, reduced personnel burden, and reduced costs. Nevertheless, completely autonomous platforms are still not fault-free and could experience decision-making errors or ethical problems due to data traps. As such, it is paramount that the level of autonomy should be controlled, and the system risk/security processes should be under human supervision. Since both humans and machines could be prone to decision-making mistakes, mutual error corrections and complementarity of humans and machines can be implemented as a sort of check and balance [48].
3. **Digital twin simulations**: FIS is a complex system that places high demands on system design, integration, testing, operation, and maintenance. The testing and simulation are difficult, long-term, and expensive to implement. Furthermore, it is difficult to make simulations using traditional computing as it exhibits some difficulties in performing full process analyses and full cycle detections of the interactions between components. Also, it is difficult to perform real-time analysis and possesses optimization functions. Hence the importance of the digital twin technology. A digital twin connects the physical and digital objects in a precise and real-time manner, featuring interoperability, scalability, real-time operability, fidelity, and closed-loop operability. The simulation analysis technology based on digital twin monitors the current state of a physical entity based on real-time feedback information [41]. Hence, digital twin would allow actualizing a complete modular space system in a virtual environment while also understanding the impact of the modular payloads and how to

leverage FIS activities to build/disassemble the components on-orbit. It is paramount for space practitioners to pay significant attention to the development of simulation technologies such as digital twin for FIS.

In conclusion, on-orbit, on-demand servicing, assembly, and manufacturing ability can enhance space exploration in several ways. FIS has implications for mission planning and logistics, vehicle design, manufacturing, assembly and integration, and product quality control and assurance. However, the harsh and isolated conditions outside the Earth's atmosphere make its feasibility challenging. Therefore, enabling technologies such as modular spacecraft need to be researched and leveraged to fully realize the potential of FIS.

Acknowledgments. This study was carried out within the MICS (Made in Italy – Circular and Sustainable) Extended Partnership and received funding from the European Union Next-Generation EU (PIANO NAZIONALE DI RIPRESA E RESILIENZA (PNRR) – MISSIONE 4 COMPONENTE 2, INVESTIMENTO 1.3 – DD 1551.11-10-2022, PE00000004). This manuscript reflects only the authors' views and opinions, neither the European Union nor the European Commission can be considered responsible for them.

Disclosure of Interests. The authors have no competing interests to declare that are relevant to the content of this article.

References

1. Deloitte Space: The commercialization of low earth orbit the commercialization of low earth orbit volume 5: providing the foundation for growth. (2022)
2. Executive Office of the President of the United States: In-space servicing, assembly, and manufacturing national strategy (2022)
3. Harris, D.J.: Hubble's quarter century in orbit has opened a universe of possibilities. Proc. Natl. Acad. Sci. **112**, 3176–3177 (2015). https://doi.org/10.1073/pnas.1500872112
4. Ma, B., Jiang, Z., Liu, Y., Xie, Z.: Advances in space robots for on-orbit servicing: a comprehensive review. Adv. Int. Syst. (2023). https://doi.org/10.1002/aisy.202200397
5. Chhabra, A., Kim, D.: Ground robotic platform for simulation of on-orbit servicing missions. J. Aerosp. Inf. Syst. **19**, 480–493 (2022). https://doi.org/10.2514/1.I011008
6. Kringer, M., Böhrer, C., Frey, M., Pimpi, J., Pietras, M.: Direct Robotic Extrusion of Photopolymers (DREPP): Influence of microgravity on an in-space manufacturing method. Front. Space Technol. (2022). https://doi.org/10.3389/frspt.2022.899242
7. Cavaciuti, A.J., Heying, J.H., Davis, J.: In-space servicing, assembly, and manufacturing for the new space economy (2022)
8. Piskorz, D., Jones, K.L.: On-orbit assembly of space assets: a path to affordable and adaptable space infrastructure center for space policy and strategy (2018)
9. Piskorz, D., O'Quinn, C.: Setting the standard: launch units for the smallsat era. (2018)
10. Davis, J.P., Mayberry, J.P., Penn, J.P.: On-orbit servicing: inspection, repair, refuel, upgrade, and assembly of satellites in space (2019)
11. NASA: NASA's Dragonfly project demonstrates robotic satellite assembly critical to future Space Infrastructure development (2017)

12. Werner, D.: Satlets: crazy idea or ingenious concept? This week's test on ISS will offer clues (2017). https://spacenews.com/satlets-crazy-idea-or-ingenious-concept-this-weeks-test-on-iss-will-offer-clues/.
13. Made In Space Inc.: Fiber optics: best in class fluoride-based fiber for medical, telecom, and research (2018)
14. Kagan Pittman: Manufacturing in space with robots and 3D printing (2017). https://www.engineering.com/manufacturing-in-space-with-robots-and-3d-printing/.
15. Corbin, B.A., Abdurrezak, A., Newell, L.P., Roesler, G.M., Lal, B.: Global trends in on orbit servicing, assembly and manufacturing (OSAM) (2020)
16. Clinton, R.G.: AM in Space: ISM and IRMA NASA initiatives manufacturing technology centre coventry UK, Coventry (2019)
17. Deloitte Space: The commercialization of low earth orbit the commercialization of low earth orbit volume 4: bringing earth to space. (2022)
18. Hoffmann, M., Elwany, A.: In-space additive manufacturing: a review (2023). https://doi.org/10.1115/1.4055603
19. Kulu, E.: In-space manufacturing-2022 industry survey and commercial landscape. (2022)
20. Abdulhamid, F., Terzi, S., Sullivan, B.P.: Factory in space - shaping the scenario. In: Proceedings of the Summer School Francesco Turco (2023)
21. Watson, J.J., Collins, T.J., Bush, H.G.: A history of astronaut construction of large space structures at NASA langley research center. In: Proceedings, IEEE Aerospace Conference. pp. 7–3569–7–3587. IEEE. https://doi.org/10.1109/AERO.2002.1035334
22. Rapley, C.G., Sylwester, J., Phillips, K.J.H.: New results from the solar maximum mission/bent crystal spectrometer. Sol. Phys. **292**, 50 (2017). https://doi.org/10.1007/s11207-017-1070-y
23. National Aeronautics and Space Administration: Servicing history and long- term plans. NASA Facts (1993)
24. Stockman, H.S.: Highlights of the hubble space telescope (1994). https://doi.org/10.1007/978-94-011-0794-5_10
25. Bradley, S., Col peter garretson: In-Space manufacturing: a roadmap to the future (2018). https://apps.dtic.mil/sti/pdfs/AD1055025.pdf
26. Friend, R.B.: Orbital express program summary and mission overview (2008). https://doi.org/10.1117/12.783792
27. Xue, Z., Liu, J., Wu, C., Tong, Y.: Review of in-space assembly technologies. Chin. J. Aeronaut. **34**, 21–47 (2021). https://doi.org/10.1016/j.cja.2020.09.043
28. Kaiser, C., Sjöberg, F., Delcura, J.M., Eilertsen, B.: SMART-OLEV—an orbital life extension vehicle for servicing commercial spacecrafts in GEO. Acta Astronaut. **63**, 400–410 (2008). https://doi.org/10.1016/j.actaastro.2007.12.053
29. Chen, P.-Y., Wang, Y.-F., Chen, L.-Y., Wang, G.-L., Wang, E.-C.: Reaction of o-(oxiranylmethyl) benzonitriles with sodium borohydride or Grignard reagent/CuI: a new synthesis of substituted 3-alkyl-3,4-dihydroisocoumarins. ARKIVOC **2013**, 119–128 (2013). https://doi.org/10.3998/ark.5550190.0014.310
30. Li, W.-J., et al.: On-orbit service (OOS) of spacecraft: a review of engineering developments. Prog. Aerosp. Sci. **108**, 32–120 (2019). https://doi.org/10.1016/j.paerosci.2019.01.004
31. Zocca, A., et al.: Challenges in the technology development for additive manufacturing in space. Chin. J. Mech. Eng. Addit. Manufact. Front. **1**(1), 100018 (2022). https://doi.org/10.1016/j.cjmeam.2022.100018
32. NASA: Mission update: OSAM-1 successfully passes key decision point-C. https://nexis.gsfc.nasa.gov/05292020_osam1_update.html. Last accessed 07 Nov 2023
33. The National Aeronautics and Space Administration: On-orbit servicing, assembly,andmanufacturing2(OSAM-2). https://www.nasa.gov/mission/on-orbit-servicing-assembly-and- manufacturing-2-osam-2/. Last accessed 11 Nov 2023

34. Jackson, M., Joseph, S.: Manufacturing in space. In: Space Systems: Emerging Technologies and Operations. pp. 233–284. New Prairie Press, Manhattan, KS (2021)
35. The Manufacturing Technology Centre Limited: Final report summary - AMAZE (Additive Manufacturing Aiming Towards Zero Waste & Efficient Production of High-Tech Metal Products). (2018)
36. Mies, D., Marsden, W., Warde, S.: Overview of additive manufacturing informatics: "a digital thread." Integr. Mater. Manuf. Innov. **5**, 114–142 (2016). https://doi.org/10.1186/s40192-016-0050-7
37. Makaya, A., Pambaguian, L., Ghidini, T., Rohr, T., Lafont, U., Meurisse, A.: Towards out of earth manufacturing: overview of the ESA materials and processes activities on manufacturing in space. CEAS Space J. **15**, 69–75 (2023). https://doi.org/10.1007/s12567-022-00428-1
38. Ministry of National Defense of the People's Republic of China: China conducts 1st 3D printing experiment in space. http://eng.mod.gov.cn/xb/News_213114/TopStories/4864760.html,Last accessed 10 Nov 2023
39. Santomartino, R., et al.: Toward sustainable space exploration: a roadmap for harnessing the power of microorganisms. Nat. Commun. (2023). https://doi.org/10.1038/s41467-023-37070-2
40. Ye, Y., Huo, B., Zhang, M., Wang, B., Zhao, X.: The impact of modular designs on new product development outcomes: the moderating effect of supply chain involvement. Supply Chain Manage. Int. J. **23**, 444–458 (2018). https://doi.org/10.1108/SCM-01-2018-0021
41. Zhang, Z., et al.: Modularity, reconfigurability, and autonomy for the future in spacecraft: a review, (2023). https://doi.org/10.1016/j.cja.2023.04.019
42. SpaceWorks: The business imperative for modularity
43. Tal Azoulay, Giulia Federico, Ran Qedar: Modular satellite manufacturing to enhance space assets resiliency
44. Paulauskas, K.: Space: NATO's latest frontier (2020)
45. The Aerospace Corporation: Building the modular future of space
46. Pirat, C., Ribes-Pleguezuelo, P., Keller, F., Zuccaro Marchi, A., Walker, R.: Toward the autonomous assembly of large telescopes using cubesat rendezvous and docking. J. Spacecr. Rockets **59**, 375–388 (2022). https://doi.org/10.2514/1.A34945
47. Pyrak, M., Anderson, J.: Performance of northrop grumman's mission extension vehicle (MEV) RPO imagers at GEO. (2021)
48. Nahavandi, S.: Trusted Autonomy between humans and robots: toward human-on-the-loop in robotics and autonomous systems. IEEE Syst. Man Cybern. Mag. **3**, 10–17 (2017). https://doi.org/10.1109/MSMC.2016.2623867

Integrating Digital Twins for Enhancing Smart Agriculture Within a Circular Economy Framework

Komalaporn Kummalue[1], Suepphong Chernbumroong[2(✉)], Chitaporn Pratan[3], Sulaiya Piemchaiwat[3], Sri Devi Ravana[4], and Subrata Dutta[5]

[1] SunSpace Farm, Chiang Mai, Thailand
[2] Department of Digital Technology Management, College of Arts, Media and Technology, Chiang Mai University, Chiang Mai, Thailand
suepphong.c@cmu.ac.th
[3] The Halal Science Center, Chulalongkorn University, Bangkok, Thailand
[4] Department of Information Systems, Faculty of Computer Science and Information Technology, University of Malaya, Kuala Lumpur, Malaysia
sdevi@um.edu.my
[5] Department of Computer Science and Engineering, National Institute of Technology, Jamshedpur, India
sdutta.cse@nitjsr.ac.in

Abstract. This research paper explores the integration of Digital Twins (DTs) into smart agriculture within a Circular Economy (CE) framework. The study investigates the potential of Digital Twins to simulate and optimize agricultural processes, thereby enhancing resource efficiency and reducing waste. It delves into the system design architecture required for robust Digital Twins implementation in agriculture. Through a multidisciplinary approach, the paper aims to provide a comprehensive understanding of how Digital Twins can transform agricultural practices, promote sustainability, and contribute to a more efficient agricultural sector. Additionally, the research proposes the development of a training platform using Digital Twins technology to enhance the digital literacy and skills of farmers, facilitating the adoption of smart farming practices across Asia.

Keywords: Digital Twins · Smart Agriculture · Circular Economy · System Design Architecture

1 Introduction

1.1 Background and Motivation

Agriculture is a crucial sector that sustains human life by providing food, raw materials, and employment. However, modern agriculture faces several challenges, including resource depletion, environmental degradation, and climate change. To address these issues, innovative solutions that enhance sustainability and efficiency are necessary.

P. Sureephong et al. (Eds.): PLM 2024, IFIP AICT 740, pp. 409–422, 2025.
https://doi.org/10.1007/978-3-031-93319-6_31

Digital Twins (DTs) offer a promising avenue for transforming agriculture by providing a dynamic virtual model that simulates and optimizes agricultural processes. Smart Agriculture (SA) leverages advanced technologies such as the Internet of Things (IoT), artificial intelligence (AI), and big data to improve farming practices. Integrating Digital twins within a Circular Economy (CE) framework can significantly reduce waste, enhance resource reuse, and increase process efficiency.

1.2 Smart Farming in Asia

The adoption of smart farming technologies in Asia has the potential to revolutionize agriculture, making it more efficient, sustainable, and resilient. Continued research, innovation, and collaboration are essential to fully realize the benefits of smart farming and address the unique challenges faced by the agricultural sector in different Asian countries. Smart farming, integrating advanced technologies like IoT, AI, and robotics, is transforming agriculture globally. Asia, with its diverse agricultural practices and rapid technological adoption, provides a unique landscape for the implementation of smart farming technologies. While smart farming technologies offer numerous benefits, challenges such as high initial costs, lack of technical expertise, and data integration issues persist. Collaborative efforts between governments, tech companies, and farmers are crucial to overcoming these barriers and ensuring the widespread adoption of smart farming practices across Asia [1].

Regional Case Studies China: The integration of smart farming technologies in China has shown significant improvements in efficiency and sustainability. The use of AI and IoT for monitoring and managing large-scale farms has reduced labor costs and increased crop yields [2]. India: Smart farming in India focuses on addressing water scarcity and climate change impacts. Innovations such as automated irrigation systems and climate-resilient crops are being developed and implemented to ensure sustainable agricultural practices [3].

1.3 Circular Economy in Agro-Food Systems

The implementation of Circular Economy in agro-food systems has shown promising results in reducing environmental impacts and enhancing economic benefits. And the integration of digital twins into smart agriculture, aligned with the principles of the circular economy, offers a promising pathway towards sustainable and efficient agricultural practices. By leveraging advanced technologies and innovative business models, it is possible to address the challenges of resource scarcity, environmental degradation, and food waste in the agro-food sector. Future research should focus on developing scalable solutions and fostering collaboration among stakeholders to enhance the adoption of these technologies and achieve sustainable agricultural development. Digital Twins in Smart Agriculture represent a significant technological advancement that can enhance smart agriculture by creating a virtual replica of physical agricultural systems. This technology enables real-time monitoring, predictive analysis, and optimization of agricultural processes [4]. Digital Twins can improve resource efficiency by providing precise data on crop health, soil conditions, and weather patterns, allowing farmers to make informed decisions that reduce waste and enhance productivity [4].

Integrating Digital Twins with Circular Economy principles can further promote sustainability by enabling the efficient use of resources and minimizing waste through predictive maintenance and optimized supply chain management. The adoption of Digital Twins in agriculture faces challenges such as high initial costs, the need for technological expertise, and data integration issues. However, the potential benefits, including increased efficiency, reduced environmental impact, and improved economic outcomes, make it a worthwhile investment [4]. Collaboration among stakeholders, including farmers, technology providers, and policymakers, is essential to overcome these challenges and fully realize the benefits of Digital Twins in enhancing smart agriculture within a Circular Economy framework [5].

1.4 Objectives

The primary objectives of this paper are to investigate how digital twins can enable smart agriculture, evaluate their impact on resource efficiency and waste reduction in a circular economy perspective, and propose a robust system design architecture for agricultural digital twins. By achieving these objectives, the study aims to provide valuable insights and recommendations for the sustainable transformation of the agricultural sector.

The secondary objective of this paper is to design a training platform for farmers by using the digital twin technology. The platform will allow farmers to learn and practice various aspects of smart agriculture, such as crop management, irrigation, and fertilization, in a virtual environment that mimics the real-world conditions. The platform will also provide feedback and guidance to farmers on how to optimize their performance and reduce their environmental impact. By achieving these objectives, the study aims to enhance the digital literacy and skills of farmers and facilitate the adoption and scaling of smart farming in Asia.

2 Literature Review

Digital Twins are increasingly recognized for their transformative potential in agriculture. By integrating Digital Twins with digital and physical assets, and ensuring seamless communication, farmers can optimize resource use, reduce waste, and enhance sustainability. This review focuses on the integration of Digital Twins in smart agriculture within a Circular Economy framework, emphasizing digital and physical assets and communication technologies. Integrating Digital Twins with digital and physical assets, along with robust communication technologies, creates a comprehensive system architecture for smart agriculture within a Circular Economy framework. This approach leverages real-time data, advanced analytics, automation, and sustainability practices to enhance efficiency and productivity. By providing farmers with the tools and insights needed to optimize their operations, this integration supports a more sustainable agricultural future.

2.1 Physical Twins in Agriculture

Physical twins refer to the real-world counterparts of digital twins, including farm equipment, sensors, and control systems that interact directly with the environment to collect data and perform tasks. These components are crucial for providing accurate and timely data to their digital counterparts, enabling precise monitoring and control of agricultural processes. The integration of physical twins with digital platforms forms the foundation for the development of Digital twins in agriculture, supporting data- driven interventions and simulations [6].

2.2 Digital Platforms

Digital platforms are essential for managing the data collected from physical twins and facilitating integration with digital twins. These platforms provide the infrastructure for device connectivity, data processing, and automation within an interconnected digital ecosystem. They leverage IoT technologies to ensure seamless communication between sensors, actuators, and control systems, supporting real-time data analytics and decision-making processes [7].

2.3 Digital Twins in Agriculture

Digital Twins are virtual models that replicate the characteristics, behavior, and state of physical systems. In agriculture, Digital Twins simulate various farming processes, allowing for better decision-making and operational efficiency. Key components of Digital Twins include virtual models, simulation tools, and continuous data integration from physical counterparts. These components enable farmers to perform simulations, predict outcomes, and optimize operations based on data-driven insights [8, 9].

2.4 Digital and Physical Assets

Digital Assets: Digital assets in agriculture include sensors, control systems, and data management platforms that collect and process data from the physical environment. These assets are critical for implementing Digital Twins, as they provide the necessary data for accurate modeling and simulation. Sensors play a crucial role in collecting real-time data on various environmental and operational parameters, including soil moisture, temperature, humidity, and nutrient levels [10, 11].

Physical Assets: Physical assets refer to the actual farm equipment, sensors, and control systems that interact directly with the environment to collect data and perform tasks. These components are essential for providing accurate and timely data to their digital counterparts, enabling precise monitoring and control of agricultural processes. The integration of physical assets with digital platforms forms the foundation for the development of Digital Twins in agriculture, supporting data-driven interventions and simulations [8] (Figs. 1 and 2).

2.5 Circular Economy Framework

Integrating Digital Twins within a Circular Economy framework focuses on optimizing resource use, reducing waste, and enhancing sustainability. Key strategies include:

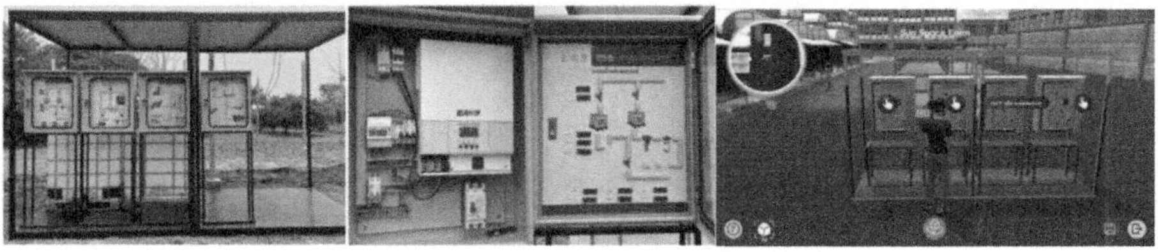

Fig. 1. Farm's Digital and Physical Assets transform to Digital Twin

- Closed-Loop Systems: Utilizing data from digital twins to optimize the use of water, fertilizers, and energy [10].
- Sustainable Practices: Precision agriculture to minimize resource overuse.
- Enhanced Productivity: Using simulations to optimize crop yields and mitigate risks such as pests or adverse weather conditions [8].

2.6 Model-Driven Engineering (MDE) in Agriculture

Model-Driven Engineering plays a crucial role in the development and implementation of digital twins in agriculture. Model-Driven Engineering emphasizes the use of high-level models to drive the development process, enhancing productivity, quality, and maintainability. Key aspects of Model-Driven Engineering relevant to digital twins in agriculture include:

- **Metamodels:** Defining the structure and semantics of modeling languages used to create digital twins.
- **Model Transformation:** Converting high-level models into executable code or other models, facilitating automation.
- **Code Generation:** Automatically producing executable code from models, reducing manual coding efforts and errors.

By applying Model-Driven Engineering principles, the development of digital twins can be streamlined, ensuring consistency, standardization, and scalability in their implementation [12].

3 Methodology

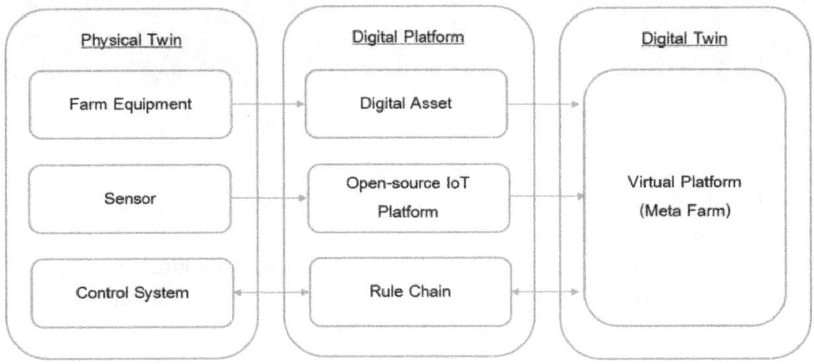

Fig. 2. Designing an integrated system architecture for agriculture.

The integration of Physical Twins, Digital Platforms, and Digital Twins into a cohesive system architecture can significantly enhance smart agriculture, particularly within a Circular Economy framework. This approach leverages advanced technologies to optimize resource use, improve productivity, and ensure sustainable agricultural practices.

Physical Twin

A physical twin refers to the real-world counterpart of a digital twin, which is a virtual model that mirrors the characteristics, behavior, and state of a physical object or system. In the context of agriculture, physical twins include farm equipment, sensors, and control systems that interact with the physical environment to collect data, perform tasks, and implement actions based on real-time information and analysis from their digital counterparts.

Components: Farm Equipment, Sensors, Control Systems

Functionality: Data Collection, Task Execution

Digital Platforms

Digital platforms are essential frameworks that enable the creation, deployment, and management of various digital services and applications. They provide a foundation for connecting devices, processing data, and automating processes in an interconnected digital ecosystem.

Components: IoT Hub, Data Analytics, User Interface:

Functionality: Device Connectivity, Data Processing, Automation

Digital Twin

A Digital Twin is a virtual representation of a physical object, system, or process, designed to replicate and simulate real-world functions and behaviors. It serves as a bridge between the physical and digital worlds, enabling data-driven insights, predictive maintenance, and optimization of operations. Digital twins are used across various industries, including manufacturing, healthcare, smart cities, and agriculture, to enhance efficiency, reduce costs, and improve decision-making.

Components: Virtual Models, Simulation Tools, Data Integration

Functionality: Simulation and Analysis, Predictive Maintenance, Operational Optimization (Fig. 3)

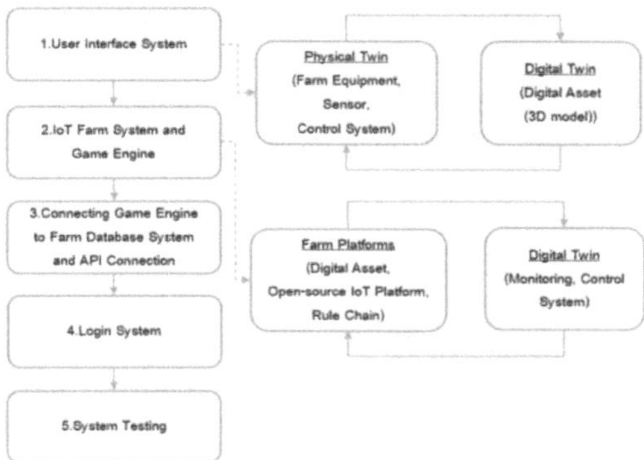

Fig. 3. Research Framework

In developing the platform for this experiment called Meta Farm, the following steps will be taken.

3.1 User Interface System

Designing a User Interface (UI) system that integrates with the physical twin of a farm involves creating a digital platform that allows users to interact with and control the physical aspects of the farm through real-time data visualization and control mechanisms. This integration enhances monitoring, decision-making, and automation processes in agriculture.

Farm Equipment

Farm Equipment encompasses a wide range of machinery and tools used in agricultural operations to enhance productivity, efficiency, and precision. These can include tractors, harvesters, planters, and irrigation systems, among others. Modern farm equipment often integrates advanced technologies such as GPS, IoT, and automation to perform tasks with high precision and minimal human intervention.

Sources: SunSpace Farm's Farm Equipment (e.g., Greenhouse & Control Station, Equipment Dome & Training Unit, Smart Distillation, Smart Dehydrator)

Sensors

Sensors are critical components in the agricultural sector, providing real-time data on various environmental and operational parameters. These devices measure factors such as soil moisture, temperature, humidity, light intensity, and nutrient levels, enabling farmers to make data-driven decisions to enhance crop management and productivity.

Sources: SunSpace Farm's sensors

1. Greenhouse (e.g., Light intensity sensor, Air temperature sensor, Air humidity sensor, Soil moisture sensor, Soil temperature sensor, EC value in soil, Soil pH)
2. Equipment Dome (e.g., Light intensity sensor, Air temperature sensor, Air humidity sensor, Soil moisture sensor, Soil temperature sensor)
3. Smart Distillation (e.g., Water temperature sensor, Steam temperature sensor, Coolant temperature sensor, Pressure sensor)
4. Smart Dehydrator (e.g., Temperature sensor, Humidity sensor)
5. Vertical Hydroponic (e.g., Temperature sensor, Humidity sensor)
6. Weather Station

Control Systems

Control System in agriculture are used to manage and automate various processes to improve efficiency and productivity. These systems can control irrigation, lighting, temperature, and nutrient delivery in both open fields and controlled environments like greenhouses. They often rely on data from sensors and are integrated with farm equipment to execute precise actions.

Sources: SunSpace Farm's Control systems

1. Greenhouse & Control Station (e.g., Air stirring fan, Exhaust fan, Water Integration for Mist spray head, Water Integration for Dripping water head)
2. Vertical Hydroponic & Control Station (e.g., Watering control system, Fertilizer control system)

3.2 IoT Farm System and Game Engine

Integrating IoT farm systems and game engines with digital platforms enhances the overall efficiency, monitoring, control, and user engagement in farming operations. This integration leverages real-time data collection, advanced analytics, automation, and gamification to create a more interactive and productive farming environment.

Open-source IoT Platform

An open-source IoT platform is a type of digital platform that supports the development and deployment of Internet of Things (IoT) applications. These platforms offer various tools and services to connect, manage, and analyze data from IoT devices. Open-source IoT platforms are customizable and scalable, making them suitable for a wide range of applications, from smart homes to industrial automation. Key features often include device management, real-time data processing, and integration with various communication protocols like MQTT, CoAP, and HTTP.

Sources: SunSpace Farm's Open-source IoT Platform

1. Greenhouse Monitoring (e.g., Light intensity sensor, Air temperature sensor, Air humidity sensor, Soil moisture sensor, Soil temperature sensor) and Control (e.g., Air stirring fan, Exhaust fan, Water Integration for Mist spray head)

2. Equipment Dome Monitoring (e.g., Light intensity sensor, Air temperature sensor, Air humidity sensor, Soil moisture sensor, Soil temperature sensor) and Control (e.g., Water Integration for Mist spray head, Water Integration for Dripping water head)
3. Vertical Hydroponic Monitoring (e.g., Power System Voltage, Electric Power, Temperature sensor, Humidity sensor) and Control (e.g., Water Integration for Hydroponic Plant, Water Integration for Water Tank)
4. Weather Station Monitoring (e.g., Air Temperature, Air Humidity, Light Intensity, Wind Speed, Rainfall, Atmospheric Pressure, CO2 Concentration, UV index)

Digital Asset
A digital asset refers to any content or media in a digital format that provides value to an organization or individual. This can include documents, images, videos, software, and other types of digital content. Digital assets are often stored and managed using digital asset management (DAM) systems, which provide tools for organizing, retrieving, and securing these assets. Effective management of digital assets is crucial for maintaining their integrity, accessibility, and usability.

Sources: SunSpace Farm's Digital Asset.

Equipment (Controller, Fan Ceiling, Fan Wall, Hydrosol, Water Pump)

Rule Chain
A rule chain is a sequence of logical rules or conditions used to process and analyze data within a system. In the context of IoT platforms, rule chains enable the automation of data processing tasks by defining specific actions to be taken when certain conditions are met. This can include data filtering, transformation, alert generation, and more. Rule chains are essential for implementing complex workflows and ensuring that IoT systems can respond dynamically to real-time data inputs.

Sources: SunSpace Farm's Rule Chain.

ThingsBoard Rule Engine Allows users to define rule chains to process incoming data from IoT devices, triggering actions like alerts, data transformation, or external API calls.

Virtual Platform
A Virtual Platform in the context of Digital Twins refers to the digital environment where the twin resides. This platform enables the creation, simulation, and analysis of digital twins, providing a comprehensive suite of tools for data integration, visualization, and interaction. Virtual platforms for digital twins often incorporate advanced technologies such as IoT, AI, machine learning, and big data analytics to ensure accurate and real-time modeling of their physical counterparts.

Sources: SunSpace Farm's Virtual Platform.

1. Digital Asset (3D model) Equipment (e.g., Base Controller, Controller, Barrier, Customer Service, Fan Ceiling, Fan Wall, Hydroponic Box) Building (e.g., Bathroom, Container, Dome, Meeting Table, Wall, Greenhouse) Plant (e.g., Empty Pots, Plant Pots, Flower Wall Panel, Cabbage, Salad)
2. Monitoring, Control System (Figs. 4, 5, 6, 7 and 8)

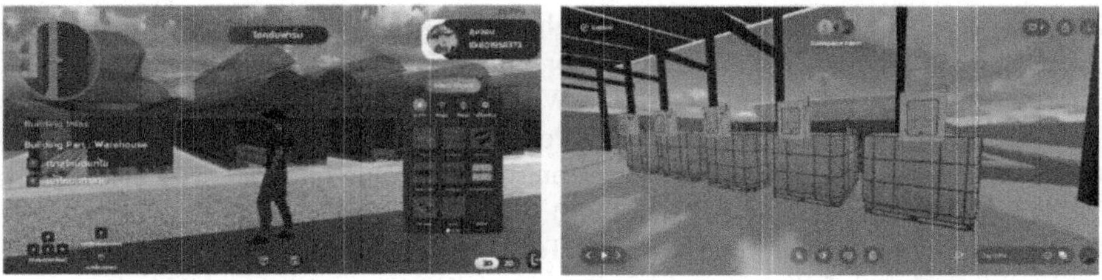

Fig. 4. Digital Asset (3D model)

3.3 Connecting Game Engine to Farm Database System and API Connection

Connecting Game Engine to Farm Database System and API Connection.

Integrating a game engine with the farm database system and API connection can significantly enhance the user engagement and interactivity of the digital platform. This integration leverages real-time data from the physical farm and digital twin to create immersive and educational gaming experiences. Farm Database System: A central repository for storing all collected data from physical twins, including sensor readings, equipment status, and environmental conditions. API (Application Programming Interface): A set of protocols and tools that enable different software applications to communicate with each other. In this case, APIs allow the game engine to interact with the farm database and digital platform.

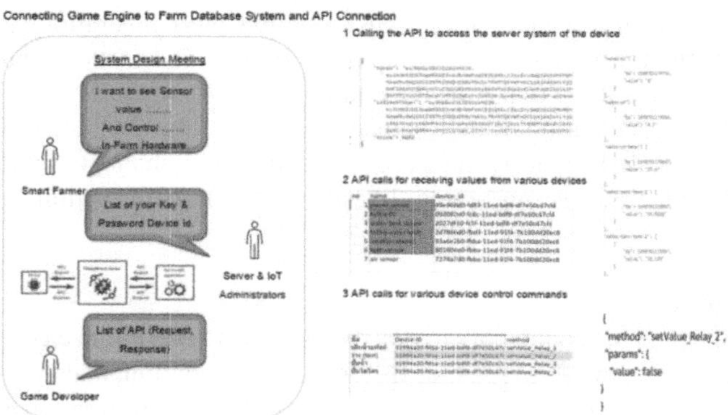

Fig. 5. Connecting Game Engine to Farm Database System and API Connection.

3.4 Login System and System Testing

Login following: Administrator role, Editor role, Client role.

Testing by: Administrator role, Editor role (SunSpace Farm Administrator), Client role.

Fig. 6. System Testing by farm administrators and volunteers

4 Results

4.1 User Interface System

Designing a user interface (UI) system that integrates with a farm's physical twin involves creating a digital platform.

Meta Farm 3D Model: Building, Farm Equipment, Plants

Fig. 7. Meta Farm 3D Model: Building, Farm Equipment, Plants compare between Physical twin and Digital twin

Meta Farm sensor and Control system

Fig. 8. Meta Farm Control: Set Equipment Controllers.

4.2 Connecting Game Engine to the Farm Database System and Connecting to the API

Users can interact and control the physical aspects of the farm through real-time data visualization and control mechanisms.

This integration enhances monitoring, decision making and automation in agricultural processes.

1. Monitoring sensor data such as water level sensor. Temperature sensor soil moisture sensor.

2. Control systems such as turning on and off water pumps (Fig. 9).

Fig. 9. Monitoring sensor data and Control systems

5 Conclusion

5.1 Creating a Virtual Working System to Control Equipment in Real Time

Developing a robust virtual system that can manage and monitor the usage of agricultural equipment in real-time is essential for modern farming. This system would lever- age advanced technologies like IoT and AI to ensure that equipment operates efficiently and only, when necessary, thereby reducing wear and tear and extending the lifespan of machinery. Real-time control also enables quick responses to any issues, minimizing downtime and maintaining continuous operation.

5.2 Integrating the System to Visualize Environmental Data for Timely Assessment

The integration of environmental data visualization into the virtual system is crucial for optimizing plant cultivation. By collecting data from various sensors (e.g., soil moisture, temperature, humidity), the system can provide farmers with a comprehensive view of the current conditions affecting their crops. This real-time visualization allows for immediate assessment and adjustments, such as altering irrigation schedules or applying fertilizers, ensuring that crops receive optimal care tailored to the prevailing environmental conditions.

5.3 Studying, Designing, and Compiling API Commands for IoT Device Control

A critical component of the system involves the study, design, and compilation of API (Application Programming Interface) commands that are suitable for controlling various IoT devices used on the farm. This involves ensuring compatibility and seamless communication between the central system and peripheral devices, such as sensors, actuators, and drones. Well-designed APIs facilitate the automation of farming processes, allowing for precise and coordinated actions based on real-time data inputs.

6 Future Works

6.1 Train 100 Farmers Using Digital Twin Technology

The objective is to organize training for 100 farmers in the upper northern region of Thailand to demonstrate the practical applications and benefits of Digital Twin technology in agriculture. And improve understanding and adoption of Digital Twin technology among farmers. Improve decision-making and productivity through real-time data and simulation.

6.2 Integrate Circular Economy Knowledge with Digital Twin for Smart Agriculture

Its objective is to embed circular economy principles and practices into Digital Twin models for smart agriculture to promote sustainability. And increase awareness and adoption of circular economy practices among farmers. Reducing waste and increasing resource reuse in agricultural processes.

6.3 Combining Game Elements with Digital Twin for Smart Agriculture

Its objective is to increase user engagement and learning by combining game elements with a Digital Twin system for smart agriculture. And increase participation and motivation among farmers in using Digital Twin technology. Improve learning outcomes and practical application of Digital Twin insights through gamification.

Acknowledgments. We would like to extend our heartfelt gratitude to the following organizations and individuals.

SunSpace Farm, Chiang Mai, Thailand: We sincerely appreciate your generous provision of farm equipment and platform data, which have greatly contributed to the practical aspects of our study.

The Halal Science Center, Chulalongkorn University, Bangkok, Thailand: Your support in maintaining the project's farming standards has been crucial in ensuring the integrity and quality of our research. We are deeply grateful for your partnership and collaboration.

Research Participants, Collaborators and Project Team Members: Our deepest appreciation goes to all the farmers, technology providers, and policymakers who participated in this study. Your insights, feedback, and collaboration have been invaluable in shaping the outcomes of this research. And hard work throughout the project. Your multi- disciplinary approach and commitment have been key to our success.

We are sincerely thankful for your partnership and contributions, which have significantly advanced our study on integrating digital twins for enhancing smart agriculture within a circular economy framework.

References

1. Temmen, N., Schilling, J.: Smart farming technology in japan and opportunities for EU companies. EU-Japan Centre for Industrial Cooperation Report (2021)
2. Yunlong, Z.: Advancements in smart farming in china. Chin. Agric. Rev. (2023)
3. Singh, R., et al.: Climate-smart agriculture in India: innovations and impacts. Agric. Syst. (2023)
4. Jones, D., et al.: Digital twins in smart agriculture: implementation challenges and opportunities. Agric. Syst. **178**, 102736 (2020)
5. Hamam, M., et al.: Circular economy models in agro-food systems: a review. Sustainability **13**(3453), 1–18 (2021)
6. Verdouw, C., Tekinerdogan, B., Beulens, A., Wolfert, S.: Digital twins in smart farming. Agric. Syst. **189**, 103046 (2021). https://doi.org/10.1016/j.agsy.2020.103046
7. IoF2020 Project: European union. (2021)
8. Peladarinos, N., Piromalis, D., Cheimaras, V., Tserepas, E., Munteanu, R.A., Papageorgas, P.: Enhancing smart agriculture by implementing digital twins: a comprehensive review. Sensors **23**(17), 7128 (2023). https://doi.org/10.3390/s23167128
9. Nasirahmadi, A., Hensel, O.: Toward the next generation of digitalization in agriculture based on digital twin paradigm. Sensors **22**, 498 (2022). https://doi.org/10.3390/s22020498
10. Zhang, B., Qiao, Y.: AI, sensors, and robotics for smart agriculture. Agronomy **14**(11), 1180 (2024). https://doi.org/10.3390/agronomy14061180
11. MacPherson, J., et al.: Future agricultural systems and the role of digitalization for achieving sustainability goals. a review. Agron. Sustain. Dev. (2022). https://doi.org/10.1007/s13593-022-00792-6
12. Bordeleau, F., Combemale, B., Eramo, R., van den Brand, M., Wimmer, M.: Towards Model-Driven Digital Twin Engineering: Current Opportunities and Future Challenges. Int. Conf. Syst. Model. Manage. (2020). https://inria.hal.science/hal-02946949

Digital Twins for Waste Management in Manufacturing: Preliminary Investigation on Possible Uses

Laila El Warraqi(✉), Daniele Perossa, and Paolo Rosa

Department of Management, Economics and Industrial Engineering, Politecnico di Milano, via Lambruschini 4b, 20156 Milan, Italy
laila.elwarraqi@polimi.it

Abstract. The increase of demand and resources consumption is significantly impacting the sustainability of the manufacturing industry which needs to operate responding to the demand growth while efficiently managing the waste generated. Over recent years, different strategies have emerged to allow addressing waste-related challenges. One of these strategies is the implementation of circular economy (CE) principles, encompassing initiatives such as resource consumption, recycling, and reuse. CE is indeed a comprehensive approach that needs to be investigated, specifically integrating it with digital technologies. The role of digital technologies is recognized not only for enhancing manufacturing productivity, but also for promoting environmental sustainability. Among these technologies, the digital twin (DT) stands out as promising thanks to its ability to replicate systems in real-time, recognized for its potential in advancing the manufacturing sector. The DT can be explored to support also waste management strategies in manufacturing. In particular, this paper aims to analyze the usage of this digital technology in supporting waste management in manufacturing, identifying strategies that support waste reduction and recycling practices. Specifically, these strategies will be examined concerning their application to equipment, products, processes and production systems, aligning waste management issues with the scope of the DT applications and the lifecycle stages where it is used.

Keywords: Digital Twin · Waste Reduction · Recycling · Circular Economy · Industry 4.0 · Manufacturing

1 Introduction

1.1 The Waste Management Issue in Manufacturing

In today's dynamic environment, resource consumption has seen a great increase that continues to impact the manufacturing industry. Indeed, this growth of consumption is also affected by the significant increase of the population around the world [1]. It started therefore to be recognized the relevance of this theme which resulted in continuous research and investigation on strategies for resources saving which are relevant as they

Published by Springer Nature Switzerland AG 2025
P. Sureephong et al. (Eds.): PLM 2024, IFIP AICT 740, pp. 423–438, 2025.
https://doi.org/10.1007/978-3-031-93319-6_32

allow the reduction of GHG emissions, increasing sustainability and keeping the Earth green and clean [1]. Waste management is a great challenge that is continuously faced by mankind in this century [2]. Such problem emerges from the following main causes.

- During production; as manufacturing accounts for a quarter of worldwide employment [3] and includes a broad range of products, therefore, a great amount of waste is generated in the different production phases. Manufacturing waste can in particular be defined as materials left unused or left over after the completion of manufacturing processes, these materials are generated at every stage of production and could be in a solid or liquid form [2]. Manufacturing in fact is essential for sustaining the economic growth and the generation of goods and services has expanded drastically in order to align with the global demand. Indeed, a great range of products - including for instance household needs, electronics/electrical items, eatables, etc. – are generated every day and linked to the increasing production rate, there is also the increase of materials waste which occurs for several reasons such as scraps, overproduction, obsolete inventory, packaging waste or energy losses that all companies are called to manage.
- When products reach their end-of-life (EoL), and thus their usage becomes less functional. In this case, products can be subject to different routes. EoL management is a great challenge that many practitioners and scholars are taking into consideration. At this phase of the lifecycle it is needed to understand products status in order to decide whether to proceed on discarding them or to identify other EoL management strategies. In particular, recently the focus has been devoted to circular economy (CE) which at its center considers the closed flow of materials, reducing and minimizing resources usage. Moving towards CE requires priority order in waste management strategies [4]. In particular, the EU Commission provided the Waste Framework Directive, a five-step waste hierarchy which establishes the priority orders for managing and disposing waste including: prevention, preparing for reuse, recycling, recovery and disposal[1]. Additionally, in 2017 Potting et al. have proposed a comprehensive CE approach which considers the so called 9R strategies (refuse, rethink, reduce, reuse, repair, refurbish, remanufacture, repurpose, recycle, recover) [5] that group all the possible waste management strategies requiring deep investigations on their application.

From the highlighted points it can be noticed how in manufacturing environments the waste management issues are starting to be regarded and, besides the already mentioned 9R solutions, other different approaches including the zero-waste manufacturing (ZWM) started to be adopted by both producers and consumers to tackle the waste and minimize its impact on the natural resources [3]. Additionally, policy instruments play a relevant role; in particular, at the European level the EU Waste Policies contribute to the circular economy by extracting high-quality resources from waste as much as possible while protecting the environment and human health. Particularly, different specific policies have been drafted to manage specific wastes such as: i) batteries and accumulators, ii) biodegradable waste, iii) construction and demolition waste, iv) end-of-life vehicles, v) landfill waste, vi) packaging waste, etc. In this sense, guidance documents and information are needed to manage waste efficiently.

[1] https://environment.ec.europa.eu/topics/waste-and-recycling/waste-framework-directive_en

1.2 The Role of Digital Technologies

Besides the approaches introduced in the previous section, a relevant role is also played by digital technologies. Specifically, the Industry 4.0 (I4.0) concept opened the era of technological advancement and recently technologies started to be used not only to improve production in terms of time and quality, but also to drive the transition towards sustainability. In manufacturing context, the sustainable development goes hand in hand with the technological advancement. Indeed, the digital technologies introduced by I4.0 – which include additive manufacturing, artificial intelligence, big data, advanced analytics, internet of things, virtual and augmented reality, digital twins, etc. – offer improvements in terms of reduced material and energy consumption and waste and emissions generation [6]. The role of I4.0 technologies is continuously studied in order to assess how their implementation can support the transition toward sustainability. In particular, recently such tools have started to be implemented to enhance the CE strategies [7]. Addressing all these strategies through digital technologies is relevant and indeed a great part of research is dedicated to the study of the role of I4.0 technologies in supporting these concepts. CE however is a huge theory that is continuously evolving and requires both researchers and practitioners to understand how to measure and quantify its impact in a real context [8]; all the approaches that it proposes need to be fully investigated and linked to the usage of digital technologies. One of these promising technologies is the digital twin (DT), which has a great relevance in supporting the transition of manufacturing towards sustainable approaches [9].

1.2.1 Digital Twin

The DT is defined as the virtual representation of a system and is characterized by the synchronization between the virtual and real system thanks to sensed data and connected smart devices, mathematical models and real time data elaboration [10, 11]. DT plays a relevant role for the manufacturing industry, and besides simulation, its application can support decision-making in autonomous systems, which even addresses both the product and process levels for parts, machines and factories [12, 13]. The application of DTs is established in different areas and is relevant as it can allow to understand and/or predict the behavior of a considered system and its components. Additionally, as stated in [14] such technology can support different lifecycle stages of a product or system including: i) design phase through strategies such as iterative optimization where the DT can be applied to realize design optimization between static configuration and dynamic execution [15] and improve the accuracy and efficiency for product material selection [16]. Additionally, at this phase the DT can be applied to provide data integrity as well as to allow virtual evaluation and verification to reduce inconsistencies between the actual and expected behavior [14, 17], ii) manufacturing phase for real-time monitoring, production control, workpiece performance prediction, process evaluation and optimization, production planning, iii) service phase for predictive maintenance, fault detection and diagnosis, state monitoring, performance prediction and virtual test, and iv) retire phase. Such lifecycle phases can be grouped under three macro-ones, mainly: beginning of life (BoL) including the design phase and manufacturing phase of a product, middle of life (MoL) including the service phase of a product (i.e., when the product is used) or of a production system (i.e., when the system performs the production in manufacturing), and end of life (EoL) including the retire stage.

In particular, in the case of waste management, since there is still lack of research that links the application of DT to such issue, the authors believe that the DT can support studies across the lifecycle of a considered system or product in: i) the BoL through design evaluations, to improve products development and manufacturing or to optimize the design and development of production processes or systems which can support the extension of their lifecycle and limit possible defects; ii) the MoL by simulating the usage stage of a product or production equipment to evaluate optimization strategies and/or can be used for the optimization of production processes or systems (i.e., optimizing the system during its manufacturing phase and optimizing maintenance operations to reduce errors in production); and iii) the EoL – the DT can allow the study and optimization of EoL strategies (e.g., recycling flows) for products that are no longer usable, and additionally the usage of DT in this phase can support the CE strategies presented in Sect. 1.1.

2 Objective and Methodology

From the previous paragraph it emerges the efficacy of DTs application throughout the lifecycle of an equipment, products, process or system to encourage sustainable waste management solutions. Section 1 revealed that under the term "waste" there are different concepts which require different strategies to cope with it. In particular, the 9R strategies require a detailed analysis. Indeed, it is still missing in literature an approach that links the usage of DT with the different strategies of the 9R framework. Accordingly, this paper will focus specifically on waste reduction and recycling approaches for two primary reasons: firstly, investigating strategies for waste reduction is a first step for preventing from waste creation, and secondly, in order to consider a great part of the 9Rs approaches, recycling is a primary step to be taken into consideration.

Thus, the research objective of this paper can be stated as "*analyzing possible applications of the digital twin to support waste reduction and recycling practices*". Therefore, to cope with the challenge of manufacturing wastes, this paper will provide an analysis on the scope of application of DTs and will link it to the lifecycle stages of an equipment, product, process or system. To this aim, a literature review is conducted targeting application cases focused on the purpose of this research. In particular, after a brief introduction, in Sect. 1, on the waste issues that the manufacturing sector faces and after outlining the objective and methodology in Sect. 2, Sect. 3 will propose the methodology of the literature review and its results, Sect. 4 will provide some discussions on the results, finally Sect. 5 will outline the main conclusions of the paper.

3 Literature Review

3.1 Methodology

To respond the objectives of the paper presented in Sect. 2, a research question has been defined to guide the literature review:

"How can digital twins be effectively used to support waste reduction and recycling practices across the different stages of the lifecycle of a product or process or system in manufacturing contexts?"

To this aim, the research has been conducted in Scopus database - considering that it includes an extensive collection of academic papers - using the following keywords and logical connectors:

"manufacturing" AND "digital twin" AND ("waste" OR "recycl")*

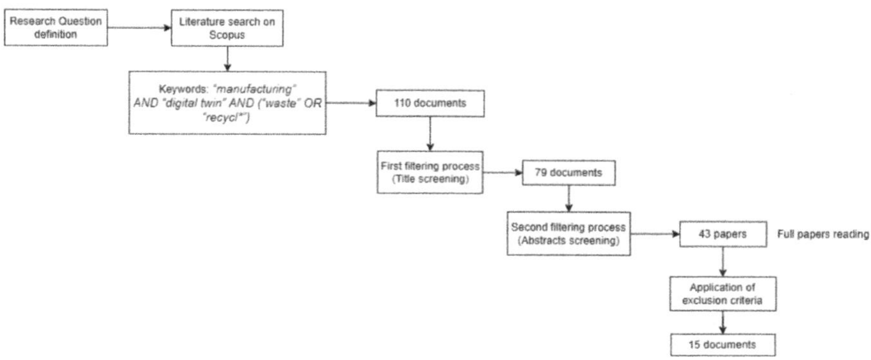

Fig. 1. Papers Selection Process.

This research resulted in a total of 110 documents, and after the assessment of their relevance according to the purpose of this paper, 15 documents have been included in the analysis. Figure 1 summarizes the papers selection process. In particular, the inclusion criteria can be stated as follows.

- Only English papers have been considered for the analysis.
- After reading the title, papers focusing on waste management practices have been considered.
- After reading the abstract, only papers applied in the manufacturing industry have been fully read.
- Finally, after fully reading the papers, application cases relevant for waste reduction and/or recycling purpose have been analyzed.

The exclusion criteria can be summarized as follows:

- Papers that did not propose a practical application of DT in manufacturing industry have been excluded.
- Papers that could not be classified in Table 1, have not been included in the analysis.

3.2 Results

The selected solutions are presented in Table 1 and, according to the different columns, each paper will be described in detail in the next sections.

Table 1. Literature review – Papers' analysis.

Reference	Application area	Application objective	Modelling scope				Lifecycle stage				DT services
			What does the DT replicate?	Equipment piece	Pro-cess/station/single machine	Production system	Product	BOL	MOL	EOL	
[18]	Electronic products	Waste reduction: product recovery – disassembly, remanufacturing and recycling	Product recovery process		x					x	Decision support, predict behavior, assess the optimal recovery strategy
[13]	Electronics	Support the recycling and remanufacturing process of the product	Electronic device				x	x	x	x	Decision support – support for the best recycling route, allow end-user read product specification and update product status
[19]	Automotive industry – Electric battery vehicles	Recycling, extending the lifecycle of the product	Battery				x	x	x	x	Optimization of batteries performance – Batteries health estimation
[20]	Textile industry	Waste reduction, recycling	Physical product, including its material, shapes, colors, textures				x	x			Design evaluation
[21]	Food industry	Waste reduction	Dairy production process		x				x		Predictive maintenance, quality control
[22]	Food retail sector	Waste reduction	Dynamics of the value chain in the grocery retail			x			x		Anomaly detection, variations in supply and demand
[23]	Food packaging	Waste reduction	Manufacturing process of smooth-wall aluminum food packaging trays		x				x		Analysis of the manufacturing process of smooth-wall aluminum food packaging trays – Elimination of defects in the rims of the tray
[24]	Food industry	Waste reduction	Food production process		x				x		Ensuring or enhancing sensory qualities of the food while preserving nutritional quality
[25]	Turbomachinery	Waste reduction – lifecycle extension of the equipment	Turbomachinery equipment	x				x	x	x	Reduction of carbon emissions, resources management, predictive maintenance
[26]	Renewable energy sector	Recycling of composite wind turbine blade	Replication of the wind turbine blade	x						x	Data gathering and analysis, design strategies
[27]	Energy - Proton exchange membrane fuel cell	Waste reduction – extend the lifecycle	Proton exchange membrane fuel cell system			x			x		Health state monitoring, remaining useful life prediction

(continued)

Table 1. (*continued*)

	(PEMFC) systems								
[28]	Additive manufacturing	Waste reduction – life extension of the 3D printer, waste reduction	Virtual replica of the 3D printer		x			x	Real time monitoring and fault detection – Fault detection – Monitor the performance of a 3D printer for improved efficiency and reduction of waste in additive manufacturing applications
[29]	Wastewater treatment process	Waste reduction	Wastewater treatment processes		x			x	Simulation of fault conditions, fault detection and health monitoring
[30]	Hazardous waste landfill	Waste reduction – optimization of the hazardous waste landfill system performance	Physical entities within the landfill, including waste pretreatment facilities, landfill facilities			x		x	Condition monitoring, optimization, fault analysis, prediction, and knowledge evolution within the hazardous waste landfill system
[31]	Cutting tools in manufacturing	Waste reduction – extend the lifecycle	Cutting tool	x			x	x	Real time monitoring, optimization, failure analysis and prediction

3.2.1 Application Area

As represented in Figure 2, the proposed solutions have been clustered according to the area of application which can be divided into a) electronics [13, 18], b) automotive [19], c) textile [20], d) food industry [21–24], e) energy [25–27], f) additive manufacturing [28], g) wastewater treatment [29], h) hazardous waste landfill [30], i) cutting tools in manufacturing systems [31].

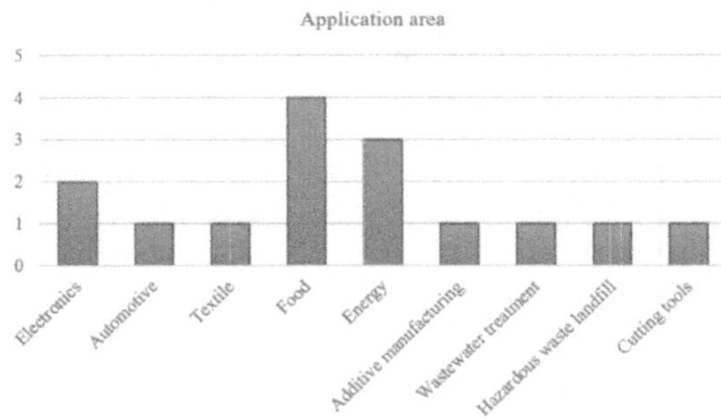

Fig. 2. Visual representation of the number of solutions per each application area.

3.2.2 Application Objective

In the *application objective* it is briefly reported the aim of the application of the DT. In line with the purpose of the present paper all the solutions focus on waste issues, some presenting general approaches that can contribute to the 9R strategies, while others presenting solutions specifically linked to the recycling approach. In particular, in [18] the focus is dedicated to electronics waste reduction proposing strategies such as product recovery by proposing remanufacturing and recycling solutions. Here the application of the DT provides decision support for these recovery strategies. Similarly, in [13] the application objective is dedicated to supporting the recycling and remanufacturing process of electronics products. In [19] recycling and lifecycle extension objectives are proposed. In this case the DT aims at improving the performance, safety, and cost-effectiveness of an electric vehicle battery system. Specifically, in this document the operation of the entire battery value chain is considered in order to offer advantages such as enhancing the economy of manufacturing, repurposing, and recycling processes. In [20] it is proposed an approach to maintain the sustainability in the fashion industry. Therefore, it is presented a DT that evaluates the design elements and characteristics of the physical product, allowing for a more efficient and sustainable production process. In particular, the waste reduction and the recycling approaches are emphasized through the study on the usage of clothing waste material. Indeed, in this paper, the DT facilitates the recycling of clothing waste since it enables the visualization and comparison of products made from recycled materials. In [21] the DT is applied in order to limit wastes and improve the quality and efficiency of dairy production. In [22] the DT is aimed at addressing supply chain variabilities to achieve zero-waste value chains in the grocery retail sector. In particular the DT is designed to predict and detect variabilities across the value chain, prevent food surplus, and facilitate reallocation of resources to minimize waste. In [23] the DT is applied to eliminate defects in the rims of smooth-walled aluminum trays used for food packaging. This is in line with the principle of waste reduction. In [24] the DT wants to ensure sensory quality of the food and product safety in order to reduce waste generation. In [25] the main objective is to optimize the equipment performance, extending its lifecycle and thus it can be understood that also the waste linked to it can be reduced. This general approach can be exploited to also support tools recycling strategies. In [26] the DT is applied in the context of recycling wind turbine blades to contribute to enabling circularity in the energy sector. In [28] the DT is used for the lifecycle extension of 3D printers to contribute to improved productivity and reduced downtime in additive manufacturing processes. This certainly goes hand in hand with the principle of waste reduction in 3D printing. In [29] it is proposed a general approach for waste management aimed at reducing and treating wastewater to improve accurate fault detection and health monitoring in wastewater treatment processes. In [30] the focus is devoted to waste pretreatment facilities and wants to optimize the hazardous waste landfill system. In [27] the application of the DT aligns with the waste reduction principle by extending the product lifespan, predicting the remaining useful life of the considered system. Finally, in [31] the application of the DT model for cutting tools is designed to monitor the tools wear status, predict the residual life, and provide services such as tool grinding, delivery, and for recycling of the cutting tools.

3.2.3 Modelling Scope

The level of detail replicated by the DT gives an idea on the system features to be considered to cope with the objectives of waste management. Table 1 highlights what is specifically modelled by the system and additionally four main levels of replication are highlighted: equipment piece, process or station or single machine, production system and product.

- In [18, 21, 23, 24, 29] are replicated the production processes, station or machine; in particular [18] replicates the process for electronics recovery, [21] replicates the dairy production process, [23] replicates the manufacturing process of smooth-wall aluminum food packaging trays, in [24] the food production process is replicated, [28] proposes the virtual replica of a 3D printer, finally in [29] it is replicated the wastewater treatment process.
- In [25, 26, 31] are replicated the production equipment; in particular, [25] considers the turbomachinery equipment, [26] replicates the wind turbine blade, while [31] models the cutting tool.
- In [22, 27, 30] are modelled the production systems. Specifically, [22] replicates the dynamics of the value chain in the grocery retail, [30] replicates all the physical entities within the landfill, including waste pretreatment facilities and landfill facilities, while [27] models the proton exchange membrane fuel cell system.
- Finally, in [13, 19, 20, 28] are considered the DT of the products. [13] proposes the DT of electronic devices, [19] considers the DT of the electric vehicle battery, finally [20] presents the DT of the physical product including its material, shapes, colors and textures.

3.2.4 Lifecycle Stage

As already presented in Section 1, the DT can be applied at different lifecycle stages of a considered item, process or system according to the purpose of the analysis. The considered stages are three – BoL, MoL, EoL – under which different propositions are included. Indeed, it is relevant to highlight the lifecycle stages, since, according to it, different waste reduction strategies can be applied. Specifically, under BoL are grouped the design, development and manufacturing applications, under MoL is grouped the usage stage of the considered item, and the manufacturing processes or systems during their operation, and under EoL are grouped all the strategies that manage the end-of-life of a product and/or system.

- *Papers considering all the three lifecycle phases.* In [13, 19, 25] the application of the DT considers BoL, MoL, EoL. In [13] it is monitored all the knowledge and information about the product as it moves from BoL to MoL and then to EoL. The product design is first modelled through computer-aided design (CAD), where all the functionalities and features of the product are considered, then the computer-aided engineering (CAE) helps to simulate the performance to validate and improve the product design. The DT can be further enhanced by the strong computing power from the cloud. Additionally, it is considered the environmental performance and impact via lifecycle assessment module where different design strategies, such as design for recycling and design for disassembly are considered. In the manufacturing phase, operations are instead controlled by the manufacturing execution system (MES) and the DT is strengthened by the related operation and product performance data collected from the factory. Once the product is purchased, the end user can interact with the DT via multiple I4.0 enablers (e.g., RFID). Each product is given a unique ID, making also the DTs of each product unique and distinguishable and therefore allows supporting different recycling and remanufacturing operations in the EoL. In [19] the DT for electric vehicle battery systems is applied across the three battery lifecycle phases; based on the considered stage three levels of DT are proposed: DT prototype which considers the design and manufacturing of the batteries, performance DT which addresses the operation of batteries in their execution life, and DT instance which considers the entire lifecycle of the batteries from design and manufacturing to their recycling. In [25] it is proposed a DT applied across multiple stages of the turbomachinery. In particular the DT can be used during the design phase to optimize the design of the equipment piece before its production, and additionally the DT can be applied in the manufacturing and assembly phase of the turbomachinery to optimize such operations, and also for maintenance operations. Finally, the proposed DT can be applied at the EoL for recycling strategies. In this paper it is highlighted the importance of storing, through the DT, design information that would allow improving the recycling approach.
- *Papers considering both BoL and MoL phases.* In [31] the DT is applied at the BoL and MoL phases for cutting tools where design, development, testing, production, monitoring, residual life management, and maintenance of the tool are performed.
- *Papers considering the BoL phase.* In [20] the DT is used at the BoL stage to allow the integration of sustainability, waste reduction, and circularity objectives into the

product development process to support the recycling and sustainability of clothing waste materials in the fashion industry.

- *Papers considering the MoL phase.* In [21] the DT allows monitoring the dairy production processes in real-time, allowing for predictive maintenance, quality control, optimization of operations, and the prevention of defects in dairy products, thus preventing from production of products that would be wasted. Similarly, in [23] the DT is applied at the MoL phase, during the execution of the manufacturing process of smooth-wall aluminum food packaging trays to optimize the manufacturing process, eliminating defects in the tray. In [22] the DT monitors the production system's operations enhancing decision making to support waste reduction in the grocery retail sector. In [24] the food production process is replicated during its execution to support circularity during production. This solution supports waste reduction in food production processes using also tools for sensory qualities of the food and product safety. All the data gathered through the sensors allow indeed the optimization of decisions through the DT. In [28] the DT is used to detect faults during the operational stage of a 3D printer. Indeed, this solution supports the reduction of defects (which then become waste) in additive manufacturing. [29] focuses on wastewater treatment processes and wants to detect faults that contribute to wastes in such process. In [30] the DT is applied in the operational phase of the lifecycle of hazardous waste landfill systems to address its efficiency and reduce the environmental risks associated with the hazardous waste disposal facilities. In [27] the DT is used during the operational stage of the proton exchange membrane fuel cell system to monitor its health and extend its lifecycle.
- *Papers considering the EoL phase.* In [18] the DT allows to recycle end-of-life products, reducing the need for resource extraction and minimizing the environmental impact of electronic waste. In [26] the DT allows the usage of data on blade's condition and supports, thanks to robotic fabrication techniques, to disassemble the blade and prepare the materials for recycling or repurposing.

3.2.5 DT Services

The DTs proposed in each of the presented solutions provide overall the following different services to support their application's objectives.

In [18] the DT is applied for decision support, to predict the process' behavior and assess the optimal recovery strategy. In [13] it supports optimal recycling root decisions and additionally it allows the end-user to read products specifications and update the product status allowing the DT to be updated on recent data. In [19] batteries health performance is estimated through the DT. In [20] the DT allows products' design evaluations. In [21] the DT supports predictive maintenance and quality control of the dairy production system. In [22] to support zero-waste value chains, the DT ensures anomaly detection and monitors the variations in the supply and demand. In [23] the DT allows the elimination of defects in the rims of the tray. In [24] the DT ensures food sensors' quality while also preserving food nutritional quality. In [25] the usage of DT allows the reduction of carbon emissions, resources management and predictive maintenance operations. In [26] the DT allows data analysis and design improvement strategies. In [28] the DT allows real-time monitoring of the performance of the 3D printer and fault detection. In [29] the DT simulates the fault conditions in wastewater treatment process and allows health monitoring. In [30] the DT allows condition monitoring, optimization, fault analysis, prediction and knowledge evolution within the hazardous waste landfill system. In [27] the DT provides an analysis of the health state and allows the prediction of the remaining useful life of a proton exchange membrane fuel cell system. In [31] the DT supports real-time monitoring, optimization, failure analysis and prediction for cutting tools.

4 Discussion

From the analysis of the different proposed solutions, several key points emerged. As presented in Fig. 3, the purpose of application of the DTs is strictly linked to the lifecycle stage.

i. The considered papers indicate that when the objective specifically involves recycling approaches, the DT needs to be applied also at the EoL stage, as it is in this phase where it is necessary to manage all the end-of-life strategies, including recycling. However, it should not be neglected the relevance of applying DTs also at BoL and MoL. In the analyzed papers, recycling strategies mainly apply to equipment pieces, products or production processes focused on products recycling. The DT therefore allows studying methodologies to improve recycling management at the EoL; however, the analysis shows that few solutions are currently applied at the EoL, indicating the need for further research on EoL management strategies through DT applications.
ii. Overall, waste reduction and treatment strategies should not be considered only at the EoL. Indeed, strategies to prevent from waste generation need also to be applied at earlier lifecycle phases. The analysis allowed assessing that integrating sustainable practices into the BoL phase, such as product development practices that support lifecycle extension, can reduce the waste generation from products that eventually would have been no longer useful. Indeed, at the BoL stage, the DT can be applied to test design practices that improve product lifecycle and that will also simplify EoL management. However, the literature review shows that DT applications for design purposes (therefore at the BoL phase) are still lacking. Additionally, for waste reduction, DTs at the MoL phase i.e., during the operational stage – are mainly used for defects detection. Indeed, defects can damage products and/or production systems reducing their effectiveness and the duration of their lifecycle. For instance, defective products will cause wastes, while inefficient production systems will negatively impact the production, leading to waste generation.
iii. Using DTs to link data of the different lifecycle phases can optimize decisions taken in the single lifecycle stages. For example, at the BoL phase, considering that once a product reaches the EoL will need to be recycled, the usage of the DT can test product design strategies, through for instance the so-called *design for recycling* approach, to facilitate the EoL management. In this sense, data from the DT of the EoL stage will improve the DT of the BoL phase. This approach is promising and could bring to many benefits, however applications in this sense are still lacking and need further research.

Besides the insights presented through the above points, the analysis emphasizes the importance of considering not only products performance – as usually the waste is linked to products – but it is also crucial to consider the performance of the manufacturing processes and systems. In this sense, the ability of DTs in virtually replicating physical systems, can be exploited to optimize production processes and/or systems to detect failures or defects that could affect the final product. The proposed study is a proof of the relevance of real-time monitoring and fault detection also at process and system level. Indeed, considering the product and not the production process or system means considering only how to dispose or recycle the products or their materials, while considering the production process or system means regarding waste prevention strategies before the product reaches the disposal stage. During the operational life of a production process or system, the DT can be implemented to monitor in real-time the process or system considered. Gathering data from the real system can allow failure analysis and prediction and could allow the prediction of future behaviors thanks to the analysis of historical data.

Additionally, in the context of waste management, a proper DT implementation should address all the lifecycle stages, each one for a specific purpose, but the data from the different lifecycle stages should be integrated. DTs at the BoL phase can support the design of manufacturing processes and systems that reduce waste generation during the operational life (at the MoL), and additionally can support the design of products that facilitate their EoL management, therefore data from all phases should be integrated.

Researchers and practitioners implementing DTs for waste management should define clear objectives for each lifecycle phase of the considered element (i.e., equipment, product, process/station/machine or system). DTs can be leveraged to simulate and test prototypes, identifying potential waste reduction opportunities (e.g., preventing from defective products or inefficient systems, or ensuring efficient materials recovery at the EoL). Nevertheless, current research does not give enough relevance to data integration across the different lifecycle stages. DTs should create a comprehensive view of the element's lifecycle, enabling informed decision-making at each stage.

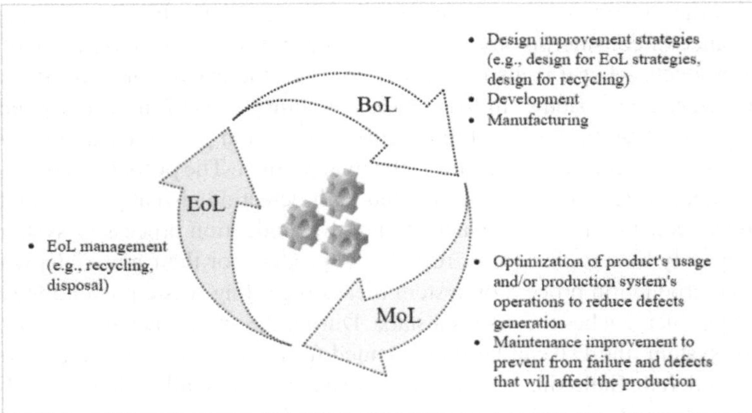

Fig. 3. DT application purpose across the lifecycle for waste reduction and recycling.

5 Conclusions

The paper explored the issue of waste creation in manufacturing and aimed at analyzing possible sustainability practices. Great improvements can be supported through the application of digital technologies. The I4.0 paradigm allowed the introduction of different digital technologies, and it started to be agreed that their role in supporting sustainability practices is crucial in today's very dynamic environment. The focus of this paper has been devoted to the DT application for waste reduction and recycling, as this technology, allowing the creation of virtual replicas, can support studies on optimization strategies for the analyzed issue. The conducted review allowed linking the lifecycle stages where DTs are applied with the scope of application and allowed assessing the relevance of applying the DT not only at equipment and product levels but also at production process and system levels. As emerged from the literature, the application of DTs for waste management purpose is still lacking and specifically, considerations at the BoL and EoL phases are still limited. Therefore, further investigations are needed. Future works should discuss the application of DT for supporting CE objectives in a broad sense to evaluate general approaches that can support DTs better development and application for all the CE strategies.

Acknowledgments. This study was carried out within the MICS (Made in Italy – Circular and Sustainable) Extended Partnership and received funding from the European Union Next- GenerationEU (PIANO NAZIONALE DI RIPRESA E RESILIENZA (PNRR) – MISSIONE 4 COMPONENTE 2, INVESTIMENTO 1.3 – D.D. 1551.11-10-2022, PE00000004). This manuscript reflects only the authors' views and opinions, neither the European Union nor the European Commission can be considered responsible for them.

This work has received funding from the European Union's Horizon Europe Research and Innovation programme under grant agreement No 101091490. In any case, the present work cannot be considered as an official position of the supporting organisation, but it reports just the point of view of the authors.

Disclosure of Interests. The authors have no competing interests to declare that are relevant to the content of this article.

References

1. Sinha, S., Modak, N.M.: A systematic review in recycling/reusing/re-manufacturing supply chain research: a tertiary study. Int. J. Sustain. Eng.
2. Awogbemi, O., Von Kallon, D.V., Bello, K.A.: Resource recycling with the aim of achieving zero-waste manufacturing. Sustain. (2022)
3. Singh, S., Ramakrishna, S., Gupta, M.K.: Towards zero waste manufacturing: a multidisciplinary review. J. Clean. Prod. **168**, 1230–1243 (2017)
4. Kosacka-Olejnik, M.: How manage waste from End-of-Life Vehicles? - method proposal. IFAC-PapersOnLine **52**(13), 1733–1737 (2019)
5. Liu, Q., Trevisan, A. H., Yang, M., Mascarenhas, J.: A framework of digital technologies for the circular economy: digital functions and mechanisms. Bus. Strateg. Environ
6. Laskurain-Iturbe, I., Arana-Landín, G., Landeta-Manzano, B., Uriarte-Gallastegi, N.: Exploring the influence of industry 4.0 technologies on the circular economy. J. Clean. Prod. **321**, 128944 (2021). https://doi.org/10.1016/j.jclepro.2021.128944
7. Haupt, M., Vadenbo, C., Hellweg, S.: Do we have the right performance indicators for the circular economy?: Insight into the swiss waste management system. J. Ind. Ecol. **21**(3), 615–627 (2017). https://doi.org/10.1111/jiec.12506
8. Sassanelli, C., Rosa, P., Rocca, R., Terzi, S.: Circular economy performance assessment methods: a systematic literature review. J. Clean. Prod. **229**, 440–453 (2019). https://doi.org/10.1016/j.jclepro.2019.05.019
9. He, B., Bai, K.J.: Digital twin-based sustainable intelligent manufacturing: a review. Adv. Manuf. **9**(1), 1–21 (2021)
10. E. Negri, L. Fumagalli, and M. Macchi, "A Review of the Roles of Digital Twin in CPS-based Production Systems," *Procedia Manuf.*, 2017
11. Kritzinger, W., Karner, M., Traar, G., Henjes, J., Sihn, W.: Digital Twin in manufacturing: a categorical literature review and classification. IFAC-PapersOnLine **51**(11), 1016–1022 (2018)
12. Posada, J., et al.: Visual computing as a key enabling technology for Industrie 4.0 and industrial internet. IEEE Comput. Grap. Appl. **35**(2), 26–40 (2015). https://doi.org/10.1109/MCG.2015.45
13. Wang, X.V., Wang, L.: Digital twin-based WEEE recycling, recovery and remanufacturing in the background of Industry 4.0. Int. J. Prod. Res. **57**(12), 3892–3902 (2019). https://doi.org/10.1080/00207543.2018.1497819
14. Liu, M., Fang, S., Dong, H., Xu, C.: Review of digital twin about concepts, technologies, and industrial applications. J. Manuf. Syst. **58**, 346–361 (2021). https://doi.org/10.1016/j.jmsy.2020.06.017
15. Liu, Q., Zhang, H., Leng, J., Chen, X.: Digital twin-driven rapid individualised designing of automated flow-shop manufacturing system. Int. J. Prod. Res. **57**(12), 3903–3919 (2019). https://doi.org/10.1080/00207543.2018.1471243
16. Xiang, F., Zhang, Z., Zuo, Y., Tao, F.: Digital twin driven green material optimal-selection towards sustainable manufacturing. Procedia CIRP **81**, 1290–1294 (2019). https://doi.org/10.1016/j.procir.2019.04.015
17. Tao, F., et al.: Digital twin-driven product design framework. Int. J. Prod. Res. (2019)
18. Tozanlı, Ö., Kongar, E., Tozanlı, Ö., et al.: Evaluation of waste electronic product trade-in St.pdf. (2020)

19. Naseri, F., et al.: Digital twin of electric vehicle battery systems: comprehensive review of the use cases, requirements, and platforms. Renew. Sustain. Energy Rev. **179**, 113280 (2023). https://doi.org/10.1016/j.rser.2023.113280
20. Lee, H.: The dual strategy for textile and fashion production using clothing waste. Sustain. **15**(15), 22–24 (2023)
21. Konstantinidis, F.K., et al.: Achieving zero defected products in Diary 4.0 using digital twin and machine vision. ACM Int. Conf. Proc. Ser. (2023)
22. Sengupta, S., Dreyer, H.: Realizing zero-waste value chains through digital twin-driven S and OP: a case of grocery retail. Comput. Ind. **148**, 103890 (2023). https://doi.org/10.1016/j.compind.2023.103890
23. Mahmoodi, P., Elrefaey, A., Hassan, O.I., Morgan, H.D., Sienz, J., Belblidia, F.: Towards a wickless smooth–Wall aluminium food packaging tray mould tool digital twin- Advanced computational modelling supported by experimental validation. Appl. Math. Model. **105**, 375–386 (2022)
24. Udugama, I.A., Kelton, W., Bayer, C.: Digital twins in food processing: a conceptual approach to developing multi-layer digital models. Digit. Chem. Eng. **7**, 100087 (2023). https://doi.org/10.1016/j.dche.2023.100087
25. Xie, R., Chen, M., Liu, W., Jian, H., Shi, Y.: Digital twin technologies for turbomachinery in a life cycle perspective: a review. Sustainability **13**(5), 2495 (2021). https://doi.org/10.3390/su13052495
26. Geiger, R., Hannan, Y., Travia, W., Naboni, R., Schlette, C.: Composite wind turbine blade recycling - value creation through Industry 4.0 to enable circularity in repurposing of composites. IOP Conf. Ser. Mater. Sci. Eng. **942**(1), 012016 (2020). https://doi.org/10.1088/1757-899X/942/1/012016
27. Meraghni, S., Terrissa, L.S., Yue, M., Ma, J., Jemei, S., Zerhouni, N.: A data-driven digital-twin prognostics method for proton exchange membrane fuel cell remaining useful life prediction. Int. J. Hydrogen Energy (2021)
28. Rachmawati, S.M., Putra, M.A.P., Lee, J.M., Kim, D.S.: Digital twin-enabled 3D printer fault detection for smart additive manufacturing. Eng. Appl. Artif. Intell. **124**, 106430 (2023). https://doi.org/10.1016/j.engappai.2023.106430
29. Liu, W., He, S., Mou, J., Xue, T., Chen, H., Xiong, W.: Digital twins-based process monitoring for wastewater treatment processes. Reliab. Eng. Syst. Saf. **238**, 109416 (2023). https://doi.org/10.1016/j.ress.2023.109416
30. Wu, Y., Li, Y.: Digital twin-driven performance optimization for hazardous waste landfill systems. Math. Probl. Eng. **2022**, 1–10 (2022). https://doi.org/10.1155/2022/7778952
31. Xie, Y., Lian, K., Liu, Q., Zhang, C., Liu, H.: Digital twin for cutting tool: modeling, application and service strategy. J. Manuf. Syst. **58**, 305–312 (2021). https://doi.org/10.1016/j.jmsy.2020.08.007

Named Entity Recognition for Building Information Model Verification Using Large Language Models

Insaf Nahri[1,2](✉), Romain Pinquié[3], Philippe Véron[1], Nicolas Bus[2], and Mathieu Thorel[2]

[1] LISPEN EA 7515, Arts et Métiers Institute of Technology, 13100 Aix-en-Provence, France
{insaf.nahri,philippe.veron}@ensam.eu, insaf.nahri@cstb.fr

[2] Information System and Applications Division, CSTB, 06560 Sophia Antipolis, France
{nicolas.bus,mathieu.thorel}@cstb.fr

[3] Univ. Grenoble Alpes, CNRS, Grenoble INP, G-SCOP, Grenoble, France
romain.pinquie@grenoble-inp.fr

Abstract. Building Information Modeling (BIM) boosts collaboration and efficiency in the construction industry by digitalizing data exchange throughout a building's lifecycle. However, traditional manual methods for extracting information from unstructured documents for BIM validation are inefficient and error-prone. This study proposes integrating BIM with Natural Language Processing (NLP), specifically Named Entity Recognition (NER), utilizing pre-trained Large Language Models (LLMs) like GPT-4 and Mistral 7B, coupled with prompt engineering techniques to automate extracting requirements from Building Technical Specifications (BTS) documents. The capability of GPT-4 and Mistral 7B to extract BIM entities in zero-shot and few-shot scenarios was compared with *Camembert_base*, a BERT Transformer-based language model fine-tuned on a dataset of 72 BTS documents. The comparison aimed to assess LLMs' viability as alternatives to traditional NER models and the impact of domain-specific fine-tuning. Findings indicate GPT4 and Mistral 7B show potential in recognizing BIM entities, yet their effectiveness varies, highlighting the limits of direct LLMs application without further adaptation. *Camembert_bas*e outperformed in few-shot scenarios, achieving over **90% F1-score** across all entities. This underscores the importance of fine-tuning LLMs with domain-specific data for optimized NER tasks in BIM validation.

Keywords: BIM · NLP · NER · LLM · GPT-4 · Mistral · CamemBERT · BTS · Construction Industry

1 Introduction

BIM transforms the construction industry by enhancing lifecycle management, collaboration, cost-efficiency, and quality through digital data exchange [1]. The key to fully leveraging BIM lies in accurately validating BIM models, a process heavily reliant on

Published by Springer Nature Switzerland AG 2025
P. Sureephong et al. (Eds.): PLM 2024, IFIP AICT 740, pp. 439–452, 2025.
https://doi.org/10.1007/978-3-031-93319-6_33

extracting precise information from unstructured documents to ensure digital representations align with the necessary technical specifications and requirements[2]. Thechallenge of automatically extracting information from these unstructured documents is significant, especially in an industry characterized by specialized terminology and various data sources [3]. Information Extraction (IE) is crucial in construction for interpreting vast amounts of data in documents like BTS[1], which details the technical requirements for project execution. These documents are critical for the French construction sector, dictating performance expectations and project scope, thereby playing a central role in public procurement processes.

Traditionally, NER, a fundamental technique for IE, has relied on creating extensive labeled datasets or developing rule-based systems. These methods, although effective, require considerable resources and continuous updates to handle domain-specific language changes and new information [4]. The labor-intensive nature of these approaches highlights the industry'sneed formore efficient methods.

The advent of LLMs like GPT-4 and Mistral 7B offers a new avenue for addressing these challenges. With their ability to understand and generate humanlike text, LLMs represent a significant opportunity to automate IE tasks, reducing the need for manual effort associated with traditional methods [5]. This shift could alleviate the reliance on constructing labeled datasets and rule-based systems, presenting an appealing solution for the construction industry's efficiency and accuracy demands.

This study investigates the feasibility of using LLMs, specifically GPT-4 and Mistral 7B, to extract entities from BTS documents automatically. It assesses these models in zero-shot and few-shot learning scenarios to evaluate their ability to identify and classify domain-specific entities effectively. A comparison with Camembert_base was made to benchmark their performance. This comparison explores whether LLMs, through prompt engineering, can match or exceed the accuracy of fine-tuned language models specifically adapted for the construction domain, such as Camembert_base fine-tuned on 72 BTS documents.

The paper is organized into a literature review on IE techniques relevant to construction (Sect. 2), a methodology section detailing the use of LLMs and the fine-tuning of Camembert_base (Sect. 3), a comparative analysis of experimental results (Sect. 4), and a conclusion summarizing the findings, limitations, and suggestions for future research (Sect. 5).

2 Related Work

Section 2.1 presents an overview of IE, followed by a comprehensive review of NER approaches, a critical subtask of IE, in Sect. 2.2. Section 2.3 details the examination of the application of LLMs in the construction industry.

[1] BTS: known in French as the " Cahier des Clauses Techniques Particulières (CCTP)" is a document outlining the technical requirements, performance expectations, and scope of a construction project. Created by the project manager, it breaks the project into work packages and is essential for the public procurement process.

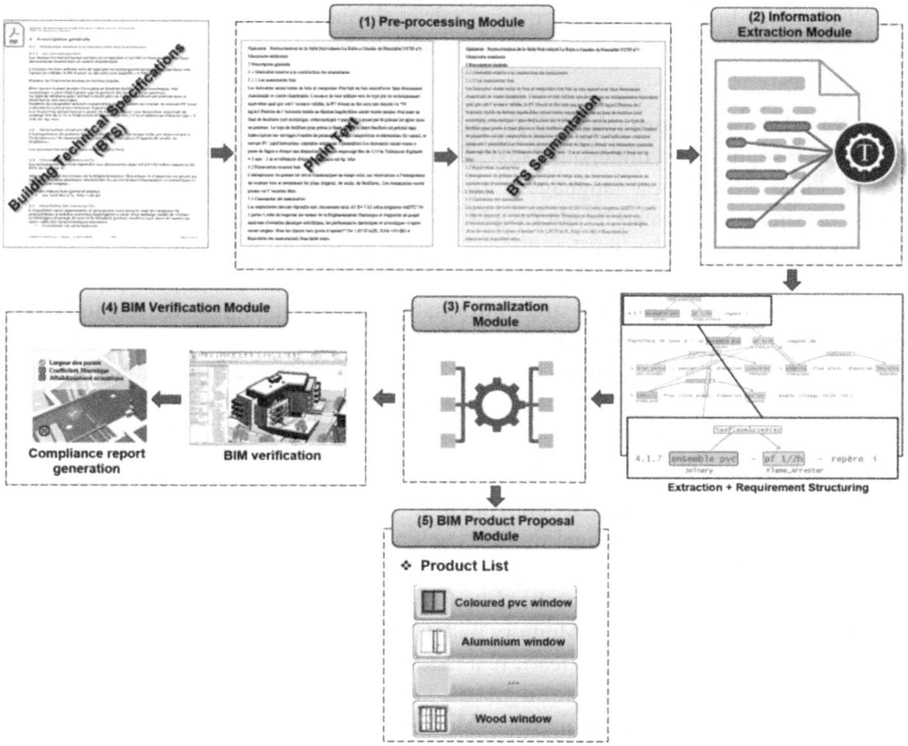

Fig. 1. Use cases for requirements extraction from BTS documents: (1) BIM verification use case (Module 4) ensuring compliance with BIM and BTS, and (2) product BIM proposal use case (Module 5) facilitating construction product recommendations.

2.1 Information Extraction (IE)

IE in NLP automatically derives structured insights from unstructured or semistructured textual data, identifying and capturing relevant information such as entities, relations, and events within the text. This process finds applications in various domains, including healthcare [6], the automotive industry [7], and the chemical sector [8]. Approaches to IE range from traditional rule-based systems that rely on hand-crafted rules and patterns [9] to more advanced machine learning and deep learning models [10].

2.2 Named Entity Recognition (NER)

NER is a crucial task within IE, focused on identifying and categorizing named entities into predefined categories like person names, organization names, and locations. Given an input sentence X of n tokens $x1, x2,..., xn$, and a set of predefined entity types $E = e1, e2,..., ek$, where k represents the number of predefined entity categories, NER aims to identify and classify entities within the unstructured text into these predefined categories.

NER aids in tasks such as information retrieval, question answering, and document classification, utilizing approaches ranging from rule-based systems to machine learning algorithms.

Rule-Based NER Approaches. Rule-based methods use specific rules created by experts to recognize and understand text. These rules are based on the expert's domain-specific knowledge of how language is used, and they can be in different formats, such as text parsing rules or search patterns [11]. They work well for clear-cut cases but can be hard to set up and may not handle unexpected information well. Many fields have successfully used these methods. In crime analysis, they combine dependency parsing and Part-Of-Speech tagging to extract information [12]. In the medical domain, regular expressions identify peripheral arterial disease [13]. In construction, dependency trees and suffix-based matching extract key mechanical, electrical, and plumbing information [4]. Additionally, these methods often rely on domain-specific dictionaries or lists that include important words or phrases to look for in the text. This makes the process straightforward and effective in various areas [14].

Machine Learning-Based NER Approaches. NER has undergone significant evolution, transitioning from statistical models such as Hidden Markov Models and Conditional Random Fields (CRF), which require extensive manual feature definition [15], to advanced deep learning approaches like Convolutional Neural Networks (CNN) and Bidirectional Long Short-Term Memory with CRF (BILSTM-CRF) models. These deep learning methods automatically learn features from large datasets but depart from earlier reliance on manual feature crafting [16]. The advent of transformer-based language models, particularly Masked Language Models (MLMs) like BERT and its variants, including CamemBERT for French, based on the RoBERTa model [28], has changed the landscape. These models capture contextual nuances, allowing for the accurate extraction of domain-specific entities by predicting words masked in sentences. Attention mechanisms further enhance their effectiveness, streamlining training compared to traditional Recurrent Neural Networks(RNNs), making them adept at understanding intricate details in texts without extensive labeled data [17, 18]. In addition to MLMs, Causal Language Models (CLMs) like GPT-4 [20] and Mistral 7B [27] have introduced innovative methodologies for NER through "Incontext learning" (ICL) [21], enabling both zero and few-shot learning scenarios. This approach, which relies less on labeled data, is contrasted with MLMs by generating text from prompts and is currently a researchfocusfor its effectiveness in NER tasks [22]. The ongoing exploration into the comparative performance of MLMs and CLMs in NER, especially within specialized domains, underscores the rapid development in the field of NER and its application across a wide range of sectors.

Few-Shot NER. Can be approached using both pre-trained MLMs and CLMs, leveraging their distinct architectures and training paradigms for effective entity recognition with minimal training examples. Few-shot learning facilitates entity recognition with limited examples, ranging from zero-shot (without examples) to K-shot (using K examples) [22].

Few-Shot NER with Pre-trained MLMs. MLMs are trained to predict masked tokens in a text, acquiring deep contextual representations [18]. For few-shot NER, these models are fine-tuned on a small set of labeled examples, adapting their rich pre-trained embeddings to the task with limited data. This approach benefits from the contextual understanding developed during pre-training, allowing for effective entity recognition even with few examples. However, performance can be limited by entities' domain specificity and the training examples' representativeness.

Few-Shot NER with CLMs. CLMs predict the next token in a sequence using a generative approach [10]. In few-shot settings, CLMs can use a few examples as prompts, guiding the model to generate entity labels directly without explicit task-specific fine-tuning. This method leverages the model's generative capabilities and the flexibility of prompt-based learning. The primary challenge here is the dependency on the quality and format of the prompt, which can significantly influence the model's performance on the NER task.

2.3 LLMs in the Construction Industry

In the construction industry, the use of LLMs like GPT and Mistral is emerging, with existing research highlighting the effectiveness of MLMs such as BERT for tasks like extracting requirements from regulatory documents and identifying risk clauses [19, 24]. The application of CLMs in construction has been less explored, but it shows promise for task-specific applications through zero-shot and few-shot learning, evidenced by the use of GPT for material selection guidance and accident cause analysis in highway construction [23, 25].

These models offer significant advantages in handling tasks without extensive labeled datasets, simplifying the IE process. However, challenges remain in adapting these models to the domain-specific knowledge required in construction [23, 26]. This research evaluates the potential of LLMs in enhancing the accuracy and efficiency of requirements extraction from BTS documents, with a dual focus: 1) ensuring BIM and BTS compliance for optimal lifecycle management in the construction sector (Module 4, illustrated in Fig. 1), and facilitating construction product recommendations aligned with BTS-derived requirements (Module 5, illustrated in Fig. 1). Achieving these objectives relies on BIM's integral role across a building's lifecycle, from conception to demolition or renewal. The methodology underpinning this research, particularly the NER as an initial step for IE, is detailed in Sects. 3 and 4.

3 LLMs in Extracting BIM Requirements Entities: A Use Case on Joinery Work Package

Figure 2 illustrates the method for extracting BIM requirements from unstructured BTS documents, starting with NER as the initial step in IE. The approach commences with the collection of BTS documents (Step 1, Fig. 2), followed by preprocessing to remove irrelevant content that could hinder the IE algorithm (Step 2, Fig. 2). A segmentation algorithm then isolates raw requirements by identifying entities like BIM concepts and properties, utilizing dictionaries of BIM (Step 3, Fig. 2). Next, an annotation process is applied via an annotation tool (Step 4, Fig. 2), and the labeled dataset is divided into

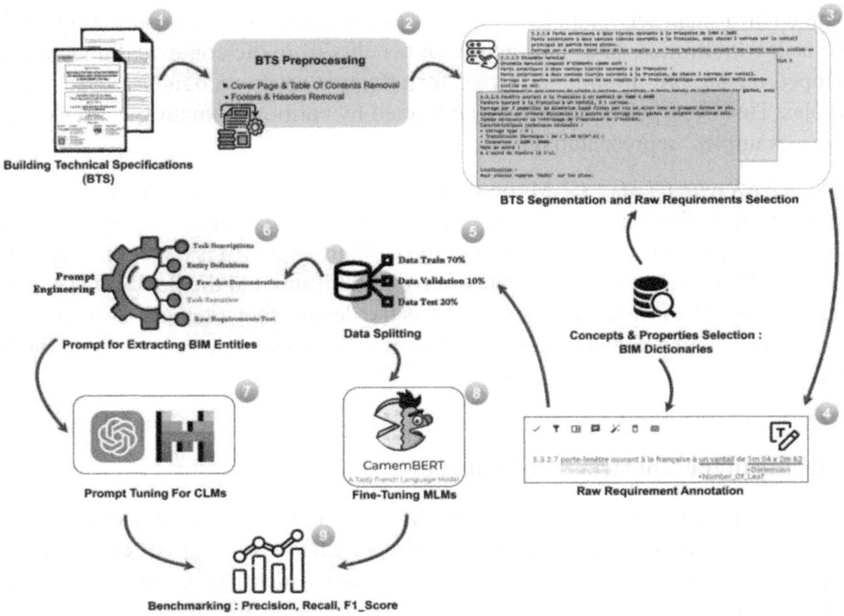

Fig. 2. Summary of the proposed methods for extracting BIM requirements entities from BTS

training, testing, and validation subsets for LLMs (Step 5, Fig. 2). CLMs are guided (Step 7) using the prompt detailed in Step 6, followed by fine-tuning MLM with the training and validation subsets in Step 8. Finally, the performance of CLMs and MLMs is assessed using precision, recall, and F1-score metrics and testing subset (Step 9).

3.1 BTS Gathering

The dataset gathers 72 BTS in XML-based PDF format from 61 online platforms, corresponding to a total of 1505 pages and 508,456 words. These documents, produced by 61 contractors and building owners, exhibit diverse writing styles due to their varied authorship to aid in developing a model for extracting formal requirements irrespective of writing style. Subsequent manual analysis of 10 documents of the dataset (chosen randomly) aimed to decipher the structure of requirements and evaluate the expression diversity. This analysis highlighted the extensive range of requirements detailed in BTS, employing varied terminology and encompassing entities from broad categories like “Door” to more specific items such as “Handle”, “Strap hinge”, and “Frame”, etc.

3.2 BTS Preprocessing

Thedataset curation process entailedselectively removingnon-essential elements from BTS documents, including cover pages, tables of contents (TOC), footers, and headers. This removal was necessary because these elements introduced noise to the proposed algorithms. For instance, entities in the TOC could result in the extraction of additional,

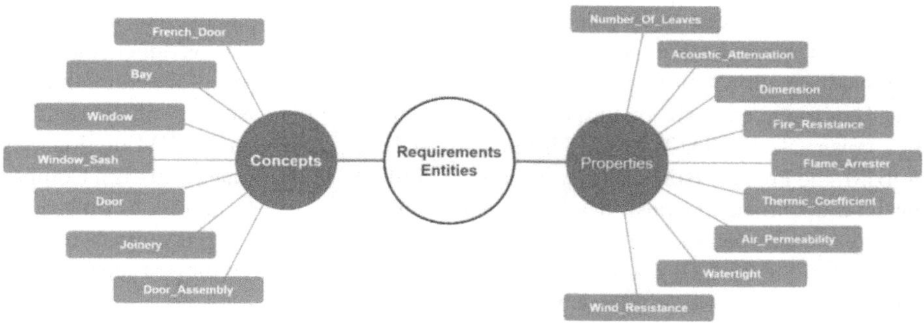

Fig. 3. Chosen entities within the joinery work package

unnecessary entities. The variability and lack of standardization in the inclusion and formatting of cover pages and TOC necessitated their manual exclusion. A systematic approach for footers and headers was adopted by breaking down each page's text into sentences to pinpoint and discard those recurring throughout the BTS documents. This technique relied on cross-page sentence pattern analysis to detect and eliminate the consistently positioned and similar headers and footers across documents.

3.3 BTS Segmentation and Raw Requirements Selection

BTS documents underwent segmentation into paragraphs, leveraging the document's numbering system (e.g., "1, 1.1, 1.1.1, 2" or Roman numerals "I, II, III") to extract raw requirements. Regular expressions were employed to recognize this structured format, facilitating the extraction of elements like chapters and sections. This method allowed for the automatic identification of headlines, with the subsequent text treated as separate paragraphs. This strategy streamlined the analysis of extensive BTS documents, typically over 75 pages, containing detailed process and technical requirements.

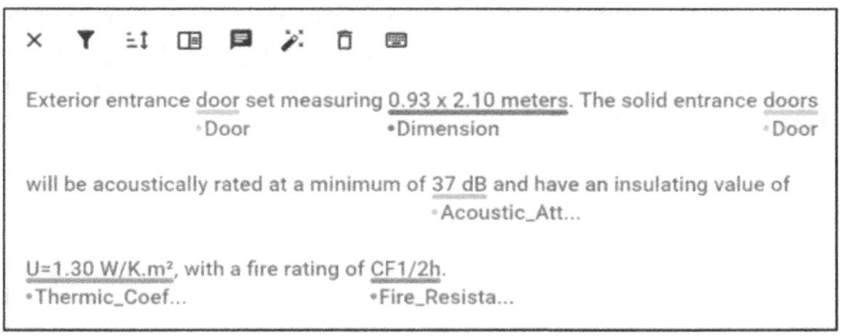

Fig. 4. Doccano interface with raw text data during the annotation process

By segmenting BTS documents, the focus was on extracting technical requirements relevantto BIM, specifically those segments containing BIM concepts and properties. In

BIM, "concepts" refer to the fundamental ideas or building elements like walls, while "properties" describe their specific characteristics or attributes such as dimensions. This segmentation allows for targeted analysis of technical aspects over process-related content. Due to the extensive nature of BTS documents and the vast array of BIM entities they cover (Sect. 3.1). The study limited its analysis to three main BIM dictionaries: POBIM, the Product BIM Dictionary, and the Model BIM Dictionary [30–32]. Comparing BTS documents against these dictionaries enabled to isolate and focus on specific concepts and properties (Fig. 3), leading to the identification of 233 relevant raw requirements from the 72 BTS documents that were collected, the three longest requirements containing up to 2,600 words each.

3.4 Raw Requirements Annotation and Splitting

Data annotation involvesmarking up the concepts and properties within selected raw requirements. Doccano [29] was chosen as a tool for text annotation because of its user-friendly interface and the convenient JsonLines format it produces. This format includes the text and a collection of annotated entities, with each entity accompanied by an identifier, start and end offsets, and a designated label.

Table 1. Doccano output showing annotated entities with start and end offsets

Id	Entity Type	Offsets	Entity
I1	Door	18 22	door
I2	Dimension	37 55	0.93 x 2.10 m
...	...	...	...
I5	Fire_Resistance	200 206	CF1/2h

Figure 4 demonstrates the Doccano interface during the labeling process, while Table 1 presents the resulting annotated data structure. After annotating 233 raw requirements, they were partitioned into three datasets: 70% for training, 20% for testing, and 10% for validation. These datasets have been utilized to train and evaluate the LLMs. Figure 5 depicts the distribution of entity labels across the training, testing, and validation datasets.

3.5 CLMs Prompt for Extracting BIM Entities

An English prompt was utilized for GPT-4 and Mistral 7B, taking advantage of English's widespread presence in their training datasets. This implies the LLMs likely encountered similar elements during their training, aiding their understanding. To facilitate entity recognition from the French text, precise translations of each concept and property were provided. Thisapproach simplifies entity recognition by guiding the models with familiar terms in their training language (English or French). The prompt includes five elements as depicted in Fig. 6:

Fig. 5. Entity label distribution within the training, testing, and validation data sets

1) Task Descriptions to explain the task; 2) Entity Definitions for model clarity; 3) Few-shot Demonstrations with ten raw requirement examples from the training dataset, ensuring representation of each entity type at least once; 4) Task Execution Instructions with specific guidelines; 5) The raw requirement testfor the CLMs to handle.

The study employs a tagging prompt strategy, one of three main methods identified in the literature [26], wherein specific tags encapsulate each entity. This study utilizes “ < *Entity_Label* > “ and “ < */Entity_Label* > “ tags to highlight entities. This method aids precise entity identification and simplifies the process, facilitating accurate entity localization within texts. This strategy is critical for automatically evaluating CLMs against MLMs like *camembert_base* by accurately determining entity positions. The results were inaccurate when attempting to have the CLMs directly return entities with their start and end offsets. This outcome highlights that LLMs struggle with precise counting of the entity positions. Furthermore, to minimize errors, exact string matching was avoided. This method prevented issues such as retrieving incorrect information that the CLM didn't identify.

4 Experimentation with Joinery Work Package

This section compares GPT-4 and Mistral 7B, employing zero-shot and 10-shot learning approaches to recognize BIM entities from BTS documents, following the template outlined in Fig. 6. Their performance was evaluated on a labeled test dataset using precision, recall, and F1-score metrics.

Additionally, the *camembert_base* model was fine-tuned on the training dataset with a learning rate of 2e-5, a batch size of 16, over 4 epochs, using the AdamW optimizer and a linear warmup schedule. Hyperparameter selection was based on empirical evidence and best practices for fine-tuning transformer models. The AdamW optimizer was chosen for its effectiveness in preventing weight overfitting, and a linear warmup schedule was used to stabilize early training dynamics. Validation against a pre-described dataset helped monitor performance and prevent overfitting. The model was tested on the same test dataset used for GPT-4 and Mistral 7B, ensuring consistent comparison. The distribution of entities in the training, validation, and testing datasets is shown in Fig. 5.

To tailor the *camembert_base* model for the construction domain, domainspecific pre-training was conducted on a corpus of approximately 500,000 construction related tokens. During fine-tuning, the training data included annotated examples of construction entities, enhancing the model's recognition capabilities. The BILOU tagging scheme was implemented to improve precision in recognizing entity boundaries, assigning specific prefixes to each token to denote its position. This approach improved the model's compatibility with machine learning models and facilitated detailed evaluation. Evaluation metrics included precision, recall, and F1-scores segmented by entity type, providing insights into the model's performance across different construction entity categories.

4.1 Evaluation Process

In evaluating language models' capability to identify and extract BIM entities, the study employed a method that transforms text tagging into a precise delineation of entities marked by their start and end positions within the text.

This method is designed to replicate precisely the entity categorization method used in the labeled dataset. Subsequently, the BILOU scheme is utilized to refine the calculation of essential metrics such as **P**recision (P), **R**ecall (R), and the **F1**-score (F1), which are standard in the field of information retrieval. Given the dataset includes multiple types of entities, the metrics are calculated as a weighted average to account for the variety and imbalance of entity classes. This weighting approach, based on the distribution insights from the labeled dataset, as illustrated in Fig. 5, ensures that the impact of each class on the overall metric is proportional to its size. This methodology provides a balanced and nuanced evaluation, allowing us to assess the models' effectiveness in entity extraction with high accuracy and fairness.

Table 2 presents comparative results of different language models on some selected entities. The highest P, R, and F1 are highlighted in bold. Table 3 showcases the extraction performance of all BIM concepts and properties extracted from BTS documents using zero and few-shot GPT-4, Mistral 7B, and fine-tuned Camembert_base, measured by weighted P, R, and F1.

4.2 Discussion

Table 3 shows MLMs perform better than CLMs in domain-specific NER, especially in fields like construction that require in-depth knowledge. CLMs, due to their generative nature, often include unnecessary contextual information, leading to a need for further post-processing and affecting **P**. Their difficulty is recognizing specific entities, as shown in Table 2, results in lower **R**. However, CLMs can identify certain entities with contextual clues, such as "*Acoustic insulation Rw + Ctr > = 30dB* ", aiding in the recognition of "Acoustic_Attenuation" even in zero-shot scenarios. Yet, not all BTS documents provide such context explicitly, posing challenges in property value extraction like "Rw + Ctr > = 30dB" without direct clues. Despite this, CLMs can recognize entities like "Dimensions" without contextualclues.

Table 3 emphasizes Few-shot Mistral 7B's advantage over Few-shot GPT-4, with Mistral 7B scoring higher in concept identification **(86% vs. 84%)** and property extraction **(71% vs. 63%)**. This highlights Mistral 7B's better performance in domain-specific

Task Descriptions:

You are an experienced building owner from France, with a keen ability to recognize various knowledge entities related to French building technical specifications. These entities encompass **'Dimension'**, **'Door'**, **'Door assembly'**, **'Window'**, **'Window sash'**, **'French door'**, **'Number of leaves'**, **'Acoustic attenuation'**, **'Fire resistance'**, **'Flame arrester'**, **'Thermal coefficient'**, **'Air permeability'**, **'Watertight'**, and **'Wind resistance'**. Your task involves Named Entity Recognition (NER). For each entity, a definition will be provided to aid in their identification and labeling. Your objectives are as follows:

1. Retain the original text format but label each entity mentioned by enclosing it with **'<Entity_Label>'** and **'</Entity_Label>'**, where Entity_Label is the identified label of the entity, selected based on the provided definition.
2. Within the text, instances of these entities will occur, sometimes without their explicit names. You are required to identify and extract all instances of these entities, relying on the provided definitions and the textual context. Examples of both input and the corresponding output are provided below for clarification.

Entity Definitions:

1. **Door assembly**: Or *'Bloc-porte'* in French, A complete unit comprising a door frame, hinges, handle, and any other necessary components for installation.
2. **Window**: Or *'Fenêtre'* in French, An opening in a wall or roof of a building, typically glazed and framed, to admit light or air.
3. **French Door** : Or *'Porte-fenêtre'* in French, A style of door with glass panes extending for most of its length.

...

n-1. **Acoustic attenuation**: Or *'Affaiblissement acoustique'* in French, The reduction in intensity of sound as it passes through a medium, typically measured in decibels (dB).

n. **Fire resistance**: Or *'Indice coupe-feu'* in French, The ability of a material or assembly to withstand fire or prevent its spread for a specified period.

Few-shot Demonstrations:

1. **Input** : "Fixed glazed window sash directly in its frame, 1.04m x 0.80m. Fixed glazed window sash directly in its frame, with a single pane. Extensions necessary to compensate for the thickness of the insulation. Minimum technical specifications: • Type of glazing: 4; • Thermal transmission: Uw < 1.40 W/m².K; • Dimensions: 1.04m x 0.80m."

   Output : "Fixed glazed *<Window sash>window sash</window_sash>* directly in its frame, *<Dimension>1.04m x 0.80m</Dimension>*. Fixed glazed *<Window_sash>window sash</window_sash>* directly in its frame, with a single pane. Extensions necessary to compensate for the thickness of the insulation. Minimum technical specifications: • Type of glazing: 4; • Thermal transmission: *<Thermic_coefficient>Uw < 1.40 W/m².K</Thermic_coefficient>*; • Dimensions: *<Dimension>1.04m x 0.80m</Dimension>*."

...

Task Execution:

1. The task involves identifying and extracting building technical specification entities from French text. Do not extract entities that are not included in the provided entities list.
2. Refrain from creating entities that are absent from the given text, and avoid altering the original text. Work with the text as it is provided.
3. The provided text may not contain any entities from the list given. If no entities exist, return the text without labeling any other entity.
4. Allocate sufficient time for analyzing and verifying all entities before finalizing them, ensuring that all specified entities have been accurately extracted.

Raw Requirements Test:

1. **Input** : "03.1.1 exterior joinery wood / lacquered aluminum\n\n\nSupply and installation of mixed joinery in wood/aluminum with exterior in solid wood of local species to\npaint inside and lacquered aluminum exterior:\n- Lacquered aluminum, color chosen by the client from a color chart\n- Frame in solid ash, redwood, or fir: non-exotic wood to be painted\n- Glazed parts in double glazing insulating: with burglary-resistant glass on the internal face and safety glass\nfor the spandrel parts\n- Insulating double glazing type "argon" climatplus 4 s protect, 33.2/16/44.2, UG=1.1W/m2°K provide\nCEKAL certificate -\nLaminated glass on the exterior\n- Thermal performance of the frame UF= 1.2W/m2°C refer to the study"

   Output : " "

...

Fig. 6. The components utilized to construct the prompt for extracting BIM entities from BTS documents using CLMs

Table 2. Comparative results of models on selected **Concepts** and **Properties**

Models	Concepts											
	Door_Assembly			Window_sash			Door			Joinery		
	P	R	F1	P	R	F1	P	R	F1	P	R	F1
Zero-shot-GPT4	0.39	0.60	0.47	0.27	0.70	0.39	0.20	0.37	0.26	0.25	0.47	0.32
10-shot-GPT4	0.74	**0.9**	0.83	**1.00**	0.83	0.90	0.86	0.89	0.87	**0.93**	0.69	0.79
zero-shot-Mistral 7B	0.17	0.85	0.28	0.22	0.62	0.33	0.06	0.47	0.10	0.10	0.17	0.12
10-shot-Mistral 7B	0.97	**0.95**	0.96	**1.00**	0.91	0.95	0.84	0.88	0.86	0.92	0.78	0.85
Camembert_base	**0.99**	0.93	**0.96**	**1.00**	**1.00**	**1.00**	**0.95**	**0.99**	**0.97**	0.91	**1.00**	**0.95**
Models	Properties											
	Number_Of_Leaves			Acoustic_Attenuation			Air_Permeability			Thermic_Coefficient		
	P	R	F1	P	R	F1	P	R	F1	P	R	F1
Zero-shot-GPT4	0.00	0.00	0.00	0.23	0.19	0.21	0.00	0.00	0.00	0.82	0.67	0.73
10-shot-GPT4	0.79	0.47	0.59	0.57	0.68	0.62	0.25	0.75	0.38	0.98	0.60	0.75
zero-shot-Mistral 7B	0.32	0.12	0.17	0.51	0.63	0.56	0.02	0.17	0.03	0.51	0.50	0.51
10-shot-Mistral 7B	0.52	0.48	0.50	0.81	0.81	0.81	0.12	0.20	0.15	0.99	**0.99**	**0.99**
Camembert_base	**0.97**	**0.98**	**0.98**	**0.98**	**0.98**	**0.98**	**1.00**	**0.86**	**0.92**	**1.00**	0.96	0.98

Table 3. Assessment of NER performance through a comparative analysis using weighted **P**, **R**, and **F1** metrics for all concepts and properties extraction

Model		Concepts			Properties	
	P	R	F1	P	R	F1
Zero-shot-GPT4	0.28	0.51	0.36	0.41	0.37	0.39
10-shot-GPT4	0.86	0.84	0.84	0.67	0.64	0.63
Zero-shot-Mistral 7B	0.11	0.59	0.18	0.52	0.39	0.44
10-shot-Mistral 7B	0.89	0.84	0.86	0.71	0.74	0.71
Camembert_base	**0.93**	**0.97**	**0.95**	**0.98**	**0.96**	**0.97**

tasks, particularly in identifying BIM entities from French BTS documents. Such findings underline the importance of careful model and training strategy selection to achieve the best NER results in specialized domains like BIM.

10-shot learning enhances GPT-4 and Mistral 7B's performance, showing their potential for adaptability. Specifically, 10-shot Mistral 7B achieves a **99% F1** for "Thermic_Coefficient" as shown in Table 2, but *Camembert_base* outperforms all, consistently achieving over **95% F1** in concept and property extraction, due to their cross-domain familiarity, as shown in Table 3.

5 Conclusion and Perspectives

This study automated the extraction of information from BTS documents critical to the French construction industry. It utilized LLMs, focusing on CLMs such as GPT-4 and Mistral 7B through prompt strategies and few-shot learning. This approach enabled CLMs to generate results without relying on labeled datasets. Nevertheless, it did not capture all necessary BIM entities. The research also explored MLMs by fine-tuning Camembert_base, which surpassed CLMs in domain-specific NER, achieving a 95% weighted F1 score in extracting BIM concepts and properties. The study identified limitations, including the partial effectiveness of CLMs' prompt strategies for domain-specific NER. The potential for enhancing CLMs through refined prompts and improved few-shot examples was noted. Future efforts will aim to refine these strategies and investigate the fine-tuning of CLMs with labeled datasets to improve performance in specialized NER tasks. This is intended to contribute to the digital transformation in the construction industry by facilitating advancements in building lifecycle management.

References

1. Doumbouya, L., Gao, G., Guan, C.: Adoption of the Building Information Modeling (BIM) for construction project effectiveness: the review of BIM benefits. Am. J. Civil Eng. Arch. **4**(3), 74–79 (2016)
2. Ahn, D.,Cha, H.: Integration of building maintenance data in application of Building Information Modeling (BIM). J. Build. Constr. Plan. Res. **2014** (2014)
3. Soibelman, L., Wu, J., Caldas, C., Brilakis, I., Lin, K.Y.: Management and analysis of unstructured construction data types. Adv. Eng. Inf. **22**(1), 15–27 (2008)
4. Wu, L.T., Lin, J.R., Leng, S., Li, J.L., Hu, Z.Z.: Rule-based information extraction for mechanical-electrical-plumbing-specific semantic web. Autom. Constr. **135**, 104108 (2022)
5. Min, B., et al.: Recent advances in natural language processing via large pretrained language models: a survey. ACM Comput. Surv. **56**(2), 1–40 (2023)
6. Zhu, E., Sheng, Q., Yang, H., Liu, Y., Cai, T., Li, J.: A unified framework of medical information annotation and extraction for Chinese clinical text. Artif. Intell. Med. **142**, 102573 (2023)
7. Ahaggach, H., Abrouk, L., Lebon, E.: Information extraction and ontology population using car insurance reports. In International Conference on Information Technology-New Generations, pp. 405–411. Springer, Cham (2023)
8. Nguyen, D.Q., et al.: ChEMU: named entity recognition and event extraction of chemical reactions from patents. In Advances in Information Retrieval: 42nd European Conference on IR Research, ECIR 2020, Lisbon, Portugal, 14–17 April 2020, Proceedings, Part II 42, pp. 572–579. Springer, Heidelberg (2020)
9. Kluegl, P., Toepfer, M., Beck, P.D., Fette, G., Puppe, F.: UIMA Ruta: rapid development of rule-based information extraction applications. Nat. Lang. Eng. **22**(1), 1–40 (2016)
10. Radford, A., Narasimhan, K., Salimans, T., Sutskever, I.: Improving language understanding by generative pre-training (2018)
11. Hobbs, J.R., Riloff, E.: Information Extraction. Handb. Nat. Lang. Process. **15**, 16 (2010)
12. Rahma, F., Romadhony, A.: Rule-based crime information extraction on Indonesian digital news. In: 2021 International Conference on Data Science and Its Applications (ICoDSA), pp. 10–15. IEEE (2021)

13. Savova, G.K., Fan, J., Ye, Z., Murphy, S.P., Zheng, J., Chute, C.G., Kullo, I.J.: Discovering peripheral arterial disease cases from radiology notes using natural language processing. In AMIA Annual Symposium Proceedings, vol. 2010, p. 722. American Medical Informatics Association (2010)
14. Akanbi, T., Zhang, J.: Design information extraction from construction specifications to support cost estimation. Autom. Constr. **131**, 103835 (2021)
15. Morwal, S., Jahan, N., Chopra, D.: Named entity recognition using hidden Markov model (HMM). Int. J. Nat. Lang. Comput. (IJNLC) **1** (2012)
16. Shi, L., Jianping, C., Jie, X.: Prospecting information extraction by text mining based on convolutional neural networks–a case study of the Lala copper deposit, China. IEEE Access **6**, 52286–52297 (2018)
17. Zheng, Z., Lu, X.Z., Chen, K.Y., Zhou, Y.C., Lin, J.R.: Pretrained domain-specific language model for natural language processing tasks in the AEC domain. Comput. Ind. **142**, 103733 (2022)
18. Devlin, J., Chang, M.W., Lee, K., Toutanova, K.: Bert: pre-training of deep bidirectional transformers for language understanding. arXiv preprint arXiv:1810.04805 (2018)
19. Schönfelder, P., König, M.: Deep learning-based entity recognition in construction regulatory documents. In ISARC. Proceedings of the International Symposium on Automation and Robotics in Construction, vol. 38, pp. 387–394. IAARC Publications (2021)
20. Achiam, J., et al.: Gpt-4 technical report. arXiv preprint arXiv:2303.08774 (2023)
21. Min, S., Lewis, M., Zettlemoyer, L., Hajishirzi, H.: Metaicl: learning to learn in context. arXiv preprint arXiv:2110.15943 (2021)
22. Zhang, J., et al.: 2iner: instructive and in-context learning on few-shot named entity recognition. In Findings of the Association for Computational Linguistics: EMNLP 2023, pp. 3940–3951 (2023)
23. Saka, A., et al.: GPT models in construction industry: opportunities, limitations, and a use case validation. Dev. Built Environ., 100300 (2023)
24. Moon, S., Chi, S., Im, S.B.: Automated detection of contractual risk clauses from construction specifications using bidirectional encoder representations from transformers (BERT). Autom. Constr. **142**, 104465 (2022)
25. Smetana, M., Salles de Salles, L., Sukharev, I., Khazanovich, L.: Highway Construction Safety Analysis Using Large Language Models. appl. Sci. **14**(4), 1352 (2024)
26. Naguib, M., Tannier, X., Névéol, A.: Few shot clinical entity recognition in three languages: masked language models outperform LLM prompting. arXiv preprint arXiv:2402.12801 (2024)
27. Jiang, A.Q., et al.: Mistral 7B. arXiv preprint arXiv:2310.06825 (2023)
28. Martin, L., et al.: CamemBERT: a tasty French language model. arXiv preprint arXiv:1911.03894 (2019)
29. Nakayama, H., Kubo, T., Kamura, J., Taniguchi, Y., Liang, X.: doccano: Text Annotation Tool for Human (2018). https://github.com/doccano/doccano
30. plan bim 2022 - Le dictionnaire national de propriétés POBIM | Plan BIM 2022'. Accessed 25 Nov 2022. https://plan-bim-2022.fr/actions/ptnb-axe-c-le-dictionnaire-national-de-proprietes-pobim/le-dictionnaire-national-de-proprietes-pobim/
31. Didier balaguer. Le CEN, pour une norme européenne en matière d'interopérabilité', datBIM. Accessed 12 Sept 2023. https://www.datbim.com/2017/03/14/le-cen-pour-une-norme-europeenne-en-matiere-dinteroperabilite/
32. Delcambre, B.: Mission numérique bâtiment. Repéré sur le site du Ministère du logement et de l'égalité des territoires et de la ruralité (2014)

ESG Digital Twin

Thepparit Sinthamrongruk[1(✉)] and Napat Harnpornchai[2]

[1] College of Arts, Media and Technology, Chiang Mai University, Chiang Mai, Thailand
thepparit.s@cmu.ac.th

[2] Faculty of Economics, Chiang Mai University, Chiang Mai University, Chiang Mai, Thailand
napateconcmu@gmail.com

Abstract. This research proposes a digital twin system based on ESG principles to improve organizational performance in planning and support services. The system integrates environmental, social, and governance factors for real-time data collection on CO2 emissions and workplace safety. It offers efficient and accurate ESG performance measurement with seamless visualization across various devices. The implemented system demonstrates reliable data transmission with minimal packet loss. The successful application within a controlled environment highlights its potential for real-world use in smart cities. Future research could explore broader integration and user experience in large-scale deployments. This paves the way for utilizing digital twin technology for sustainable and socially responsible urban development.

Keywords: Digital Twins · ESG · Sensor

1 Introduction

Thailand is highly committed to developing smart cities and establishing a Smart City Office under government supervision [1] to drive a comprehensive shift towards a digital city system. ESG (Environmental, Social, and Governance) is critical to this development. Adopting environmentally, socially, and responsibly managed systems as a part of the Digital Government initiative has become a pivotal step toward achieving these goals. These systems, designed to ensure the city's development is sustainable, socially inclusive, and governed by ethical principles, directly apply ESG principles in the Smart City transformation process.

ESG has emerged as a critical issue for modern organizations, serving as a comprehensive framework for assessing a company's commitment to sustainability [2] by evaluating its performance across three key dimensions:

- **Environmental:** This dimension focuses on a company's impact on the natural environment. It encompasses practices that minimize environmental harm, such as reducing air, water, and land pollution through efficient processes and cleaner technologies, and resource conservation, conserving precious resources like water, energy, and raw materials to promote environmental sustainability.

Published by Springer Nature Switzerland AG 2025
P. Sureephong et al. (Eds.): PLM 2024, IFIP AICT 740, pp. 453–464, 2025.
https://doi.org/10.1007/978-3-031-93319-6_34

- **Social:** This aspect emphasizes a company's social responsibility and positive societal impact. It highlights practices that promote fair labor practices, ensure fair wages, provide safe working conditions, and ensure adherence to labor rights for all employees. It also fosters diversity and Inclusion, creating an inclusive workplace that embraces diversity, equity, and equal opportunities.
- **Governance:** This pillar focuses on a company's corporate governance practices and ethical decision-making. It emphasizes transparency and accountability, maintaining transparency in financial reporting, corporate activities, and decision-making processes.

The digital twin, a potent tool of digital transformation techniques, can revolutionize operations [3, 4]. It simulates real-world scenarios by installing sensors and feeding data into a model, often in a three-dimensional format, to visualize, analyze, and significantly enhance operations, making them more efficient. This can be seen in various industries, from simulating the design process of a car to modeling logistics systems. In ESG integration, the digital twin can monitor and optimize the performance of the 3Ps: people, places, processes, and ESG. For instance, it can help measure and report CO2 emissions from operations, create suitable working environments, and assist organizations in achieving their ESG goals.

Integrating ESG principles into the digital twin system practices presents a significant and multifaceted challenge. However, it also holds immense potential for positive impact. This potential can lead to a more sustainable and inclusive urban environment when harnessed effectively. To successfully address this challenge, it is crucial to consider the following key aspects: 1. Holistic system framework, 2. Data-driven decision- Making, 3. Stakeholder engagement, and 4. Cost-effective solutions [5].

Therefore, this paper presents the development of a digital twin system based on ESG principles that focuses on enhancing the efficiency and delivery of planning and support services to optimize organizational performance. This system differs from previous technologies in the following ways:

- Enables measuring ESG performance efficiently and accurately: digital twin technology captures insights from physical systems or processes in real-time, dimension by dimension. Focus on reporting CO2 emissions from operations.
- The developed digital twin technology can report suitable working environments to create safety and promote employee workplace standards.

The following section will explain the literature review, the proposed system's details and structure, the performance evaluation results, and the conclusion.

2 Digital Twin and ESG

Digital twin technology is one of the key components of a smart city, which utilizes IoT systems as the foundation for connecting various devices into a network to facilitate ease of use and command [3]. Digital twin focuses on creating cities with sustainable development [4], including energy conservation, using clean energy, addressing environmental issues, environmental balance, and transportation efficiency [6].

The real-world impact of digital twins is evident in cities like Seattle, Amsterdam, Barcelona, and Stockholm [7]. This technology bridges the natural and virtual worlds, incorporating mixed reality (MR) technology to visualize simulated environments. It presents images, videos, audio data, and other information [8] on various devices, from computers and mobile phones to tablets and laptops [9].

The concept of digital twins, first introduced in 2005, aimed to create realistic models in the digital world [10] that could seamlessly connect with the real world [11]. As digital twins gained recognition, they were increasingly incorporated into research and experimentation [12]. Today, digital twins are widely used in the foundational aspects of various products related to technologies such as IoT, AI, network edge, and Cloud [13], which are crucial in addressing environmental sustainability challenges.

The digital twin also manages the environment and various objects within homes or buildings. This includes monitoring weather conditions and inspecting usage statistics of objects inside a particular room. This is achieved through collaboration with multiple sensors, such as vibration, light, and noise [14], in real-time and non-real-timeformats [15].

In the context of ESG, it's important to highlight that greenhouse gas emissions, including CO2 and methane (CH4), are critical contributors to global warming, which drives climate change. These emissions profoundly impact the environment, leading to various issues such as rising sea levels, extreme weather events like hurricanes and heat waves, and ecosystem degradation. The 2017 Global Status Report [16] underscores that the built environment, encompassing buildings, homes, and infrastructure, is a significant source responsible for over 40% of global greenhouse gas emissions.

In the United Kingdom, housing is reported to be the largest energy consumer for heating and hot water, accounting for 25% of total energy use [17]. This consumption also contributes significantly to CO2 emissions and their detrimental impact on climate change. Consequently, the UK has set an ambitious target of achieving net-zerogreenhouse gas emissions by 2035 [18].

Building Information Modeling (BIM) and digital twins offer promising solutions to address these challenges by providing real-time insights into space utilization and indoor environmental quality. Factors such as materials, appliances, cooking activities, other occupant activities, and inadequate ventilation can significantly impact indoor air quality and, consequently, human health, even more so than outdoor air [19]. Therefore, continuous monitoring of Total Volatile Organic Compounds (TVOCs) and equivalent carbon dioxide (eCO2) levels across different spaces is crucial. This data proves invaluable in identifying issues, tracking ventilation system performance [20], and alerting occupants to their behavior, which can be linked to energy consumption patterns. This holistic approach is critical to achieving net-zero buildings, designed to produce as much energy as they consume, thereby minimizing their carbon footprint and contributing to environmental sustainability [21].

Despite significant research efforts in recent years to utilize digital twins for real-time building performance monitoring and optimization, paving the way for practical implementation among industry professionals, the development of digital twin systems remains in its early stages.

This research presents a pioneering application of digital twins for real-time monitoring indoor pollutant emissions from occupied assets. The study focuses on integrating IoT data, processing it through a digital twin Engine, and visualizing the results on a single platform. The user-friendly web-based interface enables real-time indoor air quality monitoring and automated analysis of historical time-series data. Guided by the ESG principle, the system aims to forecast future emission trends.

The following section will present the details of the proposed digital twin technology, which consists of three main components: 1) ESG IoT system, 2) Server and Digital Twin engine, and 3) Real-time visualization.

3 The Proposed Framework of ESG Digital Twin

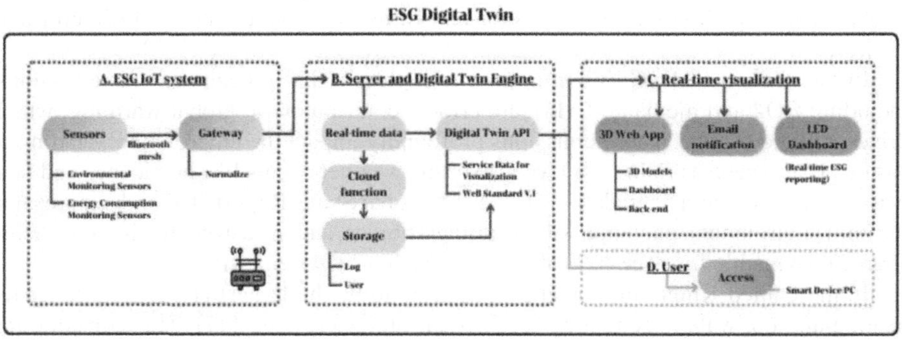

Fig. 1. The proposed framework of ESG Digital Twin.

The system consists of three main components: ESG IoT System, Server and Digital Twin Engine, and Real-time visualization system, as illustrated in Fig. 1. The ESG IoT System is a comprehensive network of sensors that monitor the environment and electricity consumption. These sensors diligently collect data and transmit it via a Bluetooth Mesh network to the Gateway, the central hub of our system. The Gateway then formats and dispatches the data to the Server and Digital Twin Engine.

Upon receiving real-time data, the server and digital twin engine store and process a portion based on environmental quality. It also converts electricity consumption values into carbon dioxide emissions. It reformats the data for data services to the Real-time visualization system, which consists of a 3D web application for easy user access through web browsers. This email alert system notifies users when environmental quality poses a risk to occupants (e.g., PM2.5 levels exceed the threshold). An LED display in the hallway presents ESG performance reports to the public.

3.1 ESG IoT System

The system comprises 26 sensors that cover the laboratory and office spaces, as shown in Fig. 2. These sensors collect ESG data and then transmit the data via a Bluetooth Mesh network to a gateway. The data is then normalized and packaged into small packets before being sent to the server via 4G mobile signals to the digital twin engine. The goal is to achieve the fastest possible data transmission speed and the smallest possible data size for rapid rendering of 3D models. Overall, the sensor system can be divided into two main subsystems:

Fig. 2. ESG IoT system

Subsystem 1: Environmental Monitoring Sensors
This subsystem consists of 12 sensors strategically placed throughout the laboratory and office spaces to monitor various environmental parameters, including temperature, humidity, light levels, noise levels, TVOC (total volatile organic compounds), and CO2. These sensors collect real-time data and transmit it to the gateway via the Bluetooth Mesh, as illustrated in Fig. 3.

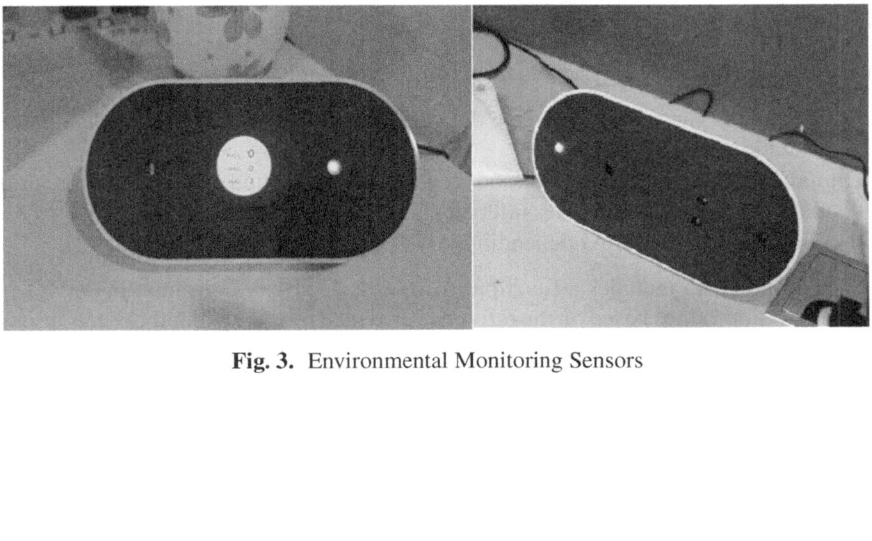

Fig. 3. Environmental Monitoring Sensors

Subsystem 2: Energy Consumption Monitoring Sensors

This subsystem has 14 sensors installed on air conditioners and other energy-consuming devices. These sensors monitor the real-time energy consumption of these devices, as depicted in Figs. 4 and 5. The gateway receives data from both subsystems and performs the following tasks:

- Data Normalization: The gateway normalizes the data from different sensors to ensure consistency and comparability.
- Data Packaging: The gateway packages the normalized data into small packets for efficient transmission.
- Data Transmission: The gateway transmits the data packets to the server via 4G mobile signals.

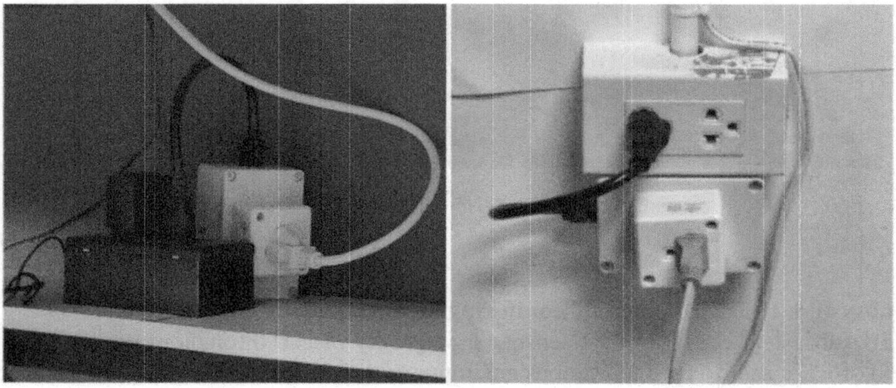

Fig. 4. Energy Consumption Monitoring Sensors (Plug)

3.2 Server and Digital Twin Engine

The server receives the data packets from the gateway and performs the following tasks:

- Data Storage: The server stores the data in a secure database for future analysis and visualization, as shown in Fig. 6.
- Digital Twin engine: Process data and create JSON-formatted data services for various displays. This part also involves inferring the facility engineering knowledge to create the Engine's rule-based AI depending on WELL Building Standard v1 [22].

The system also includes a function to convert electricity consumption values into carbon emission values based on the reference provided by the Thailand Greenhouse Gas Management Organization (Public Organization) or TGO, where 1 kWh is equivalent to 0.5986 kgCO2e.

Fig. 5. Energy Consumption Monitoring Sensors (Air Conditioner)

Fig. 6. Real-timeserver

3.3 Real-Time Visualization

This part is developed to display cross-platform visualizations and provide error alerts based on the engine's sensor detections and conditions the facility engineer sets. This real-time visualization system uses entirely open-source tools for three-dimensional representation, including TypeScript and the Three.js library. It has been tested on iOS, Android, and Windows operating systems.

4 Experimental Results

4.1 IoT Performance

In this research, we have developed code to check the performance measurement values of data transmission in an IoT system by checking the "callback" data to the source device. We use the following values:

Throughput

Throughput measures the amount of data transmitted per unit of time. Measuring Throughput is essential to evaluating system performance because throughput indicates how much data the system can handle. To measure throughput, we first define the type and size of the data. We then test by sending data from the source to the destination, using the timing from the start of data transmission to the completion of all data reception. Finally, we calculate the throughput value.

$$\text{Throughput} = \text{Data size} / \text{Timeperiod} \tag{1}$$

Latency

Latency measures the time taken from the start of data transmission to the point where the data is ultimately received. To find latency, we start sending data, record the time, and then wait for the data response.

$$\text{Latency} = \text{Responding time} - \text{Starttime} \tag{2}$$

Packet Loss

Packet loss is the process of determining the amount of data that has not been delivered to the destination. It can be expressed as a percentage of lost packets to the total number of packets sent. Measuring packet loss indicates a system's reliability in terms of data transmission accuracy.

To test packet loss, we define the type and number of packets to be tested. Then, we send all the data, count the number of packets received at the destination, and calculate the number of packets lost.

$$\text{Packet Loss} = \text{Total packet} - \text{Receivedpacket} \tag{3}$$

$$\text{Packet Loss } (\%) = (\text{Packet Loss} / \text{Total packet}) \times 100 \tag{4}$$

Table 1 presents the estimated performance of the ESG IoT system, providing precise parameters such as throughput, latency, and packet loss.

For the throughput value, the system's efficiency is calculated as the total amount of data (Data size) from the gateway sent out to the server according to Eq. (1). The gateway efficiently sends out a total of 708 Bytes per time, which is derived from 12 environmental sensors; each sensor uses 24 Bytes of data (for a total of 288 Bytes)

Table 1. The performance evaluation table for ESG IoT systems with throughput, latency, and packet loss

Metric	*Desired Range*
Throughput	0.000708 Mbps or 708 Bytes per second
Latency	78 ms (ms)
Packet Loss	0%

combined with 14 energy consumption monitoring sensors, each using 30 Bytes of data (for a total of 420 Bytes).

The latency value is calculated from the Bluetooth Mesh specification used to transmit data and the latency value between the gateway and the server; it takes an average time of 78 ms per data transmission according to Eq. (2) from 216 sample packages. Regarding Packet Loss, the experiment found an error of 0% calculated according to Eqs. (3) and (4) in the testing area.

For the system reliability testing, the research team established a controlled testing environment within the laboratory and office spaces of the main building of the College of Arts, Media and Technology, Chiang Mai University. They combined an area of 70 square meters inside an office building at Chiang Mai University, and the testing period spanned nine months (240 days), operating 24 h a day, seven days a week (24/7), to monitor the effects of changes and evaluate the system's reliability. Throughout the testing period, the system functioned seamlessly, with only a single instance of a sensor reporting an error due to a power outage caused by a concert event in front of the building.

4.2 Visualization Performance

For evaluating the data retrieval speed and frame rate for display, the system was tested using cross-platform display devices, including:

- **Windows Operating System:** PC with Intel® Core™ i7-12700F 4.90 GHz processor, 64 GB RAM, and an Nvidia RTX4060ti graphics card with 8 GB of memory.
- **OSX**: MacBook M1 8-core CPU and 7-core GPU
- **iOS System**: iPad Gen 9 (2021) with A13 Bionic chip.
- **Android System:** Samsung Galaxy Tab S9 FE with Exynos 1380 and 6 GB RAM.

The developed digital twin was tested for display on popular web browsers, including Chrome, Firefox, and Safari. It involved rendering a large 3D model with over 90,000 polygons and displaying a user interface, as shown in Fig. 7. A live demo of the visualization and functionality can be viewed at this link: https://cmu.to/esgdigitalt-win.

Table 2 presents the frame rate measurements obtained using an embedded library within the program to evaluate the frame rate across various devices, operating systems, and browsers. It is evident that for display on personal computers, the presented digital twin can achieve an average frame rate of 123.67 frames per second (FPS) in real time. The data access speed is breakneck, requiring only 0.01 s to retrieve data from the digital twin engine. The OSX system's average frame rate is 49.17 FPS, with the same data access speed of 0.01 s as on PC systems.

Regarding display on smart devices, the system can deliver an average frame rate of 58.5 FPS on Android devices, with an access time of 0.15 s. For iOS devices, the system operates with an average frame rate of 48.83 FPS and an access time of 0.14 s. Additionally, Chrome on Android exhibits a minimum FPS that is twice as high as iOS. iOS gradually increases the frame rate to maintain stability and conserve device power.

Table 2. The performance evaluation table for ESG IoT systems with throughput, latency, and packet loss

Personal Computer (PC)			Smart Devices		Overall
FPS	Windows OS	OSX	Android	iOS	
*Maximum**	144	60	60	60	88
*Minimum**	103.33	38.33	57	37.67	68.11
*Average**	123.67	49.17	58.5	48.83	**73.89**
*Access time***	0.01	0.01	0.15	0.14	0.06

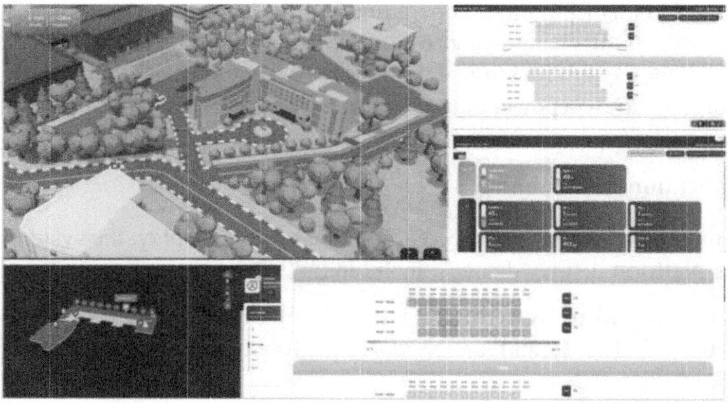

Fig. 7. Web-based visualization

In addition, the system can also display real-time ESG results for carbon dioxide emissions from electricity consumption in work activities and workplace environments, as shown in Fig. 8.

Fig. 8. ESG report on LEDdashboard

5 Conclusion

This research successfully developed a digital twin system based on ESG principles to optimize organizational performance in planning and support services. The system integrates environmental, social, and governance factors into its operations, offering several advantages over previous technologies.

Here are the key findings:

- **Efficient and Accurate ESG Performance Measurement:** The digital twin technology captures real-time data on various aspects, allowing for accurate measurement of ESG performance. This includes factors like CO2 emissions and workplace safety.
- **Improved Visualization:** The system offers seamless visualization across various devices and browsers, with a high frame rate and fast data retrieval speed. Users can access real-time data on comfort, energy usage, and air quality.
- **Reliable Data Transmission:** The system demonstrates reliable data transmission with zero packet loss (0%). Throughput and latency values, which fall within the desired 78 ms, ensure efficient data transfer.

The system's potential for real-world application in similar settings is highlighted by successful implementation within a controlled environment, such as a pilot city or a specific organization.

Future research could explore integrating this system with existing infrastructure and its impact on broader urban environments. Additionally, further studies could investigate user experience and data security aspects in a large-scale deployment scenario.

Overall, this research paves the way for utilizing digital twin technology to enhance sustainability and social responsibility within intelligent cities. By integrating ESG principles, organizations can achieve a more efficient and responsible approach to urban development.

References

1. Digital Economy Promotion Agency (depa) Thailand: Smart city office [Website] (2024). https://www.depa.or.th/th/smart-city-plan/smart-city-office.

2. Ioannou, I., Serafeim, G.: The consequences of mandatory corporate sustainability reporting. Harvard Business School Research Working Paper. pp. 11–100 (2017)
3. Barricelli, B.R., Casiraghi, E., Fogli, D.: A survey on digital twin: definitions, characteristics, applications, and design implications. IEEE Access **7**, 167653–167671 (2019)
4. Bonab, A.B., Fedele, M., Formisano, V., Rudko, I.: In complexity we trust: a systematic literature review of urban quantum technologies. Technol. Forecast. Soc. Chang. **194**, 122642 (2023)
5. Piras, G., Agostinelli, S., Muzi, F.: Digital twin framework for built environment: a review of key enablers. Energies **17**(2), 436 (2024). https://doi.org/10.3390/en17020436
6. Bibri, S.E.: The IoT for smart sustainable cities of the future: an analytical frame- work for sensor-based big data applications for environmental sustainability. Sustain. Cities Soc. **38**, 230–253 (2018)
7. Chen, J., Gan, W., Hu, M., Chen, C.M.: On the construction of a post-quantum blockchain for smart city. J. Inf. Secur. Appl. **58**, 102780 (2021)
8. Kritzinger, W., Karner, M., Traar, G., Henjes, J., Sihn, W.: Digital twin in manufacturing: a categorical literature review and classification. IFAC-PapersOnline **51**(11), 1016–1022 (2018)
9. Fonseca, Í.A., Gaspar, H.M.: Challenges when creating a cohesive digital twin ship: a data modelling perspective. Ship Technol. Res. **68**(2), 70–83 (2021)
10. Grieves, M.W.: Product lifecycle management: the new paradigm for enterprises. Int. J. Prod. Dev. **2**(1/2), 71 (2005). https://doi.org/10.1504/IJPD.2005.006669
11. Zhou, M., Yan, J., Feng, D.: Digital twin framework and its application to power grid online analysis. CSEE J. Power Energy Syst. **5**(3), 391–398s (2019)
12. Tao, F., Qi, Q.: Make more digital twins. Nature **573**(7775), 490–491 (2019)
13. Zhang, M., et al.: Digital twin data: methods and key technologies. Digital Twin **1**, 2 (2022)
14. Ma, X., Tao, F., Zhang, M., Wang, T., Zuo, Y.: Digital twin enhanced human- machine interaction in product lifecycle. Procedia Cirp **83**, 789–793 (2019)
15. Qi, Q., et al.: Enabling technologies and tools for digital twin. J. Manuf. Syst. **58**, 3–21 (2021)
16. Abergel, T., Dean, B., Dulac, J.: Towards a zero-emission, efficient, and resilient buildings and construction sector: global status report 2017. UN Environment and International Energy Agency, Paris, France, p. 22 (2017)
17. BEIS: UK Greenhouse gas emissions, final figure. In: E.I.S. Department for Business (Ed.) (2022)
18. HM Government: Net Zero Strategy: Build Back Greener (2021)
19. Lee, S.C., Li, W.M., Ao, C.H.: Investigation of indoor air quality at residential homes in Hong Kong—case study. Atmos. Environ. **36**(2), 225–237 (2002)
20. Fisk, W.J., De Almeida, A.T.: Sensor-based demand-controlled ventilation: a review. Energy Build. **29**(1), 35–45 (1998)
21. Zhao, L., Zhang, H., Wang, Q., Wang, H.: Digital-twin-based evaluation of nearly zero-energy building for existing buildings based on scan-to-BIM. Adv. Civ. Eng. **2021**(1), 6638897 (2021)
22. The building standard - storyblok. (n.d.). https://a.storyblok.com/f/52232/x/2453e8308e/well-v1-with-q2-2020-addenda.pdf

Managing Organizational Transformation Within PLM Approach Between Theory and Practice

Cyril Amergé[1,2], Farouk Belkadi[2](✉), and Michel Paillet[1]

[1] Cognitive Companions, Consulting, Paris, France
{cyril.amerge,michel.paillet}@cognitive-companions.com
[2] Nantes Université, École Centrale Nantes, LS2N Laboratory, UMR 6004, Nantes, France
{cyril.amerge,farouk.belkadi}@ls2n.fr

Abstract. Organizational and digital transformation projects are highly challenging and the current practices against these challenges are still limited in terms of efficiency and effectiveness. This is leading to the need of innovative decision supports to help managers conducting such strategies in optimal way. As a first step towards this perspective, this paper summarizes the results of an initial exploratory phase, involving a study of the theoretical foundations and current consulting practices of organizational transformation with PLM approach, in order to identify current limitations.

Keywords: Organizational transformation · co-evolution · PLM approach · consulting practices

1 Introduction

Companies are evolving in an increasingly competitive market, constrained by a variety of normative requirements and rapid changes of the socio-economic environment. To deal with these evolutions, today's companies are constantly restructuring their internal processes and upgrading their material and human resources for better resilience and agility. Proposing new product and/or service offers implies new organizational transformations among the entire value, but also the global supply chain and the customer relationships.

For some complex projects, companies need to build dynamic and contextual collaborations with various partners, at different levels of integration. New transactional synchronization activities are required, leading also to some modifications on the internal processes of each company. The management of such variety of processes is more complex in case of dynamic partnership, where the OEM wants to maintain strong relationships on specific projects while keeping less degree of partners' integration in other projects. The success of a contractual relationship requires the best compromise between all stakeholder's requirements and optimal ways to obtain these compromises. Conventional methods for supporting the establishment of such relationships tend to fix

Published by Springer Nature Switzerland AG 2025
P. Sureephong et al. (Eds.): PLM 2024, IFIP AICT 740, pp. 465–479, 2025.
https://doi.org/10.1007/978-3-031-93319-6_35

the definition of the integration strategy at the early stages of project development. As a result, the partners' variability and degrees of freedom become limited, depending on the level of mutual integration.

On the other hand, the need to structure and synchronize processes is coupled with the need to share information and knowledge via the digital chain, which is composed of various interconnected software and information systems. With the Industry 4.0 paradigm, the concept of the digital chain is extended to include various technologies such as IoT, cloud, digital twins and other augmented/virtual reality tools, etc.

In addition, as described by some consultants, the observation of industrial practices to support the organizational transformation shows that these actions seem to be opportunistic and not already connected to a structured methodology or standard referential. At the operational level, this weakness leads to serious difficulties in the articulation between major managerial approaches, such as program management, systems engineering, product lifecycle management, configuration management, etc.

According to the above assumptions, the realization of any digital transformation project is complex to achieve and requires a preliminary mapping of processes and resources. So, the strategy of organizational transformation has a direct impact on the digital transformation. Inversely, the organizational transformation is highly constrained by the actual capacities of the human and digital capabilities in the company. It is then important to think "co-transformation" strategy rather than individual transformations to ensure the consistency of the whole value chain and its related digital chain.

This paper is part of an ambitious research work between Centrale Nantes and Cognitive Companion (consulting company specialized on organizational and digital transformations). This research work aims to develop new management-aid approach that takes benefits from existing standards, methods and practices for supporting the emergent needs of "co-evolution" strategies, while reaching organizational agility and operational performance.

This paper focuses on the exploration stage and proposes a synthesis of literature survey to explain major reasons and contexts that lead to the need of organizational. In second time, an overview of important methods and tools that should be considered as a support of such transformations is given. Finally, two examples of consulting practices to support organizational transformation in industry and bank domains are presented to understand the gap. Finally, the conclusion will set up the future steps to build this Framework.

2 Why Organizational Transformations Are Needed?

Undertaking new organizational transformation is a risky and costly decision that should be considered with high precaution because it can lead to a destabilization of the current behaviour of the company. In consequence, delays and over costs can negatively affect the competitiveness of the company. Furthermore, resistance to change can appear internally and increases the probability of project failure.

So, it is important to analyse the balance benefit/lost and strength/weakness before moving towards such challenging projects. In this section, a study of main beneficial strategies that requests a strong organizational transformation.

2.1 Shifting Toward New Business Models

Companies are constantly facing market competitiveness, forcing them to search for new business opportunities and new ways to create added value for existing or new clients. For instance, the core idea of Product-Service System (PSS) is to create new value by providing an integrated system of product and service elements with a total care throughout their lifecycle to meet the customers' requirements while reducing resource consumption and environmental impacts [1]. PSS is not a simple addition of separate product and service components but a combination of complex processes for the definition, the assembly and the integration of mechanical, cyber-physical and IoT components, but also organizational and digital infrastructures [2]. Three main characteristics of PSS can justify the need of major changes on the internal structure of companies who intend to implement a PSS-oriented strategy in existing or new markets. Firstly, PSS value is achieved through long-term relationship with the client along the whole PSS lifecycle. Indeed, in difference with classical business model where the client is fully owner of the product after finalizing the transaction, company in case of PSS will still be owner of the system during all its life duration. It is responsible of to maintain correct conditions of working of the PSS, to update it when new technologies or standards are coming, but also to manage its end of life. New skills, roles and processes are needed to support that business model evolution and managerial rules has to be defined to go with these new organizational capabilities (i.e. in case of bike sharing PSS, it is needed to have transportation facilities with drivers to ensure the right filling balance all stations).

Secondly, the goal of PSS is to increase the usage experience of the client by adding complementary offers. This is made possible thanks to digital infrastructures that help analysing and clustering the profile of the clients and their usage experience to provide the right service (i.e. sensors embedded in machines help analysing the behaviour of the production system and predicting potential failures) [3]. Digital skills and resources are needed as part of a global digital transformation to perform intelligent data processing operations for various decision aid perspectives.

Thirdly, the PSS development process is highly collaborative because of the involvement of heterogeneous stakeholders with complementary skills and resources.

Even if companies are intending to take great benefits and competitive advantage from new PSS offers, they are not ready or able to extend their core competencies to cover all the fields requested by such development projects [4]. A concrete PSS example is the smart shoes to detect domestic and worksite accidents, as well as worker fatigue. In this case, the manufacturer specialized in creation and production of shoes calls upon other firms specialized in electronics and sensing systems to install the right devices. This collaboration effort is also needed during the exploitation and updating phases to conduct the supplying activities. The problem of organizational transformation is extended to the selection of best partners and the assessment of collaboration performance between potential partners [5].

2.2 Frugality and Regionalization of Markets

Frugal innovation is another strategy to penetrate new market or market segment. This theory was introduced to propose new business management directions, supporting the evolution of strategies in industrial companies when proposing new products dedicated to regional markets [6]. The outcomes of a frugal process are new products and services that "seek to minimize the use of resources in the complete value chain with the objective of reducing the cost of ownership while fulfilling or even exceeding certain pre-defined criteria of acceptable quality standards" [7]. The acronym "Frugal" is an aggregation of six attributes: Functional, robust, user-friendly, growing, affordable and local [8].

Following the affordable attribute, frugal innovation is not a low-cost approach where the target is solely cost minimization, even by reducing the functionalities of the product or by choosing very cheaper components. The key issue is rather to provide products and services that correctly meet the exact needs of the customer with acceptable cost regarding to the economic context of the target market in the target regional market. Similarly, local attribute suggests considering better the existing capabilities in the target market as a solution to optimize the production performance.

Focusing on technological advances as the unique source of innovation and developing products from scratch is not the best way to reach high performance and frugality. Reusing and adapting existing solutions, which have often proved their efficiency in previous projects, is a key direction to increase time and money saving through intelligent knowledge exploitation [9].

As a complement to the above strategy, it is important to align the development and production processes with mass customization, full customization and on demand perspectives. This way of doing aims at giving the clients the possibility to customize their orders according to their preferences and financial capacities [10]. The final product will match as close as possible the customer needs through alternatives of architectures that combine existing components and related production solutions.

Also, Adapting current production systems with respect to the target market capabilities, supply chain constraints and policy barriers is another complementary option. This becomes highly important when the company has to propose large variety of product versions for the same market. In this case, high flexibility is required from the production network.

Thus, frugality and mass/full customization are leading to new organizational transformations in order to update the internal processes and to build the supplier network for the realization and the management of variety product of alternatives and service offers. To ensure a good behaviour of the supplier network, it is necessary to align the information systems of different actors of this network. This conducts to a need of digital transformation as a complement of the organizational transformation.

2.3 Ramp Up and Resilience

In difference with the two previous strategies, which are generally predefined and well planned depending on the target market characteristics, resilience and ramp up are more unpredictable and request high reactivity form the organizational structure, with the possibility to anticipate various challenging situations.

Indeed, the aim of the resilience strategy is not to oppose to the malicious events, but to minimize their impacts on the global performance. Physical resilience, for instance, measures the ability of an object to return to its initial state after a particular perturbation or pressure. This concept has since inspired countless studies in various domains. In industrial domain, the resilience refers to the capacities of internal processes and resources to generate alternative and proactive scenarios face to particular situations. Decentralized Manufacturing (DM) are kinds of innovative strategies dedicated to increase the resilience of nowadays enterprises [11]. It is a paradigm that the manufacturing cooperation among different geographical-distributed enterprises and the cooperative network is self-aware and can make local decisions. Samely, the supply chain (SC) resilience addresses the capability of the supply chain to be ready for any unforeseen situation, respond to ongoing disruptions, and bounce back after facing disruptions. It has three major facets: Readiness, Response and Recovery (RRR). Readiness refers to the ability of a firm to be prepared for any unexpected SC disruptions through proactive planning. Response refers to the ability of a firm to face and encounter the SC disruptions. Whereas recovery is the capability of a supply chain to reorganize and redesign the supply chain operations at the original state [12].

Likely, the ramp up paradigm concerns the capacity of an organizational system to rapidly answer rising demand of the market. The organizational transformation in both cases is based on the combination of alternatives of scenarios to be triggered according to the characteristic of the situation. In such configuration, the company has to provide a reservoir of heterogeneous resources with coordination rules.

3 Methodological and Informational Supports for Organizational Transformation

3.1 Co-evolution Strategy and PPR Approach

One solution to perform efficient organizational transformation is to design various alternatives of scenarios, all depending to specific situations. These scenarios should be tailored to the specificities of product variants and constrained by the internal capabilities and resources. According to these directions, the co-evolution of product, processes and production systems and the consideration of their reciprocal influences with the strategic decisions of the company are often required to reach economic and social sustainability in the frugal context.

Herein, the change propagation behaves as a cause–effect wave across the enterprise, spanning all product, process and production system dimensions. According to Tolio et al., (2010) [13] the term "co-evolution" represents the ability to strategically and operationally manage the propagation of engineering changes to gain competitive advantage from the resulting market and regulatory dynamics.

The co-evolution of the production system is usually extended to the production network level that aims to create several collaborative relations between OEM and supplier companies for a better management of their distinctive skills and resources in the whole production process. Supplier selection and evaluation are among the main factors to be resolved at the earlier stage to guarantee successful results from any OEM- Supplier collaboration [14].

To realize efficient co-evolution strategy, it is important to work conjointly on the design of Product-Process-Resource structures (PPR). PPR as an approach is often used as a conceptual framework for supporting the design and deployment of information system within PLM approach. However, the application of such approach is still constrained by several aspects that should be solved:

- Product configuration management does not systematically take into account all business views on common objects. This is a consequence of the problem of data inconsistency, when representative data of each view is managed in different sources and heterogeneous formats (including difference of semantic).
- Process mapping is often carried out at a global level, as formalized by internal procedures. Day-to-day practices and interactions in real-life work situations are not sufficiently represented in agility support models.
- The resources used by processes are often considered quantitatively when planning. Individual and collective skill levels (derived from communities of practice) are partially taken into account. Also, only few tools make differences between novice and expert users, when providing useful information for decision-making.

Thus, the success of any co-evolution strategy should be based on robust models ensuring the global consistency between all development stages (from product design to production network configuration) and supporting the propagation of decisions through all managerial levels. The power of modularity for such problematics is clearly laid-out in the literature [15], since it consists of decomposing complex systems into independent but interconnected parts that can be treated as conceptual, logical, physical, or organizational/operational units [16].

3.2 Modularity and Configuration Management

Product architecture is the way by which the product functions are arranged into physical units and the way in which these units interact to implement these functions [17]. This concept is strongly connected to the concept of a module, which represents a physical or conceptual grouping of product components to form a consistent unit that can be easily identified and replaced. Based on the modularity concept, Hubka and Eder (1998) [18] define the modular design process as "connecting the constructional elements into suitable groups from which many variants of technical systems can be assembled".

As an output of this process, a generic product architecture (GPA) can bring cost savings and enable quicker introduction of multiple product variants, through the concepts of product family and configuration mechanism [19]. The whole product structure is obtained by the specification of modules' interfaces to support connections between modules in specific configurations. The GPA can be constructed by using a global methodology: Break systems into discrete modules; ensure modules interchangeability with other ones; and provide well-defined interfaces between modules in the targeted GPA. Thus, modularization consists of deciding about the characteristics used to group and deciding about the characteristics used to separate different components in one common module.

By transposing the concept of product module to different items (i.e. process and resource modules), different improvement pillars can be achieved within the organizational transformation. Firstly, the company can propose various offers to the client after considering the feasibility of each proposed offer in terms of realization processes and resources. Different product configurations can be built as an adaptation of existing products or the creation of new ones through the combination and connection of existing modules developed separately in previous projects.

Therefore, by the matching between product modules, process modules and production capabilities, it is possible to rapidly introduce new products in the market that fit with the client needs. The development of a product for a new market can be obtained through a concurrent adjustment of the design, the production strategy and the production network as a global solution. Using the modular-based approach can improve the customer satisfaction by proposing high customization possibilities based on pre-defined offers and reduce the interaction by engaging customers only in the modules which they are interested in and presenting a high potential of adaptation.

Finally, Modularity can make the internal organizational transformation easier since it will be based on existing piece of work already proved in previous projects. The modularity concept can be used to represent the completeness of knowledge as an accumulation of different reusable fragments, available at the beginning of a new development project. The innovative character depends on the completeness level of these knowledge fragments at the beginning of the project.

3.3 PLM Approach as a Support for Organizational Transformation

Organizational transformation requests strong changes of the operational processes at both internal and external levels. These processes are generally synchronous, exchange heterogeneous sets of data and information, and share various resources.

At the external level, Virtual Organisation (VO) and Extended Enterprise (EE) are the main categories of networked organisations (NO) able to provide new working processes, roles and models of collaboration between companies [20]. New rules and constraints are described as a set of implicit or contractual agreements between the members of the network to avoid conflicts of interest at managerial level as well as conflicts between resources and processes at the operational level. Furthermore, rules and procedures of the NOs should guarantee flexibility for the implementation of engineering and business processes when faced with new changes in the economic environment. This will allow each partner to collaborate strongly on some projects while working alone on other projects.

Looking to the product lifecycle management (PLM) as an approach (and not a simple tool), it is proved that this paradigm presents great advantages for implementing such kind of coordination rules at various levels of inter-enterprise collaboration [21]. Indeed, the PLM is defined as a systematic concept for the integrated management of all product related data and processes through the entire life cycle. It can be assimilated to an intelligent strategy that applies in a consistent way various process models and IT solutions to support the creation, management, and use of product information across the extended enterprise. In manufacturing industry, the PLM extends the boundaries of product data management (PDM) tools and results from the integration of heterogeneous business information systems [22]. Also, thanks to the configuration panels, the PLM is able to ensure a data sharing strategy depending to the characteristics of collaborative processes. The main logic is that a higher collaboration level implies stronger connection between internal processes and IT systems implementing the PLM approach. Saying that, it can be confirmed that PLM is an ideal support for the digital and organizational co-transformation in the enterprise.

4 Consulting Practices for Organizational and Digital Transformations

This section gives an overview of a methodological Framework developed in Cognitive Companions to support the companies when conducting their organizational and digital transformations. This framework can be used in an industrial context within a PLM approach, but it remains useful in other activity sectors like banking. It can be used for a complete transformation program or to address only the transformation of some potential processes. Two examples of application in industry and banking sectors are presented in the subsections bellow to illustrate how the Framework can be applied.

4.1 C-BPM for a Companywide Organization Based on PLM Approach

The aim of C-BPM (Cognitive-Business Process Management) is to develop a capability that enables continuous adaptation of the production system and the produced systems (product). The principle of the method is to focus on the business objects of the company, and to analyze their role in the value-creation processes. The aim is to give clear vision to decision-makers (at different levels) about the complexity of their daily activities. To do so, an important step is to describe the whole cartography of business objects, as well as their different states' changes resulting from the transformation processes. State-machine diagram can be used for this step. Business objects and transformation processes are then under explicit control. Thus, the second step will focus on the description, implementation and sharing of transformation procedures and their related data All stakeholders involved in the transformation project can develop a common vision of all business objects, processes and procedures. They can co-participate in their review and improvement. All these elements are managed in configuration management modules, enabling their traceability and sharing. The PLM (Product Lifecycle Management) backbone, combined with tools specific to each activity, is at the heart of the information system support of the transformation process. So to achieve that, the knowledge management system must be put under the control of configuration management module for all processes of product design and manufacturing, but also other core business activities of the company. So, the configuration management strategy is involved in enterprise-wide activities and not only limited to engineering domain.

In this example, the current transformation project concerns a major international automotive company and aims to work on the generalization of the use of PLM system among all entities of the engineering department. This transformation aims to achieve several strategic objectives such as: zero nonconformity, first time right, etc.

Following the logic of extended enterprise design, we started to integrate Finance, Procurement and suppliers in the process and the configuration management tools. It is designed around a unique data lake, trusted source of data with a dedicated governance, and with dedicated common interface to collect needs and represent specific activities. This is the first step towards the generalization of the proposed Framework, where all activities will be documented and controlled through a unique configuration management repository.

C-BPM is based on a concept of "requirements-driven organization", where the concept of requirements must respect the following elements:

- Be captured on a tangible support, validated and published,
- Be clear, concise and valid,
- Described through a collection of data in digital or physical format.

A company's management system must be properly defined by the collection of published data and comply with it (see Fig. 1). The requirements for a management system are determined by the company's strategic business plan and activities. Products and services must meet customer and market needs, and all of these elements (products, services, customer demands, and market needs) must be included in and conform to published data collections.

The use of this framework is highly dependent on the capability of the organization to ensure that all data collections are clear, concise, and valid in a way that serves as a useful reference for collaboration. It is therefore necessary to work on business objects and process and their definition or identification. Effective communication within and between the different functional activities is a major challenge. The ability to communicate formally and effectively is a prerequisite for growing and maintaining a high level of effectiveness. Communications about product requirements are made through data collection, objects (and thus forms), records, and metadata.

These communication tools are used during all phases of the product lifecycle. It is therefore important that all stakeholders understand the semantic used in C-BPM. It is also important to distinguish between the phases of the product/service lifecycle and the core business processes of the company, including transaction and management processes. This is the main purpose of C-BPM to address jointly the product/service lifecycle and the core business processes and all elements supporting it.

For this purpose, the role of the "Business Object Management Infrastructure" to support the common repository for the storage of data and knowledge of both product/service and core business processes in a single point of truth able to facilitate communication between stakeholders.

Products/service information is provided in the form of a collection of design data including product requirements, plans, processes and actual work instructions. Company information is provided in the form of a strategic business plan, including core processes and their owners, operational standards that define the company's requirements and objectives, and basic procedures that are how-tos for achieving the company's objectives.

All this information needs to be managed in an enterprise-wide configuration management to achieve consistent compliance while eliminating waste and reducing spending on service resources.

The main difficulties encountered by organizations in achieving this objective of reducing intervention costs is induced by the questioning of the quality of the information they disseminate. Thus, the ability to communicate requirements based on extended configuration management enables organizations to achieve consistent performance. A well-tuned management system where data collections, and therefore requirements, are clear, concise and valid, leads to constant compliance to avoid corrective actions.

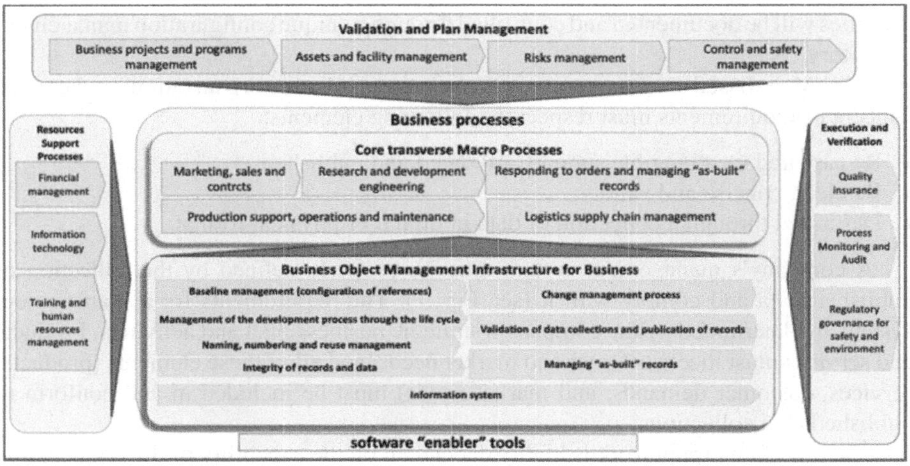

Fig. 1. Global C-BPM Approach for organizational transformation in industrial sector

C-BPM is related to extended configuration management with the objective of taking KM out of the engineering department alone and having knowledge manage-ment extended to the entire organization, and thus managing all information that has an impact on: Security; Costs; Benefits; Environment; Regulation or brand recognition.

This approach ensures that configuration management is not limited to a subset of arbitrarily chosen elements. The focus is on:

- The excellence of the operating model and in particular on the integration of processes
- Continuous improvement of this operating model

The supporting strategy provides the instruction handbook for having an adaptability that ensures the ability to:

- Constantly adapting,
- Optimize the dissemination and re-use of best practices, standards, but also product components,
- Ensure that all requirements remain clear, concise and valid,
- Communicate to all these elements in quickly and accurately to all stakeholders and in an extended enterprise principle.

Extending this scope to the entire enterprise is a major paradigm shift for most organizations, and that is why we can address it partially depending on the maturity level of the organization.

4.2 C-BPM in a Banking Sector Process Transformation

The first attempts to implement the method were carried out as part of a project in the banking sector, revealing that the introduction of an extended configuration management approach is not restricted to industrial or IT projects, and that C-BPM is in essence independent of the business sector.

A factoring company wanted to review its decision-making processes and translate business issues into a quantitative decision support system. The implementation of the method in this context relied on the definition of business objects and associated indicators to enable the requirements to be defined clearly, concisely and validly.

We have used C-BPM principle to transform this process.

As described above, these activities around the identification of business objects and processes are a preliminary work needed to ensure that everybody is aligned on the core activities and objects of interest of the company. We will thereafter describe how this work was achieved in this particular context. The process starts by clarifying the business questions, which are translated into a set of performance indicators, and Management rules based on these indicators. To be able to respond clearly to this definition, it is necessary to have a rigorous specification of the indicator according to its five aspects:

- The business object or asset to which the indicator applies.
- The nature of the indicator, which is an attribute of the business object under study.
- The Portfolio of the considered assets, which define the scope of the study.
- The statistical moment calculated for this population.
- The temporal characteristics of the estimated indicator.

However, in the present context, only two aspects of these indicators were perceived by the business actors:

- The business object in the form of asset that is the subject of the analysis.
- The generic nature of the indicator, i.e. the attribute of the business object being examined.

What's more, this business object, the asset that is the subject of the question, seems to be obvious to the business. We need to be careful, however, because this is often not the case. In this case, the business object may turn out not to be the customer but a more complex notion such as the status of the customer's contract. As a first step, therefore, it is necessary to identify precisely the business objects that will enable the business questions to be translated rigorously. The process applied is as follows (Fig. 2):

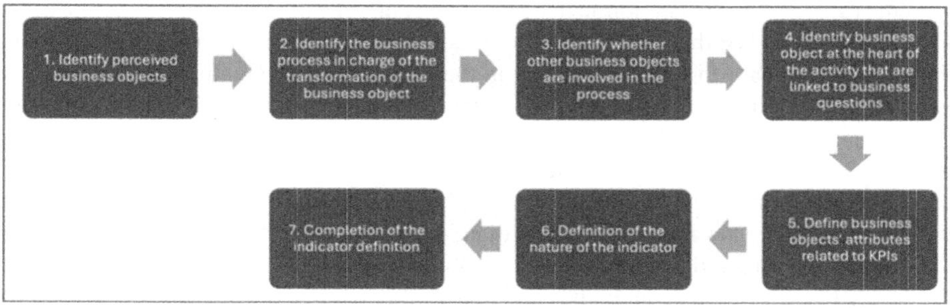

Fig. 2. C-BPM based transformation process in bank sector

1. Collection of stakeholders' perception regarding all business objects they handle. This will help to clarify all ambiguities and built a common perception.

2. Description of the business object transformation processes and complete the description of the related business objects by identifying their different states.
3. Identification of new business objects or missed ones from the analysis of transformation processes.
4. Selection of business objects at the heart of the target activity, and linked them to the goals of the organizational and/or digital transformation. If the selected objects are not sufficient to answer these questions, a new identification loop can be launched with the same order cited above.
5. Detailing the core business objects through the description of their attributes and the connection to performance indicators (i.e. consistency of the data regarding the objective of the business activity, etc.).
6. Completion of the indicator definition based on the above elements. For instance, a customer is a company asset, and the customer's turnover are attributes that defines different indicators such "the number of cancelled contracts per month for a particular product". However, missed information can be added for this definition such as: the population on which the indicator is based, the statistical moment of the population and the temporal characteristics of the indicator.
7. It is then necessary to complete the definition of the indicator with the missing elements.

This work on business processes and objects are coupled by an awareness of the need for a rigorous definition of statistical indicators. It has helped to clarify which objects are really at the heart of the factoring business, over and above the customs and habits of the service. By clarifying requirements and recording their configuration in a single repository, we were able to clarify and optimize activities and manage business risk more effectively.

The first results of using this method were to clarify team members' vision of the current situation. As a conclusion, the following aspects were modified in the new version. In the initial process, the contract went through a validation phase: a first verification of the legal quality of the contract, to check that all the regulatory elements were present and valid, and a second: a commercial verification, to check that the contract complied with the rules governing economic margins. The contract validation process is illustrated in Fig. 3. We can see that the validation tasks are carried out in parallel, and so during this validation phase the contract object in the process has no single state. This area of vagueness led to delays in contract processing, as the two teams in charge of validation could wait for each other since there was no single path, and we ended up with an object with no defined state in the state machine.

After brainstorming and validation workshops, the process has been changed into a linear version (Fig. 4) and with intermediate status to ensure the optimal execution of the process with a good appropriation of tasks by teams.

Using C-BPM in this context has had many other results, one of which is to identify which tasks are currently highly mechanical and cannot for the moment be taken over by the cognitive companions resulting from digital transformation. However, this enables us to anticipate the effects of the evolution of these tools on the organization's business and activities, and to manage their impact in advance.

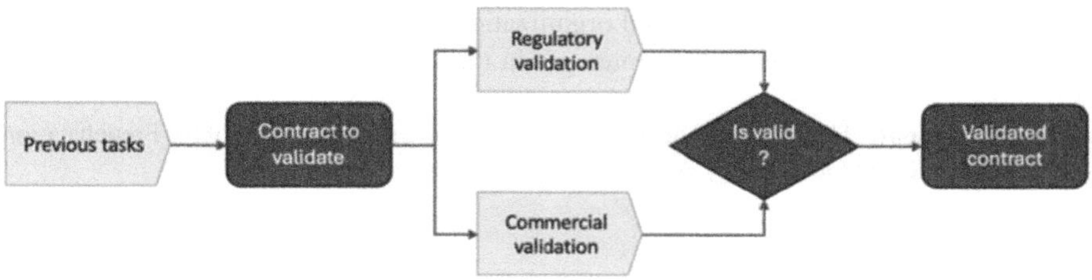

Fig. 3. Initial version of the contract validation process

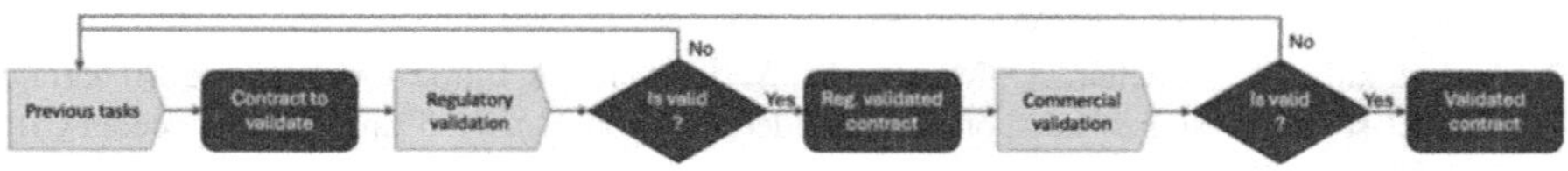

Fig. 4. Final version of the contract validation process

5 Conclusion and Perspective

This paper presents a first step of study towards the creation of a decision aid Framework to support the implementation of PLM approach at both organizational and informational levels, to cover both organizational and digital transformations. The literature survey shows the importance to work on the co-evolution of product structure and production strategy to cover all dimensions of the organizational transformation. For this reason, it is important to extend the configuration management strategy to the three aspects: Product, Process and resources. By the way, the organizational transformation for the company requests to include all processes from different departments. Even if the transformation project considers specific business departments or processes, it should integrate the big picture of the company to analyze the interactions between all organizational and operational entities.

As part of the PLM approach, the configuration management can help the connection of all data sets, and consequently supports the digital transformation in parallel to the organization one. The lesson learned from consulting practices confirm that organizational and digital transformations should not be achieved in separate way but should be deployed in a concurrent way. Mutual constraints are exchanged between the two transformations projects.

The goal of the decision system is to help the configuration of co-transformation strategies, while iintegrating mutual constraints in the problem solving process. For this reason, the aim of the future work is to go towards a unified methodology based on the C-BPM Framework to support a co-evolution strategy (organizational and digital). The current step is to extend the analysis of current consulting practices to identify advantages and limits. Based on this step, the main best practices will be integrated in a unique process, respecting the main steps of the C-BPM Framework. In the next step, principles of co-evolution and configuration management will be transformed into operational procedures and guidelines for the co-transformation support framework. The last step will be the capitalization and organization of expert knowledge as an ontology coupled with business rules to help manager in the steps of problem solving.

Acknowledgement. The presented results were conducted with the support of the French national agency of research and technology ANRT.

References

1. Brady, T., Davies, A., Gann, D.M.: Creating value by delivering integrated solutions. Int. J. Project Manag. **23**(5), 360–365 (2005)
2. Andriankaja, H., Boucher, X., Medini, K., Vaillant, H.: A framework to design integrated product-service systems based on the extended functional analysis approach. Procedia CIRP **47**, 323–328 (2016)
3. Maleki, E., et al.: Ontology-based framework enabling smart product-service systems: application of sensing systems for machine health monitoring. IEEE J. Internet things **05**(06), 4496–4505 (2018)
4. Wallin, J., Parida, V., Isaksson, O.: Understanding product-service system innovation capabilities development for manufacturing companies. J. Manuf. Technol. Manag. **26**, 763–787 (2015)
5. Harrat, M., Caverot, G., Belkadi, F., Bernard, A.: A Fuzzy logic assessment Framework for collaborative development of Product-Service Systems. Int. J. Value Chain Manag. **14**(2), 143–167 (2023)
6. Zeschky, M., Widenmayer, B., Gassmann, O.: Frugal innovation in emerging markets: the case of mettler toledo. Res. Technol. Manag. **54**(4), 38–45 (2011)
7. Tiwari, R., Herstatt, C.: Frugal innovation: a global networks' perspective. Swiss J. Bus. Res. Pract. **3**, 245–274 (2012)
8. Berger, R. 2013. Frugal Products: Study Results. Stuttgart: Berger Strategy Consultants. http://www.rolandberger.com/media/pdf/Roland_Berger_Frugal_products_20130212.pdf
9. Giovannini, A., Aubry, A., Panetto, H., El-Haouzi, H., Canciglieri Junior, O., Pierrel, L.: Knowledge representation, retrieval and reuse for product family design: an antilogicist approach. Comput. Ind. Eng. **101**, 391–402 (2016)
10. Shao, X.F.: What is the right production strategy for horizontally differentiated product: standardization or mass customization? Int. J. Prod. Econ. **223**, 107527 (2020)
11. Leng, J., et al.: Towards resilience in Industry 5.0: a decentralized autonomous manufacturing paradigm. J. Manuf. Syst. **71**, 95–114 (2023)
12. Qader, G., Junaid, M., Abbas, Q., Mubarik, M.S.: Industry 4.0 enables supply chain resilience and supply chain performance. Technol. Forecast. Social Change **185**, 122026 (2022)
13. Tolio T., et al.: SPECIES: co-evolution of products, processes and production systems. CIRP Ann. Manuf. Technol. **59**(2), 672–693 (2010)
14. Cheraghi, S.H., Dadashzadeh, M., Subramanian, M.: Critical success factors for supplier selection: an update. J. Appl. Bus. Res. **20**(2), 91–108 (2004)
15. Sako M. Modularity and outsourcing: the nature of co-evolution of product architecture and Organisation architecture in the global automotive industry. (Book chap) In: The Business of Systems Integration, pp. 229–253. Oxford University Press, New York 2005)
16. Jiao, J., Tseng, M.M.: Fundamentals of product family architecture. Integr. Manuf. Syst. **11**(7), 469–483 (2000)
17. Ulrich, K.T., Eppinger, S.D.: Product Design and Development, 3rd edn. McGraw-Hill, New York (2004)

18. Hubka, V., Eder, E.W.: Theory of technical systems, 2nd edn. Springer, Heidelberg (1998)
19. ElMaraghy, H., et al.: Product variety management. CIRP Ann. Manuf. Technol. **62**(2), 629–652 (2013)
20. Kamel, M., Benzekri, A., Barrère, F., Laborde, R.: Evaluating the virtual organizations security solutions using the ISO/IEC 17799 standard. In: International Conference on Concurrent Enterprising, Sophia-Antipolis, 4–6 June 2007 (2007)
21. Ming, X.G., Yan, J.Q., Lu, W.F., Ma, D.: Technology solutions for collaborative product life cycle management status review and future trend. Concurr. Eng. Res. Appl. **13**(4), 311–319 (2005)
22. Schuh, G., Rozenfeld, H., Assmus, D., Zancul, E.: Process oriented framework to support PLM implementation. Comput. Ind. **59**(2–3), 210–218 (2008)

Evaluating the Sustainability of Artificial Intelligence Applications Across the Product Lifecycle

Giovanni Marongiu(✉), Giulia Bruno, Paolo Chiabert, and Franco Lombardi

Politecnico di Torino, Corso Duca degli Abruzzi 24, 10129 Turin, Italy
giovanni.marongiu@polito.it

Abstract. Due to the emergence of global challenges, the European policy called Industry 5.0 demands technologies to become resilient, sustainable and humancentric; the integration of Artificial Intelligence (AI) and Knowledge Management (KM) (i.e., AI-based KM) represents an opportunity to accomplish this ambitious goal thanks to the possibility of combining humans and computers' unique skills. While the compliance of AI-based KM with human-centric requirements is apparent, its long-term sustainability is still an object of discussion. In response to I5.0's request for metrics of sustainability, this paper presents a lifecycle analysis of AI-based KM's applications retrieved from literature to establish whether they can be deemed socially, environmentally, and economically sustainable. A list of sources was collected by means of a literature search, to be later evaluated through sustainability criteria specific for the design and use phases of life. It was found that AI-based KM applications supporting fractions of the knowledge lifecycle may become fragile in the long-term, because potentially leading to repositories' indefinite growth and information overload in humans and computers; furthermore, it emerged that not enough attention was paid to AI's accountability and data governance. The mentioned pain points, complemented in this article with practical examples of the counter-measures adopted by virtuous sources, provide scholars with practical indications to lead future research efforts, and contribute to the validation of a framework to assess the longterm sustainability of AI solutions.

Keywords: Artificial Intelligence · Knowledge Management · Sustainability · Life Cycle

1 Introduction

In 2020, the European Union (EU) introduced the concept of I5.0 after recognizing the Industry 4.0 (I4.0) paradigm to be too technology-oriented, fragile, and generally unable to address the challenges Europe and humanity as a whole had to face [1]. The emergent awareness on the planet and society's needs made the EU redefine the role of industry and technology as the main instrument toward a different form of progress based on human progress and environmental well-being, rather than production and consumption. This

Published by Springer Nature Switzerland AG 2025
P. Sureephong et al. (Eds.): PLM 2024, IFIP AICT 740, pp. 480–495, 2025.
https://doi.org/10.1007/978-3-031-93319-6_36

transformative view calls for innovative ways of value creation based on regenerative features and for different applications of technologies able to complement and enhance humans' unique abilities. The achievement of such ambitious goals depends on many factors, including the availability of metrics for sustainability. The development of AI proposed by I5.0, building upon the advancements of I4.0, holds great promise. EU requires AI to become more sustainable and human-centric, a request which can be met by integrating AI and KM to further improve humans' knowledge processes. In fact, through the suggested combinations, humans are supported rather than replaced, as AI is implemented to mitigate common weaknesses such as information overload and decision-making biases.

Given the EU's prioritization of human-centricity and sustainability principles for I5.0, it is evident that AI-based KM can meet the former requirement, yet there is a lack of adequate data concerning the latter. Through an examination of AI and Knowledge repositories applications in the past decade, this article seeks to uncover the level of support they provide for knowledge activities and their potential sustainability. In order to do this, some qualitative metrics for evaluating AI's sustainability were retrieved from literature, and systematically applied to assess solutions' sustainability across their lifecycle, more specifically in the design and use phases of life.

In the next sections, after a brief overview of the state of the art, the procedure used for the literature search and analysis will be presented, followed by the presentation of results.

2 State of the Art

I5.0 is not a technological paradigm; rather, it is an attempt to redirect the technologies at the core of I4.0 toward goals of Sustainability, Human-centricity, and Resilience. According to EU, technological development should not be an end in itself or, worse, a means to increase profitability alone; instead, the ultimate goal must be to improve the whole supply chain sustainability and resilience to shocks, and to enhance workers and society's living conditions [1].

The key driver behind this transformative view is industry which, in order to fulfill its strategic role, must become responsible, regenerative, self-sustaining, and adaptive, decentralized, and digitalized with purpose. In other words, the technologies enabling the fourth industrial revolution must be repurposed toward objectives of global holistic prosperity.

2.1 The AI and Its Rise in Popularity

Generally speaking, it is possible to identify nine technologies at the core of I4.0 and, consequently, of I5.0: autonomous robots, simulation, horizontal and vertical system integration, industrial internet of things (IIoT), cybersecurity, additive manufacturing, augmented reality, big data and analytics [2].Their usage and integration is believed to be the gateway to new levels of efficiency and productivity.

One of the most popular and impacting technologies proposed by the smart revolution for big data analytics is AI: the development of computer systems able to receive and interpret external data, learn from them and use what has been learnt to accomplish certain tasks that generally require human intelligence [3]. Despite its current popularity, AI is not a recent discipline: the concept of a "thinking" machine originated during the 20th century, although with mechanisms substantially different from the ones used today[1]. In fact, after the initial enthusiasm generated by the fast progress of AI's abilities, the limitations imposed by the computational capacity induced countries and institutions to withdraw their investments in the discipline.

Nowadays, trust in AI has been renovated thanks to the recent technological advancements in computational power and to the emergence of new inference mechanisms that allow computers to truly learn (differently from rule-based systems); the combination of these two factors made AI more flexible and considerably less expensive, thus allowing widespread adoption in the industry. The advantages offered by AI seem to justify its diffusion: for instance, machine learning (one of the principal and most popular techniques) is able to analyze vast amounts of data, taking into account tens or hundreds of features to make decisions (e.g., predict a nominal or numerical value) in a fraction of time it would take a human operator. AI's methodologies are also widely used for other tasks like image and text generation (i.e., generative AI), image recognition, clustering of customers, robot navigation, and recommendations systems. Notwithstanding the benefits delivered, past AI's implementations often demonstrated the short-sighted vision typical of I4.0: they often attempted to maximize shareholders' value, neglecting its potential impacts and synergies with society and environment. This attitude is in sharp contrast with the EU's policy, which explicitly calls for AI applications designed specifically for sustainability, resilience, and human-centricity [1]. To achieve the European Union's goals, it is crucial to reintroduce the human operator in the cycle, and to develop AIs that adapt to human needs and complement their skills.

Furthermore, the inadequacy of technology alone to maximize the efficiency and effectiveness of systems has been proved in many domains, like the Zero-defect manufacturing (ZDM) [4]. For many decades, human workers have been seen as the main source of defects in manufacturing and for this reason have been progressively substituted by machines in multiple tasks. However, researchers now consider the technocentric approach of removing the human workers from the loop to be ineffective for the achievement of true ZDM; instead, establishing a synergy between engineers, operators, and cutting-edge technologies may lead to improved defective recognition, deeper root cause analysis, and more effective problem-solving, thus leading to economic, social, and environmental well-being.

2.2 Cognitive Computing: The Pursue of AI-Based KM

The creation of human-centric and sustainable AI systems is not an easy endeavor. Building on the proposition of complementing human abilities, a viable and interesting possibility is to integrate AI into KM processes.

---

[1] They mostly relied on rule-based knowledge representation, which assumes all knowledge can be represented through if-then statements [3].

KM is a multi-disciplinary field aimed at managing all knowledge related processes, including knowledge identification, creation, sharing and application [5, 6] to improve multiple dimensions of organizational performance.

The effort of integrating AI and KM to create AI-based KM or "*Cognitive computing*"[2] is justified by the issue that many benefits of KM are currently countered by obstacles such as the information overload (i.e., excessive information to be processed), the constant growth of knowledge repositories making retrieval tasks difficult (especially under time pressure) [7], and the high cost of creating and codifying knowledge [8]. In fact, despite humans' incredible capacity for innovation and creativity, they are constrained in their knowledge intensive task by the amount of data they can consider at once and by their biases and personal beliefs. Human-centric AI can potentially fill these fallacies, especially when dealing with large knowledge bases. Machines' abilities in data analysis and pattern recognition make possible to aggregate and select the most relevant features to present to human operators, thus reducing the cognitive load and, as a consequence, improving the decision-making quality. Recommendation systems and natural language processing can also facilitate knowledge retrieval and knowledge creation processes through context recognition and entity extraction. Furthermore, they can also combine and process human-generated knowledge to provide new insights, reinforcing the knowledge creation loop.

While AI-based KM's human-centricity is apparent, its sustainability is still a subject of debate. As a matter of fact, it is necessary to take into account multiple dimensions of sustainability at once in order to define an application viable in the long term. In an effort to meet the EU's requirements for measuring technological growth, this article seeks to assess the sustainability of AI-driven KM platforms during their lifecycle, employing sustainability benchmarks.

3 Literature Review Methodology

The answer to the research problem was found by means of a literature review. The literature review was divided into a literature search phase, which produced a set of articles to be analyzed, and an analysis phase, during which the documents were analyzed by means of predefined criteria. In the next two sections, the process followed for the literature search and the criteria used for the source evaluation are presented.

3.1 The Literature Search Phase

The query used for the phase of literature search was aimed at covering all possible applications of AI to Knowledge Bases and Knowledge Processes (Fig. 1). The explored time period has been limited to documents from 2013 to 2024 to gain an up to date overview of AI in KM, especially considering the growth in effectiveness and popularity this technology experienced during the last 10 years.

This initial query produced 54 results, which have been screened two times, first by reading the abstract, then by reading the full document (Fig. 2).

[2] The replication of the human way of thinking through data mining, pattern recognition, and natural language processing [29].

TITLE-ABS-KEY (("Knowledge Management" OR "Organizational learning" OR "Knowledge worker") AND "AI" AND ("Knowledge Base" OR "Knowledge repository" OR "Knowledge repositories")) AND PUBYEAR > 2013 AND PUBYEAR < 2025

Fig. 1. Query used on Scopus, last search on 18/02/2024.

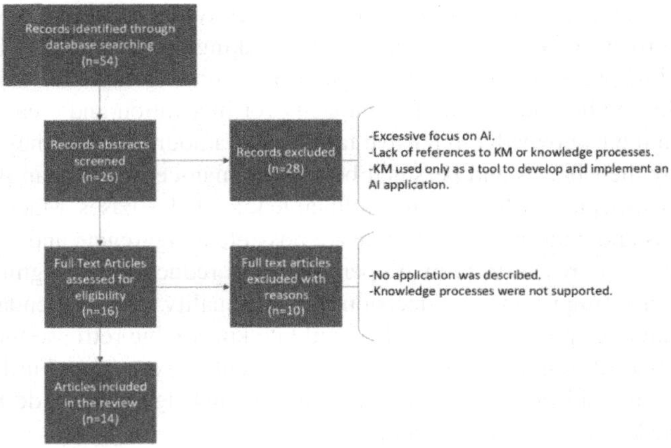

Fig. 2. The literature search process.

During the abstract screening, it has been possible to remove most of the non-relevant results from the analysis. The main principles for exclusion included an excessive focus on AI, the lack of references to KM and knowledge processes or the usage of KM as a simple tool to support the development and implementation of AI. In fact, the purpose is to understand how AI can support KM and knowledge processes, not vice versa. Using these criteria, the initial 54 results have been reduced to 26 documents.

The following step was to read the full articles in order to find the ones to be analyzed. During this step, in addition to the previous principles for exclusion, two new criteria have been used: first, articles that did not describe any application or implementation of AI have been eliminated; then, only sources describing a clear application to support a knowledge process have been maintained. Through the described screening process, 16 documents were selected for the analysis.

Finally, following the analysis process, 14 of them have been included in this literature review (Table 1).

3.2 Knowledge Processes and Application Sustainability

The analysis of the retrieved sources was twofold. During the first part, applications were categorized according to the step of the knowledge processes supported; following this analysis, the same sources were reviewed against 9 sustainability criteria encompassing different lifecycle phases.

Table 1. Articles under analysis.

Year	Authors	Title	Document Type
2019	Senda M, Iwasa D, Hayashi T, Ohsawa Y [9]	Data Classification by Reducing Bias of Domain-Oriented Knowledge Based on Data Jackets	Conference paper
2019	Berquand A, Murdaca F, Riccardi A, Soares T, Generé S, Brauer N et al. [10]	Artificial Intelligence for the Early Design Phases of Space Missions	Conference paper
2019	Lau EKW, Zhao A, Ko A [11]	Hybrid artificial intelligence B2B2C business application – Online travel services	Conference paper
2020	Sekiguchi K, Hori K [12]	Organic and dynamic tool for use with knowledge base of AI ethics for promoting engineers' practice of ethical AI design	Journal article
2021	Groza A, Toderean L, Muntean GA, Nicoara SD [13]	Agents that Argue and Explain Classifications of Retinal Conditions	Journal article
2022	Sticha A, Broussard S, Havenaar I, Vardeman C, Brenner P [14]	Hybrid Knowledge Engineering to Build a Global Assassination Dataset	Conference paper
2022	Gammack J, Akay H, Ceylan C, Kim SG [15]	Semantic knowledge management system for design documentation with heterogeneous data using machine learning	Conference paper
2023	Chee J, Kwa ED, Goh X [16]	"Vertigo, likely peripheral": the dizzying rise of ChatGPT	Journal Article
2023	Telukdarie A, Munsamy M, Katsumbe M, Maphisa X, Philbin SP [17]	Industry 4.0 Technological Advancement in the Food and Beverage Manufacturing Industry in South Africa— Bibliometric Analysis via Natural Language Processing	Journal Article
2023	Trianni V, Nuzzolese AG, Porciello J, Kurvers RHJM, Herzog SM, Barabucci G et al. [18]	Hybrid Collective Intelligence for Decision Support in Complex Open-Ended Domains	Conference paper

(continued)

Table 1. *(continued)*

Year	Authors	Title	Document Type
2023	Chen Z [19]	Artificial Intelligence-Virtual Trainer: Innovative Didactics Aimed at Personalized Training Needs	Journal article
2023	Kernan Freire S, Niforatos E, Wang C, Ruiz-Arenas S, Foosherian M, Wellsandt S et al. [20]	Lessons Learned from Designing and Evaluating CLAICA: A Continuously Learning AI Cognitive Assistant	Conference paper
2023	Lee L, Li Q, Duan Y, Zheng J[21]	Study on STBLI Knowledge Base Construction	Conference paper
2023	Yu X [22]	Design of Modern English-Chinese Bilingual Interactive AI Teaching Platform for International Shipping Based on Structured Data Sources	Conference paper

In the first part of the analysis, six "inner" phases[3] have been used for classification [23]:

- Knowledge identification (KI).
- Knowledge acquisition (KA).
- Knowledge development (KDe).
- Knowledge distribution (KDi).
- Knowledge utilization (KU).
- Knowledge preservation (KP).

Each of them contributes to a different phase of knowledge creation and sharing: first the knowledge sources are identified and mapped both internally and externally (KI), then external sources can be acquired (KA), while internal ones can be developed through internalization or combination (KDe) [24]; the created knowledge can be shared or transferred (KDi), and then used (KU). KU in this case is a wide category that includes the task of knowledge storage, retrieval, and application. The last process covers all the activities through which knowledge elements are selectively identified for retention (KP). At the end of this step of the analysis, the number of applications supporting each process was counted.

Following this classification, each application was evaluated for sustainability. Over time, the initial emphasis placed by I5.0 on industrial environmental sustainability has been expanded to also cover social and economic aspects; their balance is crucial for the achievement of humanity and planet well-being in the long term [25]. For this reason, all three dimensions have been considered in this analysis. Although a proper sustainability analysis should assess the whole AI-based KM applications' life cycle, the limited

[3] Opposed to the two outer processes, employed to monitor and control the six inner ones.

amount of information provided by the sources only allowed the evaluation of two phases through 9 criteria: 6 covering the design phase [26] and 3 referred to the use phase. For each criterion one or more questions were defined prior to the analysis in order to produce coherent answers throughout the work (Table 2). Combining criteria related to the design and to the use phase of life, it has been possible to evaluate each source on multiple dimensions and to develop a representative image of current research direction on the topic of AI and KM for sustainability.

Table 2. Sustainability criteria

Life cycle phase	Sustainability criteria	Questions
Design phase	Safety and technical robustness	Does it lead to or help to avoid harm, the establishment of vulnerable relationships, and information asymmetries between business and consumers?
	Oversight and human agency	Does it substitute humans or complement their skills?
	Accountability	Is it transparent? Can its decisions be understood and explained by the human expert? Are they justified?
Use phase	Fairness, diversity, and nondiscrimination	Is it fair, non-biased, universally accessible?
	Environmental and societal well-being	Does it harm the environment or humans in any way?
	Data governance and privacy	What is its role with respect to data privacy?
	Social sustainability	Is it aimed at humans' well-being?
	Economic sustainability	Is it aimed at competitiveness improvements?
	Environment sustainability	Is it aimed at environmental well-being?

4 Results

The number of sources on the subject published in the past 3 years demonstrates a clear positive trend. In fact, 2023 was a particularly prosperous year for AI-based KM research, especially if compared to previous years. It is also interesting to notice the absence of sources published before 2019; it could be argued that the reason behind this phenomenon is the limited appeal of KM, which might have led researchers to devise methods to facilitate knowledge processes by means of knowledge bases, without integrating them into a distinct KM framework. Notwithstanding these considerations, 2023 witnessed a sharp increase in publications, indicating the growing focus of researchers on the use of AIs in knowledge processes.

4.1 Analysis of the Supported Knowledge Process

To better understand how AIs have been applied to KM, articles were first categorized according to the knowledge process they were deployed to support (Table 3).

Table 3. Knowledge process supported, by source.

Authors	KI	KA	KDe	KDi	KU	KP	Phases supported
Senda M, Iwasa D, Hayashi T, Ohsawa Y [9]	X				X		2
Berquand A, Murdaca F, Riccardi A, Soares T, Generé S, Brauer N et al. [10]		X	X	X	X	X	5
Lau EKW, Zhao A, Ko A [11]					X		1
Sekiguchi K, Hori K [12]			X	X	X	X	4
Groza A, Toderean L, Muntean GA, Nicoara SD [13]				X	X		2
Sticha A, Broussard S, Havenaar I, Vardeman C, Brenner P [14]		X	X				2
Gammack J, Akay H, Ceylan C, Kim SG [15]			X		X		2
Chee J, Kwa ED, Goh X [16]				X	X		2
Telukdarie A, Munsamy M, Katsumbe M, Maphisa X, Philbin SP [17]	X	X					2
Trianni V, Nuzzolese AG, Porciello J, Kurvers RHJM, Herzog SM, Barabucci G et al. [18]		X		X	X		3
Chen Z [19]	X	X	X	X		X	5
Kernan Freire S, Niforatos E, Wang C, Ruiz-Arenas S, Foosherian M, Wellsandt S et al. [20]	X		X	X	X	X	5
Lee L, Li Q, Duan Y, Zheng J[21]	X	X	X				3
Yu X [22]	X	X	X	X	X		5
				Average:			3

Then, the occurrence of each process was measured to identify the most popular one among applications (Fig. 3).

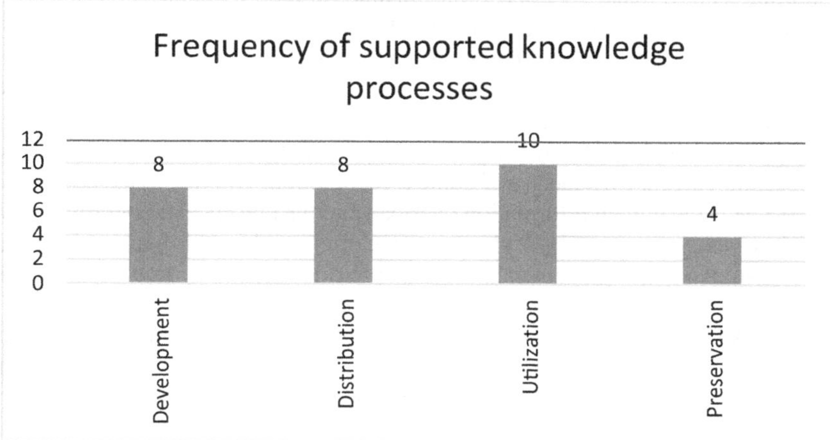

Fig. 3. Frequency of supported knowledge processes

According to the analysis, the average number of lifecycle phases managed by AI-based KM applications is three; in fact, only four of them attempted to address at least 4 phases of knowledge lifecycle. The remainder of the article will showcase how the desired outcome of the solution may require managing multiple phases of knowledge's lifespan to develop sustainable applications.

The most supported process within the knowledge lifecycle is KU. It has been already mentioned that KU covers a diverse set of subprocesses, such as storing, retrieving, reusing, and applying expertise. Since the topic of this research revolves around AI-based KM and knowledge bases, it is reasonable to expect that most solutions will address the listed activities. Following utilization, the phases of knowledge lifecycle with the highest level of support are KDe and KDi, KA, KI and, finally, KP.

Among the sources collected, AI is usually implemented in ways that complement human abilities and allow them to develop their full potential. For instance, [10, 20] assist operators by automatically collecting tacit knowledge respectively in manufacturing and in design. Tacit knowledge is a form of knowledge that is personal, embedded in people and routine and, because of this, difficult to define and articulate; furthermore, tacit knowledge is tightly linked with the context in which it was generated [24]. Companies have a marked interest in converting tacit into explicit knowledge, in order to extract, store (in a knowledge repository), share and reuse their operators' distinctive know-how. The conversion process can only be performed by the expert; however, creating an explicit (i.e., coded) knowledge repository is a challenging endeavor. Furthermore, similar initiatives by organizations are often associated with limited chances of success due to the fact that experts tend to alter the nature of their know-how to fit the characteristics of the re-user, thus penalizing the completeness and effectiveness of the repository's content [8]. To compensate for these shortcomings, researchers employed

AI to systematically extract context and tacit knowledge, which are later stored inside Knowledge graphs to make them retrievable by means of natural language queries. The possibility to automatically collect experts' knowledge as it emerges from practices is an important step to make AI for KM sustainable; in fact, it enables constant updates to the knowledge base without requiring the knowledge creator to spend any effort in the codification process, but still enabling new know-how to be implemented into the reasoning mechanism. In addition, the usage of Natural Language Processing (NLP) queries supports both experts and novices when retrieving past experiences from large knowledge bases, in particular by integrating the potentially inaccurate terminology used by this second category by means of context recognition. Another issue addressed by AI-based KM is the human information overload and the presence of cognitive biases in the decision-making process. Hybrid systems are often used for this purpose because, by combining human and AI judgments, they can provide users with insights and different perspectives to overcome commonly recognized human limitations.

An interesting aspect is the low occurrence of applications supporting KP. In a society that keeps producing data, information, and knowledge, and that plans to store and apply it, a mechanism to selectively identify which elements to maintain and which to "forget" is crucial. In fact, the problem of information overload, which has been always attributed to humans and that make searching large knowledge bases difficult, may expand to AI systems as well. Researchers are now discussing about the phenomenon of information overload in computers, which may originate from repository's growth in size and from the distribution of false and inaccurate information [27], and that will lead to lower efficiency and effectiveness in the future. This is the reason why the correct management of knowledge and meta-knowledge lays at the basis of the long-term sustainability of AI applied to knowledge bases. Despite the topic's relevance, an exiguous number of sources implemented a system to support this process. The preferred approach is a feedback loop. The primary purpose behind this mechanism's deployment is to improve results' accuracy by discriminating among high and low quality knowledge elements through users' opinions. Nevertheless, the same tool can be adapted to provide a hierarchical order of the most valued knowledge elements, assisting knowledge managers in trimming obsolete elements from the organizational intellectual assets.

4.2 Analysis Against the Lifecycle Sustainable Criteria

During the second step of the analysis, each article was judged by means of the criteria previously presented (Table 2) and the number of articles respecting each criterion has been counted, both for the design (Fig. 4) and the use phase of life (Fig. 5).

When adopting design criteria, almost all applications were judged positively under the aspects of safety, fairness, and complementarity with humans. In particular, this last result on human oversight is to be expected: the KM framework acts as a constraint to AI because it forces the design to revolve around humans, considered to be the primary driver and user of knowledge processes.

Among the criteria considered, accountability is one of the most impacting. Accountability refers to the ability to explain and justify AI's reasoning [28]. This feature plays a central role in the integration and interaction between human and computer's work. In the process of implementing AI, offering explanations for decisions and presenting

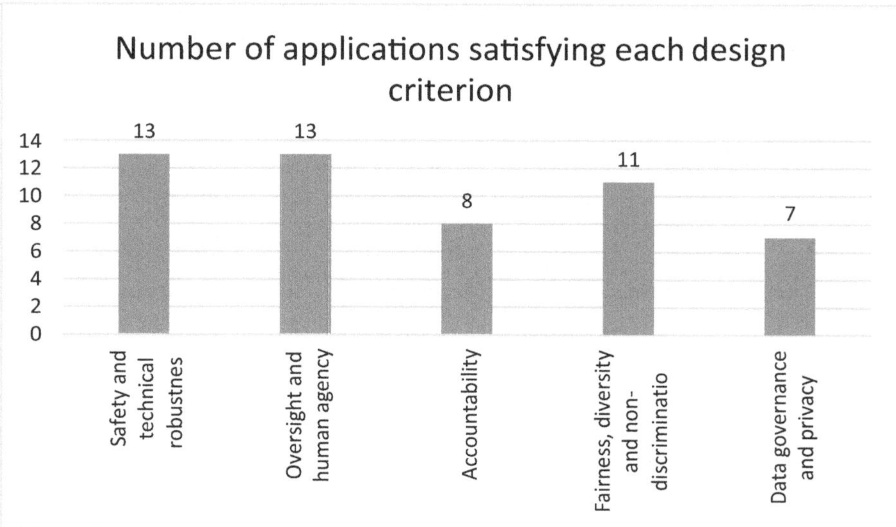

Fig. 4. Evaluation of AI applications' sustainability according to design criteria

assurance cases can greatly impact the acceptance of end users. Additionally, numerous industries, such as healthcare [13], may soon make them a compulsory requirement. Despite its importance, only half of the articles was found to be compliant with this principle. Furthermore, technology is currently moving in the direction of black box AIs like Deep Neural Networks; as the inference mechanism's complexity grows it will be essential for researchers to find ways to explain the reasons behind each judgment.

Another concerning criterion is the data governance and privacy. Many of the analyzed solutions collect users' data to further train and enlarge their knowledge bases. While this has been described as a desirable characteristic for the long term-sustainability of AI, data collection and governance adds a further layer of complexity to the evaluation of AI social sustainability, an issue recognized by authors themselves [20].

Finally, the lack of data about the specifics of each solution did not allow to evaluate their environmental sustainability.

Although less specific from the point of view of the criteria adopted, the second part of the analysis provides interesting hints on the direction of the research, thanks to the possibility to discern the end goal of the proposed solutions (Fig. 5).

It can be noticed that almost every AI solution attempts to enhance the social sustainability of the implementing organization. Once again, this result is unsurprising given the context of the research centered around the knowledge processes.

Instead, only 50% of the applications under consideration show clear elements of their intention to support long term economic sustainability of the organization.

Finally, only 3 over 14 sources aim at improving the environmental performance of the implementing organization.

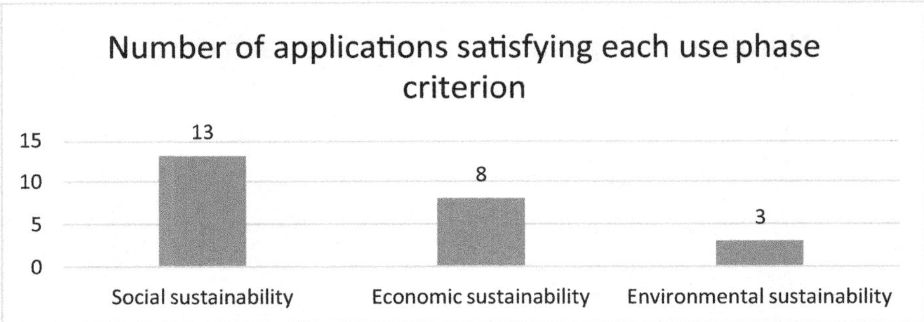

Fig. 5. Evaluation of AI applications' sustainability according to use phase of life criteria

These findings indicate a noticeable disparity in the three aspects of sustainability and suggest a need for a more equitable distribution of resources to achieve balance across all three dimensions.

5 Conclusions

The implementation of I4.0 technologies, in particular AI, for sustainable and human-centric development is made possible by AI-based KM, in which artificial reasoning performs tedious and complicate tasks, leaving humans free to approach complex problems requiring innovative and flexible solutions. While the integration of AI and KM is a clear instance of smart technologies deployed for human-centric development, there is still uncertainty regarding its long-term sustainability.

This research evaluated the lifecycle sustainability of AI applications supporting knowledge processes, leading to the provision of important recommendations for future AI design and deployment.

The first relevant outcome of the analysis is the overview of the supported knowledge processes. It was found that most applications covered the know-how usage phase of life, for instance to collect and store the operators' tacit knowledge. The possibility to automatically enlarge the knowledge base used by the AI for inference was identified as an important element to provide sustainable AI applications; however, the lack of support on knowledge disposal and preservation may represent an element of fragility in the future because of the constant growth of repositories' size and complexity, and the resulting information overload for humans and machines. Overall, both mechanisms should exist at the same time for self-sustaining AI-based KM.

The second outcome is instead the evaluation of the retrieved sources' sustainability using criteria from the design and the use phases of life. It was found that almost none of the retrieved cases explicitly aimed at reaching economic, social, and environmental sustainability in their usage phase of life; balancing all three dimensions is a mandatory requirement for an application to be sustainable in the long-term. Considering instead the design phase of life, the lack of transparency and accountability, and the problem of data governance may reflect negatively on AI social sustainability; therefore, they are principles that designers must consider to guarantee the accomplishment of I5.0's ambitious goals.

The presented analysis is affected by three limitations. The first limit is related to the low number of evaluated articles, which might prevent the gathering of enough observations to draw an accurate picture of the state of the art. The findings could be enriched by including more sources, for instance recent AI solutions which were not retrieved by means of the presented query but that can still be identified as supporting KM processes.

A second limitation is due to the criteria used to evaluate the documents in their use phase. Although functional to the current purposes, they can be furtherly broken down to accurately track the presence of elements of sustainability.

Finally, due to the lack of data regarding the carbon footprint and energy consumption attributable to the applications during their lifecycle, it was not possible to accurately evaluate the sources' environmental sustainability. Researchers can contribute by mentioning in their papers the algorithm, the general architecture and the hardware employed to train and run the model, to provide a basis for a more rigorous sustainability assessment by means of generic measures like mean energy consumption, embedded energy, and components' carbon footprint.

In conclusion, the framework for the assessment and the recommendations provided for designing and deploying AI-based KM solutions contribute to renovate and redirect the effort to make smart technologies sustainable.

Acknowledgements. The authors would like to acknowledge the company Eurodies Italia S.r.l. for the support provided for the research presented in the paper.

References

1. Renda, A., et al.: Industry 5.0, a transformative vision for Europe : governing systemic transformations towards a sustainable industry. Publications Office of the European Union (2022)
2. Rüßmann, M., Lorenz, M., Gerbert, P., Waldner, M., Engel, P., Harnisch, M., Justus, J.J.: Industry 4.0: The Future of Productivity and Growth in Manufacturing Industries. https://www.bcg.com/publications/2015/engineered_products_project_business_indus-try_4_future_productivity_growth_manufacturing_industries. Accessed 19 Feb 2024
3. Haenlein, M., Kaplan, A.: A brief history of artificial intelligence: on the past, present, and future of artificial intelligence. Calif. Manag. Rev. **61**, 5–14 (2019)
4. Wan, P.K., Leirmo, T.L.: Human-centric zero-defect manufacturing: state-of-the-art review, perspectives, and challenges. Comput. Ind. **144** (2023)
5. Yew Wong, K., Aspinwall, E.: Characterizing knowledge management in the small business environment. J. Knowl. Manag. **8**, 44–61 (2004)
6. Rhem, A.J.: The Connection between Artificial Intelligence and Knowledge Management, http and-knowledge-management. Accessed 22 Feb 2024
7. Gray, P.H., Durcikova, A.: The role of knowledge repositories in technical support environments: speed versus learning in user performance. J. Manag. Inf. Syst. **22**, 159–190 (2005)
8. Markus, M.L.: Toward a theory of knowledge reuse: types of knowledge reuse situations and factors in reuse success. J. Manag. Inf. Syst. **18**, 57–93 (2001)

9. Senda, M., Iwasa, D., Hayashi, T., Ohsawa, Y.: Data classification by reducing bias of domain-oriented knowledge based on data jackets. In: 2019 6th International Conference on Signal Processing and Integrated Networks (SPIN), pp. 404–407. IEEE, Noida (2019)
10. Berquand, A., Murdaca, F., Riccardi, A., Soares, T., Generé, S., Brauer, N., Kumar, K.: Artificial intelligence for the early design phases of space missions. In: IEEE Aerospace Conference. pp. 1–20. IEEE, Big Sky (2019)
11. Lau, E.K.W., Zhao, A., Ko, A.: Hybrid artificial intelligence B2B2C business application – online travel services. In: Uden, L., Ting, I.H., Corchado, J. (eds.) Knowledge Management in Organizations, pp. 515–524. Springer, Cham (2019)
12. Sekiguchi, K., Hori, K.: Organic and dynamic tool for use with knowledge base of AI ethics for promoting engineers' practice of ethical AI design. AI Soc. **35**, 51–71 (2020)
13. Groza, A., Toderean, L., Muntean, G.A., Nicoara, S.D.: Agents that argue and explain classifications of retinal conditions. J. Med. Biol. Eng. **41**, 730–741 (2021)
14. Sticha, A., Broussard, S., Havenaar, I., Vardeman, C., Brenner, P.: Hybrid knowledge engineering to build a global assassination dataset. In: 2022 IEEE Fifth International Conference on Artificial Intelligence and Knowledge Engineering (AIKE), pp. 32–35. IEEE, Laguna Hills (2022)
15. Gammack, J., Akay, H., Ceylan, C., Kim, S.G.: Semantic knowledge management system for design documentation with heterogeneous data using machine learning. Procedia CIRP **109**, 95–100 (2022)
16. Chee, J., Kwa, E.D., Goh, X.: "Vertigo, likely peripheral": the dizzying rise of ChatGPT. Eur. Arch. Otorhinolaryngol. **280**, 4687–4689 (2023)
17. Telukdarie, A., Munsamy, M., Katsumbe, T.H., Maphisa, X., Philbin, S.P.: Industry 4.0 technological advancement in the food and beverage manufacturing industry in South Africa—bibliometric analysis via natural language processing. Information **14**, 454 (2023)
18. Trianni, V., et al.: Hybrid collective intelligence for decision support in complex open-ended domains. In: Lukovicz, P., Mayer, S., Koch, J., Shawe-Taylor, J., Tiddi, I. (eds.) HHAI 2023: Augmenting Human Intellect, pp. 124–137. IOS Press (2023)
19. Chen, Z.: Artificial intelligence-virtual trainer: innovative didactics aimed at personalized training needs. J. Knowl. Econ. **14**, 2007–2025 (2023)
20. Kernan, F., et al.: Lessons learned from designing and evaluating CLAICA: a continuously learning AI cognitive assistant. In: IUI 2023 Proceedings of the 28th International Conference on Intelligent User Interfaces, pp. 553–568. Association for Computing Machinery (ACM), Sydney (2023)
21. Lee, L., Li, Q., Duan, Y., Zheng, J.: Study on STBLI knowledge base construction. In: 2023 2nd International Conference on Robotics, Artificial Intelligence and Intelligent Control (RAIIC), pp. 226–231. IEEE, Mianyang (2023)
22. Yu, X.: Design of modern english-chinese bilingual interactive AI teaching platform for international shipping based on structured data sources. In: 2023 IEEE 5th Eurasia Conference on IOT, Communication and Engineering (ECICE), pp. 460–464. IEEE, Yunlin (2023)
23. Fink, K., Ploder, C.: Knowledge management toolkit for SMEs. Int. J. Knowl. Manag. **5**, 46–60 (2009)
24. Nonaka, I.: A Dynamic Theory of Organizational Knowledge Creation. organ. Sci. **5**, 14–37 (1994)
25. Leng, J., et al.: Industry 5.0: prospect and retrospect. J. Manuf. Syst. **65**, 279–295 (2022)
26. Alloghani, M.A.: Criteria for sustainable AI software: development and evaluation of sustainable ai products. In: Artificial Intelligence and Sustainability, pp. 33–51. Springer, Cham (2024)
27. Reuter, L., Berndt, J.O., Ulfert, A.S., Antoni, C.H., Ellwart, T., Timm, I.J.: Intentional forgetting in distributed artificial intelligence. KI Kunstliche Intelligenz **33**, 69–77 (2019)

28. Novelli, C., Taddeo, M., Floridi, L.: Accountability in artificial intelligence: what it is and how it works. AI Soc. **39**, 1871–1882 (2023)
29. El Asri, H., Benhlima, L., Agnaou, A.: Integrating artificial intelligence in knowledge management: a primer. In: Saka, A., et al. (eds.) Advances in Integrated Design and Production. CPI 2019, pp. 516–526. Springer, Heidelberg (2021)

Author Index

Published by Springer Nature Switzerland AG 2025
P. Sureephong et al. (Eds.): PLM 2024, IFIP AICT 740, pp. 497–499, 2025.
https://doi.org/10.1007/978-3-031-93319-6